D1288507

Digital Integrated Circuit Design

THE OXFORD SERIES IN ELECTRICAL AND COMPUTER ENGINEERING

SERIES EDITORS

Adel S. Sedra, *Electrical Engineering*
Michael R. Lightner, *Computer Engineering*

SERIES TITLES

Allen and Holberg, *CMOS Analog Circuit Design*
Bobrow, *Elementary Linear Circuit Analysis, 2nd Ed.*
Bobrow, *Fundamentals of Electrical Engineering, 2nd Ed.*
Campbell, *The Science and Engineering of Microelectronic Fabrication*
Chen, *Analog & Digital Control System Design*
Chen, *Linear System Theory and Design, 3rd Ed.*
Chen, *System and Signal Analysis, 2nd Ed.*
Comer, *Digital Logic and State Machine Design, 3rd Ed.*
Cooper and McGillem, *Probabilistic Methods of Signal and System Analysis, 3rd Ed.*
Franco, *Electric Circuits Fundamentals*
Fortney, *Principles Of Electronics: Analog & Digital*
Granzow, *Digital Transmission Lines*
Guru and Hiziroglu, *Electric Machinery & Transformers, 2nd Ed.*
Hoole and Hoole, *A Modern Short Course in Engineering Electromagnetics*
Jones, *Introduction to Optical Fiber Communication Systems*
Krein, *Elements of Power Electronics*
Kuo, *Digital Control Systems, 3rd Ed.*
Lathi, *Modern Digital and Analog Communications Systems, 3rd Ed.*
Martin, *Digital Integrated Circuit Design*
McGillem and Cooper, *Continuous and Discrete Signal and System Analysis, 3rd Ed.*
Miner, *Lines and Electromagnetic Fields for Engineers*
Roberts and Sedra, *SPICE, 2nd Ed.*
Roulston, *An Introduction to the Physics of Semiconductor Devices*
Sadiku, *Elements Of Electromagnetics, 2nd Ed.*
Santina, Stubberud, and Hostetter, *Digital Control System Design, 2nd Ed.*
Schwarz, *Electromagnetics for Engineers*
Schwarz and Oldham, *Electrical Engineering: An Introduction, 2nd Ed.*
Sedra and Smith, *Microelectronic Circuits, 4th Ed.*
Stefani, Savant, Shahian, and Hostetter, *Design of Feedback Control Systems, 3rd Ed.*
Van Valkenburg, *Analog Filter Design*
Warner and Grung, *Semiconductor Device Electronics*
Wolovich, *Automatic Control Systems*
Yariv, *Optical Electronics in Modern Communications, 5th Ed.*

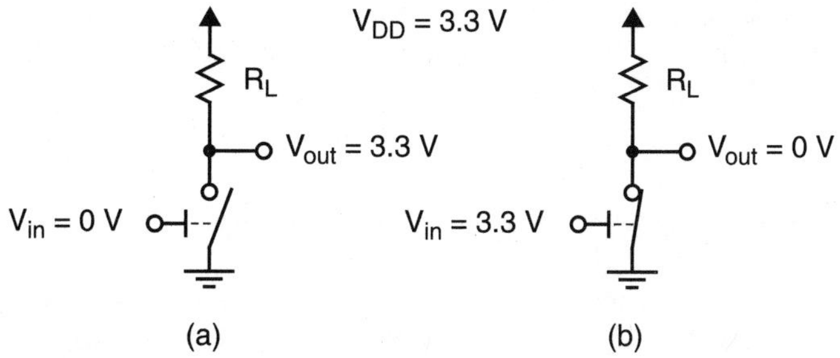

Figure 1.2 A switch-level interpretation of the inverter of Fig. 1.1 with (a) a "0" input and (b) a "1" input.

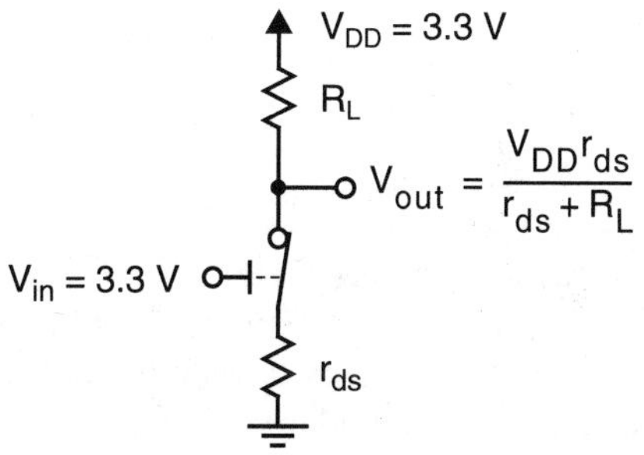

Figure 1.3 A more accurate modeling of the inverter of Fig. 1.1b.

and has a large voltage between its drain and source, or operates in the triode region, where its drain-source voltage is very small.

In a first-order analysis of the inverter of Fig. 1.1b, the **n**-*channel transistor can be considered to operate as a voltage-controlled switch; if its gate voltage is large, the switch is "on" and connects the drain to the source; if its gate voltage is low, the transistor is equivalent to an open switch and can be ignored when analyzing the circuit.* Thus, referring to Fig. 1.2a, we see that for a 0-V input signal to the drive transistor, the equivalent switch is open; therefore there is no current through the load resistor, and thus no voltage across it, and the output voltage will be equal to V_{DD} or 3.3 V.[2] Similarly, for a 3.3-V input signal, the equivalent switch is closed, which pulls the output voltage down to 0 V.

A more accurate model would be to replace the transistor by a small resistor when its gate voltage is *high* (or a "1") as is shown in Fig. 1.3. The output voltage for a "1" input would then not be 0 V, but would be given by (using simple circuit analysis)

$$V_{out} = \frac{V_{DD} r_{ds}}{r_{ds} + R_L} \tag{1.1}$$

[2]In this text, V_{DD}, the power-supply voltage, is assumed to be 3.3 V unless specifically stated otherwise.

which is slightly larger than 0 V. As long as the load resistor, R_L, is much larger than r_{ds}, then the output voltage, although not 0 V, will still be much less than the gate threshold voltage.[3]

Example 1.1

Assume that when the transistor is on, it can be modeled by a 300-Ω resistor and that the size of the load resistor R_L is 4 kΩ. What is the output voltage when the input voltage is high.

Solution: Using (1.1), we have

$$V_{out} = \frac{3.3(300)}{300 + 4000} = 0.23 \text{ V} \tag{1.2}$$

Normally, the *inverter* of Fig. 1.1b would never be used in an actual integrated circuit, primarily because the resistor, R_L, would take up too much area. In an *NMOS* technology, a type of integrated circuit processing technology that was formerly popular, an n-channel depletion transistor can be used to replace the resistor R_L as is shown in Fig. 1.4.[4] This type of transistor conducts when its gate is at the same voltage as its source, unlike an n-channel enhancement transistor.

This inverter operates very similarly to the inverter of Fig. 1.1b. Thus, the depletion transistor Q_2 operates somewhat like a resistor, but occupies much less chip area. This approximation will be made more exact in Chapter 4, but it is adequate for a first-order understanding. To obtain a low enough output voltage when the input is a "1", the width-to-length-ratio of the drive transistor Q_1 must be significantly greater than the W/L ratio of the load transistor. Normally, a factor of four is adequate and is typically used, as is the case for the inverter of Fig. 1.4.

It might also be emphasized at this time that for MOS transistors the ratio of the transistors width, W, to its length, L, is the most critical design parameter, as opposed to the absolute dimensions. Also, a transistor's W/L will sometimes be called its size and denoted as S.

A more complicated gate can be realized by adding more drive transistors (i.e., transistors that have their gates connected to an input signal). For example, the symbol for a *nor* gate and its realization using an NMOS technology are shown in Fig. 1.5a and b, respectively. This type of gate has a low output if any of the inputs are high. For the transistor realization of Fig. 1.5b, if any of the inputs is at a high voltage, the corresponding transistor can be approximated by a

[3]The gate threshold voltage V_{TH} is defined as the gate input voltage that will give an equal output voltage. This will be discussed in more detail in Chapter 4.

[4]The symbol of an n-channel *depletion* transistor is similar to that of an n-channel *enhancement* transistor, except there is an additional line between the drain and the source, which denotes that a conductive channel is present when the gate-source voltage is 0 V.

Figure 1.4 A typical inverter in an NMOS technology.

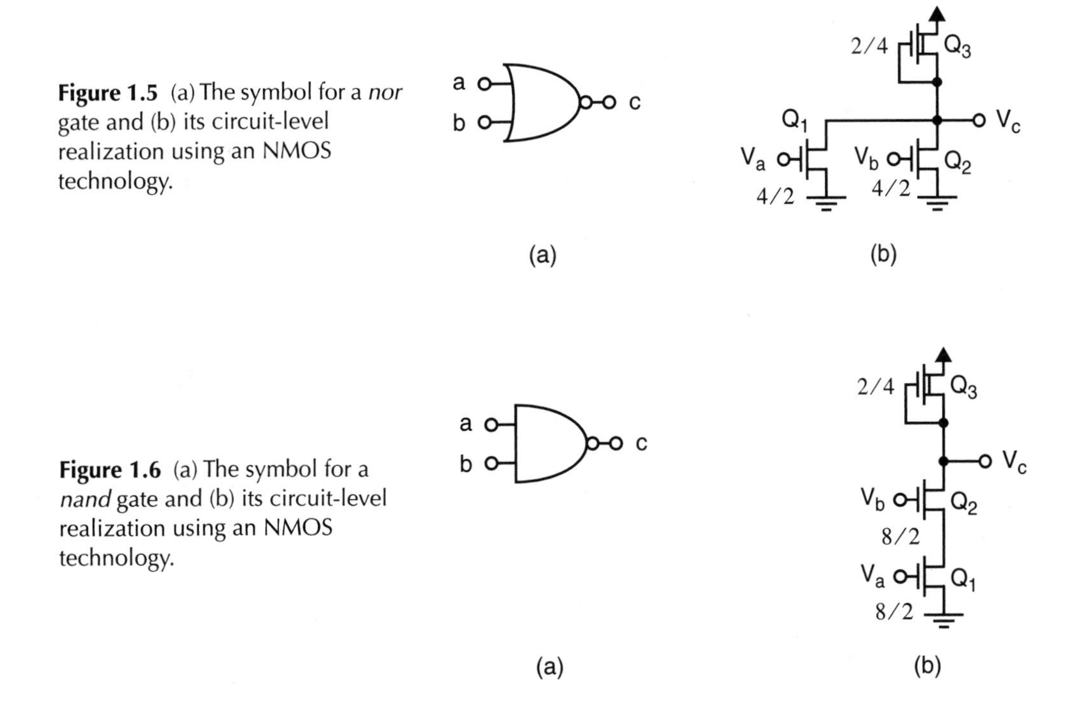

Figure 1.5 (a) The symbol for a *nor* gate and (b) its circuit-level realization using an NMOS technology.

Figure 1.6 (a) The symbol for a *nand* gate and (b) its circuit-level realization using an NMOS technology.

closed switch, and the output voltage will be pulled low, assuming the W/Ls (equivalently Ss) of the drive transistors are large enough compared to the W/Ls of the load transistor. A ratio of four is typically used, similar to what is used for the inverter. If both inputs are low, all the drive transistors will be off and can be ignored. In this case, the depletion load transistor Q_3 acts similar to a load resistor and pulls the output voltage high to V_{DD}.

Another example of an NMOS logic gate is shown in Fig. 1.6. The logic symbol of a *nand* gate is shown in Fig. 1.6a, and a transistor-level realization is shown in Fig. 1.6b. In this case, any zero input results in the output being a "1". In the transistor realization of Fig. 1.6, if either Q_1 or Q_2 has a low voltage on its gate terminal, the corresponding transistor would be off. This

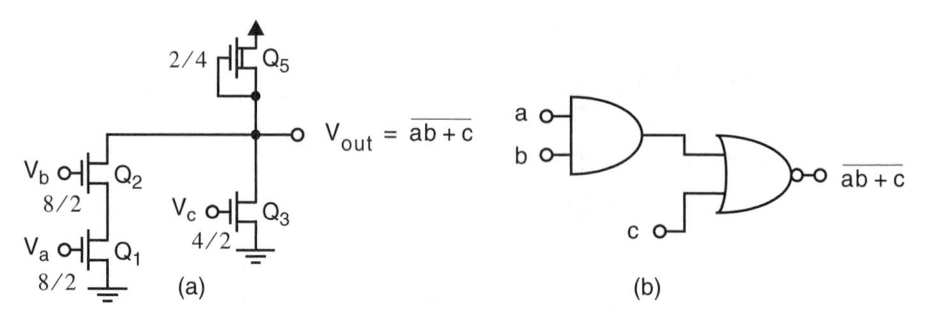

Figure 1.7 An NMOS gate that realizes the function $f = \overline{ab + c}$.

results in a high impedance (actually almost an infinite impedance) between the output-voltage node and ground, which allows the load transistor, Q_3, to pull the output voltage to V_{DD}. If both inputs are high, the impedance to ground will be much smaller. If the drive transistor's W/Ls are much larger than that of the load transistor, Q_3, then this impedance will be much less than the approximately equivalent impedance of Q_3, and the output voltage will be a small voltage that would be recognized as a "0" by any succeeding gates. Since there are two drive transistors in series, the ratio of the W/Ls of the drive transistors to the W/L of the load transistor must be larger than is necessary for an inverter or a *nor* gate. For a two-input *nand* gate, a factor of 8 is typically used.

An even more complicated transistor gate is shown in Fig. 1.7a, with its logic-gate equivalent shown in Fig. 1.7b. A general technique for generating complicated gates such as this will be presented in Chapter 4. For now, to convince oneself of the equivalence, note that the output of the transistor gate of Fig. 1.7a will be *pulled low* only if a low-impedance path through the series connection of Q_1 and Q_2, or through Q_3, exists. This will be the case only if both a and b are high, or if c is high. When this is the case, the logical output will be a "0". Thus, V_{out} is equal to $\overline{ab + c}$.

Example 1.2

Design an NMOS gate to realize the function $f = \overline{ab + cd}$.

Solution: This logic gate must have a low-impedance path between the output and ground if the inputs *a* and *b* are both high or if the inputs *c* and *d* are both high. Comparing this example to the other logic gates presented, we see that the *and* function suggests a series connection of transistors, whereas the *or* function suggests a parallel connection. Together, we are led to the gate shown in Fig. 1.8. The sizes are typical ones for this type of gate and have been chosen so that the gate threshold voltage and output *low* voltages are reasonable. This will be discussed in more detail in Chapter 4.

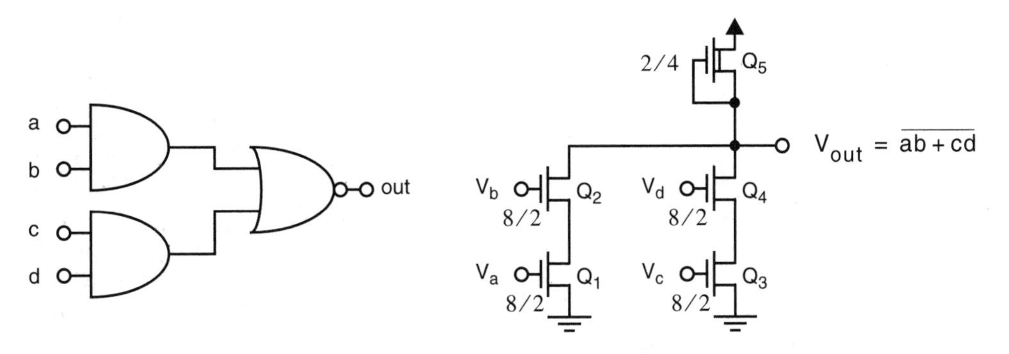

Figure 1.8 An NMOS gate that realizes the function $f = \overline{ab + cd}$.

Example 1.3

Design an NMOS gate to realize the function $f = \overline{(a + b)c}$.

Solution: This logic gate must have a low-impedance path between the output and ground if either of the inputs *a* or *b* is high and if also *c* is high. Comparing this example to the other logic gates presented previously, we see that the *or* function suggests a parallel connection of transistors, whereas the *and* function suggests a series connection. Together, we are led to the gate shown in Fig. 1.9. Again, the sizes are typical ones for this type of gate and have been chosen so that the gate threshold voltage and output *low* voltages are reasonable.

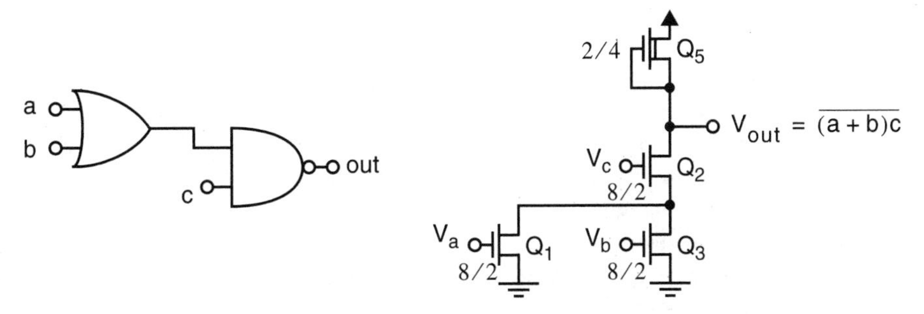

Figure 1.9 An NMOS gate that realizes the function $f = \overline{ab + c}$.

Logic gates similar to those just described, that are realized using n-channel MOS transistors only, are called NMOS logic gates. The processing steps required to make an IC composed of

NMOS gates is commonly called an NMOS technology. This technology was used by Intel Corporation to realize the 8080 microprocessor, one of the first modern microprocessors. It is a very dense technology, but unfortunately, any gate that has a "0" output dissipates power. This power dissipation limits the number of NMOS gates that can be realized in a single IC. If a moderately more complicated processing procedure is used, it is possible to realize p-channel transistors as well as n-channel transistors. By using these p-channel transistors, it is possible to realize logic gates with no d.c. power dissipation, as will be described in the next section.

In the preceding examples, the W/L ratios of the transistors were given next to them. For example, the value of L in the examples was given as 2 for the drive transistors, which would represent 2 μm. This is much larger than what is presently typical but was a representative value when NMOS gates were popular. More typical values at present are between 0.18 and 0.5 μm for minimum channel lengths. The smaller the minimum allowable channel length is, the faster the logic gates. To maximize the speed of logic gates, drive transistors are always taken with the minimum allowable channel lengths and therefore the length of a gate is often not shown in a schematic unless it is different from the minimum allowable channel length. The minimum allowable channel length is often used to describe a process technology. For example, a technology with a 0.35-μm minimum allowable channel length would be called a 0.35-μm technology.

1.2 Simple CMOS Logic Gates

A p-channel enhancement transistor works complementary to an n-channel enhancement transistor. This is the reason for the name of the logic family, *CMOS* or *complementary MOS* logic. Whereas an n-channel transistor acts like a closed switch when its gate voltage is *high*, a p-channel transistor acts like a closed switch when its gate voltage is *low*. Whereas an n-channel transistor acts like an open switch when its gate voltage is *low*, a p-channel transistor acts like an open switch when its gate voltage is *high*. Again, this is a very simplistic approximation that will be elaborated on in future chapters, but it is an adequate model to understand the logic function being implemented by many traditional CMOS gates.

There are numerous possible symbols to denote n-channel and p-channel enhancement transistors; the symbols that will be used in this text are shown in Fig. 1.10. The arrow going out of the source terminal of the transistor designates an n-channel transistor, in a similar manner to the arrow on the emitter of an npn bipolar transistor. A p-channel transistor is designated by the arrow going into the source lead of the transistor.[5] The drain junction of an n-channel transistor is defined as the junction having the higher potential; the other junction is the source junction. Whenever possible, the drain junction will be closer to the top of a schematic so current and

[5]It is not necessary to include the arrows on the sources of the n-channel transistors of *NMOS* gates because, typically, if a depletion transistor is ever used, one usually assumes that all transistors are n-channel transistors. In general, if it is not obvious, always assume the transistor is an n-channel transistor.

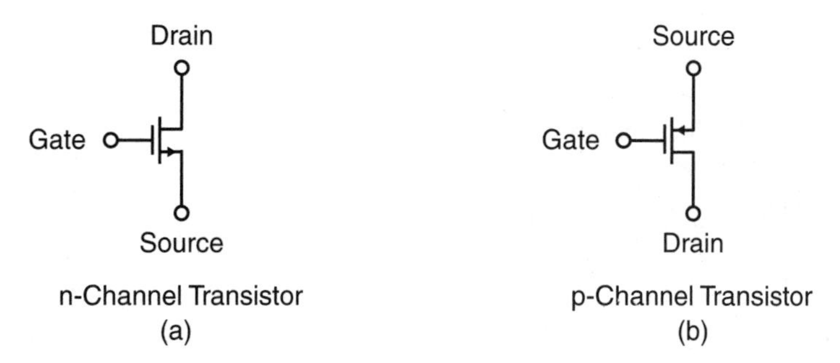

Figure 1.10 (a) An n-channel enhancement transistor and (b) a p-channel enhancement transistor.

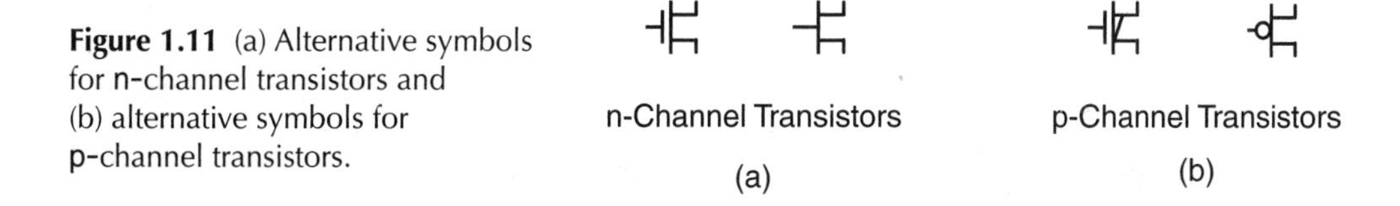

Figure 1.11 (a) Alternative symbols for n-channel transistors and (b) alternative symbols for p-channel transistors.

charge flow from the top of the schematic to the bottom, into the drain and out of the source. For p-channel transistors, the source junction is the junction with the higher potential; it will typically be shown nearer the top of the schematic and current and charge still flow from the top of the schematic to the bottom, into the source (for p-channel transistors) and out of the drain.

Some alternative symbols for n-channel and p-channel transistors are shown in Fig. 1.11. These are slightly simplified symbols that might be used to show circuits having a large number of transistors. Many other symbols have been used in the past; exactly which ones are used is not important, as long as their meanings are clear and a standard is agreed on throughout an organization.

A *traditional* CMOS logic gate consists of an n-channel network and a complementary p-channel network. The n-channel network is between the output and ground, whereas the p-channel network is between the output and V_{DD}. The inputs go to both networks. The number of transistors in each network is equal to the number of inputs. Except for the inputs, the only other connection between the two networks is at the output node. This is illustrated in Fig. 1.12. Although it might seem as if there are many requirements to remember, in practice it is quite simple. The n-channel network is designed exactly as if an NMOS logic gate were being designed; the only difference is that the transistor sizes are not as critical for the gate to function correctly. Next, the complementary p-channel network is designed. What is meant by complementary is that parallel components in the n-channel network translate into series components in the p-channel network. Correspondingly, series components in the n-channel network translate into parallel components in the p-channel network. With a little practice, designing the complementary network becomes easy.

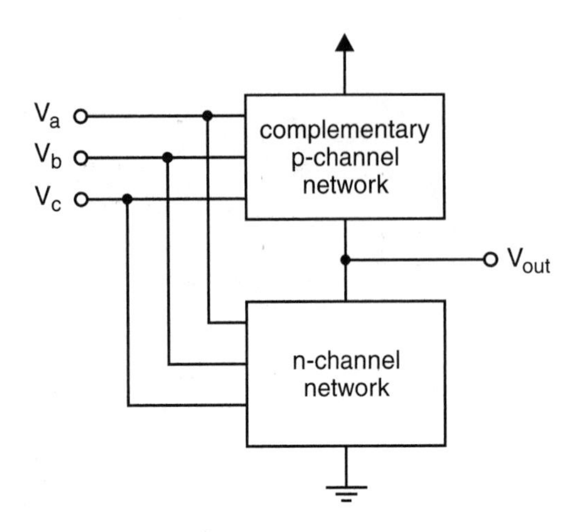

Figure 1.12 The general structure of a traditional CMOS logic gate.

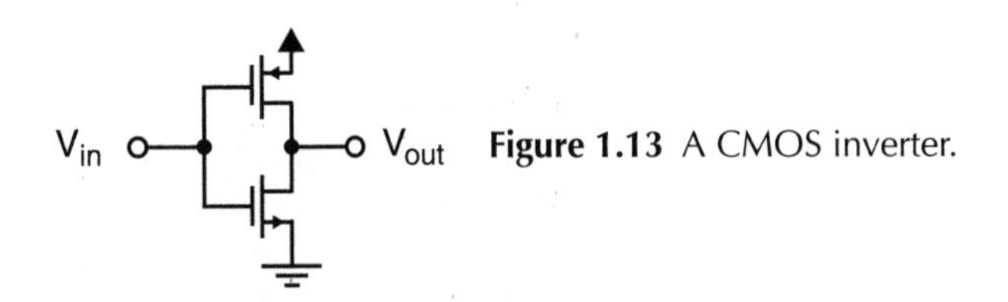

Figure 1.13 A CMOS inverter.

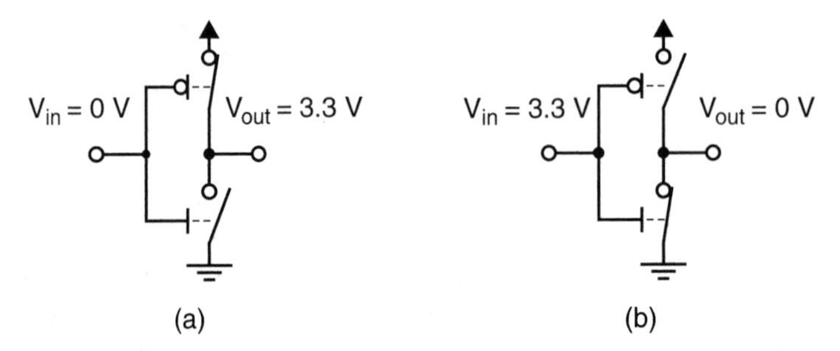

Figure 1.14 The switch-level equivalents of the CMOS inverter of Fig. 1.13 for (a) a "0" input and (b) a "1" input.

A few examples should make this procedure clear. Shown in Fig. 1.13 is perhaps the simplest possible example, a CMOS inverter. Note that the complement of a single n-channel transistor is a single p-channel transistor. The switch-level equivalent circuits for both a "0" input and for a "1" input are shown in Fig. 1.14. Note that when the n-channel transistor is on, the p-channel transistor is off; when the n-channel transistor is off, the p-channel transistor is on. When the input is a "0", the output is *pulled high* to V_{DD} by the p-channel transistor, whereas when the input is a "1", the n-channel transistor *pulls* the output low. As is typical for the low-frequency analysis of any MOS circuit, the transistors that are off can be ignored as if they

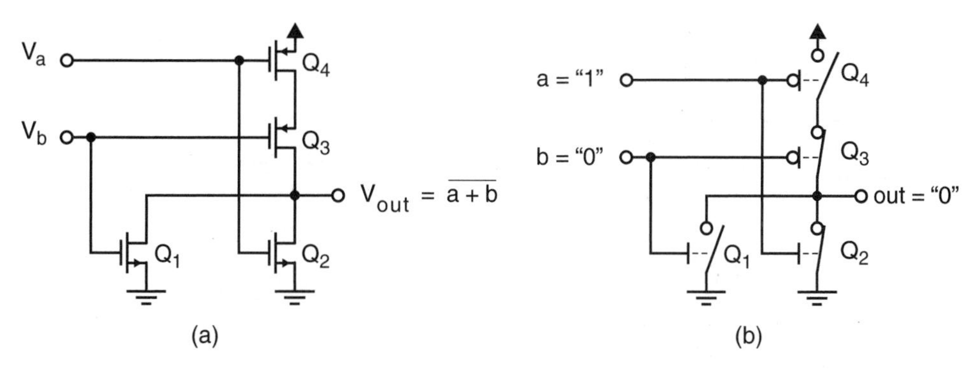

Figure 1.15 (a) A CMOS *nor* gate and (b) its switch-level equivalent for a = "1" and b = "0".

were not even in the circuit.[6] Notice that in both cases, there is no d.c. path between V_{DD} and ground, and as long as the load is purely capacitive (which is almost always the case for MOS circuits), no d.c. power is being dissipated. For traditional CMOS logic gates, irrespective of the type, no paths exist between V_{DD} and ground except during transients.

A slightly more complicated example is shown in Fig. 1.15a. This is a CMOS *nor* gate. Note first that the topology for the n-channel transistors is the same as that previously seen for the NMOS *nor* gate (not counting the load transistor) and that the complement of the two parallel n-channel transistors is two series p-channel transistors. If any input is high, at least one of the n-channel transistors will be on *pulling* the output to ground or 0 V. Also, at least one of the p-channel transistors will be off, and thus the impedance between the output and V_{DD} will be infinite. The special case of a being a "1" and b being a "0" is illustrated in the switch-level equivalent circuit of Fig. 1.15b. For this case, Q_1 is *off*, Q_2 is *on*, Q_3 is *on*, and Q_4 is *off*. Since Q_2 is *on*, the output is being *pulled low*. Since Q_4 is *off*, there is a high-impedance path between the output and V_{DD}, which results in no d.c. power dissipation.

An alternative view of a traditional CMOS gate is to think of it as an NMOS gate with a dynamic load; whenever the n-channel network represents a low-impedance path to ground, the load is high impedance; whenever the n-channel network is high impedance, the load is low impedance, *pulling* the output to V_{DD}. Since the output is never being *pulled* to V_{DD} and to ground at the same time, the gate will work correctly no matter what the width-to-length ratios (i.e., W/Ls) of the transistors are, unlike NMOS gates. *Traditional CMOS gates are called ratioless gates whereas NMOS gates are called ratioed gates.* The trade-offs when choosing the transistor dimensions for CMOS gates will be described in detail in Chapter 4; however, a reasonable choice, for now, is to choose all W/Ls equal. Note again that there never exists a d.c. path between V_{DD} and ground, and therefore the traditional CMOS *nor* gate has no d.c. power dissipation.

[6]For high-frequency or transient analysis, their parasitic capacitances must still be taken into account.

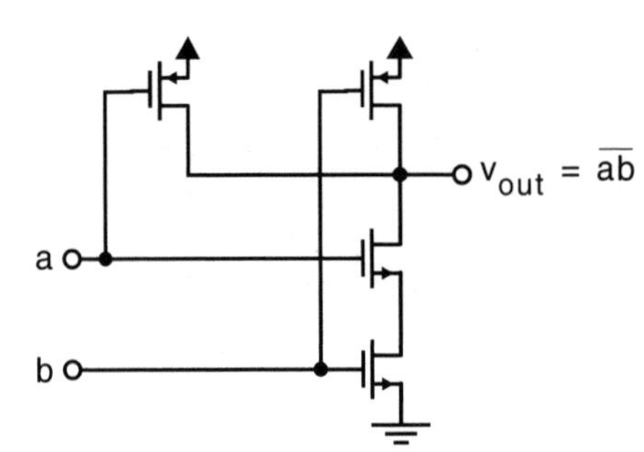

Figure 1.16 A CMOS *nand* gate.

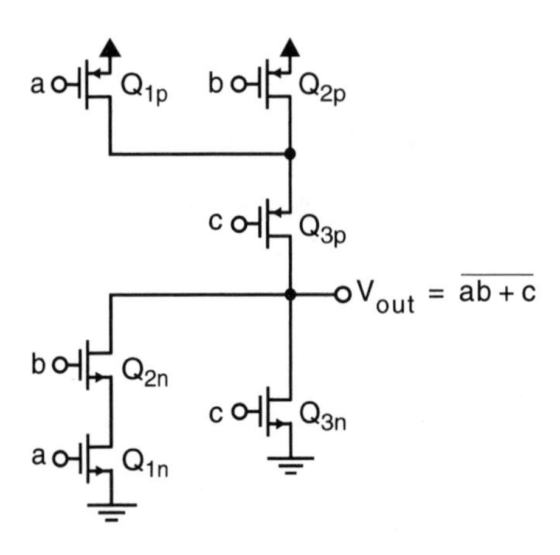

Figure 1.17 An example of a compound CMOS logic gate.

A CMOS *nand* gate is shown in Fig. 1.16. For this network, if any of the inputs is a "0", one of the n-channel transistors will be off and the n-channel network will be infinite impedance, one of the p-channel transistors will be on and the output will be pulled up to V_{DD} or a "1". Conversely, if all inputs are high, the n-channel transistors will be on, the p-channel transistors will be off, and the output will be pulled low. Thus, any "0" input guarantees a "1" output and all "1"s give a "0" output; this describes the functionality of a *nand* gate. Note that this gate is similar to a CMOS *nor* gate turned upside down, and is thought of as being a complementary gate.[7]

A final example of a CMOS gate is shown in Fig. 1.17. This is the CMOS version of the NMOS *and-or* gate shown in Fig. 1.7. Note that again the n-channel networks have identical

[7]A complementary gate is defined as one having a logic function equivalent to that obtained when the inputs and the output are all complemented.

topologies. We have used a slightly different notation to make the circuit schematic more readable. For example, the complementary transistor to Q_{1n} is Q_{1p}. These gates would be connected together in reality, but in the schematic, they are both shown being connected to the input a to make the schematic more readable. The reader should carefully note the duality between the n-channel and the p-channel networks. The n-channel network consists of two parallel subnetworks (i.e., Q_{1n} in series with Q_{2n} is one network and Q_{3n} is the other), whereas the p-channel network consists of two series networks (i.e., Q_{1p} in parallel with Q_{2p} is one network and Q_{3p} is the other).

Example 1.4

Design a traditional CMOS gate to realize the function $f = \overline{ab + cd}$.

Solution: This example is the CMOS version of the NMOS gate designed in Example 1.2. The n-channel network can be identical to that of Example 1.2, except now all the transistors would be taken with the same width to length ratios or W/Ls (for example, a 4 μm width over a 0.5 μm length). This network is shown in Fig. 1.18a with labels for the transistors. All that remains is to design the complementary p-channel network. In the n-channel network, there is a parallel connection of two subnetworks, one consisting of transistors Q_{1n} and Q_{2n} and the other consisting of transistors Q_{3n} and Q_{4n}. This transforms into a series connection of two subnetworks in the p-channel network with each subnetwork having two p-channel transistors. The n-channel subnetwork of the series connected Q_{1n} and Q_{2n} transforms into the p-channel subnetwork of parallel connected Q_{1p} and Q_{2p}. Also, the n-channel subnetwork of the series connected Q_{3n} and Q_{4n} transforms into the p-channel subnetwork of parallel connected Q_{3p} and Q_{4p}.

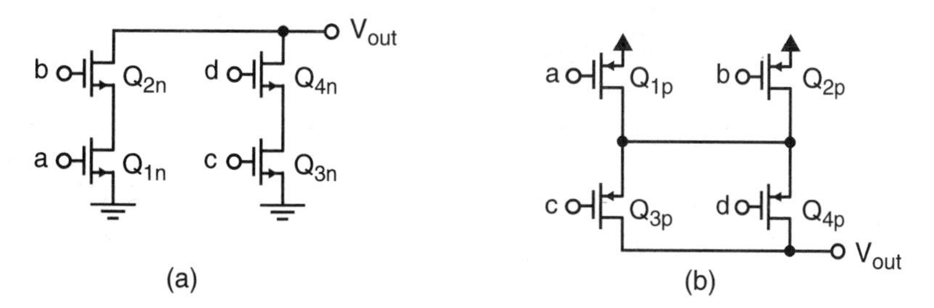

Figure 1.18 (a) The n-channel network and (b) the complementary p-channel network that when connected together realize the function $f = \overline{ab + cd}$.

Example 1.5

Design a traditional CMOS gate to realize the function $f = \overline{(a + b)c}$.

Solution: This example is the CMOS version of the NMOS gate designed in Example 1.3. The n-channel network can be identical to that of Example 1.3, except now all the transistors would be taken with the same width-to-length ratios or W/Ls (for example, a 4 μm width over a 0.5 μm length). The p-channel network is taken as the complementary network where again parallel n-channel subcircuits transform into series p-channel subcircuits and vice versa. The resulting logic gate is shown in Fig. 1.19.

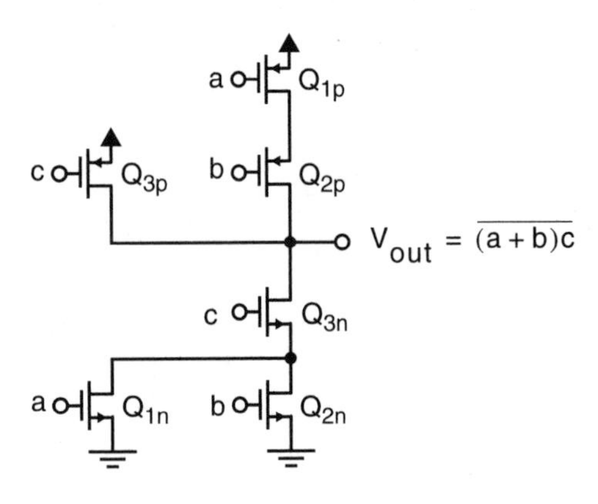

Figure 1.19 An example of a compound CMOS logic gate.

The simplistic description of NMOS and CMOS gates just presented is adequate for the design of many low- to medium-speed logic circuits.

1.3 Computer Simulation

When designing integrated circuits at the transistor level, the use of computer simulation programs has become indispensable. There are many different types of computer simulation programs that are used when developing digital integrated circuits. At the highest level, one might use a functional simulator where every logic function is simulated by the boolean functions being implemented. The logic gates might be simulated as having no delay, or possibly a fixed delay. Alternatively, one might simulate the circuits with the transistors modeled as switches. A more accurate simulation would be one in which the transistors are modeled as switches having a finite series impedance and the parasitic capacitances at every node are taken into account. However, for small circuits, the most accurate simulation is obtained using very small time steps and the I–V characteristics of each device is modeled. One of the most popular simulators that operates at this level is called *SPICE*. It was developed at the University of California at Berkeley and was originally coded by Dr. Lawrence Nagel (Nagel, 1975). It has become very popular because of its accuracy and also because it was placed in the public domain and was thus easily obtainable.

SPICE (and most of its offshoots such as H-SPICE,[8] and P-SPICE[9]) can simulate small- to medium-sized digital circuits quite accurately as long as the accurate parameters for the transistor models are used.

SPICE-like programs are not used very often to simulate *complete* integrated circuits because they are computationally intensive and somewhat prone to convergence problems. However, for the design and characterization of logic gates at the transistor level, they are indispensable. Since SPICE will be used extensively in examples and problem sets, its use will be briefly described in this section.

It is expected that most students will already be familiar with SPICE from either previous courses or laboratories, but for those students who have never used it, this section along with some first-time help from a teaching assistant and a reading of the SPICE manuals should be adequate to get started. For those students who are already familiar with SPICE, this section may be omitted.

The first step when using SPICE is to label a neatly drawn schematic with *labels* (sometimes numbers) for each node, and then to prepare an input file netlist that describes the circuit. This input file contains a line for each component along with the circuit nodes to which the components will be connected.[10] It also contains model cards that supply values for the parameters of the transistor parameters. Finally, it contains some control lines that specify the types of analyses that should be done. The order of the lines in the file is not important, although most designers settle on a consistent standard. The file must start with a title line and end with an .END line (although later versions of SPICE do not require this).

In addition to the circuit description, the parameters for the transistor models must be supplied. An example of the lines that describe the transistor *models* of an n-well CMOS process having transistor channel lengths equal to 0.6 μm might look like

```
.MODEL nch NMOS
+ LEVEL=3 PHI=0.70 TOX=1.0E-08 XJ=0.20U TPG=1
+ VTO=0.8 DELTA=2.5E-01 LD=4.0E-08 KP=1.88E-04
+ UO=545 THETA=2.5E-01 RSH=2.1E+01 GAMMA=0.62
+ NSUB=1.40E+17 NFS=7.1E+11 VMAX=1.9E+05 ETA=2.2E-02
+ KAPPA=9.7E-02 CGDO=3.7E-10 CGSO=3.7E-10 CGBO=4.0E-10
+ CJ=5.4E-04 MJ=0.6 CJSW=1.5E-10 MJSW=0.3 PB=0.99
*
```

[8]Copyright by Meta-Software.

[9]Copyright by MicroSim.

[10]In more modern versions, a graphic schematic entry program such as Viewdraw from ViewLogic might be used to prepare the input file used by SPICE.

```
.MODEL pch PMOS
+ LEVEL=3 PHI=0.70 TOX=1.0E-08 XJ=0.20U TPG=-1
+ VTO=-0.9 DELTA=2.5E-01 LD=6.7E-08 KP=4.45E-05
+ UO=130 THETA=1.8E-01 RSH=3.4E+00 GAMMA=0.52
+ NSUB=9.8E+16 NFS=6.5E+11 VMAX=3.1E+05 ETA=1.8E-02
+ KAPPA=6.3E+00 CGDO=3.7E-10 CGSO=3.7E-10 CGBO=4.3E-10
+ CJ=9.3E-04 MJ=0.5 CJSW=1.5E-10 MJSW=0.3 PB=0.95
*
```

The lines beginning with an "*" can contain anything as they are ignored. They are used for comments and spacing. Since the model statements take up more than one line, they must be continued with a "+" starting in the first position. These model parameters describe a typical 0.6 μm process.[11] These model parameters will be used throughout the book for examples and problems concerning CMOS circuits, thus, readers might carefully copy them to a special file that can be included as the first lines in their SPICE input files. The correspondence between the various parameters and the transistor models will be discussed further in Chapter 3 and in various SPICE references.

After the model specifications, the other lines that describe the circuit components and their interconnections could be included. As an example, assume the simple inverter shown in Fig. 1.20 is to be simulated for its d.c. transfer curve and transient response. The next few lines might be

```
M1 3 2 0 0 CMOSN W=4u L=0.6u AS=7.2p PS=7.6u AD=7.2p PD=7.6u
M2 3 2 1 1 CMOSP W=5.5u L=0.6u AS=9.9p PS=9.1u AD=9.9p PD=9.1u
CL 3 0 0.05pF
```

The name of the n-channel transistor is M_1. Its drain is connected to node 3, its gate is connected to node 2, and its source is connected to node 0. The fourth connection to node 0 is for the substrate. For the time being, it can be assumed that all n-channel substrates will be con-

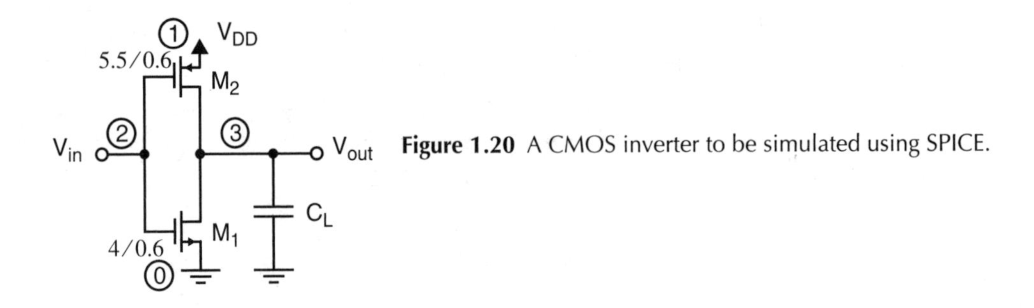

Figure 1.20 A CMOS inverter to be simulated using SPICE.

[11]This process will likely be supplanted by a more modern process having smaller dimensions by the time of publication.

nected to ground (which is always designated node 0), and that the substrates of p-channel transistors will always be connected to the positive power supply (V_{DD}), which is normally equal to 3.3 V. Thus, the p-channel connections of its drain, gate, source, and substrate are to nodes 3, 2, 1, and 1, respectively. The order of specifying the *MOS* transistor connections never changes. The next fields are references to the appropriate models (with parameters) to be used: *CMOSN* for the n-channel transistor and *CMOSP* for the p-channel transistor. The final fields are for the transistor dimensions. The most important ones are W and L, which stand for width and length, respectively. The *u* after the values designate 10^{-6}. Thus, W = 4u denotes that the width of the n-channel transistor is 4 μm. The other fields, *AS*, *PS*, *AD*, and *PD* stand for the area of the source junction, the periphery of the source junction, the area of the drain junction, and the periphery of the drain junction, respectively. For the time being, the readers may blindly assume they are given by

$$A_S = A_D = 3WL \tag{1.3}$$

and

$$P_S = P_D = W + 6L \tag{1.4}$$

These equations will be justified in Chapter 2, but for the time being it can be said that they are reasonable guesses given that the actual layout of the masks has not been done. They are included so SPICE can estimate some of the parasitic capacitances that affect the transient response times only. In some versions of SPICE, if they are not specified, they default to zero (P-SPICE); in other versions, they default to values calculated automatically using formulas similar to those given in equations (1.3) and (1.4) (H-SPICE).

The next two lines describe two independent sources, the power supply voltage (*VDD*), which is between node 1 and 0 and has a value of 3.3 V, and the step input voltage (*VIN*), which goes from 0 to 3.3 V with a delay of 0, has 100 ps rise and fall times, a 5 ns pulse width, and a 10 ns period. Other types of transient input signals such as sine waves and piecewise continuous signals are described in SPICE manuals. The descriptions of the independent voltage sources might look like

```
VDD 1 0 3.3V
VIN 2 0 PULSE(0 3.3 0 100p 100p 5n 10n)
```

Next, the desired analyses are specified. In this case, the transient response of the inverter will be analyzed from time 0 to 10 ns and the voltage of nodes 2 and 3 will be printed and plotted every 0.05 ns.

```
.TRAN 0.05n 10n
.PRINT TRAN V(2) V(3)
.PLOT TRAN V(2) V(3)
```

The next line specifies that the circuit will be analyzed for its low-frequency voltages for a variety of different input voltages. In this case, the input voltage will go from 0 to 3.3 V in increments of 0.01 V.

```
.DC VIN 0V 3.3V 0.01V
.PRINT DC V(3)
.PLOT DC V(3)
```

In addition to these types of analyses, it is possible to do small-signal a.c. frequency-response analyses about a linearized approximation to the circuit where all of the nonlinear transistors have been replaced by their small-signal linear equivalent circuits. These types of analyses are often used for analog circuit design but are not often used for digital circuit design.

The final line is an .END card that specifies there are no more lines in the input file.

```
.END
```

It should be noted that upper-case letters have been used for most names. This was required in early versions of SPICE, but is seldom required now, as most recent versions do not differentiate between upper-case and lower-case letters. Also, many of the newer versions allow for interactive simulations where simulation results may be viewed as a simulation progresses. Finally, most modern versions of SPICE include a graphic user interface for viewing voltage and current waveforms, and may include a schematic entry interface for describing the circuit. Again, the reader should consult the manuals for the version they are using. A complete listing of the input file would then look like that shown in Fig. 1.21.

```
A CMOS Inverter
*
.MODEL nch NMOS
+     LEVEL=3 PHI=0.70 TOX=1.0E-08 XJ=0.20U TPG=1
+     VTO=0.8 DELTA=2.5E-01 LD=4.0E-08 KP=1.88E-04
+     UO=545 THETA=2.5E-01 RSH=2.1E+01 GAMMA=0.62
+     NSUB=1.40E+17 NFS=7.1E+11 VMAX=1.9E+05 ETA=2.2E-02
+     KAPPA=9.7E-02 CGDO=3.7E-10 CGSO=3.7E-10 CGBO=4.0E-10
+     CJ=5.4E-04 MJ=0.6 CJSW=1.5E-10 MJSW=0.3 PB=0.99
*
.MODEL pch PMOS
+     LEVEL=3 PHI=0.70 TOX=1.0E-08 XJ=0.20U TPG=-1
+     VTO=-0.9 DELTA=2.5E-01 LD=6.7E-08 KP=4.45E-05
+     UO=130 THETA=1.8E-01 RSH=3.4E+00 GAMMA=0.52
+     NSUB=9.8E+16 NFS=6.5E+11 VMAX=3.1E+05 ETA=1.8E-02
```

Figure 1.21 A listing of an input file to be used in the SPICE simulation of a CMOS inverter.

```
+    KAPPA=6.3E+00 CGDO=3.7E-10 CGSO=3.7E-10 CGBO=4.3E-10
+    CJ=9.3E-04 MJ=0.5 CJSW=1.5E-10 MJSW=0.3 PB=0.95
*
M1 3 2 0 0 CMOSN W=4u L=0.6u AS=7.2p PS=7.6u AD=7.2p PD=7.6u
M2 3 2 1 1 CMOSP W=5.5u L=0.6u AS=9.9p PS=9.1u AD=9.9p PD=9.1u
CL 3 0 0.05pF
*
VDD 1 0 3.3V
VIN 2 0 PULSE(0 3.3 0 100p 100p 5n 10n)
*
.TRAN 0.05n 10n
.PRINT TRAN V(2)
.PLOT TRAN V(2)
*
.DC VIN 0V 3.3V 0.01V
.PRINT DC V(3) V(4) V(5)
.PLOT DC V(5)
*
.END
```

Figure 1.21 (continued).

If the name of the input file was *inverter.circ*, then on most computers SPICE could be run by using a command like "*SPICE inverter.circ* > *inverter.out.*" The details of the command line to run SPICE can be found from your computer-system administrator or a teaching assistant.

The plot of the transient response produced by SPICE version 3C1, an interactive version that runs under X-windows, is shown in Fig. 1.22. The plot of the d.c. transfer curve produced by SPICE 3C1 is shown in Fig. 1.23.

For those unfamiliar with SPICE, the example just presented should be entered and run and the results compared to those of Figs. 1.22 and 1.23. Most other SPICE simulations will use very similar input files except the component section will be different and perhaps different .MODEL cards will be required for different devices such as n-channel depletion transistors and bipolar transistors.

1.4 Transfer Curves and Noise Margins

A common way to characterize logic families is to plot their low-frequency input–output relationship through the use of transfer curves. A transfer curve is a plot of the output voltage of an inverter as a function of the input voltage. An example of a transfer curve for a CMOS inverter

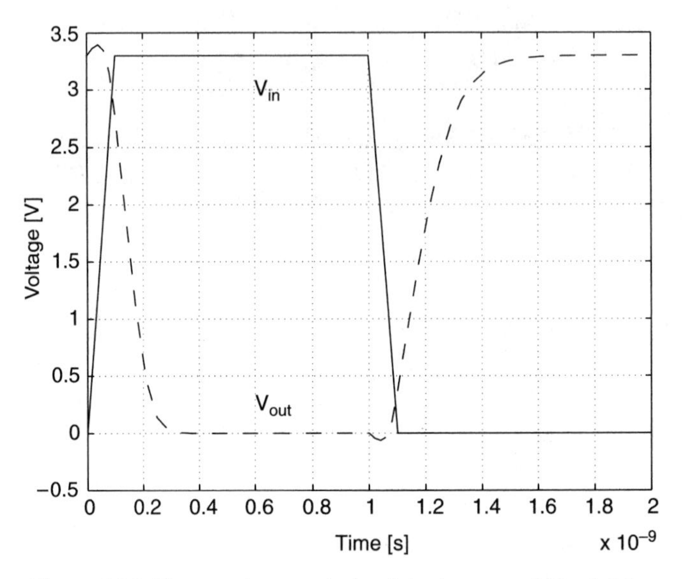

Figure 1.22 The transient analysis of the inverter of Fig. 1.20.

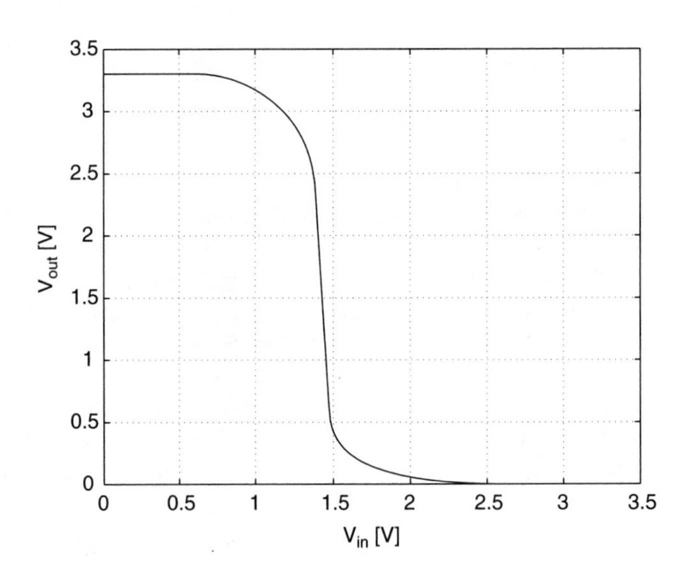

Figure 1.23 The d.c. transfer curve of the inverter of Fig. 1.20.

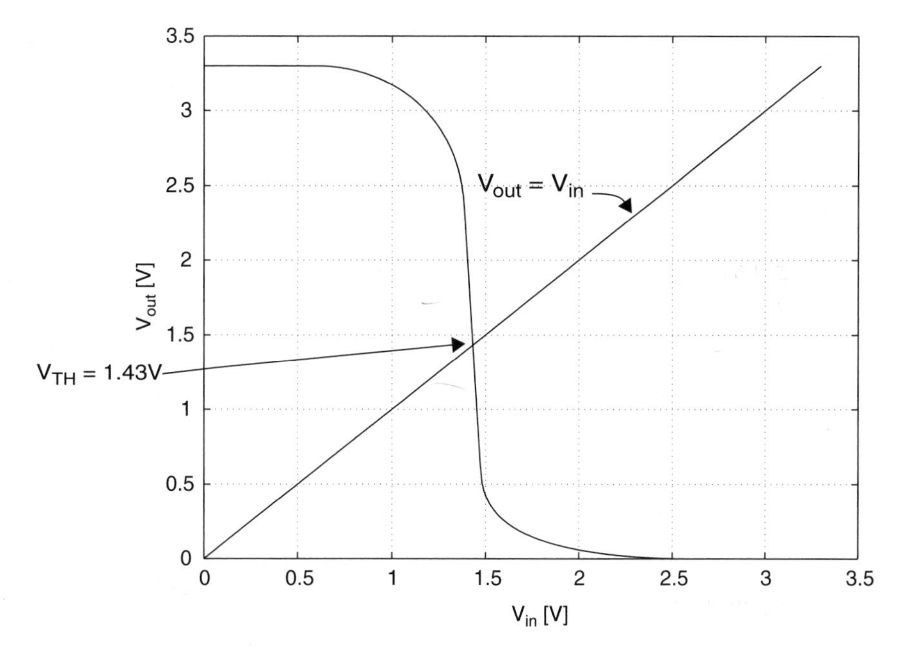

Figure 1.24 The intersection of a gate transfer curve and the line defined by $V_{out} = V_{in}$ determines the gate threshold voltage V_{TH}.

was shown in Fig. 1.23. This curve could be obtained by analyzing the gate as will be described in Chapter 4, or by simulating the gate using a program such as SPICE as was just described. The transfer curve is useful to determine if the device sizes and the circuit design are adequate to ensure *good* operation as a digital gate.

There are many parameters that can be immediately obtained from the transfer curve that are useful in making this determination. Perhaps the most useful are the gate threshold voltage, V_{TH}, the noise margins, and the small-signal gain of the gate at its threshold voltage. The threshold voltage, defined as the input voltage that gives an identical output voltage, can be found by finding the intersection of the transfer curve with a plot of the straight line resulting from plotting the function $V_{out} = V_{in}$. This has been done in Fig. 1.24 and by careful inspection it can be seen that the threshold voltage of the CMOS inverter of Fig. 1.20 is approximately 1.43 V. Alternatively, this could have been determined very accurately using the .print command from SPICE. An input voltage substantially greater than the threshold voltage V_{TH} is recognized as a "1" by the gate, whereas an input voltage substantially less than V_{TH} is recognized as a "0".

Another important characterization of a digital gate is the absolute value of the small-signal gain at the threshold voltage. The minimum allowable value has an absolute value greater than unity. If this were not the case, then as a digital signal was transmitted through more and more gates, its voltage level would get closer and closer to V_{TH} until eventually it would be *lost in noise*. The practical minimum allowable value for the absolute value of the gain of a digital gate is around $\sqrt{2}$ in order to get a reasonable amount of signal regeneration from gate to gate. Typical gains range from around two for very high-frequency circuits to around 50. The gain of

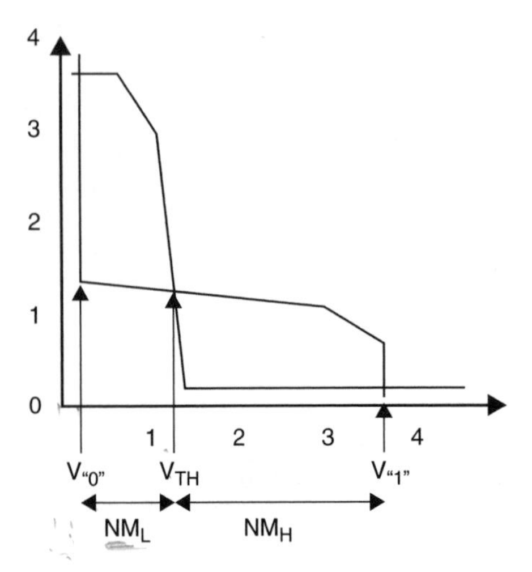

Figure 1.25 A plot of V_{out} versus V_{in} along with a plot of V_{in} versus V_{out} that can be used to obtain the noise margins.

the CMOS inverter of Fig. 1.20 is –20.1 at V_{TH}, as can be determined from the transfer curve of Fig. 1.24.

Other important parameters that can be obtained from the plot of the transfer curve are the typical gate high, $V_{“1”}$, and low, $V_{“0”}$, output voltages that correspond to a “1” and a “0” output, respectively. These voltage levels are defined as the typical gate output voltages when it is being driven by an identical gate. A formal definition would be the intersection points of a plot of V_{out} versus V_{in} with a plot of the inverse function, that is a plot of V_{in} versus V_{out}. An example of this is shown in Fig. 1.25. The transfer curves shown are for a *bipolar TTL inverter.* This type of inverter used to be quite popular for discrete logic circuits, although recently it has largely been replaced by CMOS logic. It can be seen that $V_{“0”}$ is approximately 0.4 V and $V_{“1”}$ is approximately 3.6 V. The gate threshold voltage can also be obtained as the middle intersection and is seen to be approximately 1.25 V for the TTL inverter. For most transfer curves, it is not necessary to plot V_{in} versus V_{out}, as $V_{“0”}$ and $V_{“1”}$ can usually be obtained approximately by simply inspecting the transfer curve. For example, from the transfer curve of Fig. 1.23, it is easily seen that the typical high and low voltages of a CMOS inverter are 3.3 and 0 V, respectively.

The difference between the typical high and low voltages of a logic family and the threshold voltages of a logic family are called the high and low noise margins, respectively. These are often denoted NM_H and NM_L. For the CMOS inverter of Fig. 1.20, they are approximately 1.9 and 1.4 V, as is easily seen from the transfer curve of Fig. 1.24. The noise margins of the TTL gate as obtained from the transfer curve of Fig. 1.25 are 0.85 V for NM_H and 2.35 V for NM_H. Obviously, the larger the noise margins, the less sensitive the gate should be to having an incorrect output value caused by noise being inadvertently injected at the input or into the ground or V_{DD} lines. Thus, the noise margins are an *indication* of the gate's insensitivity to noise. The noise margins of traditional CMOS gates are almost always larger than 1 V for a 3.3-V power supply. Some high-speed bipolar gates, such as current-mode-logic gates, can have noise margins as low as 0.2 V, and still operate dependably without making errors, assuming they have been carefully designed and manufactured.

It should be emphasized that the voltage noise margins are only indications of a logic family's insensitivity to noise. In some high-speed circuits, such as the bipolar current-mode logic circuits just mentioned, the gates are designed to have small voltage changes, but large current changes. For these circuits, a much better figure of merit would be power noise margins, which would take into account the impedance levels of the nodes of the circuit. Also, voltage noise margins ignore many other important details, such as whether the circuits use differential signals (which greatly reduces the noise sensitivity), the length of the ground and V_{DD} lines, the impedance of the lines, how many gates are connected to them, and how much noise is on the ground and V_{DD} lines due to external circuits such as output buffers. Thus, when comparing logic families, the noise margins should not be considered alone without considering many other factors.

There are many popular definitions for noise-margins different from the one presented here. Perhaps, the most popular definition is the difference between the output low or high voltages and the points at which the gain is –1 (Hodges and Jackson, 1988). This is a reasonable definition except for the fact that it makes the noise margins extremely difficult to calculate analytically for many gates. For this reason, and because knowing the noise margins very accurately is of minimal use in choosing logic families or in doing logic design, we will be using the much simpler noise margins introduced herein, which can almost always be obtained by inspection of the transfer curves or from straightforward analysis, as will be seen in Chapter 4.

1.5 Gate Delays and Rise and Fall Times

In the previous section, we saw how figures of merit such as the inverter threshold voltage and noise margins could be used to characterize the d.c. performance of a logic family. Similarly, inverter delay times and inverter rise and fall times can be used to characterize the transient performance of a logic family. The inverter delay time, t_D, is defined as the time from which the input signal to an inverter crosses through the inverter threshold voltage to the time at which the inverter output voltage crosses through the threshold voltage. This is shown graphically in Fig. 1.26. This delay includes delays due to the rise and fall times and also delays due to internal nodes of a gate, which may dominate. It is particularly relevant for multistage logic gates such as a bipolar TTL gate, but does not have that much relevance for CMOS inverters, as they have no internal nodes. For CMOS gates, better figures of merit are the gate rise and fall times.

Traditionally, the gate rise (or fall) time was defined as the time from which a logic gate output voltage has undergone a 10% change to the time at which it has undergone a 90% change (i.e., is within 10% of its final value). This definition has some problems with gates that start to change very quickly, but tend to move very slowly near the end of their transition. For this type of gate, it is possible that the gate output voltage will not have reached its 90% change voltage by the time a succeeding gate or two have already had their outputs change state. A better figure of merit is the time from which the gate input has changed through its threshold voltage to the time at which the gate output has undergone a 70% change. For the CMOS inverter, which has 0 and 3.3 V as the typical high and low output voltages, this translates to the rise time,

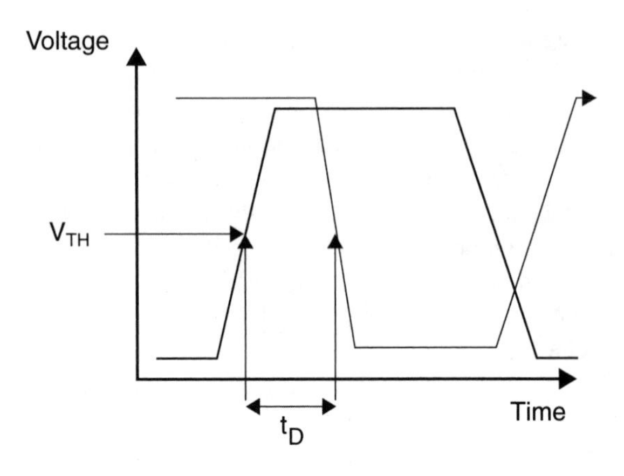

Figure 1.26 The gate delay time (t_D) is the time from which the input voltage changes through V_{TH} to the time at which the output voltage changes through V_{TH}.

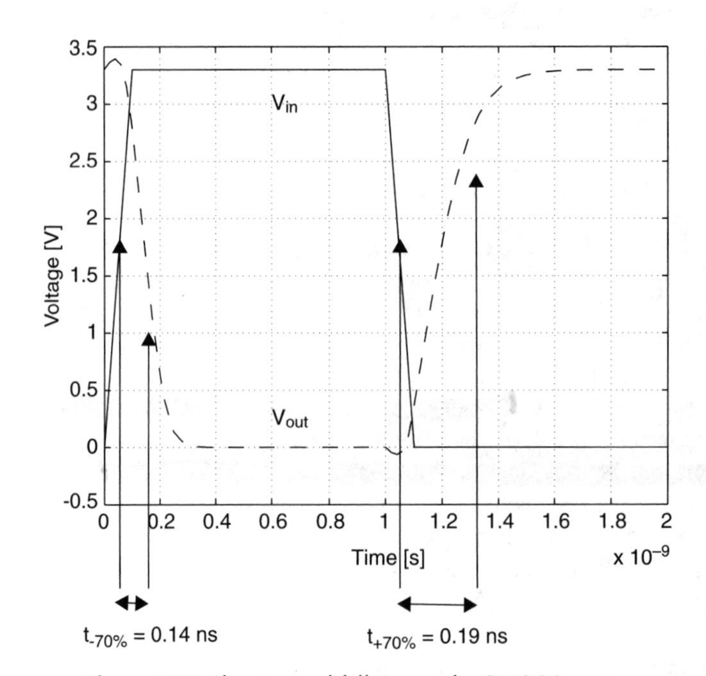

Figure 1.27 The rise and fall times of a CMOS inverter.

$t_{+70\%}$, being defined as the time it takes the input voltage to change to the time the output voltage has changed from 0 to 2.3 V. Also, the fall time, $t_{-70\%}$, is defined as the time it takes the input voltage to change to the time it takes the output voltage to change from 3.3 to 1 V. These times are shown graphically in Fig. 1.27, which is a plot of the transient response of the CMOS inverter of Fig. 1.20 that was obtained previously in Section 1.3 using SPICE. It can be seen that the rise time is 0.19 ns and the fall time is 0.14 ns.

1.6 Transient Response

The most important concern when designing a digital integrated circuit is that it operates functionally correct. The next largest concerns are maximizing speed and minimizing power and area. There are many factors that are important in determining the maximum speed of a digital circuit. One constraint is the number of digital delays through which the signal must propagate. This can be minimized by good system and logic design using techniques such as pipelining, parallel circuit design, and carry-save adders. These techniques will be covered much later in the text, and are well covered in many other books on computer architecture (Mano and Kime, 1997) and arithmetic circuits. The other way the speed of digital integrated circuits can be maximized is to minimize the delay of the gates in the critical signal paths. This subject will be dealt with extensively throughout this text.

The exact analytical determination of the transient response of a digital logic gate is completely intractable due to the high order of the system and the fact that it is highly nonlinear. Also, many critical factors such as the parasitic capacitances are not known at the time of design.[12] However, to design a good, fast, integrated circuit, it is essential that the critical nodes can be identified, that reasonably accurate delays can be approximated, and that the most important factors limiting the speeds can be identified and minimized. To do this, the designer must make approximations. In making approximations, a good knowledge of transistor operation and modeling, IC processing, and some analog design principles can be very useful.

An approximation that is almost universally used for hand analysis is that all parasitic capacitances are between nodes and ground only. Coupling capacitances are usually ignored when estimating the delay of a digital integrated circuit without the use of a computer. Another approximation that is often made is that the circuit components charging or discharging these parasitic capacitances can be approximated by either current sources or resistors. This reduces the problem to that of solving a first-order differential equation at every node.

Fig. 1.28 shows the equivalent circuit for charging a node voltage, V_N, that has previously been discharged, and has a parasitic capacitance, C_p, between that node and ground. It is also connected to the power-supply voltage V_{DD}, through a load impedance Z_L. This approximation

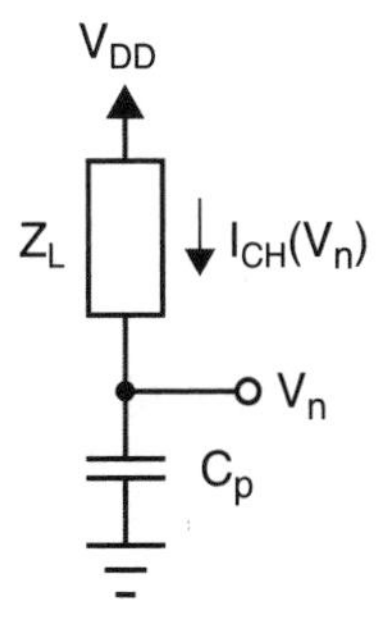

Figure 1.28 A simplified model for charging a node voltage having parasitic capacitance C_p.

[12]Primarily for this reason, digital integrated circuits should be designed so that they always function correctly *independent of the exact delay of the gates.*

assumes that the d.c. impedance between the output node and ground is infinite. The impedance between the output node and V_{DD}, Z_L, is assumed to be finite and to have an I–V characteristic described by $I_{ch}(V_n)$. Remembering that the I–V relationship of a capacitor is given by

$$I = C\frac{dV_C}{dt} \tag{1.5}$$

then the time it takes for V_n to charge from V_{n-1}, at t_1, to V_{n-2}, at t_2, is given by

$$t_2 - t_1 = \int_{V_{n1}}^{V_{n2}} \frac{C_p dV_n}{I_{ch}(V_n)} \tag{1.6}$$

When Z_L is approximated by a constant current-source, I_{ch} is independent of V_n and equation (1.6) simplifies to

$$t_2 - t_1 = \frac{C_p(V_{n-2} - V_{n-1})}{I_{ch}} \tag{1.7}$$

Rewriting equation (1.7) gives

$$\Delta t = \frac{C_p \Delta V_n}{I_{ch}} \tag{1.8}$$

where ΔV_n is the node voltage change that takes place during time Δt.

Example 1.6

For the inverter shown in Fig. 1.3, assume the input is initially a "1" and then changes to a "0". The output then changes from close to 0 to 3.3 V. Assume that during this transient the average current charging a 50 fF load capacitance is 0.4 mA. Find the time required for the output to change to 70% of its final voltage.

Solution: Assuming the output is initially at 0 V, we need to find the time it takes to change from approximately 0 V to 70% of 3.3 V, or 2.31 V. We have using equation (1.8)

$$t_{+70\%} = \frac{5\times10^{-14} 2.31}{4\times10^{-4}} = 0.29 \text{ ns} \tag{1.9}$$

Equation (1.8) is perhaps the most important equation in this text concerning the transient response of digital integrated circuits. The three golden principles to minimizing the delay of digital integrated circuits follow as a direct consequence of equation (1.8):

1. *The capacitance of critical nodes must be minimized.* This is achieved by minimizing the interconnect capacitances and the parasitic device capacitances of critical nodes. This will be discussed in the section on layout in Chapter 2.
2. *Minimize the voltage changes at all critical nodes, especially those that inherently have large capacitances, such as buses.* This is often possible by using advanced digital circuit design techniques such as sense amplifiers and common-gate or cascode amplifiers, and will be discussed in later chapters. This principle is also one of the main reasons that bipolar current mode logic (described in Chapter 8), which can have voltage changes as small as 0.4 V, is faster than traditional CMOS logic design.
3. *Maximize the currents available for charging or discharging the nodes during the transient voltage changes.* Examples of circuit design techniques to achieve this will occur throughout the text. This is especially important at IC outputs, where large load capacitances must necessarily be driven.

Although these principles appear to be quite obvious, it is surprising how often they have been ignored in many commercially available ICs.

Another commonly used approximation is to approximate Z_L of Fig. 1.28 by a resistor, R_L, as is shown in Fig. 1.29, rather than approximating it by a constant current source.

If the device charging the node is approximated by a resistor R_L, we have I_{ch} given by

$$I_{ch} = \frac{(V_{DD} - V_n)}{R_L} \tag{1.10}$$

and equation (1.6) becomes

$$t_2 - t_1 = \int_{V_{n-1}}^{V_{n-2}} \frac{C_p R_L dV_n}{V_{DD} - V_n} \tag{1.11}$$

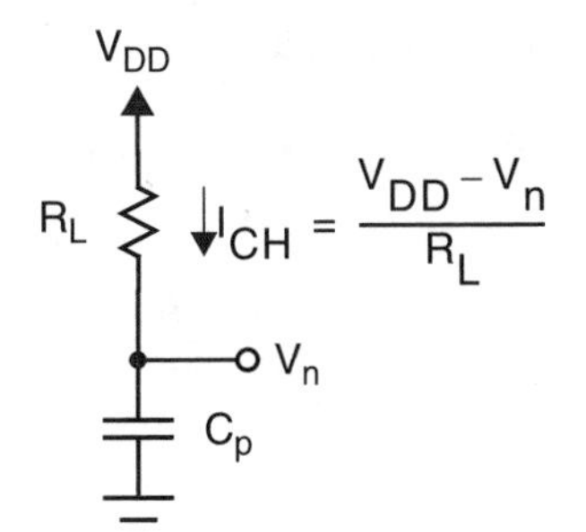

Figure 1.29 A simplified model where the load impedance is approximated by a resistor.

Evaluating the integral, we get

$$\Delta t = R_L C_p \ln\left(\frac{V_{DD} - V_{n-1}}{V_{DD} - V_{n-2}}\right) \tag{1.12}$$

Equation (1.10) could have been more simply derived by remembering that the solution for the transient response of a voltage of any first-order circuit is given by

$$v(t_2) = v(t_\infty) - [v(t_\infty) - v(t_1)]e^{-\frac{\Delta t}{\tau}} \tag{1.13}$$

where it is assumed that the input signal does not change from time t_1 on. If we solve equation (1.13) for the time Δt, we get

$$\Delta t = \tau \ln\left[\frac{v(t_\infty) - v(t_1)}{v(t_\infty) - v(t_2)}\right] \tag{1.14}$$

For the circuit of Fig. 1.29, τ is equal to $R_L C_p$, $v(t_1)$ is V_{n-1}, and $v(t_2)$ is V_{n-2}. After making these substitutions, equation (1.12) follows immediately. Equations (1.12) and (1.13) are used so often, especially for CMOS circuits, that the reader is strongly encouraged to memorize them.

Example 1.7

For the circuit of Fig. 1.3, assuming the load capacitance is 50 fF, the load resistor is 4 kΩ, and during the fall time the transistor may be approximated by a 300-Ω resistor, find the 70% rise and fall times.

Solution: During the rise time, the transistor is off and the resistance seen by the load capacitance is simply R_L, which is 4 kΩ. Therefore using (1.12), we have

$$t_{+70\%} = (4000 \times 5 \times 10^{-14}) \ln\left(\frac{3.3 - 0}{3.3 - 2.31}\right) = 2 \times 10^{-10} \times 1.20 = 0.24 \text{ ns} \tag{1.15}$$

During the fall time, we have R_L in parallel with the equivalent resistance of the transistor, R_{eq}. The parallel resistance is therefore $R_L \parallel R_{eq} = 279\ \Omega$. Assuming the final settling voltage is 0 V, we have[13] that a –70% transition occurs when V_{out} reaches 3.3 – 0.7 (3.3) = 0.99 V. Therefore

$$t_{-70\%} = (279 \times 5\times10^{-14})\ln\left(\frac{0-3.3}{0-0.99}\right) = 1.4\times10^{-11} \times 1.20 = 0.017 \text{ ns} \qquad (1.16)$$

Notice how much faster the fall time is than the rise time. In reality, the fall time would be larger due to the finite rise time of the input signal, which we implicitly assumed to be 0.

1.7 An RC Approximation to the Transient Response of a CMOS Inverter

As mentioned previously, a CMOS inverter can be modeled as a pair of switches with each having a series impedance. This is shown in Fig. 1.30.

If we assume that initially the input voltage had been at 3.3 V, and at t = 0 it changed to 0 V, the output voltage would initially be at 0 V and then at t = 0 it would start to be charged up by

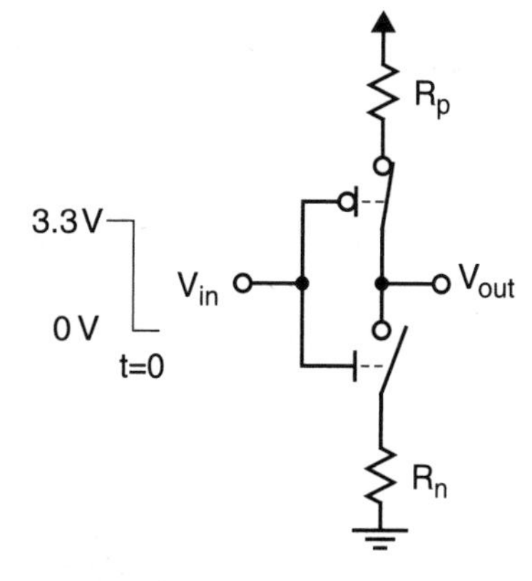

Figure 1.30 Modeling a CMOS inverter using switches and resistors.

[13]The actual final voltage is 0.23 V from Example 1.1, but this additional accuracy is not merited considering the inaccuracies of approximating a transistor by a resistor.

the p-channel transistor to 3.3 V. Assuming the p-channel transistor can be modeled by a resistor R_p, then the time it would take the output to get to 2.3 V is given by

$$\Delta t = R_p C_p \ln\left(\frac{3.3 - 0}{3.3 - 2.3}\right) = 1.2 R_p C_p \tag{1.17}$$

But in the previous section we saw using SPICE simulation that the actual gate had a rise time of $\Delta t_{+70\%} = 0.19$ ns for a 0.05 pF load capacitance. A resistor of size R_p would give the same rise time if R_p is chosen according to the formula

$$R_p = \frac{\Delta t_{-70\%}}{1.2 C_p} = 3.2\ \text{k}\Omega \tag{1.18}$$

It should be kept in mind that this size resistor gives the same rise time as a 5.5-µm-wide transistor. If a wider p-channel transistor were used, the impedance should be scaled smaller in an inversely proportional manner.

In a similar manner, the 70% fall time was found to be 0.14 ns using SPICE simulation, and this implies an equivalent resistor for a 4-µm-wide n-channel transistor given by

$$R_n = \frac{\Delta t_{-70\%}}{1.2 C_p} = 2.3\ \text{k}\Omega \tag{1.19}$$

Again, if the 4-µm-wide n-channel transistor was replaced with an n-channel transistor having twice the width, the fall times would be half as small and an equivalent resistor in this case should be 1.15 kΩ. Similarly, a 1-µm-wide transistor would be equivalent to a resistor of size 9.2 kΩ. Based on this approach, it is easy to derive the general expressions for the equivalent resistances of transistors of arbitrary width. For an n-channel transistor, we have

$$R_n = \frac{9.2\ \text{k}\Omega}{W} \tag{1.20}$$

and for p-channel transistors, we have

$$R_p = \frac{17.6\ \text{k}\Omega}{W} \tag{1.21}$$

These values are slightly high because we have assumed the load capacitance was 0.05 pF only, and ignored the parasitic junction capacitance of the transistors. In some circuits with very small loads, these capacitances can double the rise and fall times. In Chapters 2 and 4, we will see how to account for them as well.

Example 1.8

Assuming $C_L = 110$ fF, the width of the p-channel transistor is 15 μm, and the width of the n-channel transistor is 10 μm, find the 70% rise and fall times of a CMOS inverter.

Solution: Using (1.20) gives $R_n = 9.2\ \text{k}\Omega/10 = 920\ \Omega$ and using (1.21) gives $R_p = 17.6\ \text{k}\Omega/15 = 1.17\ \text{k}\Omega$. Therefore, the 70% fall time is given by

$$t_{-70\%} = 920 \times 1.1{\times}10^{-13} \times 1.2 = 0.12\ \text{ns}$$

and the 70% rise time is given by

$$t_{+70\%} = 1173 \times 1.1{\times}10^{-13} \times 1.2 = 0.15\ \text{ns}$$

These are fairly representative rise and fall times for a 0.6 μm minimum channel-length technology.

When the output load capacitance dominates, these formulas are useful for logic gates, as well, with the simple modification that if the output is being driven through a number of series transistors, than the series equivalent of the approximating resistors of the individual transistors should be used. This principle is illustrated by a number of problems.

1.8 Summary

In this chapter, we have introduced many of the basic concepts of MOS digital integrated circuits in a somewhat simplified manner. This initial description is adequate for the transistor-level design of simple circuits that do not have to operate too fast. We have introduced a number of definitions, and we have also introduced the concept of RC approximate modeling for the transients of MOS circuits. Before we can go much further than this, we must first gain a better understanding of IC manufacture, of transistor operation and modeling, and of IC-related parasitics. These will be covered in the next two chapters. Then, in Chapters 4, 5, and 6, we will use this knowledge in developing a more exact and detailed understanding of MOS logic gates. In Chapter 8, we will do the same for bipolar logic gates. After this, the text will deal with increasingly more complicated circuits and with a number of related matters that are important to digital integrated circuit design.

1.9 Bibliography

M. Annaratone, *Digital CMOS Circuit Design*, Kluwer, 1986.

M. Elmasry, ed., *Digital MOS Integrated Circuits, II,* IEEE Press, 1991.

L. Glasser and D. Dopperpuhl, *The Design and Analysis of VLSI Circuits*, Addison-Wesley, 1985.

D. Hodges and J. Jackson, *Analysis and Design of Digital Integrated Circuits*, McGraw-Hill, 1988.

M. Mano, *Digital Design,* Prentice Hall, 1984.

M. Mano and C. Kime, *Logic and Computer Design Fundamentals*, Prentice Hall, 1997.

C. Mead and L. Conway, *Introduction to VLSI Systems*, Addison-Wesley, 1980.

L. Nagel, "Spice2, A Computer Program to Simulate Semiconductor Circuits," *ERL Memorandum ERL-M520*, University of California, Berkeley, May 1975.

J. Rabaey, *Digital Integrated Circuits, A Design Perspective*, Prentice Hall, 1996.

C. Roth, *Fundamentals of Logic Design*, West, 1985.

N. Weste and K. Eshragian, *Principles of CMOS VLSI Design: A Systems Perspective*, Addison-Wesley, 1983.

1.10 Problems

For the problems in this and other chapters, the minimum transistor lengths should be assumed to be 0.5 μm unless explicitly stated otherwise. Transistor lengths in drive networks of logic gates are almost always taken equal to the minimum allowable lengths.

1.1 Design an NMOS logic gate that realizes the function $f = \overline{ab + c(d + e)f}$. Try to choose reasonable device sizes.

1.2 Repeat Problem 1.1 for a CMOS logic gate.

1.3 Design an NMOS logic gate that realizes the function $f = \overline{(ab + cd)(e + f)}$. Try to choose reasonable device sizes.

1.4 Repeat Problem 1.3 for a CMOS logic gate.

1.5 Give a CMOS realization of an *exclusive or* gate where $f = a\bar{b} + \bar{a}b$ (i.e., $f = a \oplus b$).

1.6 Give a CMOS realization of an *exclusive nor* gate where $f = ab + \bar{a}\bar{b}$ (i.e., $f = \overline{a \oplus b}$).

1.7 A single-bit full adder realizes two functions: a sum-generate function and a carry-generate function based on three inputs, a, b, and c_{in}. The sum-generate output, S, is a "1" if an odd number of inputs are "1". A logic function for this function is $S = a \oplus b \oplus c_{in}$. Give a CMOS realization of this function.

1.8 A carry-generate output, C_{out}, of a full adder has a "1" output if two or more inputs are "1". A logic function for this function is $C_{out} = c_{in}(a + b) + ab$. Give a CMOS realization of this function.

1.9 A 2-to-1 multiplexer has a select input, s, and two data inputs, d_0, and d_1. If s is a "0", the output is equal to the d_0 input, otherwise, if s is a "1", the output is equal to the d_1 input. Give a CMOS realization of this function.

1.10 A 2 to 1-of-4 decoder has two inputs and four outputs. For any possible input combination, a single output will be a "1". A different output will be a "1" for every different input combination. Give a CMOS realization of this function.

1.11 Using SPICE, find the gate threshold voltage and the noise margins of an NMOS inverter having a depletion load transistor with size of 1.6 μm/1.6 μm and an enhancement load transistor with size 3.2 μm/0.8 μm. Use the transistor model parameters from the section on SPICE, except for the depletion transistor use $V_{td} = -2.5$ V. Repeat for when the enhancement load transistor is 1.6 μm/0.8 μm and 6.4 μm/0.8 μm.

1.12 Repeat problem 1.11, but for a CMOS inverter with the p-channel transistor having a size of 5 μm/0.8 μm, and separately for the n-channel transistor having sizes 5 μm/0.8 μm, 7.5 μm/0.8 μm, and 10 μm/0.8 μm.

1.13 Repeat Example 1.8, but assume $C_L = 50$ fF, $W_p = 10$ μm, and $W_n = 5$ μm.

1.14 What is the worst-case rise time of a CMOS *nor* gate having $W_p = 15$ μm and $W_n = 5$ μm. The series p-channel transistors can be approximated by two series resistors with the size of each resistor given by (1.21).

1.15 For the circuit shown (Fig. P1.15), estimate the worst case rise and fall times when a 1-pF load capacitor is driven. Assume an n-channel transistor can be approximated by a resistor whose size is given by $R_n = 17.5$ k Ω/W (with W given in μm), and a p-channel transistor can be approximated by a resistor whose size is given by $R_p = 48.75$ kΩ/W (again, with W given in micrometers). Also, estimate the fall time if all inputs instantaneously change from '"0"s to "1"s.

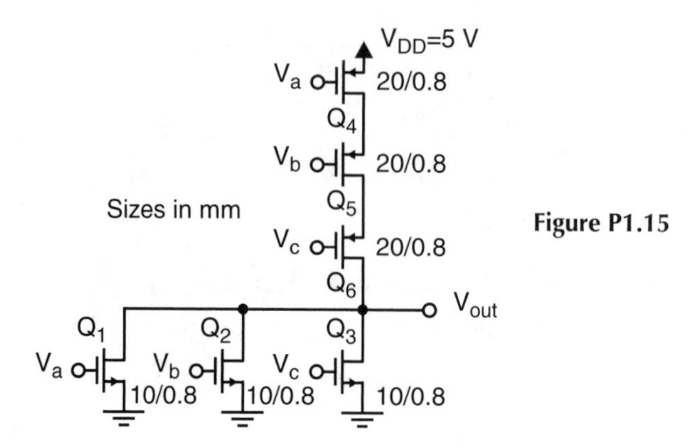

Figure P1.15

1.16 Compare your answers to those obtained using SPICE simulation. Use the same model parameters as given in the section on SPICE.

1.17 Using SPICE, find the threshold voltage of the gate shown in Problem 1.15 when only one input changes from a "0" to a "1". Repeat for all inputs changing together from "0"s to "1"s.

1.18 What is the logic function realized by the following NMOS gate (Fig. P1.18). Give an equivalent logic-gate schematic. Supply reasonable transistor sizes.

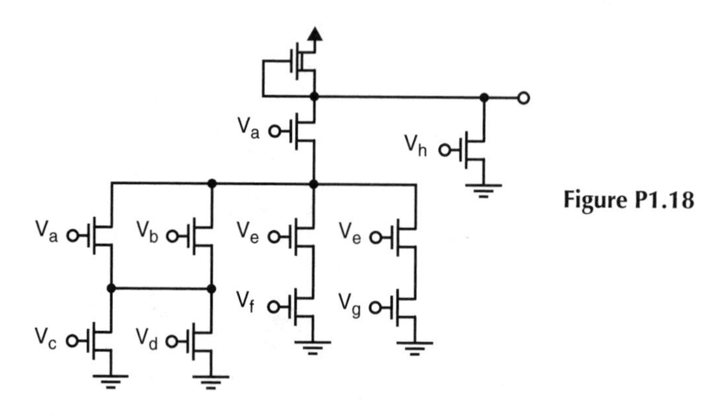

Figure P1.18

1.19 Give a CMOS equivalent circuit to the NMOS gate shown in Fig. P1.18. If possible, try to minimize the number of series p-channel transistors.

CHAPTER 2

Processing, Layout, and Related Issues

This chapter[1] describes the steps and processes used in realizing modern integrated circuits. Although emphasis is placed on CMOS processing, the technology required for bipolar-junction transistor (BJT) circuits is also described. After processing is presented, circuit layout is covered. Layout is the design phase in which the geometry of circuit elements and wiring connections is defined. This stage leads to the development of photographic masks, which are used in manufacturing a microcircuit. The concepts of design rules and the relationship between the rules and the microcircuits are emphasized. Next, circuit layout is related to the transistor models. Here, it is shown that once the layout is completed, the values of certain elements in the transistor models can be determined. This knowledge is necessary for accurate computer simulation of integrated circuits. It is also shown that by using typical design rules, reasonable assumptions to approximate transistor parasitic components can be made before the layout has been done. The chapter concludes with a brief description of more modern processing techniques.

2.1 CMOS Processing

In this section, the steps involved in processing a CMOS microcircuit are presented. For illustrative purposes, we describe here an example n-well process (with, of course, a p^- substrate)

---

[1]The authors would like to acknowledge publishers John Wiley for giving permission to copy much of the material in this chapter from Chapter 2 of *Analog Integrated Circuit Design*, Wiley, 1997.

and two layers of metal. Although the list is not complete, an attempt is also made to describe many of the possible variations during processing.

The Silicon Wafer

The first step in realizing a microcircuit is to fabricate a defect-free, single-crystalline, lightly doped wafer. To create such a wafer, one starts by creating metallurgical-grade silicon through the use of a high-temperature chemical process in an *electrode-arc furnace.* Although *metallurgical-grade silicon* is about 98% pure, it has far too many impurities for use in realizing microcircuits. Next, a gas containing silicon is formed and then reduced. Pure silicon is precipitated onto thin rods of *single-crystalline silicon.* This deposited *electronic-grade silicon* is very pure but, unfortunately, it is also *polycrystalline.* To obtain single-crystalline silicon, the silicon is melted once again and allowed to cool, as a single-crystalline *ingot* is slowly *pulled* and *turned* from the molten silicon using the *Czochralski method.* The Czochralski method starts with a seed of single crystal silicon in which the pull rate and speed of rotation determine the diameter of the crystalline rod or ingot. Typical diameters might be 4–8 in. (i.e., 10–20 cm) with typical lengths normally larger than 1 m. Producing a silicon ingot can take several days.

Normally, heavily doped silicon will be added to the *melt* before *pulling* the single-crystal ingot. After it diffuses through the molten silicon, a lightly doped silicon ingot results. In our example process, boron impurities would be added to produce a p^- ingot. The "–" superscript after the p denotes lightly doped p silicon.

The ingot is cut into wafers using a large diamond saw. A typical wafer might have a thickness as large as 1 mm. After the ingot is sawed into wafers, each wafer is polished with Al_2O_3, chemically etched to remove mechanically damaged material, and then fine polished again with SiO_2 particles in an aqueous solution of NaOH.

Very often, the company that produces the silicon wafers is not the same company that eventually patterns them into monolithic circuits. Sometimes, the surface of the wafer might be doped more heavily, and a single-crystal *epitaxial* layer of the opposite type might be grown over its surface before the wafers are delivered to the *processing* company by the *wafer-manufacturing* company. This layered approach results in an epitaxial wafer.

A starting wafer of p^- might be doped around the level of $N_A \cong 2 \times 10^{21}$ donor/m^3. Such a doping level would give a *resistivity* of 10 to 20 Ω·cm. Normally, the orientation of the crystal structure of a silicon microcircuit is "100." This orientation minimizes traps that form under the gates of transistors.

Photolithography and Well Definition

Photolithography is a technique in which selected portions of a silicon wafer can be *masked out* so that a desired processing step can be applied selectively to the remaining areas. Although photolithography is used throughout the manufacturing process of an integrated circuit, here we describe this photographic process in the context of preparing the wafer for defining the *well regions.*[2]

---

[2]Wells are doped regions that will contain one of the two types of transistors realized in a CMOS process. Wells are now normally n-type and contain p-channel transistors.

Selective coverage for well definition proceeds as follows. First, a glass mask, M_1, is created that defines where the well regions will be located. The glass mask is created by covering it in photographic materials and exposing it to an electron beam (or *e-beam*) in the regions corresponding to the well locations. Such exposure results in the well regions on the glass mask turning *opaque*, or dark. As a result, the glass mask can be thought of as a *negative* of one layer of the microcircuit. In a typical microcircuit process, 10 to 20 different masks might be required. A typical cost for these masks is currently around $100,000. Because a high degree of precision is required in manufacturing these masks, a company different from the processing company often makes them. The exposure of the opaque regions of the mask, by the electron beam, is controlled by a computer dependent on the contents of a *data base*. The data base required for the e-beam is derived from the *layout* data base produced by the designer, using a graphic terminal and a layout CAD software program.

The first step in *masking* the surface of the wafer is to thermally grow a thin layer of silicon dioxide, SiO_2, to protect the surface of the microcircuit. Details of this step are discussed later. On top of the SiO_2, a *negative photoresist*, PR_1, is evenly applied, while spinning the microcircuit, to a thickness of around 1 μm. The photoresist is a light-sensitive polymer (similar to latex).

Next, the mask, M_1, is placed in close proximity to the wafer and ultraviolet light is projected through the mask onto the photoresist. Wherever the light strikes, the polymers cross-link or *polymerize*. This change makes these regions insoluble to an organic solvent. The region in which the mask was opaque (i.e., the well regions) is not exposed. This step is shown in Fig. 2.1. The photoresist is removed in this area using an organic solvent. Next, the remaining photoresist is *baked* to harden it. After the photoresist in the well region is removed, the uncovered SiO_2 may also be removed, using an acid *etch*. (However, in some processes, where this layer is very thin, it may not be removed.) In the next step, the *dopants* to form the well are introduced into the silicon using either *diffusion* or *ion implantation* (directly through the thin oxide in cases in which it has not been removed).

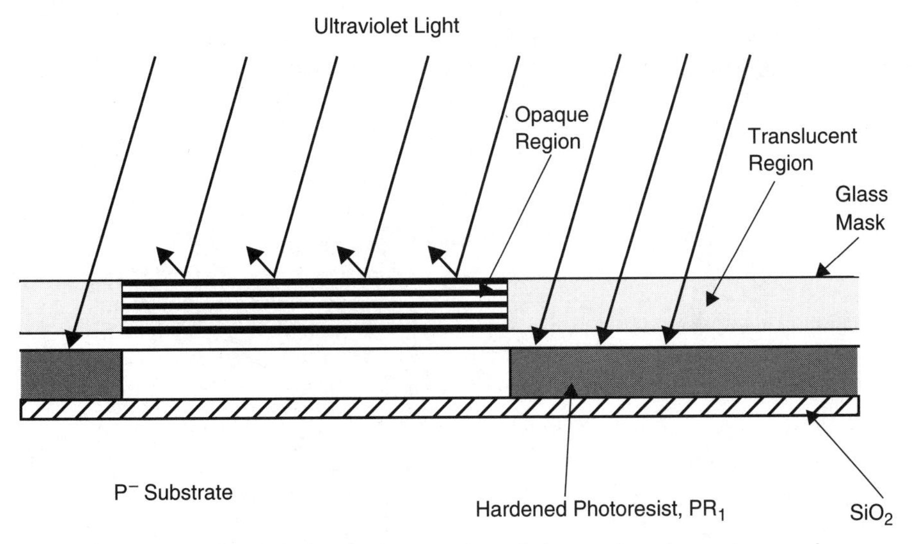

Figure 2.1 Selectively hardening a region of photoresist using a glass mask.

The procedure just described involves a *negative photoresist,* where the exposed photoresist remains after the masking. There are alternative photoresist materials, called *positive photoresists*, where the *exposed* photoresist is dissolved by the organic solvents. In this case the photoresist remains where the mask was opaque. By using both positive and negative resists, a single mask can sometimes be used for two steps, first to protect one region and implant the *complementary* region, and second, to protect the complementary region and implant the original region.

Diffusion and Ion Implantation

After the photoresist over the well region has been removed, the next step is to introduce dopants through the opening in which the well region will be located. As mentioned above, there are two approaches for introducing these dopants—diffusion and ion implantation.

In both implantation methods, normally the SiO_2 in the well region will first be removed using an acid etch. Next, the remaining hardened photoresist is *stripped* using acetone. This leaves SiO_2 that was protected by the hardened photoresist to mask all of the nonwell (i.e., substrate) regions.

In diffusion implantation, the wafers are then placed in a quartz tube that is placed in a heated furnace. A gas containing the dopant is introduced into the tube. In the case of forming an n-well, the dopant in the gas would probably be *phosphorus. Arsenic* could also be used, but it takes a much longer time to diffuse. The high temperatures of the diffusion furnace, typically 900 to 1100°C, cause the dopants to diffuse into the silicon both vertically and horizontally. The dopant concentration will be greatest at the surface and will decrease in a *Gaussian* profile further into the silicon. If a p-well had been desired, then boron would have been used as the dopant. The resulting cross section, after diffusing the n-well, is shown in Fig. 2.2.

An alternative technique for introducing dopants into the silicon wafer is ion implantation. This technique is largely replacing diffusion because it allows more independent control over the dopant concentration and the thickness of the doped region. In ion implantation, dopants are introduced as ions into the wafer as shown in the functional representation of an ion implanter of Fig. 2.3. The ions are generated by bombarding a gas with electrons from an *arc-discharge* or *cold-cathode* source. The ions are then focused and sent through a *mass separator.* This bends the ion beam and sends it through a narrow slit. Since only ions of a specific

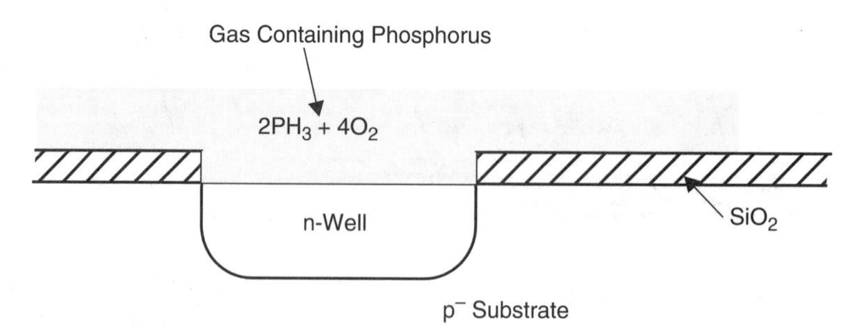

Figure 2.2 Forming an n-well by *diffusing* phosphorus from a gas into the silicon, through the opening in the SiO_2.

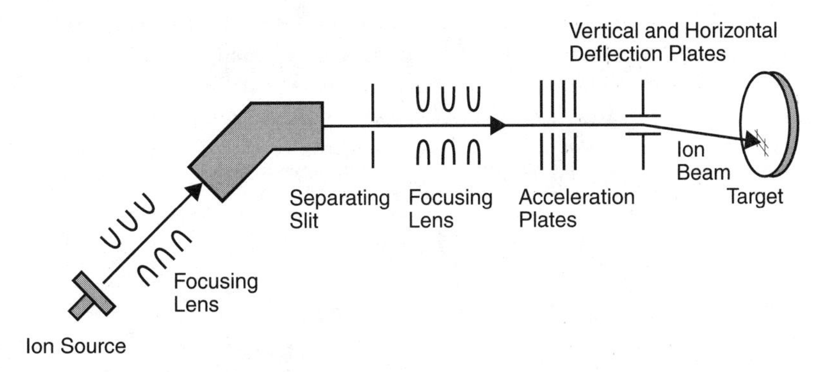

Figure 2.3 An ion implantation system.

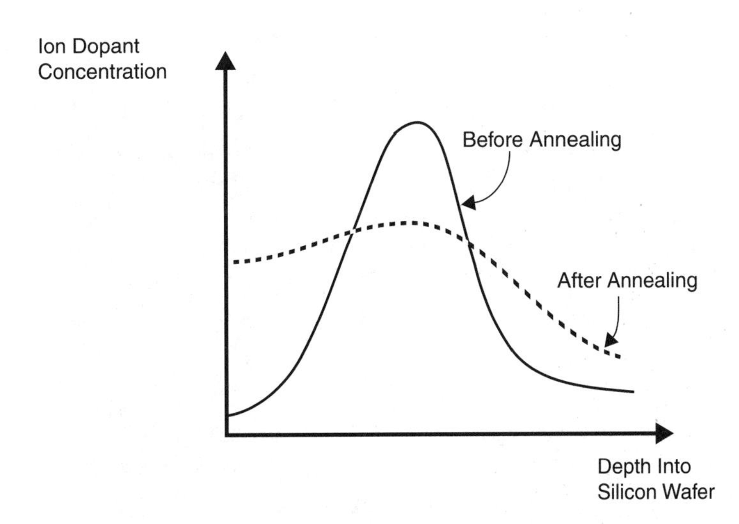

Figure 2.4 Dopant profiles after ion implantation both before and after annealing.

mass get through the slit, the beam is purified. Next, the beam is again focused and accelerated to between 10 keV and 1 MeV. The ion current might range from 10 μA to 2 mA. The deflection plates will sweep the beam across the wafer, which will often be rotated at the same time. The acceleration potential controls how deep the ions are implanted into the wafer. The beam current and time of implantation determine how heavy the dosage is. Thus, the *depth* and *dosage* are *independently controlled.*

Two problems that occur with ion implantation are *lattice damage* and a *narrow doping profile*. The lattice damage is due to nuclear collisions that result in the displacement of substrate atoms. The narrow profile results in a heavy concentration over a narrow distance, as shown in Fig. 2.4. For example, arsenic ions with an acceleration voltage of 100 keV might penetrate approximately 0.06 μm into the silicon, with the majority of ions being at 0.06 ± 0.02 μm. Both of these problems are largely solved by annealing.

Annealing is a step in which the wafer is heated to around 1000°C, for perhaps 15 to 30 mins, and then allowed to slowly cool. This heating step thermally vibrates the atoms, which allows the bonds to reform. It also broadens the concentration profile making the doping levels more uniform, as shown in Fig. 2.4. *Note that annealing is performed only once during processing after all the implantation steps have been performed but before any metal layers have been created.*[3]

For n-type dopants, arsenic will be used for shallow implantations, such as the source or drain junctions. Phosphorus might be used for the well. Boron is always used to form the p regions.

Although more expensive, ion implantation has largely replaced diffusion for forming n and p regions in a modern microcircuit due to the greater control over doping levels and depths. Another important advantage of ion implantation, as compared to diffusion, is the much smaller *sideways diffusion*. This allows devices to be more closely spaced and, more importantly for MOS transistors, minimizes the overlap between the gate and the source or drain junctions using a *self-aligned* process.

Chemical Vapor Deposition and Defining the Active Regions

The next few steps use the field-oxide mask, M_2, to form the thick *field oxide* as well as the *field implants*, which together are used to isolate transistors. This results in a thin layer of thermal SiO_2, as well as silicon nitride, Si_3N_4, everywhere the field oxide is not desired.

Often, this step will be done using positive photoresist such that wherever the mask M_2 is not opaque the photoresist will be softened. In other words, the photoresist is left intact after the organic dissolution step under the opaque regions of the mask. This region is where field oxide is not desired. A thin layer of thermal SiO_2 is then grown everywhere to protect the surface of the silicon lattice.

Next, Si_3N_4 is deposited everywhere from a *gas-phase reaction* where energy is supplied by heat (at around 850°C). This process is called *chemical vapor deposition* or *CVD*. After this step, the positive photoresist is deposited, exposed through the mask M_2, dissolved, and hardened. The hardened photoresist will be left on top of the Si_3N_4, to protect it, where the field oxide is not desired. Next, the Si_3N_4, wherever it is not protected by the photoresist, is removed by etching it away with a hot phosphoric acid. The SiO_2 is then removed with a *hydrofluoric acid etch*. Finally, the remaining photoresist is chemically removed with a process that leaves the remaining Si_3N_4 intact. The remaining Si_3N_4 will act as a mask to protect the *active regions* when the thick field oxide is being grown in the field-oxide regions.

Field Implants and the Field Oxide

The next step in our example process is to implant the *field implants* under where the *field oxide* will be grown. For example, boron will be implanted under the field oxide everywhere except in the well regions. This implant guarantees that the silicon under the field oxide will

[3] If annealing were done after deposition of a metal layer, the high temperatures required for annealing would melt the metal.

never invert (or become n) when a conductor over the field oxide has a large voltage on it. If this implant was not performed, there could be leakage currents between the junctions of separate n-channel transistors that are intended to be unconnected.

For the field oxide in the well regions, where p-channel transistors will eventually reside, an n-type implant such as As could be used. Often, it is not necessary to include field implants under the field oxide of the well regions, as the heavier doping of the well, as compared to the substrate, normally guarantees that the silicon will never invert under the field oxide in these regions.

When implanting the field implants in the substrate regions, it is necessary to first cover the wells with a protective photoresist, PR_3, so the n-well regions do not receive the p-implant. This can be done using the same mask, M_1, that was originally used for implanting the n-wells, but now a positive photoresist is used. This positive photoresist leaves photoresist where the mask is opaque (i.e., dark), which corresponds to the well regions.

After the exposed photoresist has been dissolved, we now have the cross section shown in Fig. 2.5. Notice that at this step, all the active regions, where eventually the transistors will reside, are protected from the field implants by the layers of SiO_2, Si_3N_4, and PR_2. Additionally, the complete well regions are also protected by PR_3. The field implant will be a high-energy implant with a fairly high doping level.

Growing the Field Oxide

The next step is to grow the field oxide, SiO_2. There are two different ways that SiO_2 can be grown. In a *wet process*, water vapor is introduced over the surface at a moderately high temperature. It diffuses into the silicon and, after some intermediate steps, reacts according to the formula

$$Si + 2H_2O \rightarrow SiO_2 + 2H_2 \tag{2.1}$$

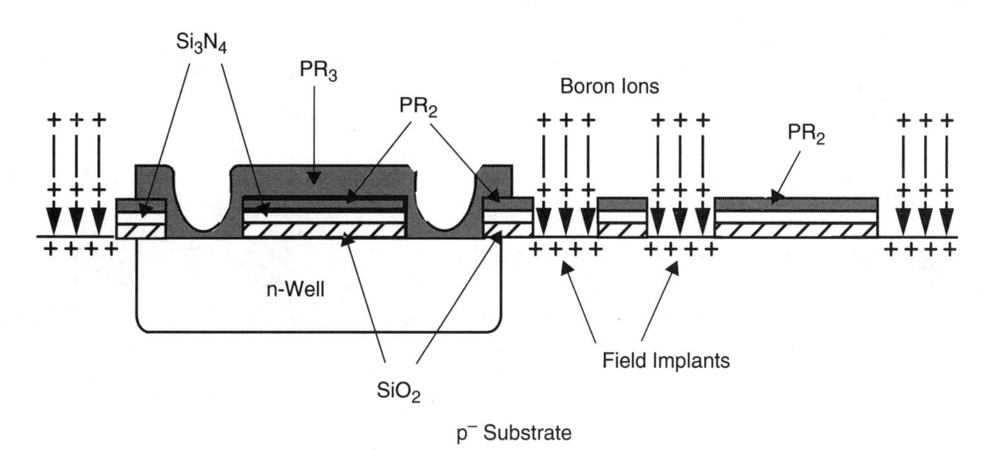

Figure 2.5 The cross section when the field implants are being formed.

In a *dry process*, oxygen is introduced over the wafer, normally at even a higher temperature, and reacts according to the formula

$$Si + O_2 \rightarrow SiO_2 \tag{2.2}$$

Since both of these processes occur at high temperature, around 800 to 1200°C, the oxide that results is sometimes called a *thermal oxide*.

Before the field oxide is grown, PR_2 and PR_3 are removed, but the silicon nitride silicon dioxide sandwich is left. The reaction does not occur wherever CVD-deposited Si_3N_4 remains, as it is relatively inert to both water and oxygen. Wherever the process does occur, the volume increases due to the additional oxygen atoms being added. SiO_2 takes up approximately 2.2 times the volume of the original silicon. This increase will cause the SiO_2 to extend approximately 45% into what was previously the surface of the silicon and 55% above what was the surface. The resulting cross section is shown in Fig. 2.6. Note that in our example process, the field oxide in the substrate region has field implants under it, whereas the field oxide in the wells does not.

When growing thermal SiO_2, the wet process is faster because H_2O diffuses faster than O_2 in silicon, but the dry process results in denser higher quality SiO_2 that is less porous. Sometimes, when growing the field oxide, the step will start with a dry process, change to a wet process, and finish with a dry process. When growing thin layers of SiO_2, as described in the next section, usually only a dry process is used.

The Gate Oxide and Threshold Voltage Adjusts

In the next step, the Si_3N_4 is removed using hot phosphoric acid. If there had been a thin layer of SiO_2 under the Si_3N_4, protecting the surface, as is shown in Fig. 2.6, it also is removed usually using hydrofluoric acid. The high-quality, thin, gate oxide is then grown using a dry process. It is grown everywhere over the wafer, perhaps to a thickness between 0.005 and 0.02 μm.

After the gate oxide has been grown, donors are implanted so that the final threshold voltages of the transistors are correct. Note that this implantation is performed directly through the thin gate oxide as it now covers the entire surface. It should be mentioned that many pro-

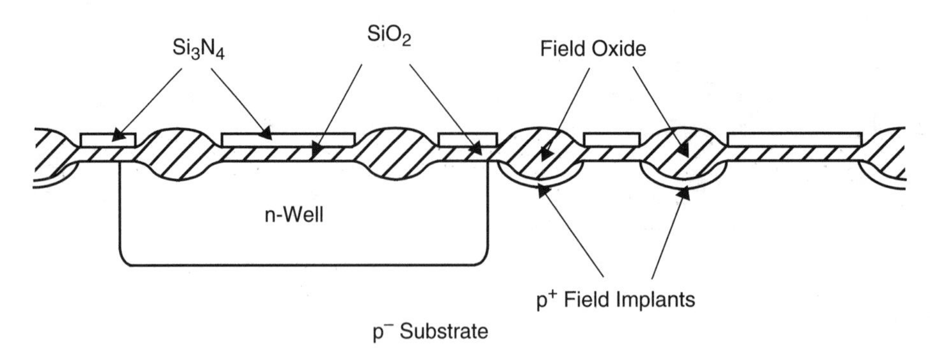

Figure 2.6 The cross section after the field oxide has been grown.

cesses differ in realizing this step. In a simple process, the threshold voltages of both the p- and n-channel transistors are adjusted at the same time. The n-channel transistors require a boron implant to change V_{tn} from its native value of around –0.1 V to its desired value of 0.7 to 0.8 V. If the n-wells are doped a little heavier than ideal, the native threshold voltage of the p-channel transistors in the well will be around –1.6 V. As a result, the same single boron threshold-adjust implant will bring it to around –0.8 to –1 V.

By using a *single threshold-voltage-adjust implant* for both n-channel and p-channel transistors, two photoresist masking steps are eliminated. If the different transistors were individually implanted, then the second one of two types would have to be protected by perhaps a negative photoresist while the first type was being implanted. Next, a positive photoresist could be used with the same mask to protect the first type while the second is being implanted. The mask used would normally be the same mask used in forming the n-wells, in other words, M_1. Thus, no additional mask is required but a number of additional processing steps are needed. *The major problem with using a single threshold-adjust implant is that the doping level of the n-well is higher than optimum.* This higher doping is necessary to achieve a reasonable final threshold voltage for the p-channel transistors. *This higher doping level increases the junction capacitances and the body effect of the transistors in the well.* A double threshold adjust allows optimum well doping. At the present time, both approaches are being commercially used, although the double threshold adjust implant is growing in favor as the dimensions decrease. The cross section at this stage is shown in Fig. 2.7.

POLYSILICON GATE FORMATION

The next step in the process is the chemical deposition of the *polysilicon* gate material. One method to create polysilicon is to heat a wafer with silane gas flowing over it so the following reaction occurs:

$$SiH_4 \rightarrow Si + 2H_2 \tag{2.3}$$

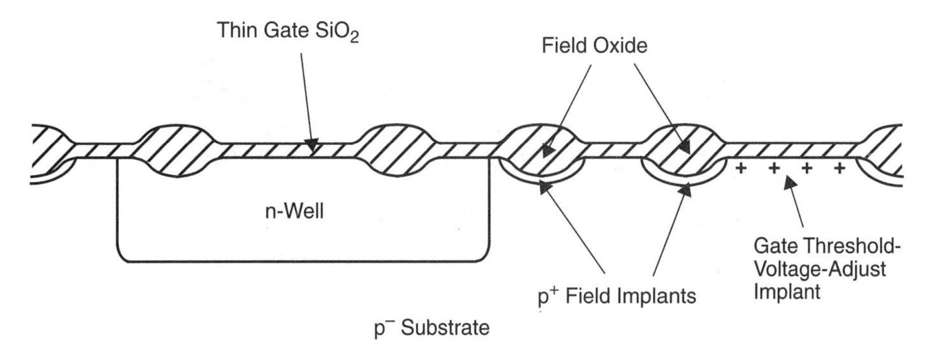

Figure 2.7 The cross section after the thin gate-oxide growth and threshold-adjust implant.

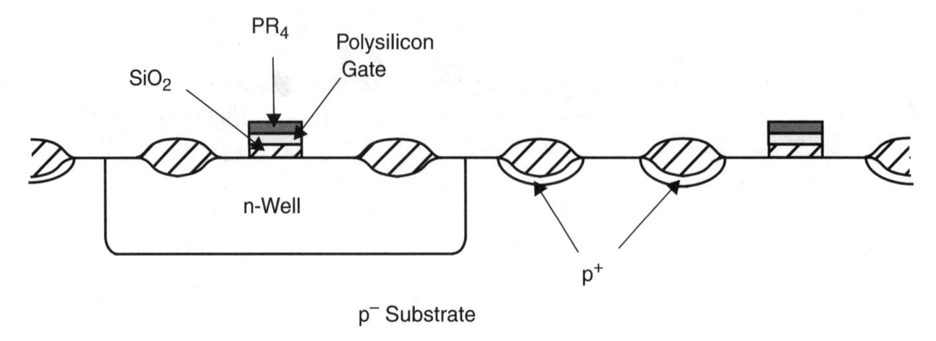

Figure 2.8 The cross section after depositing and patterning the polysilicon gates.

If this reaction occurs at high temperatures, say around 1000 to 1250°C, and the original surface of the wafer was *single crystal*, the deposited silicon would also be single crystal. This is the approach when epitaxial layers are grown in bipolar processes and some of the more-modern CMOS processes. However, when depositing the polysilicon gates, the original surface is SiO_2 and the wafer is heated only to around 650°C. As a result, the silicon that is deposited is noncrystalline or *amorphous*. It therefore is often referred to as polysilicon. Often, after the polysilicon is deposited, it will be ion implanted with arsenic to increase its conductivity. A typical final resistivity for polysilicon might be 20 to 30 $\Omega/\Box$,[4] and its thickness might be around 0.25 μm.

After the above deposition, the polysilicon gate material covers the entire wafer. This polysilicon is then patterned using a new mask, M_3, and a positive photoresist, PR_4. The mask is opaque where hardened polysilicon is to be left. After the nonhardened photoresist is removed, the polysilicon is etched away using a reactive plasma etch. This etch removes all of the polysilicon that is not protected by photoresist, but removes very little of the underlying SiO_2. This thin gate-oxide layer is used to protect the surface during the next step of junction implantation. The cross section at this phase is shown in Fig. 2.8.

Implanting the Junctions, Depositing SiO_2, and Opening Contact Holes

The next couple of steps involve the ion implantation of the junctions. In our example process, the p^+ junctions will be formed first by placing positive photoresist, PR_5, everywhere except where the p^+ regions are desired. A new mask, M_4, is used in this step. The p^+ regions are then ion implanted, possibly through a thin oxide in some processes. The cross section at this stage is shown in Fig. 2.9.

Notice that the p^+ junctions of the p-channel transistors are defined on one edge by the field oxide and, more importantly, next to the active gate area by the edge of the polysilicon gate. During the implantation of the boron, it was the gate polysilicon and the photoresist over it that pro-

[4] 20 $\Omega/\Box$ means a square area would have a resistance of 20 Ω.

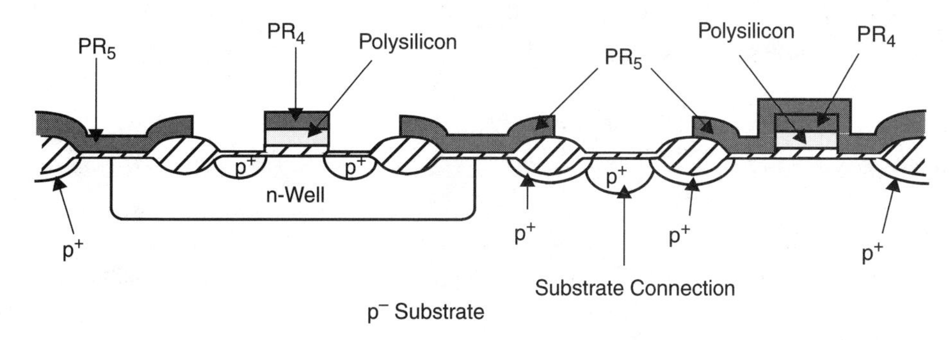

Figure 2.9 The cross section after ion implanting the p^+ junctions.

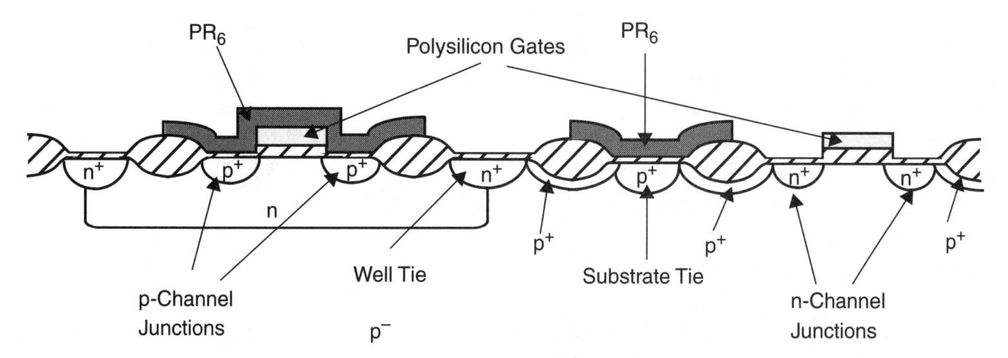

Figure 2.10 The cross section after ion implanting the n^+ junctions.

tected the channel region from the p^+ implant. Thus, the p^+ junctions are *self-aligned* to the polysilicon gates resulting in very little overlap (i.e., a small L_{ov} as defined in Chapter 3). Also, note that the effective channel area of the transistors is defined by the intersection of the gate-defining mask, M_3, and the mask used in defining the active regions, M_2 (i.e., the mask used in defining where Si_3N_4 was left). *Thus, these are the two most important masks in any MOS process.* The development of this self-aligned process has proven to be an important milestone in realizing small high-speed transistors, both in MOS processes and in BJT processes.

Also notice that a p^+ junction has been implanted in the substrate region. This junction is a *substrate tie*. It is used to connect the substrate to ground in microcircuits. These substrate ties are liberally placed throughout the microcircuit to help prevent *latch-up*, a problem discussed later. In addition, the *back side* of the wafer would normally be connected to ground, as well, through the package *header* using a *gold eutectic bonding process.*

Next, the photoresists are all removed using acetone. The p^+ active regions are then protected using the same mask, M_4, as was used for the previous step, but now using a negative photoresist, PR_6. The n^+ junctions are then implanted using arsenic. The cross section at the end of this stage is shown in Fig. 2.10.

After the junctions have been implanted, and PR_6 removed, the complete wafer is covered in CVD SiO_2. This protective *glass* layer can be deposited at moderately low temperatures of 500°C or lower. The deposited SiO_2 might be 0.25 to 0.5 μm thick.

The next step is to open contact holes through the deposited SiO_2. The contact holes are defined using mask M_5 and positive resist PR_7.

Annealing, Depositing and Patterning Metal, and Overglass Deposition

After the first layer of CVD SiO_2 has been deposited, the wafer is *annealed.* As was mentioned earlier, this might entail heating the wafer in an inert gas, nitrogen, for example, for 15 to 30 mins, at temperatures up to 1000°C. The resulting thermal vibrations heal the lattice damage sustained during all the ion implantations, broaden the concentration profiles of the implanted dopants, and increase the density of the deposited SiO_2.

Next, *interconnect metal* will be deposited everywhere. Historically, aluminum, Al, has been used for the interconnect. However, recently, alternative metals have been used that have less of a tendency to diffuse into the silicon during electrical operation of the microcircuit. The metal is deposited using evaporation techniques in a vacuum. The heat required for evaporation is normally produced by using *electron-beam bombarding*, or possibly *ion bombarding* in a *sputtering system.* After the metal is deposited on the entire wafer, it is patterned using mask M_6 and positive photoresist PR_8, and then etched.

At this time, a low-temperature annealing might take place to give better *eutectic* bonds between the metal and the silicon. The temperature of this annealing must be less than 550°C so the aluminum does not melt.

Next, an additional layer of CVD SiO_2 is deposited, additional contact holes are formed using mask M_7 and photoresist PR_9, and then a second layer of metal is deposited and etched using mask M_8 and photoresist PR_{10}. Often, the primary use of this second layer of metal might be to distribute the power-supply voltages. The bottom layer would be used more often for local interconnects in gates. In some modern processes, this process is continued once more and possibly even a fourth or a fifth time to give additional levels of metal, which allows for much denser interconnect.

After the last level of metal is deposited, a final passivation or overglass is deposited for protection. The layer would be CVD SiO_2, although often an additional layer of Si_3N_4 might be deposited because it is more impervious to moisture.

The final microcircuit processing step is to etch openings to the pads used for wire bonding. This final step would use mask M_9 and photoresist PR_{11}. Fig. 2.11 shows a cross section of the final microcircuit for our example process.

Processing Alternatives

This example CMOS process is a fairly representative, simple process. There are many variations that often involve additional masks. Some of the possible variations follow:

1. There may be two wells, one for p-channel transistors and one for n-channel transistors. This *twin-tub* process allows both wells to be optimally doped.
2. There might be an additional polysilicon layer over the first layer. This can be used to realize resistor loads in static random-access memories or poly-to-poly capacitors where a thin thermal oxide is used to separate the two layers.

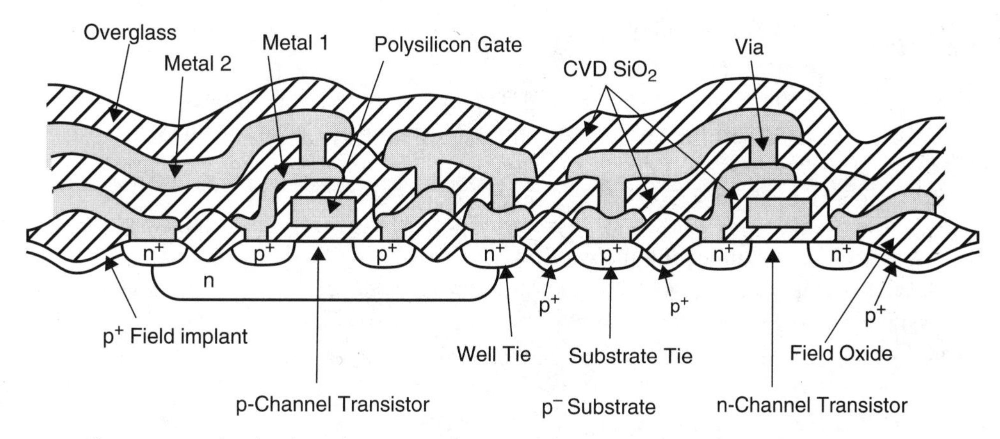

Figure 2.11 The final cross section of a CMOS microcircuit with two layers of metal.

3. There might be field implants under the field oxide in the well regions as well as the field oxide in the substrate regions.
4. Transistors might be isolated from one another by *trenches* etched into the substrate and filled with SiO_2. This allows very tight packing of the transistors.
5. Often, as was mentioned previously, there might be separate threshold-voltage-adjust implants for the n-channel and the p-channel transistors.
6. There might be three, four, or even five layers of metal for interconnect.
7. In a multimetal-layer process, it is typically necessary to add a number of additional steps whereby the surface is made smoother or *planarized* after each metal patterning step. This is normally done by some sort of reactive etching process whereby the metal is covered with SiO_2 and the *hills* are etched faster than the *valleys*.
8. There might be different metals used for the contacts than are used for the interconnect, to get better fill-in and less diffusion into the silicon surface.
9. There might be thin-film nichrome resistors under the top layer of metal.
10. There might be an epitaxial layer in which the transistors are realized. In this case, the substrate would be n^- and a p^- epitaxial layer would be grown. Before growing the epitaxial layer, the top of the substrate would be doped p^+. This type of wafer is similar to what is used in processing BJT transistors (but of the opposite type) and is starting to be commonly used in CMOS processing. Its advantages are that it is more immune to a destructive phenomenon called latch-up (latch-up will be described in the next chapter) and is also more immune to gamma radiation in space. Finally, it greatly minimizes substrate noise in microcircuits that have both analog and digital circuits (i.e., mixed-mode microcircuits).
11. A final common variation is to include additional processing steps so that BJT transistors can be included in the same microcircuit as MOS transistors. This type of process is called a *BiCMOS* process and is becoming especially popular for high-speed microcircuits, both digital and analog.

2.2 Bipolar Processing

The processing steps required for realizing BJT transistors are similar to what is used for realizing MOS transistors, with some modifications. Thus, rather than presenting the complete realization of modern BJT transistors, some of the modifications needed for realizing bipolar transistors will be briefly discussed.

A BJT process normally starts with a p^- substrate. The first masking step involves the diffusion (or ion implantation) of n^+ regions into the substrate wherever transistors are desired. These n^+ regions are used to lower the series collector resistance. Next, an n^- single-crystal epitaxial layer is deposited.

The next basic step is the formation of the field oxide for isolation. However, before the field oxide is grown and after the openings in the $Si_3N_4 - SiO_2 - PR$ sandwich have been made, the surface of the silicon is typically etched to form empty cavities. This extra etching step allows the field oxide to extend further down into the epitaxial region than would otherwise be the case. This process step, originally developed for bipolar processes, is now gaining popularity in CMOS processing as well.

After the field oxide is grown, the n^+ collector contact region is implanted. It extends from the surface down to the n^+ region buried under the transistor.

In a modern process, polysilicon is used to contact the emitter, the base, and possibly the collector. In a typical process, the base polysilicon is deposited first. It will be heavily doped p^+ so that later, during a high-temperature step, the boron dopant from the polysilicon contact diffuses into the silicon underneath the base polysilicon to make the underlying region p^+. The base polysilicon is removed in the active area of the transistor. Next, using one of a variety of possible methods, the base polysilicon is covered with a thin layer of SiO_2, perhaps 0.5 μm in thickness. This SiO_2 *spacer* allows the base contact to be very close to the emitter contact, thereby minimizing the base resistance. Next, the base will be ion implanted, after which n^+ polysilicon for the emitter will be deposited. At this point, the base and emitter are electrically separated by the thin spacer SiO_2. However, when the wafer is annealed, the n^+ from the emitter polysilicon diffuses into the base p silicon to form the true emitter region. It is at this time that the p^+ dopants from the base polysilicon also diffuse into the extrinsic base region. The final resulting structure is shown in Fig. 2.12. The important features of it are that through the use of *self-aligned contacts* and field oxide isolation, very small high-frequency BJT transistors can be realized using methods similar to those used in realizing modern MOS transistors.

2.3 CMOS Layout and Design Rules

It is the designer's responsibility to determine the geometry of the various masks required during processing. The process of defining the geometry of these masks is known as *layout* and is done using a computer and a CAD program. Here, we briefly describe some typical layout rules and the reasons for them.

When designing the layout, typically the designer does not need to produce the geometry for all of the masks because some of the masks are automatically produced by the layout program.

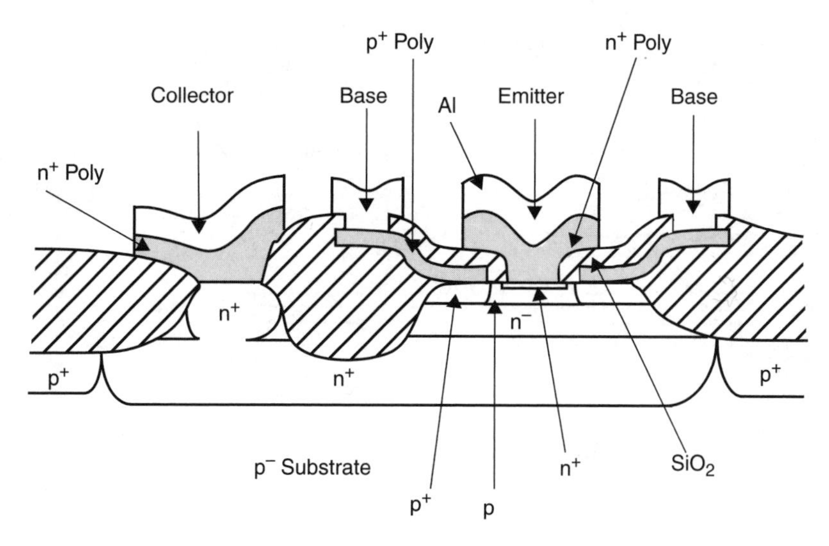

Figure 2.12 A cross section of a modern self-aligned BJT transistor with oxide isolation.

For example, typically the p^+ and n^+ masks used for the junctions are automatically generated. Also, the program might allow the designer to work in the final desired dimensions. The layout program would then automatically *size* the masks so they account for any lateral diffusions, which causes larger dimension masks, or to account for etching loss, which can cause smaller dimension masks. For example, a designer might draw a polysilicon line intending that a transistor would have a 0.5 μm length. The program might then produce a mask defining the polysilicon layer that had a 0.6 μm line width. This increased mask sizing would account for the junction overlap due to lateral diffusion and polysilicon loss due to etching.

In a modern layout program, the layout of some circuit *cells* may have already been completed and stored in a library. During overall layout, these cells would then be parametrically adapted to a required size and the corresponding geometries for every layer would be automatically generated. Often, when the cells are being connected together, they might be automatically *placed* and *routed* or connected by the program. The designer might then interactively modify this automatically generated layout. Thus, as time goes on, the layout becomes more automated. However, at the present time, the designer must still take direct control of the layout of critical cells, especially when the layout must be small or the resulting circuits must be fast. For example, it is virtually unheard of to let a computer generate the layout of a memory cell where space and capacitive loading of the connecting buses are critical. Thus, a digital microcircuit designer should be knowledgeable about the *design rules* that govern the layout required for the process used.

The two most important masks that need to be defined are those for the active region and the gate polysilicon. The intersection of these two masks will be the channel region of MOS transistors. For example, consider Fig. 2.13a, which shows a simplified view of an MOS transistor, and Fig. 2.13b, which shows the corresponding layout of the *active* mask and the *poly* (or polysilicon) mask. In Fig. 2.13b, the poly mask runs vertically. The length of the poly that intersects

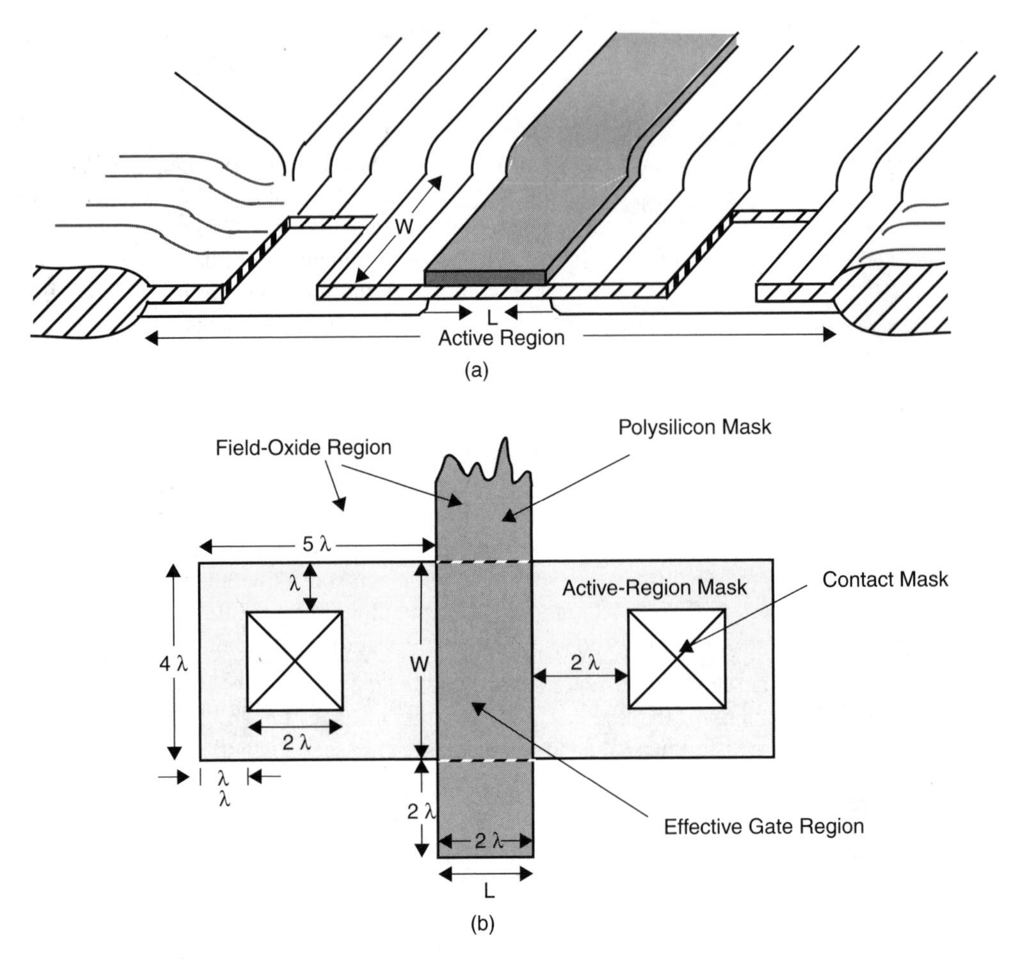

Figure 2.13 (a) A simplified view of a partially finished transistor and (b) the corresponding layout of the active, polysilicon, and contact masks.

the active-region mask is the transistor width, W. The width of the poly line is the transistor length, L, as is shown in Fig. 2.13.

The design rules for laying out transistors are often expressed in terms of a quantity λ, where λ will be equal to one-half the gate length. This generalization allows many of the design rules to be simply expressed in terms of λ and independent of actual dimensions (for example, the actual value for the minimum channel length or $2\,\lambda$). Shown in Fig. 2.13b is the smallest possible transistor realizable when a contact must be made to each junction. Also shown in Fig. 2.13b are many of the minimum dimensions in terms of λ.

When expressing design rules in terms of λ, it is implicitly assumed that each mask has a worst-case absolute alignment of under $0.75\,\lambda$. Thus, it is guaranteed that the relative misalignment between any two masks is under $1.5\,\lambda$. If an overlap between any two regions of a microcircuit would cause a destructive short circuit, then a separation between the corresponding regions in a layout of 2 λ guarantees this will never happen. For example, consider the poly mask and the contact mask in Fig. 2.13b. If, for some reason, these two regions overlapped in the microcircuit, the metal used to contact the source junction, for example, would also be short circuited to the gate poly causing the transistor to be always turned off, as is shown in Fig. 2.14. If the source happened to be connected to ground, this would also short circuit the gate to ground. To prevent the occurrence of this type of short, the contact openings must be kept at least 2 λ away from the polysilicon gates.

Another example of a catastrophic failure due to misalignment would be if a gate did not fully cross the active region, as shown in Fig. 2.14. Since the junctions are implanted everywhere in the active region except under the gate, this misalignment would cause a short circuit between the source and the drain. Thus, there is a design rule that polysilicon must always extend at least 2 λ past the active region.

Another design rule is that active regions should surround contacts by at least 1 λ. If, in reality, there was overlap between the edge of the active-region mask and the contact mask, there are no disastrous shorts that occur. The circuit will still work correctly as long as there is a sufficient overlap between the contact and the active masks so that a good connection is made between the aluminum interconnect and the junction. Since the maximum relative misalignment is 1.5 λ, this guarantees an overlap of at least 1.5 λ (assuming the minimum contact width is 2 λ).

The few design rules just described are sufficient to allow one to estimate the minimum dimensions of a junction area and perimeter before a transistor has been laid out. For example, with reference to Fig. 2.13, assuming a contact is to be made to a junction, then the active region must extend past the polysilicon region by at least 5 λ. Thus the minimum area of a small junction with a contact to it is

$$A_s = A_d = 5\,\lambda W \tag{2.4}$$

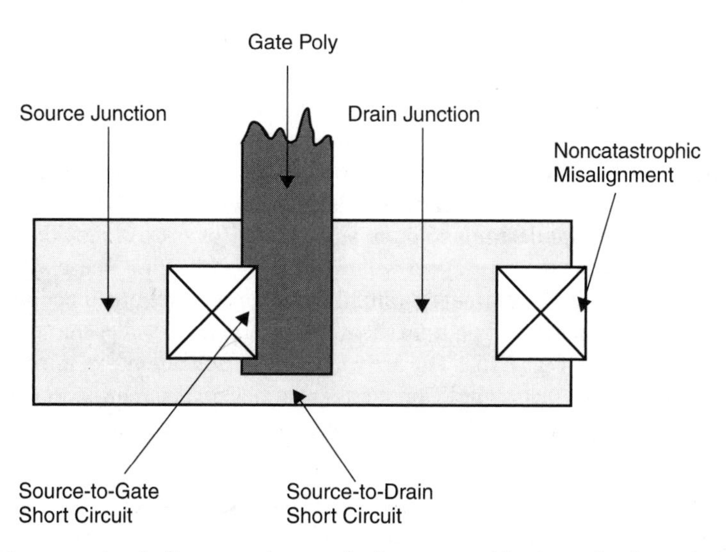

Figure 2.14 Some mask misalignment that results in catastrophic short circuits and an example of a noncatastrophic misalignment.

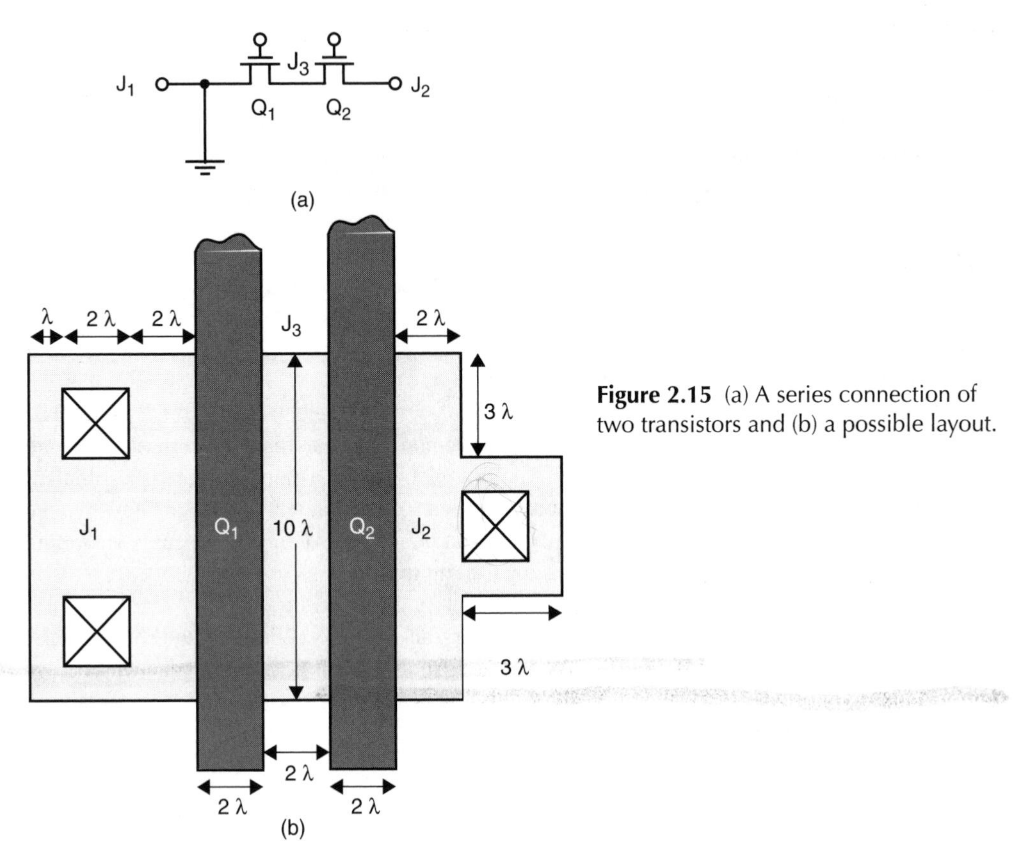

Figure 2.15 (a) A series connection of two transistors and (b) a possible layout.

where W is the transistor width. Similarly, with respect to Fig. 2.14, the perimeter of a junction[5] with a contact is given by

$$P_s = P_d = 10\lambda + W \tag{2.5}$$

These estimates may be used in the equations of Section 3.3 when estimating the parasitic capacitances in the transistor models. They may also be used in SPICE to simulate circuits so the parasitic capacitances are determined more accurately. However, it should be kept in mind that they are only estimates; the true layout will most likely differ from these estimates.

Sometimes, when it is very important to minimize the capacitance of a junction, it is possible to share a single junction between two transistors. For example, consider the series connection of two transistors shown in Fig. 2.15a. The active, poly, and contact masks might be laid out as shown in Fig. 2.15b. Notice that a single junction is shared between transistors Q_1 and Q_2. The area, and especially the perimeter of this junction, is much less than that given by equations

[5]Note that the perimeter does not include the edge between the junction and the active channel separating the junction and the gate as there is no field implant along this edge and the side-wall capacitance is therefore smaller along the edge.

(2.4) and (2.5). Also, in a SPICE simulation, the area and perimeter should be divided by 2 when it is specified in each transistor description, as it is a *shared junction*. Alternatively, all of the area and perimeter could be specified in one transistor description, and the area and perimeter of the other junction could be specified as zero.

Since the junction side-wall capacitance is directly proportional to the junction perimeter, and since the junction side-wall capacitance can be a major part of the total junction capacitance because of the heavily doped field implants, minimizing it is very important. It is of interest to note here that as transistor dimensions shrink, the ratio of the perimeter to the area increases and the side-wall capacitance becomes more important.

Example 2.1

Assuming $\lambda = 0.5\ \mu m$, find the area and perimeters of junctions J_1, J_2, and J_3 for the circuit of Fig. 2.15.

Solution: Since the width and length are shown as $10\ \lambda$ and $2\ \lambda$, respectively, and $\lambda = 0.5\ \mu m$, the physical sizes are $W = 5\ \mu m$ and $L = 1\ \mu m$.

Thus, for junction J_1, using the formulas of (2.4) and (2.5), we have

$$A_{J\text{-}1} = 5\ \lambda W = 5(0.5)5\ (\mu m)^2 = 12.5\ (\mu m)^2 \tag{2.6}$$

and

$$P_{J\text{-}1} = 10\ \lambda + W = [10(0.5) + 5]\ \mu m = 10\ \mu m \tag{2.7}$$

Since this junction is connected to ground, its parasitic capacitance is unimportant and little has been done to minimize its area. Contrast this case with junction J_2 where we have

$$\begin{aligned} A_{J\text{-}2} &= 2\ \lambda W + 12\ \lambda^2 \\ &= W + 12(0.5)^2 \\ &= 8\ (\mu m)^2 \end{aligned} \tag{2.8}$$

The perimeter is unchanged resulting in $P_{J\text{-}2} = 10\ \mu m$. Thus, it has been possible to decrease the junction area by making use of the fact that the transistor is much wider than the single contact used. However, sometimes wide transistors require additional contacts to minimize the contact impedance. For example, two contacts are used for junction J_1 to half the contact impedance.

Next, consider the shared junction. Here we have a junction area given by

$$\begin{aligned} A_{J\text{-}3} &= 2\ \lambda W \\ &= 5\ (\mu m)^2 \end{aligned} \tag{2.9}$$

Since this is a shared junction, in a SPICE simulation we would use

$$A_s = A_d = \lambda W = 2.5\ \mu m^2 \tag{2.10}$$

for each of the two transistors, which is much less than 12.5 $(\mu m)^2$. The reduction in the perimeter is even more substantial. Here we have

$$P_{J\text{-}3} = 4\lambda = 2\ \mu m \tag{2.11}$$

for the shared junction and so sharing this perimeter value over the two transistors would result in

$$P_s = P_d = 2\lambda = 1\ \mu m \tag{2.12}$$

for the appropriate junction of each transistor when simulating it in SPICE. This result is much less than the 10 μm perimeter for node J_1.

Due to the importance of minimizing the junction capacitance, one of the first steps an experienced designer takes before laying out important high-speed cells is to identify the most critical nodes and then investigate possible layouts that minimize the junction capacitance of these nodes.[6]

An additional design rule has been implicitly introduced in the previous example. Notice that for junction J_2 in Fig. 2.15, part of the active region boundary is only 2 λ away from the gate. This minimum junction area is the typical design rule for this case.

There are a number of design rules required in addition to those just mentioned. Some of these will be described with reference to the layout of a CMOS inverter shown in Fig. 2.16. Notice that the n-well surrounds the p-channel active region, and therefore the p^+ junctions of the p-channel transistors, by at least 3 λ. Notice also that the minimum spacing between the n-well and the junctions of n-channel transistors, in the substrate, is 5 λ. This large spacing is required because of the large lateral diffusion of the n-well and the fact that if the n-channel junction became short circuited to the n-well, which is connected to V_{DD}, the circuit would not

[6]Note that it is not possible to share junctions between n-channel and p-channel transistors. This limitation is one of the reasons for larger parasitic capacitances sometime encountered in CMOS microcircuits as opposed to NMOS microcircuits where only n-channel transistors are used.

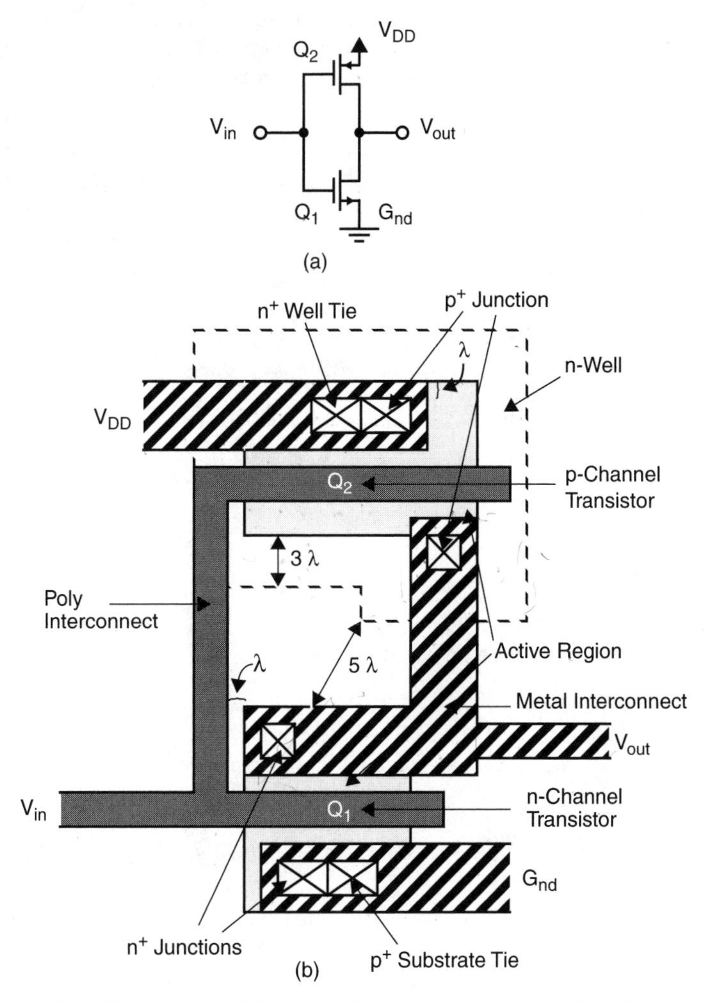

Figure 2.16 (a) A CMOS inverter and (b) a possible layout with a number of design rules illustrated.

work. Conversely, a p^+ substrate tie can be much closer to a well as it would always be connected to ground and would be separated from the well by a reverse-biased junction. A typical dimension here might be 3 λ. Since a p-channel junction must be inside the well by at least 3 λ and an n-channel junction must be outside the well by 5 λ, the closest an n-channel transistor can be placed to a p-channel transistor is 8 λ.

Notice in Fig. 2.16 that metal is used to connect the junctions of the p-channel and n-channel transistors. Normally, the metal must overlap any underlying contacts by at least λ.

A typical minimum width for first-level metal might be 2 λ, the same as the minimum width for polysilicon. However, it can be wider as is the case in Fig. 2.16, where it is 4 λ wide.

Notice also in Fig. 2.16 that a single contact opening, known as a *butting contact,* can be used to contact both the p-channel transistor source and also an n^+ well tie, as both will be connected to V_{DD}. Although the outline of the p^+ and n^+ masks are not shown in Fig. 2.16, under the contact, one half will be doped p^+ (the p-channel junction) and one half will be doped n^+ (the well tie). Also, for the n-channel transistor, a butting contact was used to connect the n-channel source to a p^+ substrate tie, which will both be connected to ground. In a typical set of design rules, a maximum distance between transistors and wells (or substrate) ties will be specified. Similarly, a maximum distance between substrate ties will also be specified. For example, it might be specified that no transistor can be more than 100 λ from a substrate tie. These rules are necessary to prevent a phenomenon called latch-up, which is described in the next chapter.

As a final example, the layout of a large transistor is described. *Normally, a very wide transistor would be composed of many smaller transistors that are connected in parallel.* A simplified layout of this approach is shown in Fig. 2.17a where four transistors, having a common

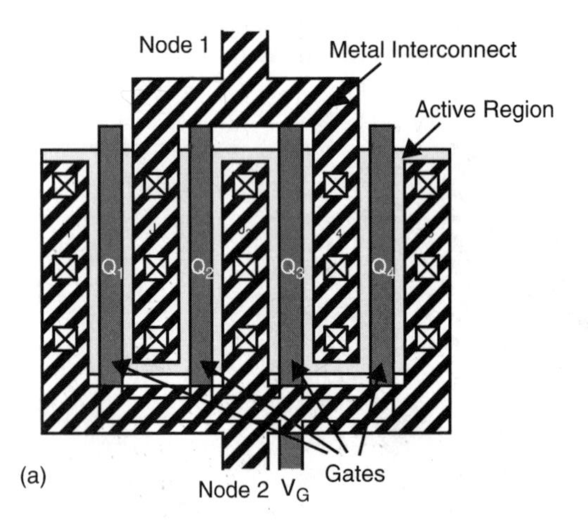

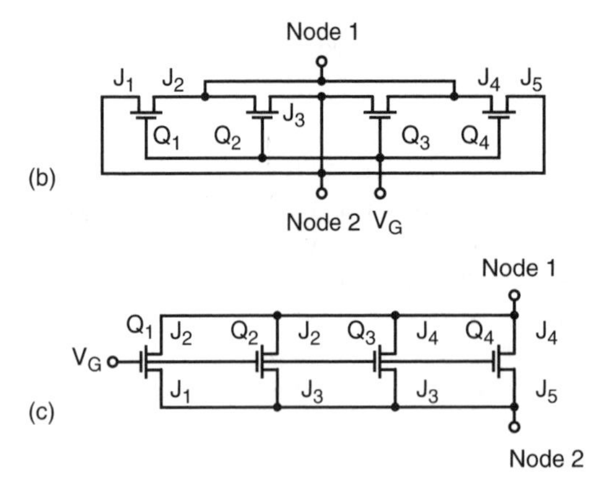

Figure 2.17 Connecting four transistors in parallel to realize a single large transistor: (a) the layout, (b) the schematic drawn in the same relative positions as the layout, and (c) the schematic redrawn to make the parallel transistors more obvious.

gate, are connected together in parallel. Shown in Fig. 2.17b is the circuit corresponding to the layout of Fig. 2.17a, where the transistors have been drawn in the same relative positions. Also shown in Fig. 2.17c is the same circuit redrawn differently where it is clear that the circuit consists of four transistors connected in parallel. Notice the second and fourth junction regions are connected by metal to node 1, whereas the first, third, and fifth junction regions are connected together by metal to realize node 2. Due to the larger total junction area, and especially perimeter, node 2 will have a larger junction capacitance than node 1. Thus, when connecting the equivalent transistor to a circuit, node 1 should be connected to the more critical node. Also notice the large number of contacts used to minimize the contact impedance. Normally, some of these would be butting contacts if either node 1 or node 2 were connected to an appropriate power supply. The use of many contacts in junction regions that are wide minimizes voltage drops that would otherwise occur due to the relatively high resistivity of silicon junctions as compared to the metal that overstraps the junctions and connects them.[7]

There are also design rules that specify the minimum pitch between polysilicon interconnects, metal 1 interconnects, and metal 2 interconnects. These might be 2λ, 2λ, and 3λ, respectively. The larger minimum pitch of metal 2 is required because it resides further from the silicon surface where the topography is less even. The minimum widths of poly, metal 1, and metal 2 might also be 2λ, 2λ, and 3λ, respectively.

This concludes our brief introduction to layout and design rules. In a modern process, many more design rules are used than those just described. However, the reasons for them and the application of the rules are normally similar to what has been described and fairly straightforward. Finally, it should be mentioned that when doing modern microcircuit layout, the design rules are usually available to the layout program and automatically checked as layout progresses.

Example 2.2

Consider the transistor shown in Fig. 2.17 where the total transistor width equals 80λ, its length equals 2λ, and $\lambda = 0.5\ \mu\text{m}$. Assuming node 2 is the source, node 1 is the drain, and the device is in the active region, find the source-bulk and drain-bulk capacitances given the parameters $C_j = 2.4 \times 10^{-4}\ \text{pF}/(\mu\text{m})^2$, and $C_{jsw} = 2.0 \times 10^{-4}\ \text{pF}/(\mu\text{m})$. Also find the equivalent capacitances if the transistor were realized as a single device with source and drain contacts similarly evenly placed.

[7]The use of metal overstrapping a higher resistivity interconnect such as polysilicon or heavily doped silicon, to lower the resistivity of the interconnect, is a recommended practice.

Solution: Starting with node 1 (i.e., the drain), we find that the areas of the junctions are equal to

$$A_{J-2} = A_{J-4} = 6\lambda \times 20\lambda = 120\lambda^2 = 30\ (\mu m)^2$$

with the perimeters given by (ignoring the gate side)

$$P_{J-2} = P_{J-4} = 6\lambda + 6\lambda = 12\lambda = 6\ \mu m$$

As a result, C_{db} can be estimated to be

$$C_{db} = 2(A_{J-2}C_j + P_{J-2}C_{jsw}) = 0.017\ pF$$

For node 2 (i.e., the source), we have

$$A_{J-1} = A_{J-5} = 5\lambda \times 20\lambda = 100\lambda^2 = 25\ (\mu m)^2$$

and

$$A_{J-3} = A_{J-2} = 30\ (\mu m)^2$$

The perimeters are found to be

$$P_{J-1} = P_{J-5} = 5\lambda + 5\lambda + 20\lambda = 30\lambda = 15\ \mu m$$

and

$$P_{J-3} = P_{J-2} = 6\ \mu m$$

resulting in an estimate for C_{sb} of

$$C_{sb} = (A_{J-1} + A_{J-3} + A_{J-5} + WL)C_j + (P_{J-1} + P_{J-3} + P_{J-5})C_{j-sw} = 0.036\ pF$$

It should be noted here that even without the additional capacitance due to the WL gate area, node 1 has less capacitance than node 2 since it has less area and perimeter.

In the case in which the transistor is a single wide device (rather than four in parallel), we find

$$A_J = 5\lambda \times 80\lambda = 400\lambda^2 = 100\ (\mu m)^2$$

and

$$P_J = 5\,\lambda + 5\,\lambda + 80\,\lambda = 90\,\lambda = 45\ \mu m$$

resulting in $C_{db} = 0.033\ pF$ and $C_{sb} = 0.043\ pF$. Note that in this case, C_{db} is nearly doubled over the case in which four parallel transistors are used.

2.4 Advanced CMOS Processing

Lightly Doped Drains

A modern processing variation that is becoming very popular is to separate the transistor junctions from the actual channel region by lightly doped regions. A typical cross section of a *lightly doped drain (LDD)* device is shown in Fig. 2.18. The use of lightly doped regions between the channel and the junctions causes the depletion width at the drain end of the transistor to extend much further into the junction region. This lowers the electric fields at the drain and minimizes the generation of hot electrons. This in turn improves long-term reliability and helps to prevent latch-up. A negative aspect of this process is the series drain and source impedances are increased.

The steps taken in realizing LDD transistors are shown in Fig. 2.19a to c. First, about 0.5 μm of polysilicon is deposited and patterned. Next, the lightly doped n^- region is ion implanted to a final depth of around 0.2 μm using phosphorus. This region is self-aligned to the polysilicon gates.

The next few steps in the LDD process result in the realization of the oxide side-wall spacers. First, high-quality oxide, such as *tetraethyl orthosilicate*, or *TEOS*, is deposited to a thickness of perhaps 0.3 μm. This is then *anisotropically* etched using reactive ion etching, or *RIE*. An anisotropic etch is a directional etch, as opposed to an *isotropic* etch, which occurs for *wet etches*, where all directions are uniformly etched. In the anisotropic etch, a plasma of ions is accelerated against the wafer. This supplies new material constantly for the chemical reaction to surfaces perpendicular to the direction of ion travel. This allows the

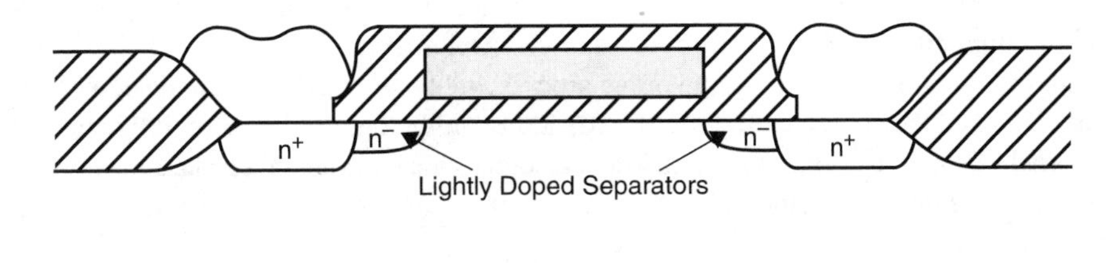

Figure 2.18 A cross section of a lightly doped drain (LDD) transistor.

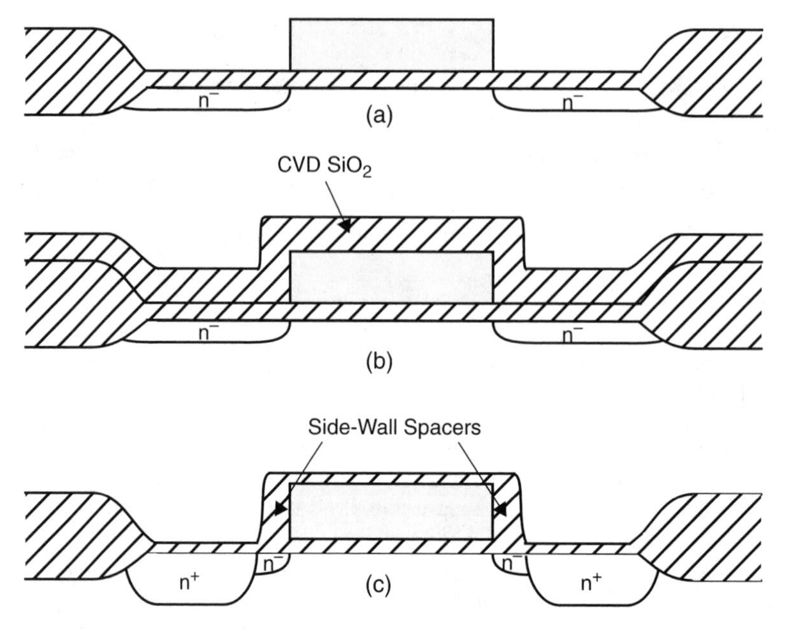

Figure 2.19 The steps in forming LDD transistors: (a) after polysilicon and CVD SiO_2 deposition and patterning, (b) after depositing high-quality thermal SiO_2, and (c) after selectively etching SiO_2 using an isotropic reactive ion etching and n^+ junction implantation.

chemical reaction involved in etching to proceed much more quickly at surfaces perpendicular to the IC as opposed to the SiO_2 sides. The net result is the side-wall spacers shown in Fig. 2.19c. In the next step, the n^+ junctions are ion implanted to a depth of 0.4 μm or so, using a mixture of phosphorus and arsenic. The mixture is used to get a more gradual concentration change that results during the later annealing due to the different diffusion constants (phosphorus diffuses much faster than arsenic). The final structure is shown in Fig. 2.19c where the shallow n^- regions are seen on either side of the channel region. The use of LDD processing allowed 0.35 μm and smaller channel lengths to be used with 3.3 V power supplies without reliability problems due to hot-electron effects.

It should be mentioned that the name of the process, lightly doped drain, is a misnomer because both the drain and the source are lightly doped. Although the lightly doped drain is beneficial in minimizing hot-electron effects, the lightly doped sources are detrimental because they give larger series source and drain resistances. This in turn causes source degeneration, which gives lower effective transistor transconductances, or effective g_ms. This can be particularly harmful in analog circuits. This series source resistance can be traded off with the hot-electron insensitivity by adjusting the phosphorus–arsenic mixture used when ion implanting the n^+ junctions.

There are many other modifications to the basic CMOS process that are used in current 0.25 and 0.35 μm processes. These might include low-impedance silicided junctions and gates using

TiS_2; local interconnect using titanium nitride, TiN, to save space; planarized oxide allowing for multiple metals using conformal oxides and etch back; tungsten contacts and interconnect; and very recently, the use of copper interconnect, to name only a few. The discussion of these subjects is beyond the scope of this text.

2.5 Bibliography

A. Glaser and G. Subak-Sharpe, *Integrated Circuit Engineering, Design, Fabrication, and Applications*, Addison-Wesley, 1977.

R. Haken, R. Haveman, R. Ekund, and L. Hutter, "BiCMOS Process Design," in *BiCMOS Technology and Applications*, A. Alvarez, ed., Kluwer Academic Publishers, 1989.

R. Haveman, R. Ekund, R. Haken, D. Scott, H. Tran, P. Fung, T. Ham, D. Favreau, and R. Virkus, "A 0.8μm 256K BiCmos SRAM Technology," *Digest of Technical Papers, 1987 International Electron Devices Meeting,* 84–843, December 1987.

S. Muraka and M. Peckerar, *Electronic Materials,* Academic Press, 1989, p. 326.

D. Reinhard, *Introduction to Integrated Circuit Engineering*, Houghton Mifflin, 1987.

S. Wolf, *Silicon Processing for the VLSI Era—Volume 3: The Submicron MOSFET,* Lattice Press, 1995.

2.6 Problems

Unless otherwise stated, assume the following:

- npn bipolar transistors: $\beta = 100$

  $V_A = 80 \text{ V}$

  $\tau_b = 13 \text{ ps}$

  $\tau_s = 4 \text{ ns}$

  $r_b = 330\ \Omega$

- n-channel MOS transistors:

  $\mu_n C_{ox} = 92\ \mu\text{A}/\text{V}^2$

  $V_{tn} = 0.8 \text{ V}$

  $\gamma = 0.5 \text{ V}^{1/2}$

  $r_{ds}(\Omega) = 8000\ L(\mu\text{m})/I_D\ (\text{mA})$ in active region

  $C_j = 2.4 \times 10^{-4}\ \text{pF}/(\mu\text{m})^2$

$C_{j\text{-}sw} = 2.0 \times 10^{-4}\ \text{pF}/\mu\text{m}$

$C_{ox} = 1.9 \times 10^{-3}\ \text{pF}/(\mu\text{m})^2$

$C_{gs(overlap)} = C_{gd(overlap)} = 2.0 \times 10^{-4}\ \text{pF}/\mu\text{m}$

- p-channel MOS transistors:

$\mu_p C_{ox} = 30\ \mu\text{A}/\text{V}^2$

$V_{tp} = -0.9\ \text{V}$

$\gamma = 0.8\ \text{V}^{1/2}$

$r_{ds}(\Omega) = 12{,}000\ \text{L}(\mu\text{m})/I_D\ (\text{mA})$ in active region

$C_j = 4.5 \times 10^{-4}\ \text{pF}/(\mu\text{m})^2$

$C_{j\text{-}sw} = 2.5 \times 10^{-4}\ \text{pF}/\mu\text{m}$

$C_{ox} = 1.9 \times 10^{-3}\ \text{pF}/(\mu\text{m})^2$

$C_{gs(overlap)} = C_{gd(overlap)} = 2.0 \times 10^{-4}\ \text{pF}/\mu\text{m}$

2.1 Discuss briefly the relationships between an ion beam's acceleration potential, the beam current, and the time of implantation on the resulting doping profile.

2.2 Place the following processing steps in their correct order: metal deposition and patterning, field implantation, junction implantation, well implantation, polysilicon deposition and patterning, field-oxide growth.

2.3 What are the major problems associated with a single threshold-voltage-adjust implant?

2.4 What is the reason for using a field implant and why is it often not needed in the well regions?

2.5 What are the major trade-offs in using a wet process or a dry process when growing thermal SiO_2?

2.6 Why is polysilicon used to realize gates of MOS transistors rather than metal?

2.7 Why can a microcircuit not be annealed after metal has been deposited?

2.8 What minimum distance, in terms of λ, would you expect that metal should be separated from polysilicon? Why?

2.9 Find the circuit that the layout shown in Fig. P2.9 realizes. Try to simplify the circuit if possible. Also, give the sizes of all transistors. Assume $L = 2\lambda$ where $\lambda = 1\ \mu m$.

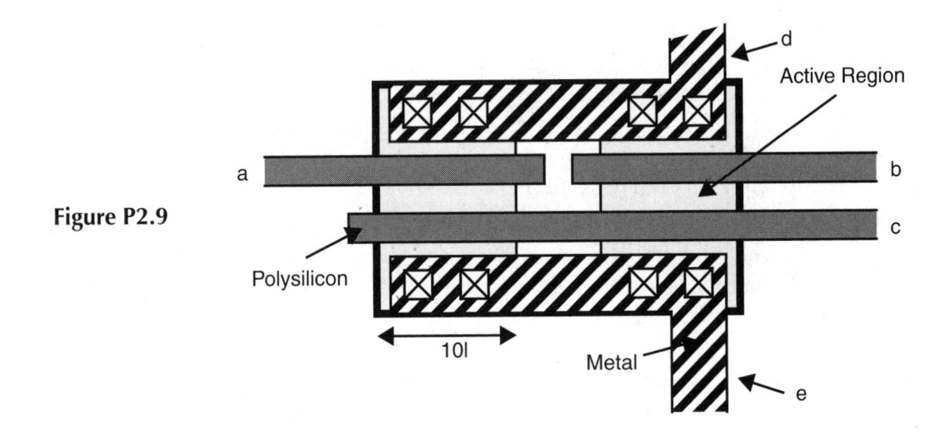

Figure P2.9

2.10 Find the transistor schematic for the CMOS logic circuit realized by the layout shown in Fig. P2.10. Give the widths of all transistors. Assume $L = 2\lambda$ where $\lambda = 0.4\ \mu m$. In tabular form, give the area and perimeter of each junction that is not connected to V_{DD} or ground.

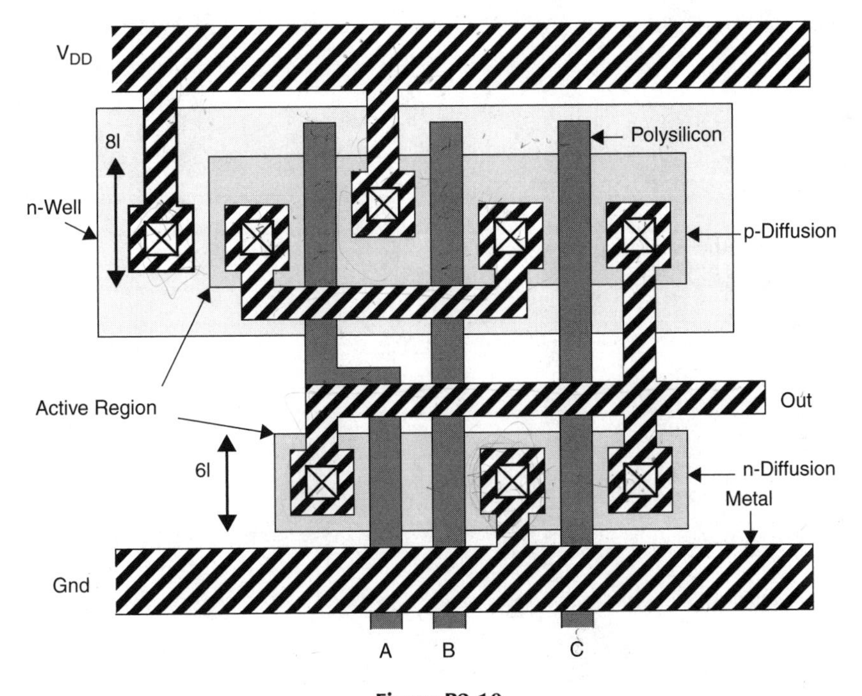

Figure P2.10

2.11 Repeat Example 2.1 for the case in which the two transistors do not physically share any junction between them but each junction is realized similar to junction J_2.

2.12 Repeat Example 2.2 where an overall transistor width of $80\ \lambda$ is still desired but assume eight parallel transistors are used (each of width $10\ \lambda$).

2.13 Repeat Example 2.2 where an overall transistor width of $80\ \lambda$ is still desired but assume two parallel transistors are used (each of width $40\ \lambda$).

2.14 Show a layout to realize two transistors in series (as in Fig. 2.15) but where each transistor has dimensions ($W = 240\ \lambda$, $L = 2\ \lambda$) and it is desired to minimize the capacitance of the node joining the two transistors (remember to consider the cases in which parallel transistors are used to realize each of the two transistors).

2.15 Given that a polysilicon layer has $7\ \Omega/\square$, what is the resistance of a long line that is $2\ \mu m$ wide and $1000\ \mu m$ long (ignore any contact resistance)?

CHAPTER 3

Integrated-Circuit Devices and Modeling

In this chapter,[1] the operation and modeling of semiconductor devices are described. It is possible to do much digital integrated-circuit design with a relatively basic knowledge of semiconductor device modeling; for a first course in digital integrated circuits, the simplified modeling presented in Section 3.1 suffices assuming students have encountered MOS transistors in a previous course; other sections need be covered only as the need arises. For more advanced courses, or for those intending to design high-speed state-of-the-art circuits, more in-depth understanding of the second-order effects of device operation and their modeling is considered critical and the whole chapter might be studied.

Section 3.2 describes pn junctions (or diodes). This section is important in understanding the parasitic capacitances in many device models, such as junction capacitances. Section 3.3 covers MOS transistors and modeling. It should be noted that it relies to some degree on the previous material presented in Section 3.2, where depletion capacitance is covered. Section 3.5 covers bipolar-junction transistors and modeling. A summary of device models and important equations are presented in Section 3.11. This summary is particularly useful for a reader who already has a good background in transistor modeling in which case the summary can be used to follow the notation used throughout the remainder of this book. Finally, this chapter concludes with an Appendix in which the more physical-based device equations are derived and in which some subjects that are less relevant to digital integrated-circuit (IC) design, such as small-signal modeling, are covered.

[1]The authors would like to acknowledge publishers John Wiley for giving permission to copy much of the material in this chapter from Chapter 2 of *Analog Integrated Circuit Design*, Wiley, 1997.

3.1 Simplified Transistor Modeling

MOS Transistors

The symbols used in this text for *enhancement* MOS transistors are shown in Fig. 3.1. MOS transistors are actually four-terminal devices; the fourth terminal is a substrate connection. For digital circuits, the substrate connection of n-channel transistors is almost always the most-negative IC voltage (i.e., ground or V_{ss}) and is not shown explicitly. Similarly, the substrate connection for p-channel transistors will be assumed to be the most positive IC voltage which is labeled V_{DD}. This will always be assumed to be the case unless stated otherwise, and therefore substrate connections will not be shown.

For an n-channel transistor, the source junction is the junction having the lower potential voltage; the drain junction is the other junction. The opposite is true for complementary p-channel transistors. If an enhancement MOS transistor has its gate-source voltage near zero, then it will not be conducting and no channel will be present. This means that the transistor can normally be ignored as if it did not exist when considering how logic gates function. However, there will be parasitic capacitances between its junctions and ground for an n-channel transistor (between its junctions and V_{DD} for p-channel transistors) due to the reverse-biased junction-substrate depletion regions. These junction capacitances need only be considered when estimating transient delays when gates are changing states. These parasitic capacitances are due to both the periphery and the area of the junctions. The junction capacitance decreases as the reverse-bias voltage of the junction increases in a nonlinear fashion (approximately inversely proportional to the square root of the reverse-bias junction voltage). A reasonable approximation is to use 0.6 times the total junction capacitance when the junction is biased at 0 V. The periphery and area of the junction can be estimated using the simple layout rules introduced in the previous chapter and also discussed in the next chapter. *The fact that the junction capacitance of* p*-channel transistors is between the junction and* V_{DD}*, rather than between the junction and ground, is inconsequential for digital circuits; the junction capacitances may still be approximated by parasitic capacitances between the junctions and ground.*

When the effective gate-source voltage is greater than zero, a channel will form for an MOS transistor. For an n-channel transistor, this means $V_{eff} = V_{GS} - V_{tn} > 0$ where $V_{tn} \approx 0.7$ V.

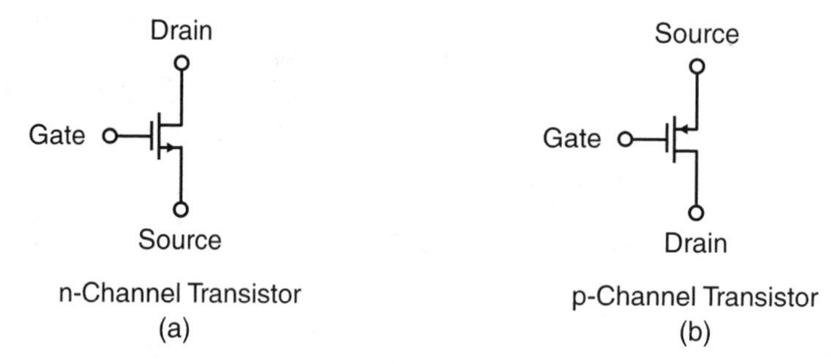

Figure 3.1 An n-channel (a) and a p-channel (b) enhancement transistor.

For a p-channel transistor, this means $V_{eff} = V_{SG} + V_{tp} > 0$ where $V_{tp} \approx -0.7$ V. For this case, the large-signal equations describing the approximate I–V relationship of n-channel MOS transistors are

$$I_D = \mu_n \frac{W}{L} C_{ox}\left[(V_{GS} - V_{tn})V_{DS} - \frac{V_{DS}^2}{2}\right] \tag{3.1}$$

for $V_{DS} < V_{eff}$ and

$$I_D = \frac{\mu_n C_{ox} W}{2 L}(V_{GS} - V_{tn})^2 \tag{3.2}$$

for $V_{DS} > V_{eff}$. In these equations, μ_n is the mobility of electrons and is approximately equal to $0.05\ m^2/V \cdot s$. C_{ox} is the gate capacitance per unit area, which is a process-dependent parameter. A typical number might be $3.5\ fF/\mu m^2$ for a $0.6\ \mu m$ process. W is the effective transistor width and L is the effective transistor length. In digital circuits, almost all transistors will have the minimum-allowable transistor length. This means that besides choosing a topology, the only other design choices, typically, are the transistor widths.

For p-channel transistors, the same equations apply but with negative signs inserted in front of each voltage variable (except for V_{eff}). For example, the inequality $V_{DS} > V_{eff}$ becomes $V_{SD} > V_{eff}$ for defining when a transistor is in the active or saturation region. The mobility of p carriers or holes is around $0.02\ m^2/V \cdot s$. With experience, one need only remember the equations for the n-channel transistors and then transform them to the complementary equations for p-channel transistors remembering that current (and charge) always flows from the source to the drain for p-channel transistors (as opposed to from the drain to the source for n-channel transistors).

When the drain-source voltage of an n-channel transistor (or the source-drain voltage of a p-channel transistor) is large, then equation (3.2) applies and a simple model for the transistor is just a current source as shown in Fig. 3.2a, when only d.c. operation is of interest.

When transient operation is of interest, then the model of Fig. 3.2b that includes parasitic junction capacitances and a gate-source capacitance can be used. The capacitor C_{gs} is approximately equal to $(2/3)WLC_{ox}$ for a transistor having a large drain-source voltage. When the

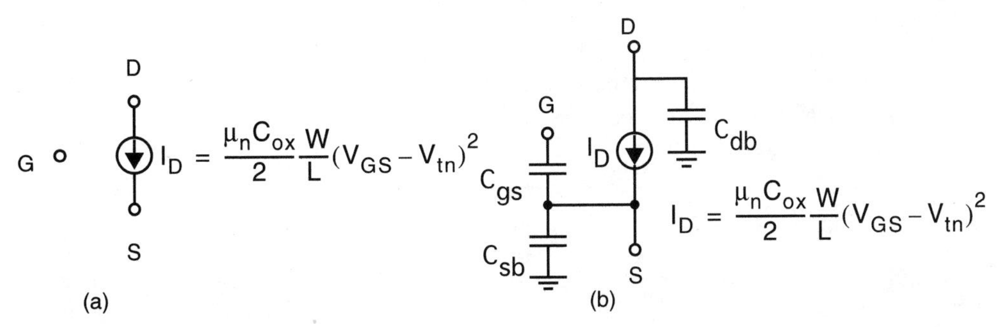

Figure 3.2 Simplified models for an n-channel transistor having a large drain-source voltage and therefore operating in the active region (a) for low frequencies and (b) for transient analyses.

drain-source voltage of a transistor is not large, then the total capacitance *seen* by a logic gate connected to a transistor gate is usually approximated by the area of the transistor gate multiplied by C_{ox}, that is $C_g \approx WLC_{ox}$. When the drain-source voltage is not known or is changing, then this latter upper bound is usually used as an approximation. These transistor gate capacitances, and the junction capacitances mentioned previously, are often the largest factors in parasitic capacitive loads in integrated circuits. Parasitic capacitances due to metal interconnects may also be important when the interconnect wires are long.

The capacitors C_{sb} and C_{db} are the junction capacitances. These capacitors are highly nonlinear with respect to the reverse-bias voltages across a junction.[2] They are roughly proportional to the junction areas. Often, the junction capacitance per unit area, $C_{j\text{-}0}$ will be specified for a reverse-bias voltage of 0 V for each type of junction for a given IC process. A commonly used approximation to account for the nonlinearity of the junction capacitance is to then multiply $C_{j\text{-}0}$ by the junction area by the factor 0.6, as stated previously. A more accurate calculation of the junction capacitances and explanations of how to estimate junction areas are discussed in later sections of this chapter and in the next chapter. This task has been largely automated in modern CAD systems, which can *extract* their sizes automatically once the layout is complete.

When the drain-source voltage of an n-channel transistor (or the source-drain voltage of a p-channel transistor) is *very* small, then equation (3.1) can be approximated by

$$I_D = \mu_n \frac{W}{L} C_{ox}(V_{GS} - V_{tn})V_{DS} \tag{3.3}$$

and

$$r_{ds} = \frac{V_{DS}}{I_D} \approx \frac{1}{\mu_n \frac{W}{L} C_{ox}(V_{GS} - V_{tn})} \tag{3.4}$$

Therefore, a simple model for the transistor in this region of operation is just a resistor as shown in Fig.3.3a when only d.c. operation is of interest. When transient operation is of interest, then the more complicated model of Fig. 3.3b, which includes parasitic junction capacitances and gate-source and gate-drain capacitances, can be used. These latter capacitances are often approximated using the formulas $C_{gs} = C_{gd} = (1/2)WLC_{ox}$. When this level of modeling is necessary, then the simplified models discussed in this section may be inadequate and more accurate modeling described later in the chapter is often required.

When a transistor has a drain-source voltage that changes from very small to very large or vice versa, then the transistor may be approximated by a resistor with a value given by

$$r_{ds} \approx \frac{2.5}{\mu_n \frac{W}{L} C_{ox}(V_{GS} - V_{tn})} \tag{3.5}$$

This approximation will be justified in the next chapter. It is only very roughly approximate (to perhaps 25% accuracy), but is useful in that it allows for very simple analyses of transient

---

[2]In an integrated circuit, practically all junctions are reverse biased and effectively have no current through them except for the junctions in bipolar transistors.

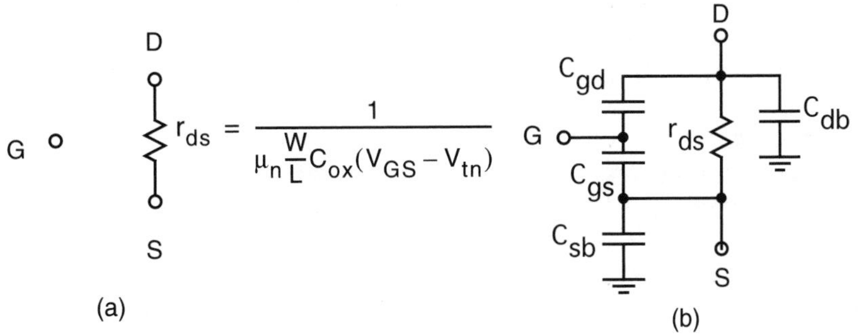

D
G
S
$r_{ds} = \frac{1}{\mu_n \frac{W}{L} C_{ox}(V_{GS} - V_{tn})}$
(a)
D
C_{gd}
C_{db}
G
r_{ds}
C_{gs}
C_{sb}
S
(b)

Figure 3.3 Simplified models for an n-channel transistor having a small drain-source voltage and therefore in the triode region (a) for low frequencies and (b) for transient analyses.

Collector
I_C
I_B
Base
$I_C = \beta I_B$
$\approx 100 I_B$
(typical)
IE
Emitter

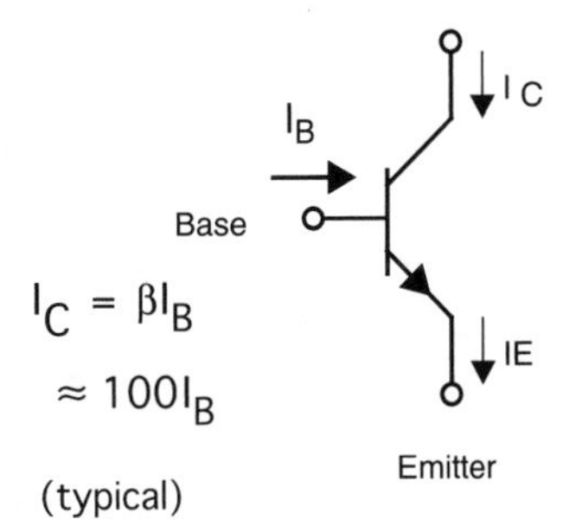

Figure 3.4 The symbol representing an npn bipolar-junction transistor.

responses assuming parasitic capacitances can be estimated (as discussed in the next chapter). Note the similarity and differences between equations (3.4) and (3.5).

It should be obvious that many approximations are necessarily used when modeling MOS transistors. These approximations are necessary because of the nonlinear operation of the transistor and the many second-order effects that are significant, but are difficult to accurately predict analytically, especially for modern submicron transistors. For these reasons, the speed of digital circuits can seldom be predicted to better than 10 to 20%; prudent designers almost always use conservative design methodologies and ensure large margins to allow for inaccurate modeling.

BIPOLAR TRANSISTORS

Normally, only *npn* bipolar transistors are used to realize *bipolar* digital integrated circuits as typical integrated *pnp* bipolar transistors are too slow. The symbol used for npn transistors is shown in Fig. 3.4. A model that is adequate for understanding the d.c. operation of most bipolar logic gates is shown in Fig. 3.5. The base current is given by

$$I_B = \frac{I_C}{\beta} = \frac{I_{CS}}{\beta} e^{V_{BE}/V_T} = I_{BS} e^{V_{BE}/V_T} \tag{3.6}$$

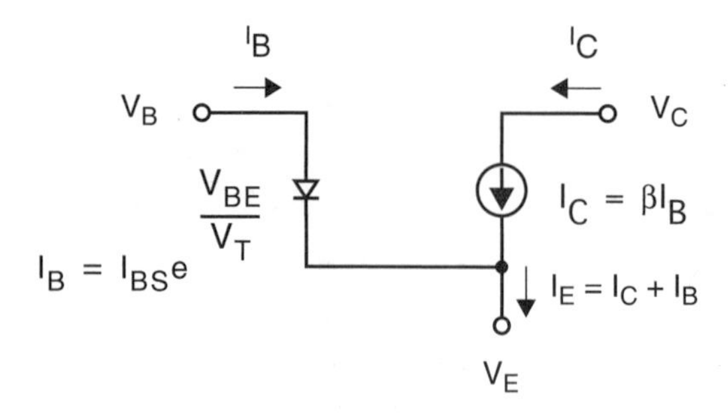

Figure 3.5 A large-signal model for a BJT.

where I_{CS} and $I_{BS} = I_{CS}/\beta$ are constants. β is the current gain of the transistor and is normally between 50 and 200. This model applies when the base-emitter voltage is greater than around 0.5 V and the collector-emitter voltage is greater than 0.3 V. If the first requirement is not true, then the transistor will not be conducting and can be ignored when considering low-frequency operation. If the latter case is not true, then the transistor will be *saturated*. This condition results in very slow operation of digital circuits due to excessive charge storage in the base and collector regions and therefore it should never occur in *good* bipolar logic circuits.

The prediction of the transient operation of bipolar logic circuits requires much more in-depth modeling than presented in this section. However, a partial understanding begins with the realization that the largest parasitic capacitances of bipolar logic circuits are the base-emitter capacitances and these are approximately given by

$$C_d = \tau_b \frac{I_C}{V_T} = g_m \tau_b \tag{3.7}$$

where $V_T = kT/q \cong 25$ mV at room temperature (and is proportional to absolute temperature) and τ_b is a constant that is approximately the inverse of the unity-gain frequency of the transistor (in radians). Thus, the major parasitic capacitances are proportional to bias currents; since the currents for charging and discharging these capacitances are also proportional to bias currents, the speed of bipolar logic is approximately independent of bias currents. This statement is approximately true at higher bias currents; at medium to lower bias currents, the junction capacitances become significant and speeds deteriorate as bias currents are reduced.

In a first-order d.c. analysis, it is often assumed that the base-emitter voltage is 0.7 V for a conducting bipolar transistor and that the collector current is approximately equal to the emitter current. These two facts are adequate for understanding the functional operation at d.c. of the bipolar gates described in this text.

The simplified and cursory approach just described for modeling MOS and bipolar transistors is adequate for doing much general-purpose noncritical digital IC design. For state-of-the-art digital IC design, knowledge of more accurate and in-depth IC modeling as discussed in the rest of this chapter is necessary. Also, for readers without previous knowledge in transistor modeling, understanding much of the material presented in the rest of this chapter is necessary, but the material is also repeated in succeeding chapters where needed. For those in a first-level course, the remaining material in this chapter is useful but not essential, and may be consulted as the need arises.

3.2 Semiconductors and pn Junctions

A semiconductor is a crystal lattice structure that can have free electrons (which are negative carriers) and/or free holes (which are an absence of electrons and are equivalent to positive carriers). The type of semiconductor typically used is silicon (commonly found in sand). This material has a valence of four, implying that each atom has four free electrons to share with neighboring atoms when forming the covalent bonds of the crystal lattice. *Intrinsic* silicon (i.e., undoped silicon) is a very pure crystal structure having equal numbers of free electrons and holes. These free carriers are those electrons or holes that have gained enough energy due to thermal agitation to escape their bonds. At room temperature, there are approximately 1.5×10^{10} carriers of each type/cm^3, or equivalently 1.5×10^{16} carriers/m^3. The number of carriers approximately doubles for every 11°C increase in temperature.

If silicon is doped with a pentavalent impurity (i.e., atoms of an element having a valence of five, or equivalently five electrons in the outer shell available when bonding with neighboring atoms), there will be almost one extra free electron for every impurity atom.[3] These free electrons can be used to conduct current. A pentavalent impurity is said to *donate* free electrons to the silicon crystal, and thus the impurity is known as a *donor.* Examples of donor elements are phosphorus, P, and arsenic, As. These impurities are also called n-type dopants since the free carriers resulting from their use have a negative charge. When an n-type impurity is used, the total number of negative carriers or electrons is almost the same as the doping concentration, and is much greater than the number of free electrons in intrinsic silicon. In other words,

$$n_n = N_D \tag{3.8}$$

where n_n denotes the free-electron concentration in n-type material (denoted by the subscript) and N_D is the doping concentration (with the subscript D denoting donor). On the other hand, the number of free holes in n-doped material will be much less than the number of holes in intrinsic silicon and can be shown (Sze, 1981) to be given by

$$p_n = \frac{n_i^2}{N_D} \tag{3.9}$$

Here, n_i is the carrier concentration in intrinsic silicon.

[3]In fact, there will be slightly fewer mobile carriers than the number of impurity atoms since some of the free electrons from the dopants have recombined with holes. However, since the number of holes of intrinsic silicon is much less than typical doping concentrations, this inaccuracy is small.

Similarly, if silicon is doped with atoms having a valence of three, for example, boron (B), the concentration of positive carriers or holes will be approximately equal to the *acceptor* concentration, N_A,

$$p_p = N_A \tag{3.10}$$

and the number of negative carriers in the p-type silicon, n_p, is given by

$$n_p = \frac{n_i^2}{N_A} \tag{3.11}$$

Example 3.1

Intrinsic silicon is doped with boron at a concentration of 10^{26} atoms/m^3. At room temperature, what are the concentrations of holes and electrons in the resulting doped silicon? Assume that $n_i = 1.5 \times 10^{16}$ carriers/m^3.

Solution: The hole concentration, p_p, will approximately equal the doping concentration ($p_p = N_A = 10^{26}$ holes/m^3). The electron concentration is found from (3.11) to be

$$n_p = \frac{(1.5 \times 10^{16})^2}{10^{26}} = 2.3 \times 10^6 \text{ electrons/m}^3 \tag{3.12}$$

Such doped silicon is referred to as p type since it has many more free holes than free electrons.

DIODES

To realize a diode, or, equivalently, a pn junction, one part of a semiconductor is doped n type, and a closely adjacent part is doped p type, as shown in Fig. 3.6. Here the diode, or junction, is formed between the p^+ region and the n region. Note that the superscripts indicate the relative doping levels. For example, the p^- bulk region might have an impurity concentration of 5×10^{21} carriers/m^3, whereas the p^+ and n^+ regions would be doped more heavily to a value around 10^{25} to 10^{27} carriers/m^3. Also note that the metal contacts to the diode (in this case, aluminum) are connected to a heavily doped region as opposed to a lightly doped region; otherwise a *Schottky diode* would occur. (Schottky diodes are discussed on p. 85.) To not make a Schottky diode, the connection to the n region is actually made via the n^+ region.

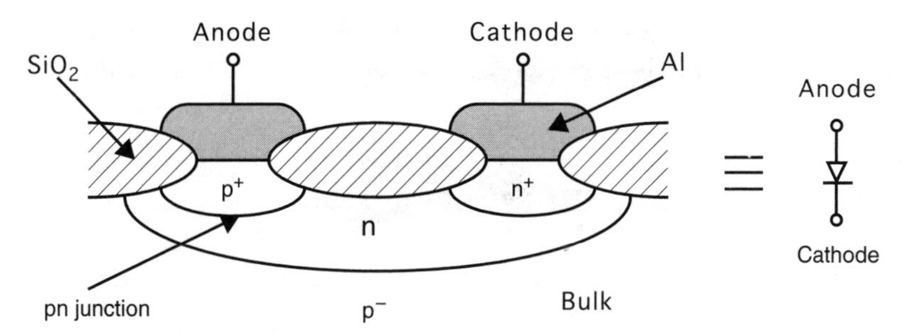

Figure 3.6 A cross-sectional view of a pn diode.

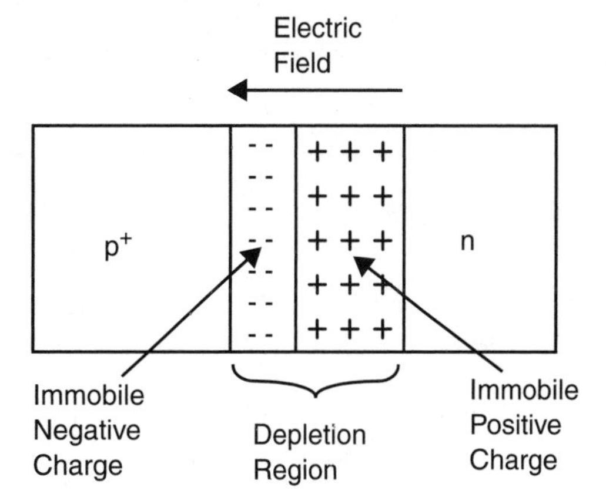

Figure 3.7 A simplified model for a diode. Note that a depletion region exists at the junction due to diffusion and extends further into the more lightly doped side.

In the p^+ side, a large number of free positive carriers are available, whereas in the n side, many free negative carriers are available. The holes in the p^+ side will tend to disperse or *diffuse* into the n side, whereas the free electrons in the n side will tend to *diffuse* to the p^+ side. This process is very similar to two gases randomly diffusing together. This diffusion lowers the concentration of free carriers in the region between the two sides. As the two types of carriers diffuse together, they *recombine*. Every electron that diffuses from the n side to the p side leaves behind a *bound* positive charge close to the transition region. Similarly, every hole that diffuses from the p side leaves behind a bound electron near the transition region. The end result is shown in Fig. 3.7. This diffusion of free carriers creates a *depletion region* at the junction of the two sides where no free carriers exist, and which has a net negative charge on the p^+ side and a net positive charge on the n side. The total amount of exposed or bound charge on the two sides of the junction must be equal for charge neutrality. This requirement causes the depletion region to extend farther into the more lightly doped n side than into the p^+ side.

As these bound charges are exposed, an electric field develops going from the n side to the p side. This electric field is often called the built-in potential of the junction. It opposes the diffusion of free carriers until there is no net movement of charge under open-circuit and steady-state conditions. The built-in voltage of an open-circuit pn junction is given by Sze (1981) as

$$\Phi_0 = V_T \ln\left(\frac{N_A N_D}{n_i^2}\right) \tag{3.13}$$

where

$$V_T = \frac{kT}{q} \tag{3.14}$$

with T being the temperature in degrees Kelvin ($\cong 300°\text{K}$ at room temperature), k being Boltzmann's constant ($1.38 \times 10^{-23}\ \text{J K}^{-1}$), and q being the charge of an electron ($1.602 \times 10^{-19}\ \text{C}$). At room temperature, V_T is approximately 26 mV.

Example 3.2

A pn junction has $N_A = 10^{25}\ \text{holes/m}^3$ and $N_D = 10^{22}\ \text{electrons/m}^3$. What is the built-in junction potential? Assume that $n_i = 1.5 \times 10^{16}\ \text{carriers/m}^3$.

Solution: Using (3.13), we obtain

$$\Phi_0 = 0.026 \times \ln\left[\frac{10^{25} \times 10^{22}}{(1.5 \times 10^{16})^2}\right] = 0.88\ \text{V} \tag{3.15}$$

This is a typical value for the built-in potential of a junction with one side heavily doped. As an approximation, we will normally use $\Phi_0 \cong 0.9\ \text{V}$ or $\Phi_0 \cong 1.0\ \text{V}$ for the built-in potential of a junction having one side heavily doped.

REVERSE-BIASED DIODES

A silicon diode having an *anode-to-cathode* (i.e., p side to n side) voltage of 0.4 V or less will not be conducting appreciable current. In this case, it is said to be *reverse biased.* If a diode is reverse biased, current flow is primarily due to thermally generated carriers in the *depletion region,* and it is extremely small. Although this reverse-biased current is only weakly dependent on the applied voltage, *the reverse-biased current is directly proportional to the area of the diode junction.* However, an effect that should not be ignored, particularly at high frequencies, is the junction capacitance of a diode. In reverse-biased diodes, this junction capacitance is due to varying charge storage in the depletion regions and is modeled as a *depletion capacitance.*

To determine the depletion capacitance, we first state the relationship between the depletion widths and the applied reverse voltage, V_R (Sze, 1981).

$$x_n = \left[\frac{2K_s\varepsilon_0(\Phi_0 + V_R)}{q}\frac{N_A}{N_D(N_A + N_D)}\right]^{1/2} \tag{3.16}$$

$$x_p = \left[\frac{2K_s\varepsilon_0(\Phi_0 + V_R)}{q}\frac{N_D}{N_A(N_A + N_D)}\right]^{1/2} \tag{3.17}$$

Here, ε_0 is the permittivity of free space (equal to 8.854×10^{-12} F/m), V_R is the reverse-bias voltage of the diode, and K_s is the relative permittivity of silicon (equal to 11.8). These equations assume that the doping changes abruptly from the n to the p side.

From the above equations, we see that if one side of the junction is more heavily doped than the other, the depletion region will extend mostly into the lightly doped side. For example, if $N_A >> N_D$ (i.e., if the p region is more heavily doped), we can approximate (3.16) and (3.17) as

$$x_n \cong \left[\frac{2K_s\varepsilon_0(\Phi_0 + V_R)}{qN_D}\right]^{1/2} \qquad x_p \cong \left[\frac{2K_s\varepsilon_0(\Phi_0 + V_R)N_D}{qN_A^2}\right]^{1/2} \tag{3.18}$$

For this case

$$\frac{x_n}{x_p} \cong \frac{N_A}{N_D} \tag{3.19}$$

This special case is called a *single-sided diode.*

Example 3.3

For a pn junction having $N_A = 10^{25}$ holes/m^3 and $N_D = 10^{22}$ electrons/m^3, what are the depletion-layer depths for a 3.3-V reverse-bias voltage?

Solution: Since $N_A >> N_D$, and we already have found in Example 3.2 that $\Phi_0 = 0.9$ V, we can use (3.18) to find

$$x_n = \left[\frac{2 \times 11.8 \times 8.854 \times 10^{-12} \times 4.2}{1.6 \times 10^{-19} \times 10^{22}}\right]^{1/2} = 0.74\ \mu\text{m} \tag{3.20}$$

$$x_p = \frac{x_n}{(N_A/N_D)} = 0.74\ \text{nm} \tag{3.21}$$

Note that the depletion width in the lightly doped n *region is 1000 times greater than that in the more heavily doped* p *region.*

The charge stored in the depletion region, per unit cross-sectional area, is found by multiplying the depletion-region width by the concentration of the immobile charge (which is approximately equal to q times the impurity doping density). For example, on the n side, we find the charge in the depletion region to be given by multiplying (3.16) by qN_D, resulting in

$$Q^+ = \left[2qK_s\varepsilon_0(\Phi_0 + V_R)\frac{N_A N_D}{N_A + N_D}\right]^{1/2} \tag{3.22}$$

This amount of charge must also equal Q^- on the p side since *there must be charge equality.* In the case of a single-sided diode when $N_A >> N_D$, we have

$$Q^- = Q^+ \cong [2qK_s\varepsilon_0(\Phi_0 + V_R)N_D]^{1/2} \tag{3.23}$$

Note that this result is independent of the impurity concentration on the heavily doped side. Thus, we see from the above relation that the charge stored in the depletion region is dependent on the applied reverse-bias voltage. *It is this charge-voltage relationship that is modeled by a nonlinear depletion capacitance.*

Reverse-biased junctions represent a major factor contributing to parasitic capacitances in integrated circuits. An ideal capacitance has a linear relationship between the voltage across it and the charge stored in it as depicted by Fig. 3.8a. However, the Q–V curve of a reverse-biased depletion region or junction as given by (3.23) or (3.22) is nonlinear as is approximately shown in Fig 3.8b. At any given bias point, the small-signal capacitance is defined as the tangent to the Q–V curve at the bias point as shown in Fig. 3.8b. It is also possible to define a large-signal average capacitance that approximates the complete Q–V curve again as shown in Fig 3.8b. For a reverse-biased junction, the small-signal capacitance at large reverse-bias voltages is normally smaller than the average capacitance; the small-signal capacitance at small reverse-bias voltages is normally larger than the average capacitance.

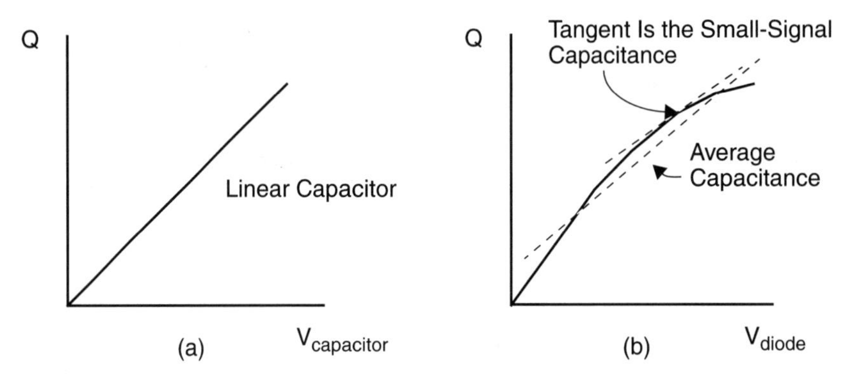

Figure 3.8 (a) The Q–V curve of an ideal capacitor and (b) the Q–V curve of a reverse-biased junction.

For small changes in the reverse-bias junction voltage, about a bias voltage, we can find an equivalent *small-signal* capacitance, C_j, by differentiating (3.22) with respect to V_R. Such a differentiation results in

$$C_j = \frac{dQ^+}{dV_R} = \left[\frac{qK_s\varepsilon_0}{2(\Phi_0 + V_R)} \frac{N_A N_D}{N_A + N_D}\right]^{1/2} = \frac{C_{j\text{-}0}}{\sqrt{1 + (V_R/\Phi_0)}} \tag{3.24}$$

where $C_{j\text{-}0}$ is the depletion capacitance per unit area at $V_R = 0$ and is given by

$$C_{j\text{-}0} = \sqrt{\frac{qK_s\varepsilon_0}{2\Phi_0} \frac{N_A N_D}{N_A + N_D}} \tag{3.25}$$

In the case of a one-sided diode with $N_A >> N_D$, we have

$$C_j = \left[\frac{qK_s\varepsilon_0 N_D}{2(\Phi_0 + V_R)}\right]^{1/2} = \frac{C_{j\text{-}0}}{\sqrt{1 + (V_R/\Phi_0)}} \tag{3.26}$$

where now

$$C_{j\text{-}0} = \sqrt{\frac{qK_s\varepsilon_0 N_D}{2\Phi_0}} \tag{3.27}$$

Many of the junctions encountered in integrated circuits are one-sided junctions with the lightly doped side being the substrate, or sometimes what is called the *well*. The more heavily doped side is often used to form a contact to interconnect metal. From (3.27), we see that *for these one-sided junctions, the depletion capacitance is approximately independent of the doping concentration on the heavily doped side, and is proportional to the square root of the doping concentration of the more lightly doped side.* Thus, smaller depletion capacitances are obtained for more lightly doped substrates—a strong incentive to strive for lightly doped substrates.

Finally, note that by combining (3.22) and (3.25), we can express the equation for the immobile charge on either side of a reverse-biased junction as

$$Q = 2C_{j\text{-}0}\Phi_0\sqrt{1 + \frac{V_R}{\Phi_0}} \tag{3.28}$$

As seen in Example 1.6, this equation is useful when one is approximating the large-signal charging (or discharging) time for a reverse-biased diode.

Example 3.4

For a pn junction having $N_A = 10^{25}$ holes/m^3 and $N_D = 10^{22}$ electrons/m^3, what is the total zero-bias depletion capacitance for a diode of area $10\ \mu m \times 10\ \mu m$? What is its depletion capacitance for a 3.3 V reverse-bias voltage?

Solution: Making use of (3.27), we have

$$C_{j\text{-}0} = \sqrt{\frac{1.6 \times 10^{-19} \times 11.8 \times 8.854 \times 10^{-12} \times 10^{22}}{2 \times 0.9}} = 304.7\ \mu F/m^2 \qquad (3.29)$$

Since the diode area is $100 \times 10^{-12}\ m^2$, the total zero-bias depletion capacitance is

$$C_{T\text{-}j\text{-}0} = 100 \times 10^{-12} \times 304.7 \times 10^{-6} = 30.5\ fF \qquad (3.30)$$

At a 3.3-V reverse-bias voltage, we have from (3.26)

$$C_{T\text{-}j} = \frac{30.5\ fF}{\sqrt{1 + (3.3/0.9)}} = 14.1\ fF \qquad (3.31)$$

As expected, we see a decrease in junction capacitance as the width of the depletion region is increased.

Graded Junctions

All of the above equations assumed an abrupt junction where the doping concentration changes quickly from p to n over a small distance. Although this is a good approximation for many integrated circuits, it is not always true. For example, the collector-to-base junction of a bipolar transistor is most commonly realized as a *graded junction*. In the case of graded junctions, the exponent 1/2 in equation (3.22) is inaccurate, and a better value to use is an exponent closer to unity, perhaps 0.6 to 0.7. Thus, for graded junctions, (3.22)is typically written as

$$Q = \left[2qK_s\varepsilon_0(\Phi_0 + V_R)\frac{N_A N_D}{N_A + N_D}\right]^{1-m} \qquad (3.32)$$

where m is a constant typically between 1/3 and 1/2.

Differentiating (3.32) to find the depletion capacitance, we have

$$C_j = (1-m)\left[2qK_s\varepsilon_0\frac{N_A N_D}{N_A + N_D}\right]^{1-m}\frac{1}{(\Phi_0 + V_R)^m} \qquad (3.33)$$

This depletion capacitance can also be written as

$$C_j = \frac{C_{j\text{-}0}}{[1 + V_R/\Phi_0]^m} \tag{3.34}$$

where

$$C_{j\text{-}0} = (1 - m)\left[2qK_s\varepsilon_0 \frac{N_A N_D}{N_A + N_D}\right]^{1-m} \frac{1}{\Phi_0^m} \tag{3.35}$$

From (3.34), we see that a graded junction results in a depletion capacitance that is less dependent on V_R than the equivalent capacitance in an abrupt junction. In other words, since m is less than 0.5, the depletion capacitance for a graded junction is more linear than that for an abrupt junction. Correspondingly, increasing the reverse-bias voltage for a graded junction is not as effective in reducing the depletion capacitance as it is for an abrupt junction.

Finally, as in the case of an abrupt junction, the depletion charge on either side of the junction can also be written as

$$Q = \frac{C_{j\text{-}0}}{1 - m}\Phi_0\left(1 + \frac{V_R}{\Phi_0}\right)^{1-m} \tag{3.36}$$

Example 3.5

Repeat Example 3.4 for a graded junction with $m = 0.4$.

Solution: Noting once again that $N_A >> N_D$, we approximate (3.35) as

$$C_{j\text{-}0} = (1 - m)[2qK_s\varepsilon_0 N_D]^{1-m} \frac{1}{\Phi_0^m} \tag{3.37}$$

resulting in

$$C_{j\text{-}0} = 81.5\ \mu\text{F/m}^2 \tag{3.38}$$

which, when multiplied by the diode's area of $10\ \mu\text{m} \times 10\ \mu\text{m}$, results in

$$C_{T\text{-}j\text{-}0} = 8.1\ \text{fF} \tag{3.39}$$

For a 3.3-V reverse-bias voltage, we have

$$C_{T\text{-}j} = \frac{8.1\text{fF}}{(1 + 3.3/0.9)^{0.4}} = 4.4 \text{ fF} \tag{3.40}$$

Notice that even for a small decrease in m, the small-signal capacitance can become significantly smaller for graded junctions. This indicates that values for small-signal capacitances, as found analytically, may have significant errors (especially for graded junctions).

Large-Signal Junction Capacitance

For digital circuits, perhaps a more relevant parameter is the large-signal capacitance. The equations for the junction capacitance given above are valid only for small changes in the reverse-bias voltage. This limitation is due to the fact that C_j depends on the size of the reverse-bias voltage instead of being a constant. As a result, it is extremely difficult and time consuming to accurately take this nonlinear capacitance into account when calculating the time to charge or discharge a junction over a large voltage change. A commonly used approximation when analyzing the transient response for large voltage changes is to use an *average size* for the junction capacitance by calculating the junction capacitance at the two extremes of the reverse-bias voltage. Unfortunately, a problem with this approach is that when the diode is forward biased with $V_R \cong -\Phi_0$, equation (3.26) *blows up* (i.e., is equal to infinity). To circumvent this problem, one can instead calculate the charge stored in the junction for the two extreme values of applied voltage [through the use of (3.28)], and then through the use of $Q = CV$, calculate the average capacitance according to

$$C_{j\text{-}av} = \frac{Q(V_2) - Q(V_1)}{V_2 - V_1} \tag{3.41}$$

where V_1 and V_2 are the two voltage extremes (Hodges and Jackson, 1988).

From (3.28), for an abrupt junction with reverse-bias voltage V_i, we have

$$Q(V_i) = 2C_{j\text{-}0}\Phi_0\sqrt{1 + \frac{V_i}{\Phi_0}} \tag{3.42}$$

Therefore,

$$C_{j\text{-}av} = 2C_{j\text{-}0}\Phi_0\frac{\left(\sqrt{1 + (V_2/\Phi_0)} - \sqrt{1 + (V_1/\Phi_0)}\right)}{V_2 - V_1} \tag{3.43}$$

One special case often encountered is charging a junction from 0 to 3.3 V. For this special case, and using $\Phi_0 = 0.9$ V, we find that

$$C_{j\text{-}av} = 0.63 C_{j\text{-}0} \tag{3.44}$$

It will be seen in the following example that (3.44) compares well with a SPICE simulation. As a rough approximation to quickly estimate the charging time of a junction capacitance from 0 to 3.3 V (or vice versa), one can use either

$$C_{j\text{-}av} \approx 0.6 C_{j\text{-}0} \quad \text{or} \quad C_{j\text{-}av} \approx \frac{2}{3} C_{j\text{-}0} \tag{3.45}$$

Example 3.6

For the circuit shown in Fig. 3.9, where a reverse-biased diode is being charged from 0 to 3.3 V, through a 10-kΩ resistor, calculate the time required to charge the diode from 0 to 2.3 V. Assume $C_{j\text{-}0} = 0.2\ \text{fF}/\mu\text{m}^2$ and the diode has an area of 20 by 5 μm. Compare your answer to that obtained using SPICE. Repeat the question for the case of the diode being discharged from 3.3 to 1 V.

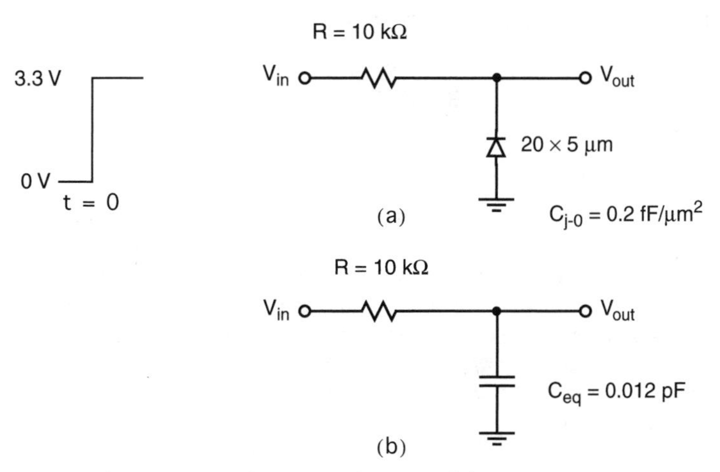

Figure 3.9 (a) The circuit used in Example 3.6 and (b) its RC approximate equivalent.

Solution: The total small-signal capacitance of the junction at 0 V bias voltage is obtained by multiplying 0.2 $fF/\mu m^2$ by the junction area to give

$$C_{T\text{-}j\text{-}0} = 0.2 \times 10^{-15} \times 20 \times 5 = 0.02 \text{ pF} \tag{3.46}$$

Using (3.44), we have

$$C_{T\text{-}j\text{-}av} = 0.63 \times 0.02 = 0.013 \text{ pF} \tag{3.47}$$

resulting in a time constant of

$$\tau = RC_{T\text{-}j\text{-}av} = 0.13 \text{ ns} \tag{3.48}$$

It is not difficult to show that the time it takes for a first-order circuit to rise (or fall) 70% of its final value is equal to 1.2τ. Thus, in this case,

$$\Delta t_{70\%} = 1.2\tau = 0.16 \text{ ns} \tag{3.49}$$

As a check, the circuit of Fig. 3.9a was analyzed using SPICE. The input data file was:

```
The Transient Response of Charging and Discharging a Diode
*
R 1 2 10k
D 0 2 DMOD
*
VIN 1 0 PULSE (0 3.3 0 10p 10p 0.99n 2.0n)
*
.MODEL DMOD D(CJO=0.02E-12, PB=0.9)
*
.OPTION NOMOD POST INGOLD=2 NUMDGT=6 BRIEF
.TRAN 0.01n 2.0n
.PRINT TRAN V(1) V(2)
.END
```

The SPICE simulation gave a 0 to 2.3 V rise time of 0.16 ns and a 3.3 to 1 V fall time of 0.13 ns. These compare favorably with the 0.16 ns predicted. The reason for the different values of the rise and fall times is the nonlinearity of the junction capacitance. For smaller bias voltages, it is larger than that predicted by (3.44), whereas for larger bias voltages it is smaller. If we use the more accurate approximation of (3.43) for the rise time with $V_2 = 2.3$ V and $V_1 = 0$ V, we find

$$C_{T\text{-}j\text{-}av} = 2 \times 0.02 \times \frac{0.9}{2.3}\left(\sqrt{1 + \frac{2,3}{0.9}} - 1\right) = 0.014 \text{ pF} \tag{3.50}$$

Also, for the fall time, we find that

$$C_{T\text{-}j\text{-}av} = 2 \times 0.02 \times \frac{0.9}{1.0 - 3.3}\left(\sqrt{1 + \frac{1.0}{0.9}} - \sqrt{1 + \frac{3.3}{0.9}}\right) = 0.011 \text{ pF} \tag{3.51}$$

These more accurate approximations result in

$$\Delta t_{+70\%} = 0.17 \text{ ns} \tag{3.52}$$

and

$$\Delta t_{-70\%} = 0.13 \text{ ns} \tag{3.53}$$

in closer agreement with SPICE. Normally, the extra accuracy that results from using (3.43) versus (3.44) is not worth the extra complication because one seldom knows the area of the junctions to better than 20% accuracy.

Forward-Biased Junctions

A positive voltage applied from the p side to the n side of a diode reduces the electric field opposing the diffusion of the free carriers across the depletion region. It also reduces the width of the depletion region. If this forward-bias voltage is large enough, the carrier *diffusion* across the junction will overcome the carrier *drift* caused by the potential drop across the depletion region, resulting in a net current flow from the *anode* to the *cathode*. For silicon, appreciable diode current starts to occur for a forward-bias voltage of around 0.5 V. For germanium and GaAs semiconductor materials, current conduction starts to occur around 0.3 and 0.9 V, respectively.

When the junction potential is sufficiently lowered for conduction to occur, the carriers diffuse across the junction due to the large gradient in the *mobile carrier concentrations*. Note that there are significantly more carriers diffusing from the more heavily doped side to the lightly doped side than from the more lightly doped side to the heavily doped side.

After the carriers cross the depletion region, they greatly increase the *minority charge concentration* at the edge of the depletion region as shown in Fig. 3.10. These minority carriers will diffuse away from the junction toward the bulk. As they diffuse, they recombine with the majority carriers, which decreases their concentration. It is this concentration gradient of the minority charge, which decreases the farther one gets from the junction, that is responsible for current flow near the junction.

The majority carriers that recombine with the diffusing minority carriers come from the metal contacts at the junctions because of the forward-bias voltage. These majority carriers flow across the bulk, from the contacts to the junction, due to an electric field applied across the bulk. This current flow is called *drift*. It results in small potential drops across the bulk, especially in the lightly doped side. Typical values of this voltage drop might be 50 mV to 0.1 V, depending primarily on the doping concentration of the lightly doped side, the distance from the contacts to the junction, and the cross-sectional area of the junction.

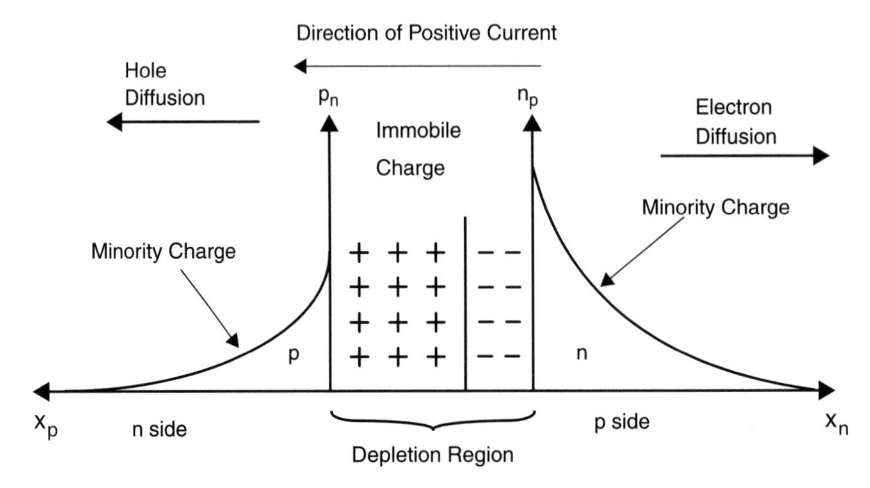

Figure 3.10 The concentration of minority carriers and the direction of diffusing carriers near a forward-biased junction.

In the forward-bias region, the current–voltage relationship is exponential and can be shown (see Appendix) to be

$$I_D = I_S e^{V_D/V_T} \tag{3.54}$$

where V_D is the voltage applied across the diode and

$$I_S \propto A_D\left(\frac{1}{N_A} + \frac{1}{N_D}\right) \tag{3.55}$$

I_S is known as the *scale current* and is seen to be proportional to the area of the diode junction, A_D, and inversely proportional to the doping concentrations.

Junction Capacitance of a Forward-Biased Diode

When a junction changes from being reverse biased (with little current through it) to being forward biased (with significant current flow across it), the charge being stored near and across the junction changes. Part of the change in charge is due to the change in the width of the depletion region and therefore the amount of immobile charge stored in it. This change in charge is modeled by the depletion capacitance, C_j, similar to when the junction is reverse biased. An additional change in charge storage is necessary to account for the change of the minority carrier concentration close to the junction required for the diffusion current to exist. For example, if a forward-biased diode current is to double, then the slopes of the minority-charge storage at the diode junction edges must double, and this, in turn, implies that the minority-charge storage must double. This component is modeled by another capacitance, called the *diffusion capacitance,* denoted C_d.

The diffusion capacitance can be shown (see Appendix) to be

$$C_d = \tau_T \frac{I_D}{V_T} \quad (3.56)$$

where τ_T is the transit time of the diode. Normally, τ_T is specified for a given technology, so that one can calculate the diffusion capacitance. *Note that the diffusion capacitance of a forward-biased junction is proportional to the diode current.*

The total capacitance of the forward-biased junction is the sum of the diffusion capacitance, C_d, and the depletion capacitance, C_j. Thus, the total junction capacitance is given by

$$C_T = C_d + C_j \quad (3.57)$$

For a forward-biased junction, the depletion capacitance, C_j, can be roughly approximated by $2C_{j\text{-}0}$. The accuracy of this approximation is not critical since the diffusion capacitance is typically much larger than the depletion capacitance.

Finally, it should be mentioned that as a diode is turned off, for a short period of time, a current will flow in the negative direction until the minority charge is removed. This behavior is greatly reduced in *Schottky diodes* since they do not have minority charge storage.

Schottky Diodes

A different type of diode, one often used in microcircuit design, is realized by contacting metal to a lightly doped semiconductor region (rather than a heavily doped region) as shown in Fig. 3.11. This diode is called a *Schottky diode*. Notice that the aluminum anode is in direct contact with a relatively lightly doped n^- region. Because the n^- region is relatively lightly doped, the work-function difference between the aluminum contact and the n^- silicon is larger than would be the case for aluminum contacting to an n^+ region, as occurs at the cathode. This causes a depletion region and, correspondingly, a diode to occur at the interface between the aluminum anode and the n^- silicon region. This diode has characteristics different from a normal pn junction

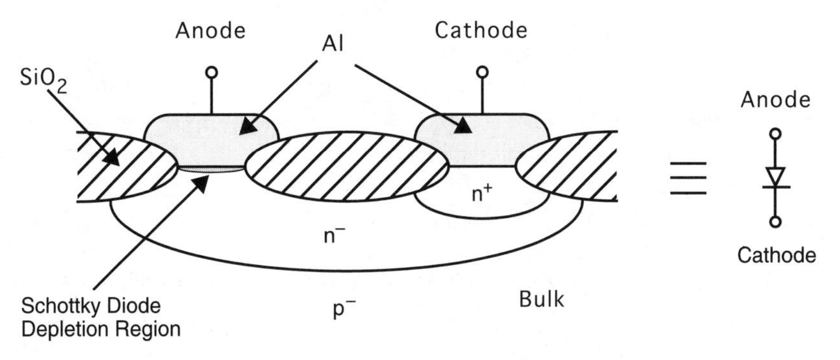

Figure 3.11 A cross-sectional view of a Schottky diode.

diode. First, its voltage drop when forward biased is smaller. This voltage drop is dependent on the metal used; for aluminum it might be around 0.5 V. More importantly, when the diode is forward biased, the minority-charge storage in the lightly doped n^+ region is greatly reduced. Thus, the small-signal model of a forward-biased Schottky diode has its diffusion capacitance equal to zero (i.e., $C_d = 0$ with reference to Fig. 3.38 in the Appendix). The absence of this diffusion capacitance makes the diode much faster. It is particularly faster when turning off, because it is not necessary to remove the minority charge first. Rather, it is necessary to discharge the depletion capacitance only through about 0.2 V.

Schottky diodes have been used extensively in bipolar logic circuits. They are also used in a number of high-speed analog circuits, particularly those realized in gallium arsenide (GaAs) rather than silicon technologies.

3.3 MOS Transistors

Presently, the most popular technology for realizing microcircuits makes use of *MOS transistors*. Unlike most *bipolar-junction transistor* (*BJT*) technologies, which make dominant use of only one type of transistor (npn transistors in the case of BJT processes[4]), MOS circuits normally use two complementary types of transistors—n-channel and p-channel. While n-channel devices conduct with a positive gate voltage, p-channel devices conduct with a negative gate voltage. Moreover, electrons are used to conduct current in n-channel transistors, whereas holes are used in p-channel transistors. Microcircuits containing both n-channel and p-channel transistors are called *CMOS* circuits, for *complementary MOS*. The acronym MOS stands for *metal-oxide semiconductor*, which historically denoted the gate, insulator, and channel region materials, respectively. However, most present CMOS technologies now utilize *polysilicon* gates rather than metal gates.

Before CMOS technology became widely available, most MOS processes made use of only n-channel transistors (NMOS). However, often two different types of n-channel transistors could be realized. One type, *enhancement* n-channel transistors, is similar to the n-channel transistors realized in CMOS technologies. Enhancement transistors require a positive gate-to-source voltage to conduct current. The other type, *depletion* transistors, conduct current with a gate-source voltage of 0 V. Depletion transistors were used to create high-impedance loads in NMOS logic gates.

A typical cross section of an n-channel enhancement-type MOS transistor is shown in Fig. 3.12. With no voltage applied to the gate, the n^+ source and drain regions are separated by the p^- substrate. The separation distance between the drain and the source is called the channel length, L. In present MOS technologies, the minimum channel length is typically between 0.18 and

[4]Most BJT technologies can also realize low-speed lateral pnp transistors. Normally these would be used only to realize current sources as they have low gains and poor frequency responses. Recently, bipolar technologies utilizing high-speed vertical pnp transistors, as well as high-speed npn transistors, have become available and are growing in popularity. These technologies are called complementary bipolar technologies.

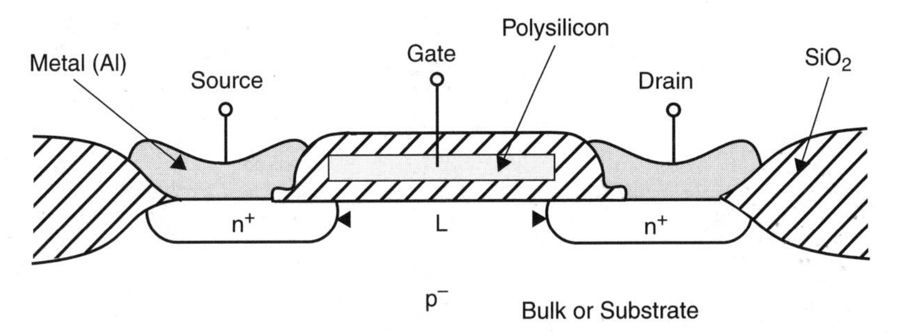

Figure 3.12 A cross-sectional view of a typical n-channel transistor.

0.5 μm. It should be noted that there is no physical difference between the drain and the source.[5] *The source terminal of an n-channel transistor is defined as whichever of the two terminals has a lower voltage.* For a p-channel transistor, the source would be the terminal with the higher voltage. When a transistor is turned *on*, current flows from the drain to the source in an n-channel transistor and from the source to the drain in a p-channel transistor. In both cases, the actual carriers travel from the source to drain, but the current directions are different because n-channel carriers (electrons) are defined negative, whereas p-channel carriers (holes) are defined positive.

The gate is normally realized using *polysilicon*, which is heavily doped noncrystalline (or amorphous) silicon. *Polysilicon gates* are used now (instead of metal) because using polysilicon gate material allows the dimensions of the transistor to be realized much more accurately during the patterning of the transistor, which involves what is called a *self-aligned process.* This higher geometric accuracy results in smaller, faster transistors.

The gate is physically separated from the surface of the silicon by a thin insulator made of silicon dioxide (SiO_2). Thus, the gate is electrically isolated from the channel and affects the channel (and hence, the transistor current) only through electrostatic coupling, similar to capacitive coupling. A typical thickness for the SiO_2 insulator between the gate and the channel is presently from 7 to 20 nm. Since the gate is electrically isolated from the channel, it never conducts d.c. current. Indeed, the excellent isolation results in leakage currents being almost undetectable. However, because of the inherent capacitances in MOS transistors, transient gate currents do exist when gate voltages are quickly changing.

Normally, the p^- substrate (or bulk) is connected to the most negative voltage in a microcircuit. In analog circuits, this might be the negative power supply, but in digital circuits it is normally ground or 0 V. This connection results in all transistors placed in the substrate being surrounded by reverse-biased junctions, which electrically isolate the transistors and thereby prevent conduction through the substrate between transistors (unless, of course, they are intentionally connected together using metal interconnect).

[5] Large MOS transistors used for power applications might not be realized with symmetric drain and source junctions.

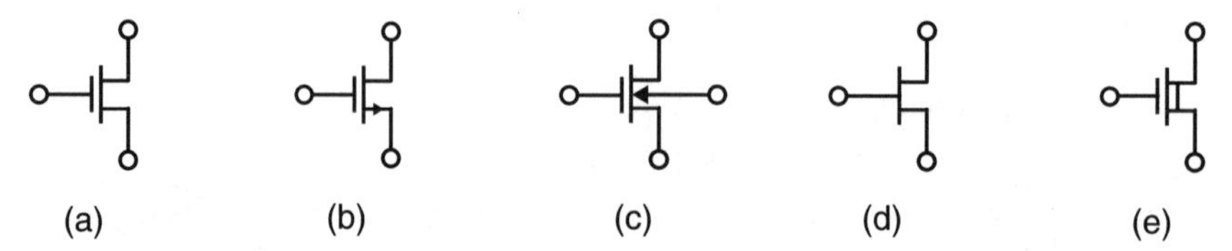

Figure 3.13 Commonly used symbols for n-channel transistors.

SYMBOLS FOR MOS TRANSISTORS

Many symbols have been used to represent MOS transistors. Figure 3.13a shows some of the symbols that have been used to represent n-channel MOS transistors. The symbol of Fig. 3.13a is often used; note that there is nothing in the symbol to specify whether the transistor is n-channel or p-channel. A common rule is to assume, when in doubt, an n-channel enhancement transistor. Figure 3.13b is the most commonly used symbol for an n-channel enhancement transistor and is used throughout this text. The arrow pointing outward on the source indicates that the transistor is n-channel, similar to the convention used for npn transistors, and usually indicates the direction of positive current.

MOS transistors are actually four-terminal devices, with the substrate being the fourth terminal. *In n-channel devices, the p^- substrate is normally connected to the most negative voltage in the microcircuit, whereas for p-channel devices, the n^- substrate is normally connected to the most positive voltage.* In these cases, the substrate connection is normally not shown in the symbol. However, for CMOS technologies, at least one of the two types of transistors will be formed in a *well* substrate that need not be connected to one of the power supply nodes. For example, an n-well process would form n-channel transistors in a p^- substrate encompassing the entire microcircuit, whereas the p-channel transistors would be formed in many n-well substrates. In this case, most of the n-well substrates would be connected to the most positive power supply, whereas some might be connected to other nodes in the circuit (often the well is connected to the source of a transistor that is not connected to the power supply). In these cases, the symbol shown in Fig. 3.13c can be used to show the substrate connection explicitly. Note that this case is not encountered often in digital circuits and is more common in analog circuits.

Sometimes, in the interest of simplicity, the isolation of the gate is not explicitly shown, as is the case for the symbol of Fig 3.13d. This simple notation is more common for digital circuits in which a large number of transistors are present. Since this symbol is also used for *JFET* transistors, it will never be used to represent MOS transistors in this text. The last symbol, shown in Fig. 3.13e, denotes an n-channel depletion transistor. The extra line is used to indicate that a physical channel exists for a 0 V gate-source voltage. Depletion transistors were used in older NMOS technologies but are not typically available in current CMOS processes.

Figure 3.14 shows some commonly used symbols for p-channel transistors. In this text, the symbol of Fig. 3.14a will be most often used. The symbol of Fig. 3.14c is sometimes used in digital circuits where the circle indicates that a low voltage on the gate turns the transistor on, as opposed to a high voltage for an n-channel transistor (Fig. 3.13a). The symbols of Fig. 3.14d

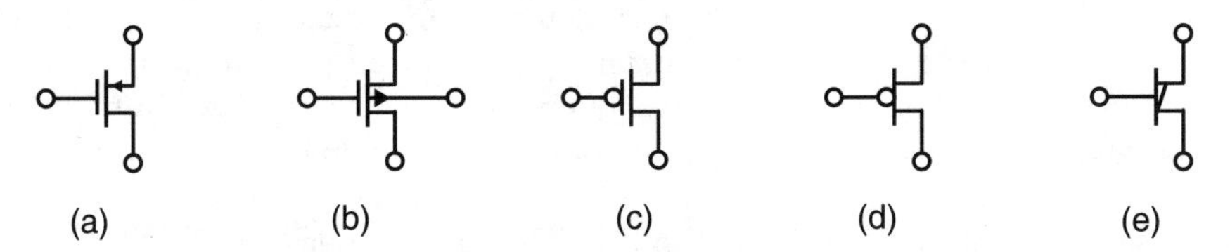

Figure 3.14 Commonly used symbols for p-channel transistors.

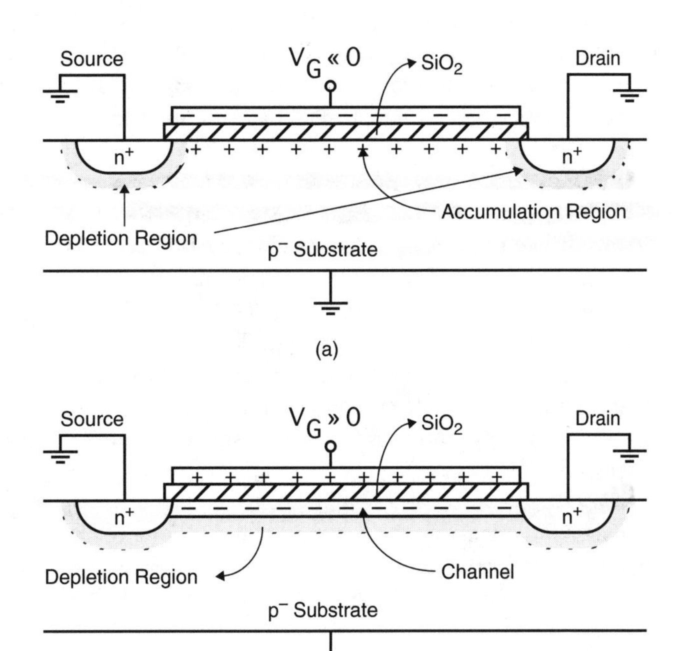

Figure 3.15 An n-channel MOS transistor. (a) $V_G \ll 0$ resulting in an accumulated channel (no current flow). (b) $V_G \gg 0$ resulting in an inverted channel (current flow).

or e might be used in larger circuits where many transistors are present, to simplify the drawing somewhat. They will not be used in this text.

Basic Operation

The basic operation of MOS transistors will be described with respect to an n-channel transistor. First, consider the simplified cross sections shown in Fig. 3.15 where the source, drain, and substrate are all connected to ground. In this case, the MOS transistor operates similarly to a capacitor. The gate acts as one plate of the capacitor, and the surface of the silicon, just under the thin insulating SiO_2, acts as the other plate.

If the gate voltage is very negative, as shown in Fig. 3.15a, positive charge will be attracted to the channel region. Since the substrate was originally doped p^-, this negative gate voltage has the effect of simply increasing the channel doping to p^+, resulting in what is called an *accumulated channel.* The n^+ source and drain regions are separated from the p^+-channel region by depletion regions, resulting in the equivalent circuit of two back-to-back diodes. Thus, only leakage current will flow even if one of the source or drain voltages becomes large (unless the drain voltage becomes so large as to cause the transistor to break down).

In the case of a positive voltage being applied to the gate, the opposite situation occurs, as shown in Fig. 3.15b. For small positive gate voltages, the positive carriers in the channel under the gate are initially repulsed, and the channel changes from a p^- doping level to a depletion region. As a more positive gate voltage is applied, the gate attracts negative charge from the source and drain regions, and the channel becomes an n region with mobile electrons connecting the drain and source regions.[6] In short, a sufficiently large positive gate-source voltage changes the channel beneath the gate to an n region, and the channel is said to be *inverted.*

The gate-source voltage, for which the concentration of electrons under the gate is equal to the concentration of holes in the p^- substrate far from the gate, is commonly referred to as the *transistor threshold voltage* and denoted V_{tn} (for n-channel transistors). For gate-source voltages larger than V_{tn}, there is an n-type channel present, and conduction between the drain and the source can occur. For gate-source voltages less than V_{tn}, it is normally assumed that the transistor is *off* and no current flows between the drain and the source. However, note that this assumption of zero drain-source current for a transistor that is *off* is only an approximation. In fact, for gate voltages around V_{tn}, there is no abrupt current change, and for gate-source voltages slightly less than V_{tn}, small amounts of *subthreshold current* will flow, as discussed in Section 3.4.

When the gate-source voltage, V_{GS}, is larger than V_{tn}, the channel is present. As V_{GS} is increased, the density of electrons in the channel increases. Indeed, the carrier density, and therefore the charge density, is proportional to $V_{GS} - V_{tn}$, which is often called the *effective gate-source voltage* and denoted V_{eff}. Specifically, define

$$V_{eff} \equiv V_{GS} - V_{tn} \tag{3.58}$$

The charge density of electrons is then given by

$$Q_n = C_{ox}(V_{GS} - V_{tn}) = C_{ox}V_{eff} \tag{3.59}$$

Here, C_{ox} is the gate capacitance per unit area and is given by

$$C_{ox} = \frac{K_{ox}\varepsilon_0}{t_{ox}} \tag{3.60}$$

[6]The drain and source regions are sometimes called diffusion regions or junctions for historical reasons. This use of the word *junction* is not synonymous with our previous use, in which it designated a pn interface of a diode.

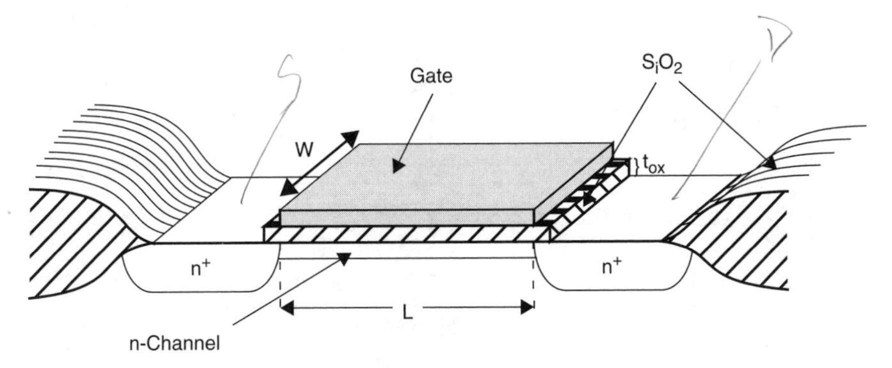

Figure 3.16 The important dimensions of an MOS transistor.

where K_{ox} is the relative permittivity of SiO_2 (approximately 3.9) and t_{ox} is the thickness of the thin oxide under the gate. A point to note here is that (3.59) is accurate only when both the drain and the source voltages are zero. Equation (3.59) is used often when analyzing MOS circuits.

To obtain the total gate capacitance, (3.60) should be multiplied by the effective gate area, WL, where W is the gate width and L is the effective gate length. These dimensions are shown in Fig. 3.16. Thus, the total gate capacitance, C_{gs}, is given by

$$C_{gs} = WLC_{ox} \tag{3.61}$$

and the total charge of the channel, $Q_{T\text{-}n}$, is given by

$$Q_{T\text{-}n} = WLC_{ox}(V_{GS} - V_{tn}) = WLC_{ox}V_{eff} \tag{3.62}$$

The gate capacitance, C_{gs}, is one of the major load capacitances that circuits must be capable of driving. Gate capacitances are also important when one is calculating *charge injection,* which occurs when an MOS transistor is being turned off because the channel charge, $Q_{T\text{-}n}$, must flow from under the gate out through the terminals to other places in the circuit.

Next, if the drain voltage is increased above 0 V, a drain-source potential difference exists. This difference results in current flowing from the drain to the source.[7] The relationship between V_{DS} and the drain-source current, I_D, is the same as for a resistor, assuming V_{DS} is small. This relationship is given (Sze, 1981) by

$$I_D = \mu_n Q_n \frac{W}{L} V_{DS} \tag{3.63}$$

where $\mu_n \cong 0.06\ m^2/V \cdot s$ is the mobility of electrons near the silicon surface, and Q_n is the charge concentration of the channel per unit area (looking from the top down). Note that as the

[7]The current is actually conducted by negative carriers (electrons) flowing from the source to the drain. Negative carriers flowing from source to drain results in a positive current from drain to source, I_{DS} by definition.

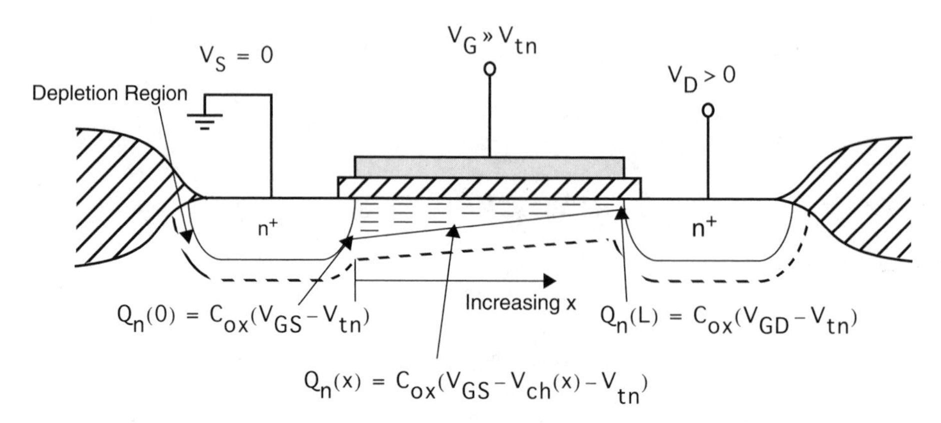

Figure 3.17 The channel charge density for $V_{DS} > 0$.

channel length increases, the drain-source current decreases, whereas this current increases as either the charge density or the transistor width increases. Using (3.62) and (3.63) results in

$$I_D = \mu_n C_{ox} \frac{W}{L} (V_{GS} - V_{tn}) V_{DS} = \mu_n C_{ox} \frac{W}{L} V_{eff} V_{DS} \tag{3.64}$$

where it should be emphasized that this relationship is valid only for drain-source voltages near zero (i.e., V_{DS} much smaller than V_{eff}).

As the drain-source voltage increases, the channel charge concentration decreases at the drain end. This decrease is due to the smaller gate-to-channel voltage difference across the thin gate oxide as one moves closer to the drain. In other words, since the drain voltage is assumed to be at a higher voltage than the source, there is an increasing voltage gradient from the source to the drain, resulting in a smaller gate-to-channel voltage near the drain. Since the charge density at a distance x from the source end of the channel is proportional to $V_G - V_{ch}(x) - V_{tn}$, as $V_G - V_{ch}(x)$ decreases, the charge density also decreases.[8] This effect is illustrated in Fig. 3.17.

Note that at the drain end of the channel, we have

$$V_G - V_{ch}(L) = V_{GD} \tag{3.65}$$

For small V_{DS}, we saw from (3.64) that I_D was linearly related to V_{DS}. However, as V_{DS} increases, and the charge density decreases near the drain, the relationship becomes nonlinear. In fact, the linear relationship for I_D versus V_{DS} flattens for larger V_{DS} as shown in Fig. 3.18.

[8] $V_G - V_{CH}(x)$ is the gate-to-channel voltage drop at distance x from the source end, with V_G being the same everywhere in the gate, since the gate material is highly conductive.

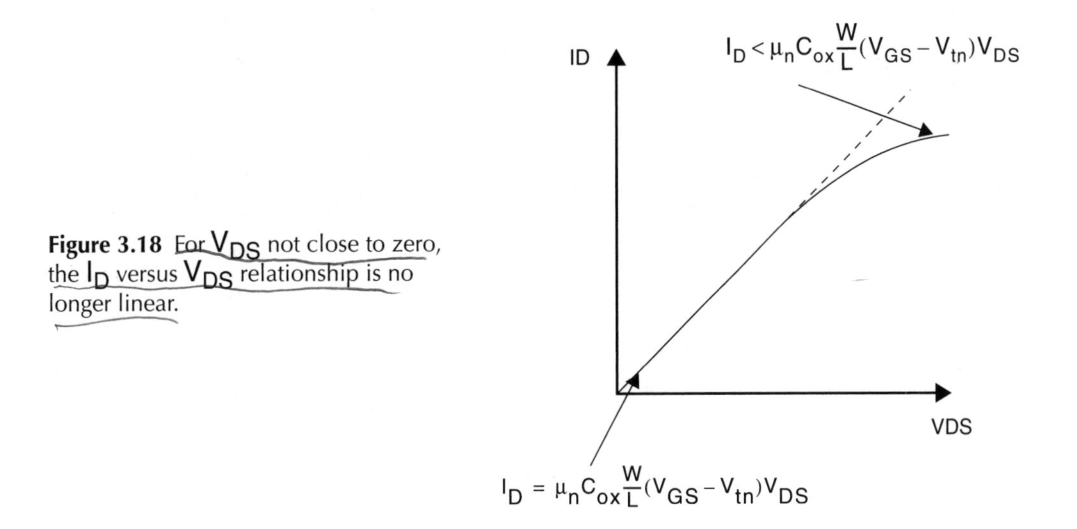

Figure 3.18 For V_{DS} not close to zero, the I_D versus V_{DS} relationship is no longer linear.

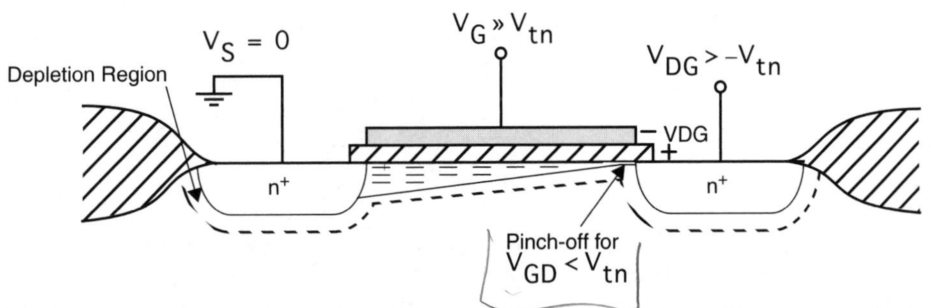

Figure 3.19 When V_{DS} is increased such that $V_{GD} < V_{tn}$, then the channel becomes pinched off at the drain end.

As the drain voltage is increased, at some point the gate-to-channel voltage at the drain end will decrease to the threshold value V_{tn}— the (approximate) minimum gate-to-channel voltage needed for n carriers to exist in the channel. Thus, at the drain end, the channel becomes *pinched off*, as shown in Fig. 3.19. This *pinch-off* occurs at $V_{GD} = V_{tn}$, since the channel voltage at the drain end is simply equal to V_D. Thus, pinch-off occurs for

$$V_{DG} > -V_{tn} \tag{3.66}$$

Denoting $V_{DS\text{-sat}}$ as the drain-source voltage when the channel becomes pinched off, we can substitute $V_{DG} = V_{DS} - V_{GS}$ into (3.66) and find an equivalent pinch-off expression

$$V_{DS} > V_{DS\text{-sat}} \tag{3.67}$$

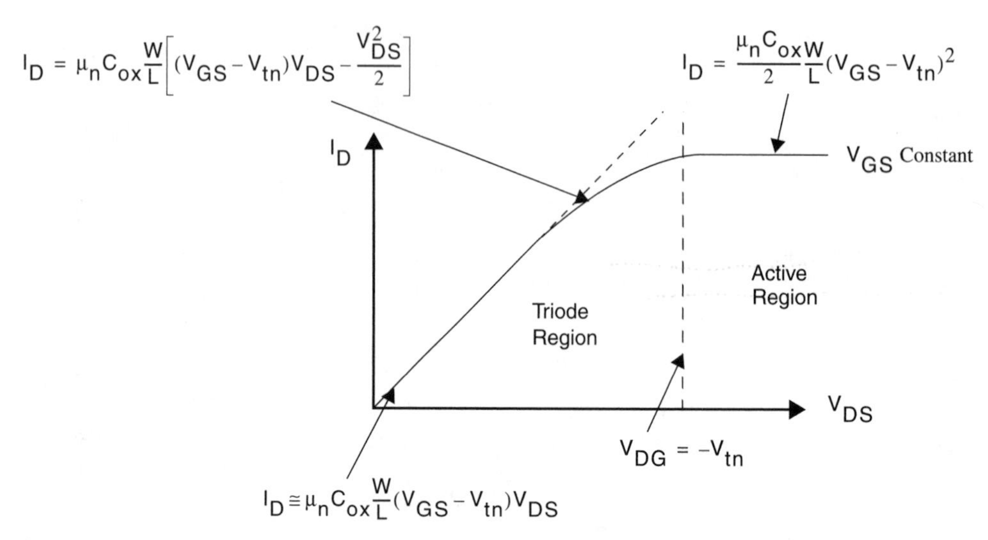

Figure 3.20 The I_D versus V_{DS} curve for an ideal MOS transistor. For $V_{DG} > -V_{tn}$, I_D is approximately constant.

where $V_{DS\text{-}sat}$ is given[9] by

$$V_{DS\text{-}sat} = V_{GS} - V_{tn} = V_{eff} \tag{3.68}$$

The electron carriers traveling through the pinched off drain region are *velocity saturated*, similar to a gas under pressure traveling through a very small tube. If the drain-gate voltage rises above this critical pinch-off voltage of $-V_{tn}$, the charge concentration in the channel remains constant (to a first-order approximation) and the drain current no longer increases with increasing V_{DS}. The result is the current–voltage relationship shown in Fig. 3.20 for a given gate-source voltage. In the region of operation where $V_{DS} > V_{DS\text{-}sat}$, the drain current is independent of V_{DS} and is called the *active region*.[10] The region where I_D changes with V_{DS} is called the *triode region*. When MOS transistors are used in analog amplifiers, they are almost always biased in the active region. When they are used in digital logic gates, they typically operate in both regions.

Before proceeding, it is worth discussing the terms weak, moderate, and strong inversion. As just discussed, a gate-source voltage greater than V_{tn} results in an inverted channel, and drain-source current can flow. However, as the gate-source voltage is increased, the channel

[9]Because of the body effect, the threshold voltage at the drain end of the transistor is increased, resulting in the true value of $V_{DS\text{-}sat}$ being slightly lower than V_{eff} (Tsvidis, 1987).

[10]Historically, the *active region* was called the *saturation region*, but this led to confusion because in the case of bipolar transistors, the saturation region occurs for small V_{CE}, whereas for MOS transistors it occurs for large V_{DS}. The renaming of the saturation region to the active region is becoming widely accepted.

does not become inverted (i.e., n-region) suddenly, but rather gradually. Thus, it is useful to define three regions of channel inversion with respect to the gate-source voltage. In most circuit applications, *noncutoff* MOSFET transistors are operated in *strong inversion*, with $V_{eff} > 100$ mV (many prudent circuit designers use a minimum value of 200 mV). As the name suggests, strong inversion occurs when the channel is strongly inverted with many channel carriers available. Note that all the equation models in this section assume strong inversion operation. *Weak inversion* occurs when V_{GS} is approximately 100 m V or more below V_{tn} and is discussed as subthreshold operation in Section 3.4. Finally, *moderate inversion* is the region between weak and strong inversion.

LARGE-SIGNAL MODELING

The *triode region equation* for an MOS transistor relates the drain current to the gate-source and drain-source voltages. It can be shown (see Appendix) that this relationship is given by

$$I_D = \mu_n \frac{W}{L} C_{ox}\left[(V_{GS} - V_{tn})V_{DS} - \frac{V_{DS}^2}{2}\right] \tag{3.69}$$

As V_{DS} increases, I_D increases until the drain end of the channel becomes pinched off, and then I_D no longer increases. This pinch-off occurs for $V_{DG} = -V_{tn}$, or approximately,

$$V_{DS} = V_{GS} - V_{tn} = V_{eff} \tag{3.70}$$

Right at the edge of pinch-off, the drain current resulting from (3.69) and the drain current in the active region (which, to a first-order approximation, is constant with respect to V_{DS}) must have the same value. Therefore, the *active region equation* can be found by substituting (3.70) into (3.69) resulting in

$$I_D = \frac{\mu_n C_{ox}}{2} \frac{W}{L}(V_{GS} - V_{tn})^2 \tag{3.71}$$

For $V_{DS} > V_{eff}$, the current stays constant at the value given by (3.71), ignoring second-order effects such as the finite output impedance of the transistor. This equation is perhaps the most important one that describes the large-signal operation of an MOS transistor. Note that (3.71) represents a squared current–voltage relationship for an MOS transistor in the active region. In the case of a BJT transistor, an exponential current–voltage relationship exists in the active region.

As just mentioned, (3.71) implies that the drain current, I_D, is independent of the drain-source voltage. This independence is true only to a first-order approximation. The major source of error is due to the channel length shrinking as V_{DS} increases. To see this effect, consider Fig. 3.21, which shows a cross section of a transistor in the active region. A pinched off region with very little charge exists between the drain and the channel. The voltage at the end of the channel closest to the drain is fixed at $V_{GS} - V_{tn} = V_{eff}$. The voltage difference between the drain and the near end of the channel lies across a short depletion region often called the *pinch-off*

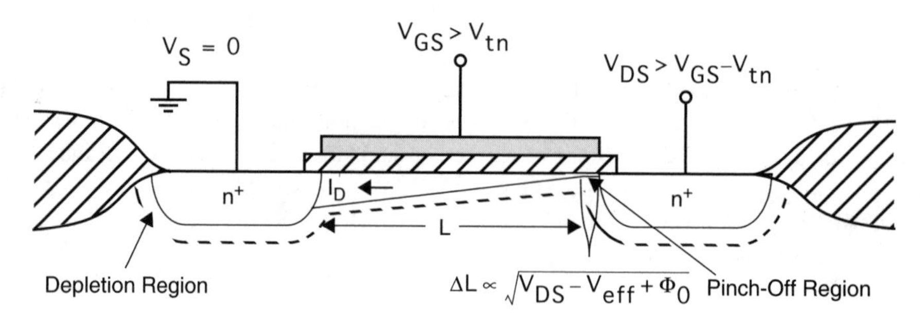

Figure 3.21 Channel length shortening for $V_{DS} > V_{eff}$.

region. As V_{DS} becomes larger than V_{eff}, this depletion region surrounding the drain junction increases its width in a square-root relationship with respect to V_{DS}. This increase in the width of the depletion region surrounding the drain junction decreases the effective channel length. In turn, this decrease in effective channel length increases the drain current, resulting in what is commonly referred to as *channel-length modulation*.

To derive an equation to account for channel-length modulation, we first make use of (3.18) and denote the width of the depletion region by x_d, resulting in

$$\begin{aligned} x_d &\cong k_{ds}\sqrt{V_{D\text{-}ch} + \Phi_0} \\ &= k_{ds}\sqrt{V_{DG} + V_{tn} + \Phi_0} \end{aligned} \tag{3.72}$$

where

$$k_{ds} = \sqrt{\frac{2K_s\varepsilon_0}{qN_A}} \tag{3.73}$$

and has units of $m/\sqrt{V}$. Note that N_A is used here since the n-type drain region is more heavily doped than the p-type channel (i.e., $N_D >> N_A$). By writing a Taylor approximation for I_D around its operating value of $V_{DS} = V_{GS} - V_{tn} = V_{eff}$, we find I_D to be given by

$$I_D = I_{D\text{-sat}} + \left(\frac{\partial I_D}{\partial L}\right)\left(\frac{\partial L}{\partial V_{DS}}\right)\Delta V_{DS} \cong I_{D\text{-sat}}\left[1 + \frac{k_{ds}(V_{DS} - V_{eff})}{2L\sqrt{V_{DG} + V_{tn} + \Phi_0}}\right] \tag{3.74}$$

where $I_{D\text{-sat}}$ is the drain current when $V_{DS} = V_{eff}$, or equivalently, the drain current when the channel-length modulation is ignored. Note that in deriving the final equation of (3.74), we have used the relationship $\partial L/\partial V_{DS} = -\partial x_d/\partial V_{DS}$. Usually, (3.74) is written as

$$I_D = \frac{\mu_n C_{ox} W}{2}\frac{W}{L}(V_{GS} - V_{tn})^2[1 + \lambda(V_{DS} - V_{eff})] \tag{3.75}$$

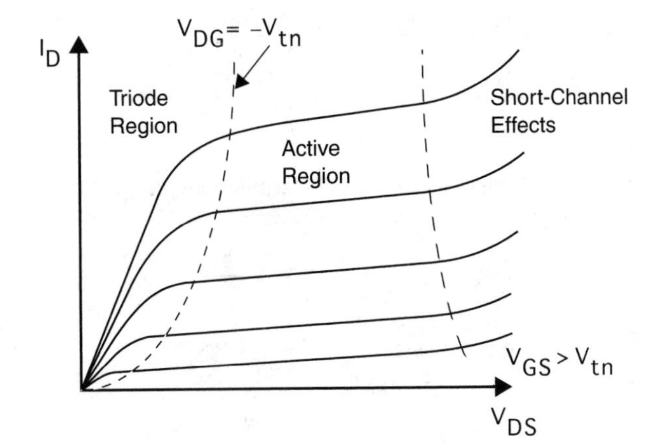

Figure 3.22 I_D versus V_{DS} for different values of V_{GS}.

where λ is the output impedance constant (in units of V^{-1}) given by

$$\lambda = \frac{k_{ds}}{2L\sqrt{V_{DG} + V_{tn} + \Phi_0}} = \frac{k_{ds}}{2L\sqrt{V_{DS} - V_{eff} + \Phi_0}} \tag{3.76}$$

Equation (3.75) is accurate until V_{DS} is large enough to cause other second-order effects, often called *short-channel effects*. For example, it implicitly assumes that current flow down the channel is not velocity saturated (i.e., increasing the electric field no longer increases the carrier speed). *Velocity saturation* commonly occurs in new technologies having very short channel lengths and therefore large electric fields. If V_{DS} becomes large enough so short-channel effects occur, I_D increases more than is predicted by (3.75). Of course, for quite large values of V_{DS}, the transistor will eventually break down.

A plot of I_D versus V_{DS} for a number of different values of V_{GS} is shown in Fig. 3.22. Note that in the active region the small (but nonzero) slope indicates the small dependence of I_D on V_{DS}.

Example 3.7

Find I_D for an n-channel transistor having a substrate concentration of $N_A = 1.4\times10^{23}/m^3$ with $\mu_n C_{ox} = 188\ \mu A/V^2$, $W = 6\ \mu m$, $L = 0.6\ \mu m$, $\Phi_0 = 0.99\ V$, $V_{GS} = 1.2\ V$, $V_{tn} = 0.8\ V$, and $V_{DS} = V_{eff}$. Assuming λ remains constant, estimate the new value of I_D if V_{DS} is increased by 0.5 V.

Solution: From (3.73), we have

$$k = \sqrt{\frac{2 \times 11.8 \times 8.854 \times 10^{-12}}{1.6 \times 10^{-19} \times 1.4 \times 10^{23}}} = 96.6 \times 10^{-9}\ m/\sqrt{V}$$

which is used in (3.76) to find λ as

$$\lambda = \frac{96.6 \times 10^{-9}}{2 \times 0.6 \times 10^{-6} \times \sqrt{0.99}} = 80.8 \times 10^{-3}\ \mathrm{V}^{-1}$$

Now, making use of (3.75), we find for $V_{DS} = V_{eff} = 0.4\ \mathrm{V}$,

$$I_{D1} = \left(\frac{188 \times 10^{-6}}{2}\right)\left(\frac{6}{0.6}\right)(0.4)^2(1) = 150\ \mu\mathrm{A}$$

In the case where $V_{DS} = V_{eff} + 0.5\ \mathrm{V} = 0.9\ \mathrm{V}$, we have

$$I_{D2} = 150\ \mu\mathrm{A} \times (1 + \lambda \times 0.5) = 156\ \mu\mathrm{A}$$

Note that this example shows almost a 5% increase in drain current for a 0.5 V increase in drain-source voltage. The SPICE-simulated values of 164 and 174 μA are greater than those calculated due to the short-channel effects of drain-induced barrier lowering.

Body Effect

The large-signal equations in the preceding section were based on the assumption that the source voltage was the same as the substrate (i.e., bulk) voltage. However, often the source and substrate can be at different voltage potentials. In these situations, a second-order effect exists that is modeled as an increase in the threshold voltage, V_{tn}, as the source-to-substrate reverse-bias voltage increases. This effect, typically called the *body effect,* is more important for transistors in a well of a CMOS process where the substrate doping is higher. The body effect is most noticeable in digital circuits in reducing the effective gate-source voltages of pass transistors and switches as discussed in Chapter 5.

To account for the body effect, it can be shown (see Appendix) that the threshold voltage of an n-channel transistor is now given by

$$V_{tn} = V_{tn\text{-}0} + \gamma\left(\sqrt{V_{SB} + |2\phi_F|} - \sqrt{|2\phi_F|}\right) \tag{3.77}$$

where V_{tn0} is the threshold voltage with zero V_{SB} (i.e., source-to-substrate voltage), and

$$\gamma = \frac{\sqrt{2qN_A K_s \varepsilon_0}}{C_{ox}} \tag{3.78}$$

The factor γ is often called the body effect constant and has units of $\sqrt{V}$. Notice that γ is proportional to $\sqrt{N_A}$,[11] so the body effect is larger for transistors in a well in which the doping is typically higher than the substrate of the microcircuit.

---

[11]For an n-channel transistor. For a p-channel transistor, γ is proportional to N_D.

Another important ramification of the body effect is to modify the equations (3.69) to (3.71). These equations, which were derived in the Appendix, are based on the assumption that the threshold voltage is constant everywhere along the channel. In reality, the threshold voltage increases closer to the drain end of the channel (for an n-channel transistor) due to the voltage drop down the channel. This causes the currents to be less than predicted by (3.69) to (3.71). Also, the drain-source voltage necessary for a transistor to be in the active region is also smaller. The I–V equations for MOS transistors have been derived taking the body effect into account in pages 119–123 [Tsvidis, 1987] and have been summarized on pg. 49 [Tsvidis, 1996]. They now become

$$I_D = \mu_n \frac{W}{L} C_{ox} \left[(V_{GS} - V_{tn}) V_{DS} - \alpha \frac{V_{DS}^2}{2} \right] \tag{3.79}$$

for

$$V_{DS} \le \frac{V_{GS} - V_{tn}}{\alpha} = \frac{V_{eff}}{\alpha} \tag{3.80}$$

and

$$I_D = \frac{\mu_n C_{ox} W}{2\alpha \; L} (V_{GS} - V_{tn})^2 \tag{3.81}$$

where

$$\alpha \cong 1 + \frac{\Upsilon}{2\sqrt{V_{SB} + |2\phi_F|}} \tag{3.82}$$

is a quantity greater than 1. Normally, α would be empirically derived.

P-Channel Transistors and Depletion Transistors

All of the preceding equations have been presented for n-channel enhancement transistors. *In the case of* p*-channel transistors, these equations can also be used if a negative sign is placed in front of every voltage variable (except* V_{eff}*, which is always positive).* Thus, V_{GS} becomes V_{SG}, V_{DS} becomes V_{SD}, V_{tn} becomes $-V_{tp}$, and so on. The condition required for conduction is now $V_{SG} > V_{tp}$, where V_{tp} is now a negative quantity for an enhancement p-channel transistor.[12] The requirement on the source-drain voltage for a p-channel transistor to be in the active region is $V_{SD} > V_{SG} + V_{tp}$. The equations for I_D, in both regions, remain unchanged, because all voltage variables are squared, except the current flow is now from the source to the drain.

[12]It is possible to realize *depletion* p-channel transistors, but these are of little value and seldom worth the extra processing involved. Depletion n-channel transistors are also seldom encountered in CMOS microcircuits, although they might be worth the extra processing involved in some applications, particularly if they were in a well.

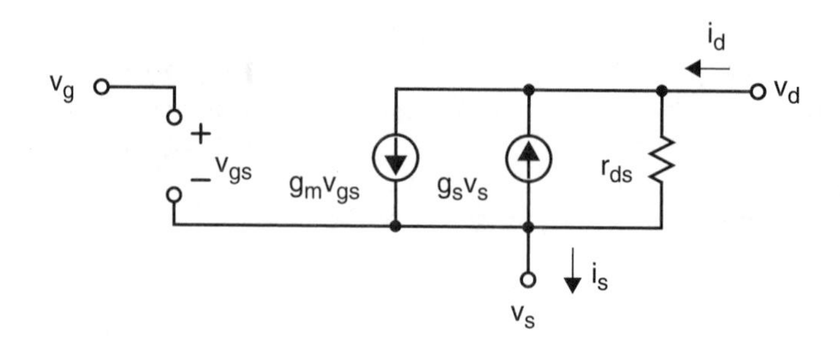

Figure 3.23 The low-frequency small-signal model for an MOS transistor.

For n-channel depletion transistors, the only difference is that $V_{td} < 0$ V. A typical value might be $V_{td} = -2$ V.

SMALL-SIGNAL MODELING

The small-signal modeling of MOS transistors is covered in the Appendix and is just briefly summarized here. A small-signal model for the active region at low frequencies is shown in Fig. 3.23, where the parameters g_m, g_s, and r_{ds} are given by

$$g_m = \sqrt{2\mu_n C_{ox}\frac{W}{L}I_D} \quad (3.83)$$

$$g_s = \frac{\gamma g_m}{2\sqrt{V_{SB} + |2\phi_F|}} \quad (3.84)$$

and

$$r_{ds} \cong \frac{1}{\lambda I_D} \quad (3.85)$$

where

$$\lambda = \frac{k}{2L\sqrt{V_{DS} - V_{eff} + \Phi_0}} \quad (3.86)$$

The parameter g_s is due to the body effect and is seldom important for digital circuits. The transistor output impedance, r_{ds}, affects the gain of logic gates only when a gate is at its threshold, and normally this also is of secondary importance. The transconductance, g_m, is important as it is a good indication of a transistor's ability to charge and discharge transistors.

A small-signal model that includes parasitic capacitances is shown in Fig. 3.24. The various capacitances are shown in a cross-sectional view of an MOS transistor in Fig. 3.25. The most-significant capacitor for digital circuits is normally the gate input capacitance. Depending on the region in which the transistor is biased, this capacitor typically varies between

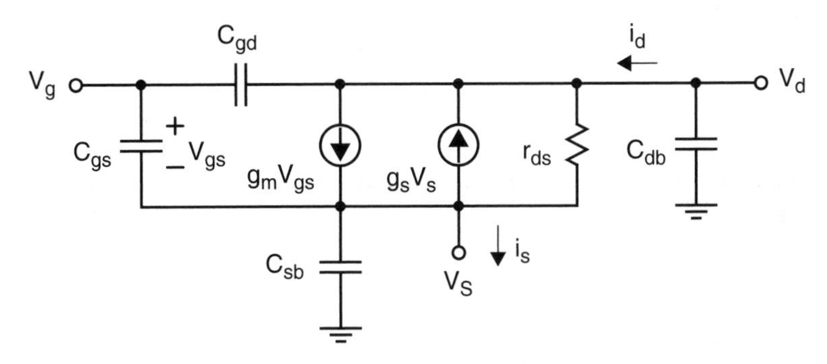

Figure 3.24 The small-signal model for an MOS transistor in the active region.

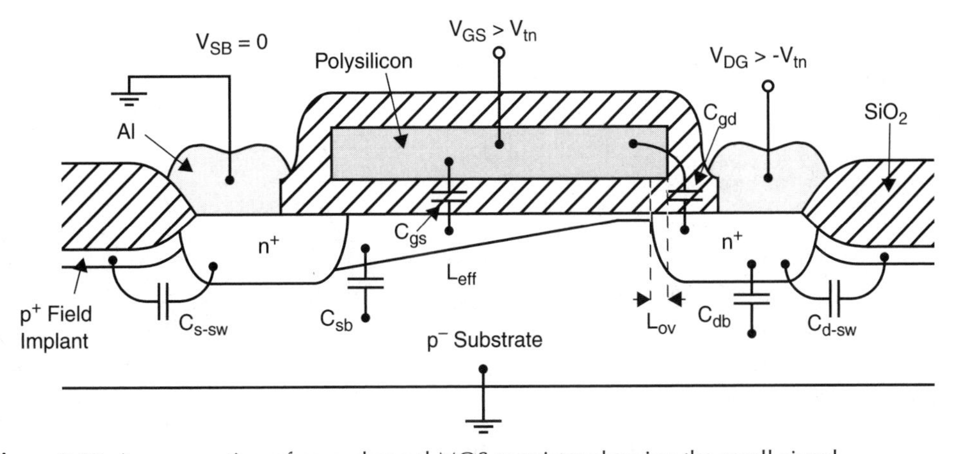

Figure 3.25 A cross section of an n-channel MOS transistor showing the small-signal capacitances.

$$\frac{WL}{2}C_{ox} < C_g < WLC_{ox} \tag{3.87}$$

It normally is conservatively estimated by its upper limit. The next most important capacitors in digital circuits are the junction capacitors, C_{sb}, and C_{db}, which are due to the reverse-biased depletion capacitances of the source and drain junctions and possibly channel-to-substrate capacitances as well. These capacitances are discussed in greater detail in the Appendix. Also, modeling in the triode and cutoff regions is covered in greater detail in the Appendix and interested readers are referred to it.

3.4 Advanced MOS Modeling

In this section, we look at some of the more advanced modeling concepts that a microcircuit designer is likely to encounter—*scaling*, *short-channel effects*, *subthreshold operation*, *leakage currents*, and *latch-up*.

SCALING

When a technology is scaled to smaller dimensions, one would like to keep the device operation unchanged, except for taking advantage of the higher speed available. This is possible, ideally, if the scaling is done to keep a constant electric field. This is theoretically possible if the voltage levels are scaled proportional to the device dimensions and the doping levels are scaled inversely proportional to device dimensions. Thus, if the topography dimensions are scaled down by a factor $1/S$ where $S > 1$, then for *constant-field scaling* one needs voltages scaled by $1/S$ and doping levels scaled by S. This constant-field scaling ideally leaves all carrier velocities unchanged. It is easy to calculate its effect on many other important characteristics of digital ICs.

For example, repeating equation (3.60), we have

$$C_{ox} = \frac{K_{ox}\varepsilon_0}{d_{ox}} \tag{3.88}$$

Since d_{ox} is inversely proportional to S, we have the gate capacitance per unit area, C_{ox}, increasing proportionately to S. But since the gate area goes down inversely proportional to S^2, we have the total gate capacitance scaling proportional to $1/S$.

Also, from equations (3.69) and (3.71), we see that the drain currents of MOS transistors are proportional to C_{ox} and also to voltages squared, but are independent of the absolute dimensions assuming W/L remains unchanged. Since C_{ox} is proportional to $1/S$ and voltages are also proportional to $1/S$, transistor currents are proportional to $1/S$.

To determine how the speed is affected by scaling, we need an equation that will not be proven until the next chapter. This equations states that the average rise and fall time of a *CMOS inverter*, t_{AV}, is given by

$$t_{AV} = \frac{1.5L^2}{(V_{DD} - V_{tn})\mu_n}\left(1 + \frac{W_p}{W_n}\right)\left(1 + \frac{\mu_n W_n}{\mu_p W_p}\right) \tag{3.89}$$

Thus,

$$t_{AV} \propto 1/S \tag{3.90}$$

as both L and $V_{DD} - V_{tn}$ are proportional to $1/S$. This inverse proportionality of delay with scaling is also true for logic gates that are more complicated than inverters. As a check, note that since a capacitor is described by the relationship

$$I = C\frac{dV}{dt} \tag{3.91}$$

we have

$$\Delta t \propto \frac{C}{I}\Delta V \tag{3.92}$$

Since the following proportionalities hold, namely $C \propto 1/S$, $\Delta V \propto 1/S$, and $I \propto 1/S$, we therefore have $\Delta t \propto 1/S$ in agreement with (3.90).

Table 3.1 The Effects of Scaling the Dimensions of an IC Proportional to 1/S While Keeping the Electric Field Constant

Parameter	Scaling factor
Device dimensions, t_{ox}, L, W, junction depth	$1/S$
Doping concentration, N_A	S
Voltage, V	$1/S$
Current, I	$1/S$
Capacitance, εA § t_{ox}	$1/S$
Delay time, VC § I	$1/S$
Power dissipation (per gate), VI	$1/S^2$
Power density, VI § A	1
Power-delay product	$1/S^3$

To determine how the power dissipation is affected by scaling, we need another result that will not be proven until the next chapter; the average power dissipation of an inverter that is switching from a "1" to a "0" and back again, once each period T, is given by

$$P_{AV} = \frac{C_L V_{DD}^2}{T} \tag{3.93}$$

Since $C_L \propto 1/S$, $V_{DD} \propto 1/S$, and $T \propto 1/S$ [from (3.71)], we therefore have

$$P_{AV} \propto 1/S^2 \tag{3.94}$$

Thus, the average power dissipation per gate goes down inversely proportional to the scaling factor squared. If the size of an IC does not change, the number of gates in an IC will go up proportionally to the scaling factor squared. Therefore, this *constant-electric-field scaling* will leave the total power dissipation and the power density of an IC unchanged. Again, this is all dependent on the power supply voltage, V_{DD}, going down inversely proportional to $1/S$. This saving in power dissipation is one of the major driving forces behind the change in the standard power-supply voltage from 5 to 3.3 to 2.5 V, which is currently occurring. The above proportionalities are summarized in Table 3.1.

Perhaps the most important insight to be gained from Table 3.1 is that constant-field scaling increases the speed and in particular minimizes the power-delay product proportional to $1/S^3$. *The power-delay product is perhaps the most important figure of merit for a logic family.*

Unfortunately, constant-field scaling is often not possible as it might not be possible to scale the voltages of the circuit proportional to $1/S$ for a variety of reasons. These might include the following: the supply voltage is determined by system considerations and cannot be lowered, the signal-to-noise ratios and noise margins become too small, the transistor threshold voltages cannot be made too close to zero without there being substantial subthreshold currents, or possibly speed is of the utmost importance and by not scaling to $1/S$, speed will go up by a proportionality greater than S. For these reasons, the voltages might be left unchanged, which is called *constant-voltage scaling*, or might be scaled approximately proportional to $1/(\sqrt{S})$, which is called *quasiconstant voltage scaling*. At present, the minimum dimensions for commercial CMOS ICs are typically between 0.18 and 0.6 μm.

Since the voltages are seldom exactly scaled proportional to $1/S$, the electric fields often increase with scaling. This higher electric field causes a number of undesirable effects. Some of these are *mobility degradation, hot-carrier effects, oxide tunneling* and resulting long-term transistor threshold-voltage shifts, drain-to-substrate currents, and decreased transistor output impedance and therefore gain. These nondesirable effects must be contended with as technology dimensions are scaled. They are often called *short-channel effects.*

SHORT-CHANNEL EFFECTS

There are a number of *short-channel effects* that degrade the operation of MOS transistors as device dimensions are scaled smaller. These effects include *velocity saturation, mobility degradation, reduced output impedance*, and *hot-carrier effects* (such as *oxide trapping* and *substrate currents*).

Transistors having short-channel lengths and large electric fields experience a degradation of the effective mobility of their carriers due to a number of factors. One of these factors is the large lateral electric field (which has a vector component in a direction perpendicular from the gate into the silicon) caused by large gate voltages and short channel lengths. This large lateral field causes the effective channel depth to change, and also causes more electron collisions, thereby lowering the *effective mobility*. Another factor is that due to large electric fields, carrier velocity begins to saturate. This saturation causes the square-law characteristic of the current–voltage relationship to be inaccurate, and the true relationship will be somewhere between linear and square.

Another important short-channel effect is due to hot carriers. These high-velocity carriers can cause harmful effects such as the generation of *electron-hole pairs* by *impact ionization* and *avalanching*. These extra electron-hole pairs can cause currents to flow from the drain to the substrate as is shown in Fig. 3.26. This effect can be modeled by a finite drain-to-ground

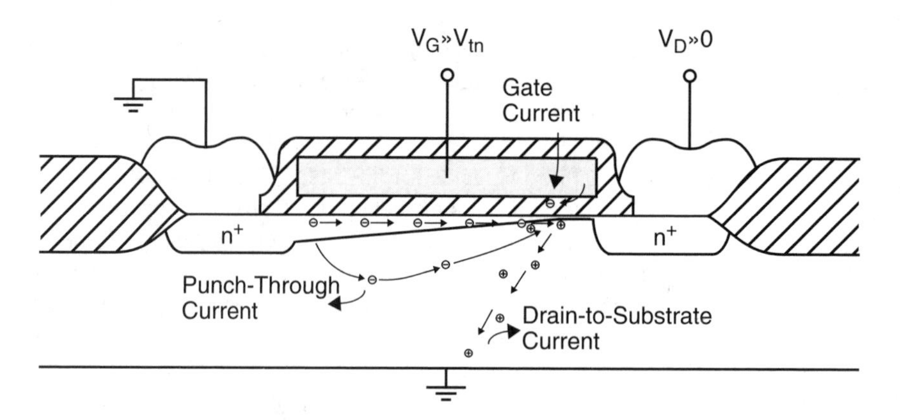

Figure 3.26 Drain-to-substrate current caused by hole-electron generation due to impact ionization at drain end of channel.

impedance. As a result, this effect is one of the major limitations on achieving very high output impedances of cascode current sources. In addition, it can cause voltage drops across the substrate and possibly cause latch-up, as described in the next section.

Another hot-carrier effect is due to electrons gaining high enough energies so they can tunnel into and possibly through the thin gate oxide. This effect can cause d.c. gate currents. However, often more harmful is the fact that any charge trapped in the oxide will cause a transistor threshold-voltage shift. As a result, hot carriers are a major factor limiting the long-term reliability of MOS transistors.

A third hot-carrier effect is when electrons with enough energy punch through from the source to the drain. As a result, these high-energy electrons no longer are limited by the drift equations governing normal conduction along the channel. This mechanism is somewhat similar to punch through in a bipolar transistor where the collector depletion region extends right through the base region to the emitter. In an MOS transistor, the channel length becomes effectively zero resulting in unlimited current flow (except for the series source and drain impedances and external circuitry). This effect is an additional cause of lower output impedance and possibly transistor breakdown.

It should be noted that all of the above hot-carrier effects are more pronounced for n-channel transistors than for their p-channel counterparts due to the larger velocities of electrons as opposed to holes.

Finally, it should be noted that short-channel transistors have much larger subthreshold currents than long-channel devices. This is partially due to what is called *drain-induced barrier lowering* or *DIBL.* This second-order effect occurs where large electric fields at the drain extend to the source region and thereby cause a lower effective threshold voltage. This effect can be largely mitigated by using a gradual diffusion concentration at the drain.

Subthreshold Operation

The device equations presented for MOS transistors above are all based on the assumption that V_{eff} (i.e., $V_{GS} - V_t$) is greater than 50 mV or so. When this is not the case, the accuracy of the square-law equations is poor, and the transistor is said to be operating in the *subthreshold region.* In this region, the transistor is more accurately modeled by an exponential relationship between its control voltage and current, somewhat similar to a bipolar transistor. In the subthreshold region, the drain current is approximately given by the exponential relationship (Geiger, et al.,1990)

$$I_D \cong I_{D\text{-}0}\left(\frac{W}{L}\right)e^{(qV_{GS}/nkT)} \tag{3.95}$$

where

$$n = \frac{C_{ox} + C_{depl}}{C_{ox}} \approx 1.5 \tag{3.96}$$

and it has been assumed $V_S = 0$ and $V_{DS} > 75$ mV, or so. The constant $I_{D\text{-}0}$ might be around 20 nA.

Although the transistors have an exponential relationship in this region, the transconductances are still small because of the small bias currents, and the transistors are slow because only small currents are available for charging and discharging capacitors. In addition, matching between transistors suffers because it is now dependent primarily on transistor-threshold-voltage matching. Normally, transistors will not be operated in the subthreshold region except in very low-frequency and low-power applications.

Leakage Currents

An important second-order device limitation in some applications is the *leakage current* of the junctions. This can be important, for example, in estimating the maximum time a sample-and-hold circuit or a dynamic memory cell can be left in the hold mode. The leakage current of a reverse-biased junction that is not close to breakdown is approximately given by (Uyemura,1988)

$$I_{lk} \approx \frac{qA_j n_i}{2\tau_0} x_d \tag{3.97}$$

where A_j is the junction area, n_i is the intrinsic concentration of carriers in undoped silicon, τ_0 is the *effective minority carrier lifetime,* and x_d is the thickness of the depletion region. τ_0 is given by

$$\tau_0 \cong \frac{1}{2}(\tau_n + \tau_p) \tag{3.98}$$

where τ_n and τ_p are the *electron* and *hole lifetimes*. Also, x_d is given by

$$x_d = \sqrt{\frac{2\varepsilon_{si}}{qN_A}(\Phi_0 + V_r)} \tag{3.99}$$

and n_i is given by

$$n_i \cong \sqrt{N_C N_V}\; e^{-\frac{E_g}{kT}} \tag{3.100}$$

where N_C and N_V are the density of states in the conduction and valence bands and E_g is the difference in energy between the two bands.

Since the intrinsic concentration n_i is a strong function of temperature (it approximately doubles for every 11°C temperature increase for silicon), the leakage current also is a strong function of temperature. Roughly speaking, it also doubles for every 11°C rise in temperature. Thus, at higher temperatures it is much larger than at room temperature. This leakage current imposes a maximum time on how long a dynamically charged output can be left in a high impedance state.

Latch-Up

One of the things that CMOS designers must be wary of, especially as dimensions shrink, is a phenomenon called *latch-up*. This effect can occur when there are substrate or well currents, or equivalently large substrate or well voltage drops, sometimes caused by capacitive coupling.

These triggering voltage drops often might occur when power is first applied to a CMOS IC. Another common cause of latch-up is substrate currents caused by capacitive coupling from output drivers.

When an IC becomes latched-up, it is equivalent to having a turned-on *silicon-controlled rectifier*, or *SCR*, between the power supply and ground. This effectively, *shorts* across the power supply on the IC, and unless the supply current is limited, it is highly likely that irreparable damage, such as a fused-open bonding wire or interconnect line, will occur.

To understand latch-up, consider the CMOS inverter cross section shown in Fig. 3.27. The *parasitic bipolar transistors*, Q_1 and Q_2, are also shown in Fig. 3.27. Q_1 is a *lateral* npn, with the base being formed by the p^- substrate. Q_2 is a *vertical* pnp, with the base being formed by the n-well region. The parasitic bipolar circuit has been redrawn in Fig. 3.28 along with some

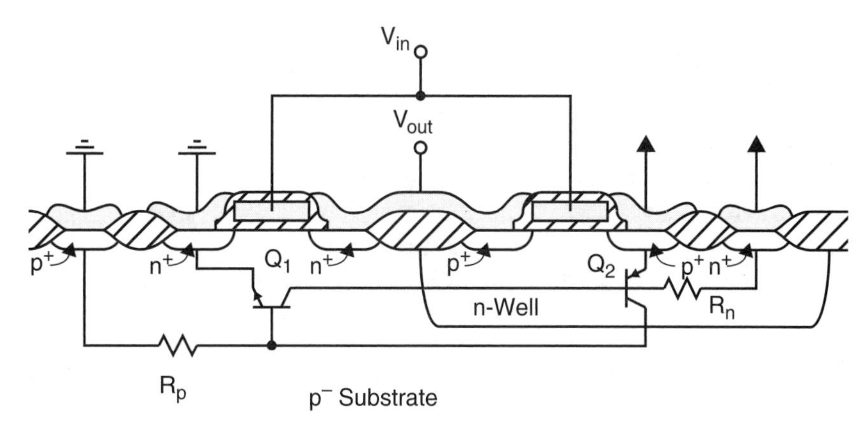

Figure 3.27 A cross-sectional view of a CMOS inverter with the schematic of the parasitic transistors responsible for the latch-up mechanism superimposed on it.

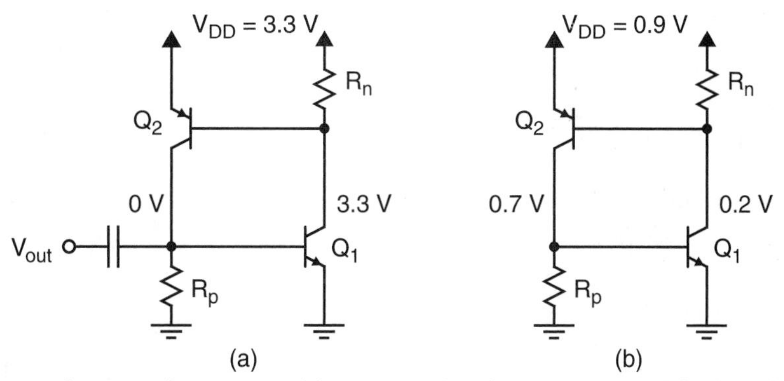

Figure 3.28 (a) The equivalent circuit of the parasitic bipolar transistors and (b) the voltages after latch-up has occurred.

of the parasitic resistances due to the lowly doped substrate and well regions. It can be seen that the circuit realizes two cross-coupled common-emitter amplifiers in a positive-feedback loop. This is the equivalent circuit of an SCR, which is sometimes called a *crowbar* switch.

Normally, the parasitic bipolar transistors are *off* and the voltages are as shown in Fig. 3.28a. However, if for some reason they turn *on*, and the loop gain is larger than unity, then the voltages will be approximately as shown in Fig. 3.28b. This effectively places a short circuit or a crowbar across the power-supply voltage and pulls V_{DD} down to approximately 0.9 V. If the power supply does not have a current limit, then excessive current will flow and a conductor will overheat and fuse open.

There are a number of ways that latch-up can be triggered. For example, the output of the CMOS inverter, V_{out} in Fig. 3.27, is capacitively coupled to the bases of the bipolar transistors by the junction-depletion capacitances of the MOS drains. If the inverter is large, such as is the case for an output buffer, then these capacitances can be large. When the output of the inverter changes, voltage *glitches* will be capacitively coupled to the base nodes of the parasitic bipolar transistors, and, if large enough, can potentially cause latch-up. Alternatively, substrate currents caused by *hot electrons* can also result in voltage drops large enough to trigger latch-up.

To prevent latch-up, the loop gain of the cross-coupled bipolar inverters must be kept less than unity. This is normally achieved by keeping the current gains of the parasitic bipolar transistors as low as is possible, and, most importantly, by keeping shunting resistors R_n and R_p as small as is possible. The current gain of the vertical pnp, Q_2, might be 50 to 100 and is difficult to minimize. The current gain of the lateral npn can be decreased by larger spacings between n-channel and p-channel transistors. However, with typically used spacings, the product $\beta_{npn}\beta_{pnp}$ is still normally greater than 1. The loop gain is primarily kept less than unity, and latch-up is prevented by decreasing R_n and R_p.

The major ways of decreasing R_n and R_p are by having low impedance paths between the substrate and well to the power supplies. One means of achieving this is by having many contacts to the substrate. For example, with an n-well technology, the design rules will normally specify a maximum distance between any place in the n-channel region of the IC and the closest p^+ junction that connects the substrate to ground. Similarly, in the p-channel regions, the maximum distance to nearest n^+ junction, which connects the wells to V_{DD}, will be specified. Also, after layout is completed, a good designer would normally fill any unused areas with extra *ties* to the substrate and well regions. In addition, any transistors that conduct large currents are normally surrounded by guard rings. These are ties to the substrate for n-channel transistors or to the well for p-channel transistors that completely surround the high-current transistors. This technique is often required for output buffers. *Finally, making sure the back of the die is connected to ground through an eutectic gold bond to the package header is very helpful.*

Perhaps one of the best ways of preventing latch-up is by using an *epitaxial process*, especially one with highly doped buried layers. For example, if a p^+ substrate has a p^- epitaxial layer, in which the transistors are placed, device performance is only marginally affected, but the highly conductive p^+ substrate has very little impedance to ground contacts and the package header. Alternatively, a p^- substrate might be used that has $n^+$and p buried regions and an intrinsic epitaxial region that is separately and optimally ion implanted to form the n-channel and p-channel regions. This self-aligned *twin-tub* technology is very immune to latch-up due to the highly conductive buried layers.

3.5 Bipolar-Junction Transistors

Historically, integrated circuits that originally dominated the market were realized using bipolar-junction transistors (BJT transistors). Then around the late 1970s, ICs realized using MOS transistors began to dominate, although BJT ICs remained somewhat popular for high-speed applications. More recently, BJT transistor understanding has become more important because of the growing popularity of BiCMOS technologies, where both BJT transistors and MOS transistors can be realized in the same IC.

Modern silicon BJT transistors can have unity-gain frequencies as high as 15 to 60 GHz, or more, as compared to a unity-gain frequency of 1 to 8 GHz for MOS transistors realized using a technology having a similar lithography resolution. However, BJT transistors do have the unfortunate characteristic that the control terminal, the base, has a finite input current when the transistor is conducting current (from the collector to the emitter for an npn transistor; for a pnp transistor the current goes from the emitter to the collector). Fortunately, at low frequencies, the base current is much less than the collector-to-emitter current—it may be only 1/100th as large as the collector current for an npn transistor. For lateral pnp transistors, the base current may be as large as 1/20th the emitter-to-collector current. Another disadvantage of bipolar transistors is they require greater area than MOS transistors realized using similar resolutions.

A typical cross section of an npn bipolar-junction transistor is shown in Fig. 3.29. Although this structure looks quite complicated, it corresponds to the approximately equivalent structure shown in Fig. 3.30. For a good BJT transistor, the distance from the n^+ emitter region to the n^- collector region, that is the base width, W, will be small, on the order of 1 μm or less. Also, as we shall see, it is necessary to have the base more lightly doped than the emitter.

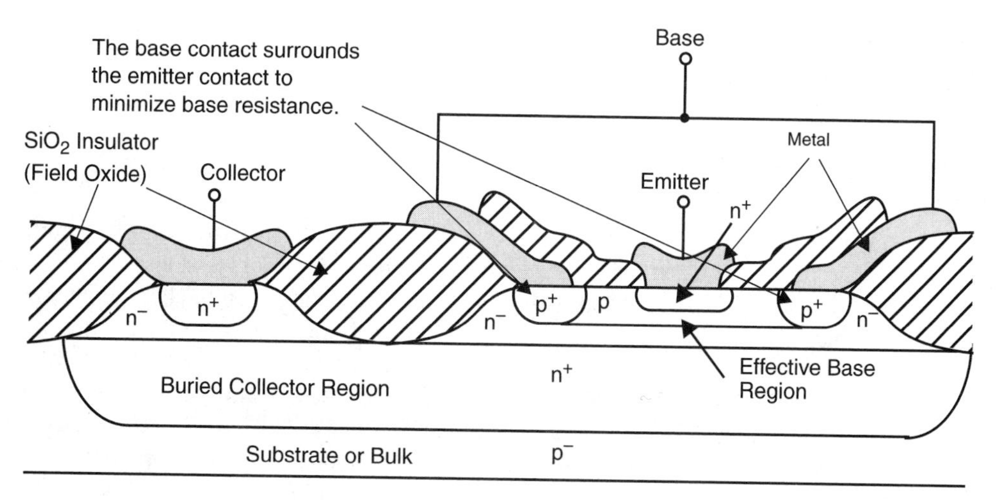

Figure 3.29 A cross-sectional view of an npn bipolar-junction transistor.

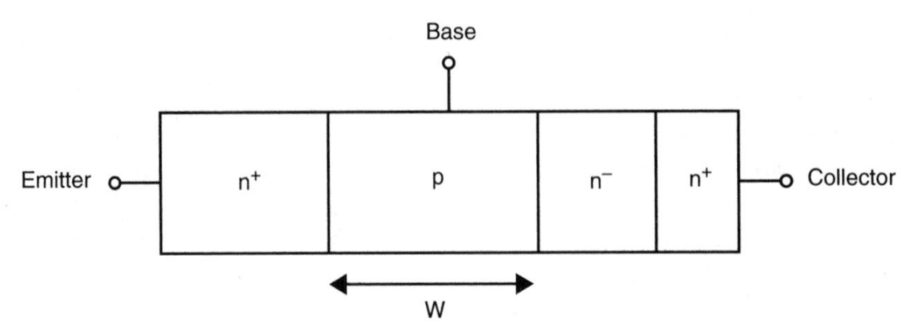

Figure 3.30 A simplified structure of an npn transistor.

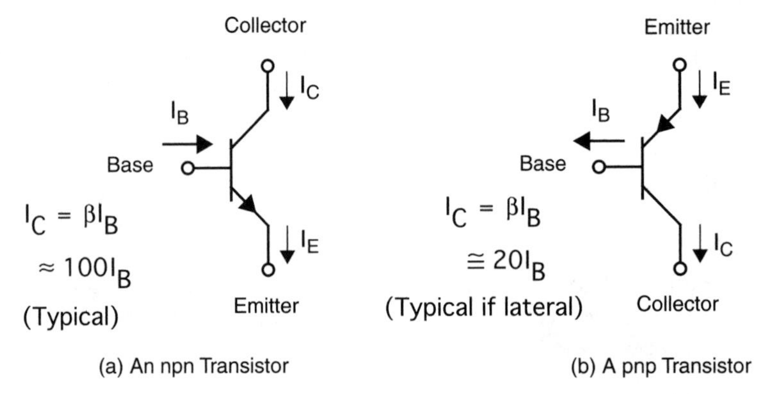

Figure 3.31 The symbol representing (a) an npn bipolar junction transistor and (b) a pnp transistor.

The circuit symbols used to represent npn and pnp transistors in electrical schematics are shown in Fig. 3.31.

Basic Operation

To understand the operation of BJT transistors, we consider an npn transistor with the emitter connected to ground as shown in Fig. 3.32. If the base voltage, V_B, is less than about 0.5 V, the transistor will be *cut off*, and no current will flow. We will see that when the base-emitter pn junction becomes forward biased, current will start to flow from the base to the emitter, but, partly because the base width is small, a much larger proportional current will flow from the collector to the emitter. Thus, the npn transistor can be considered a current amplifier at low frequencies. In other words, *if the transistor is not cut off and the collector-base junction is reverse biased, a small base current controls a much larger collector-emitter current.*

A simplified overview of how an npn transistor operates follows: When the base-emitter junction becomes forward biased, it starts to conduct, similar to any forward-biased junction.

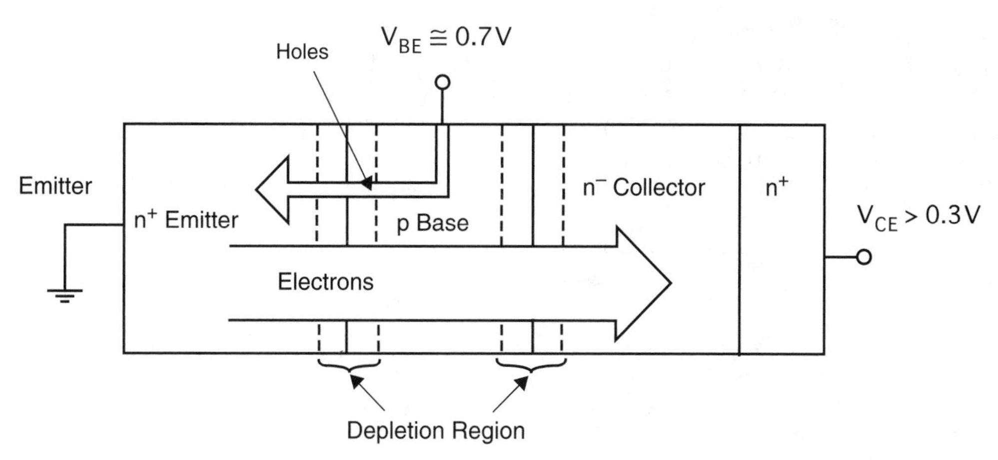

Figure 3.32 The various components of the currents of an npn transistor.

The current consists of majority carriers from the base (in this case, holes) and majority carriers from the emitter (in this case, electrons) diffusing across the junction. Because the emitter is more heavily doped than the base, there are many more electrons injected from the emitter than there are holes injected from the base. Assuming the collector voltage is large enough so that the collector-base junction is reverse biased, no holes from the base will go to the collector. However, the electrons that travel from the emitter to the base, where they are now minority carriers, diffuse away from the base-emitter junction because of the minority-carrier concentration gradient in the base region. Any of these minority electrons that get close to the collector-base junction will immediately be *whisked* across the junction due to the large positive voltage on the collector, which attracts the negatively charged electrons. In a properly designed bipolar transistor, such as that shown in Fig. 3.29, the vertical base width, W, is small, and almost all of the electrons that diffuse from the emitter to the base reach the collector-base junction and are *swept* across the junction, thus contributing to current flow in the collector. The result is that *the collector current very closely equals the electron current flowing from the emitter to the base.* The much smaller base current very closely equals the current due to the holes that flow from the base to the emitter. The total emitter current is the sum of the electron collector current and the hole base current, but since the hole current is much smaller than the electron current, the emitter current is approximately equal to the collector current (with an error of 1–2%).

Since the collector current is approximately equal to the electron current flowing from the emitter to the base, and the amount of this electron current is determined by the base-emitter voltage, it can be shown (see Appendix) that the collector current is exponentially related to the base-emitter voltage by the relationship

$$I_C \cong I_{CS} e^{V_{BE}/V_T} \tag{3.101}$$

where I_{CS} is the *scale current.* This scale current is proportional to the area of the base-emitter junction. The base current (determined by the hole current flowing from the base to emitter) is also exponentially related to the base-emitter voltage resulting in the ratio of the collector current to the base current being a constant, which, to a first-order approximation, is independent of voltage and current. This ratio is typically denoted as β, which is defined as

$$\beta \equiv \frac{I_C}{I_B} \tag{3.102}$$

where I_C and I_B are the collector and base current, respectively, for a transistor operating in the active region. Typical values of β are between 50 and 200.

Note that equation (3.101) implies that the collector current is independent of the collector voltage. This independence is an approximation that ignores second-order effects, in particular the decrease in effective base width W due to the increase in the width of the collector-base depletion region when the collector bias voltage is increased. To illustrate this point, a typical plot of the collector current, I_C, as a function of collector-to-emitter voltage, V_{CE}, for different values of I_B is shown in Fig. 3.33 for a practical transistor. The fact that the curves are not flat for $V_{CE} > V_{CE\text{-sat}}$ is indicative of the dependence of I_C on V_{CE}. Indeed, to a good approximation, the dependence is approximately linear with a slope that intercepts the V_{CE} axis at $V_{CE} = -V_A$ for all values of I_B. The intercept voltage value, V_A, is typically called the *Early voltage* for bipolar transistors. A typical value for V_A might be 50 to 100 V. This dependency results in a finite output impedance (as in an MOS transistor) and can be approximately modeled by modifying equation (3.101) to be (Sze, 1981)

$$I_C \cong I_{CS} e^{V_{BE}/V_T}\left(1 + \frac{V_{CE}}{V_A}\right) \tag{3.103}$$

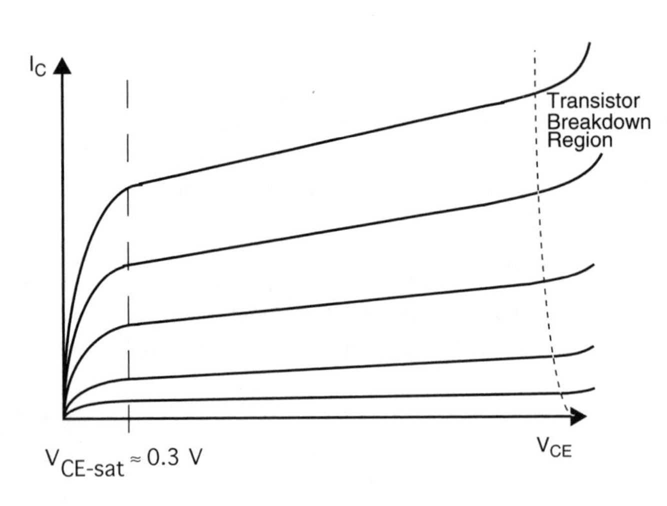

Figure 3.33 A typical plot of I_C versus V_{CE} for a BJT.

LARGE-SIGNAL MODELING

A conducting BJT having V_{CE} greater than $V_{CE\text{-}sat}$ (which is approximately 0.3 V) is said to be operating in the *active region*. Such a collector-emitter voltage is required to ensure that none of the holes from the base go to the collector. A large-signal model of a BJT operating in the active region is shown in Fig. 3.34.

Since $I_B = I_C/\beta$, we have

$$I_B = \frac{I_{CS}}{\beta} e^{V_{BE}/V_T} = I_{BS} e^{V_{BE}/V_T} \tag{3.104}$$

which is similar to the diode equation but with a multiplying constant of $I_{CS} / \beta = I_{BS}$. Since $I_E = I_B + I_C$, we have

$$I_E = I_{CS}\left(\frac{\beta+1}{\beta}\right) e^{V_{BE}/V_T} = I_{ES} e^{V_{BE}/V_T} \tag{3.105}$$

or equivalently

$$I_C = \alpha I_E \tag{3.106}$$

where α has been defined as

$$\alpha = \frac{\beta}{\beta+1} \cong 1 - \frac{1}{\beta} \cong 1 \tag{3.107}$$

If the effect of V_{CE} on I_C is included in the model, the current-controlled source βI_B, should be replaced by a current source given by

$$I_C = \beta I_B\left(1 + \frac{V_{CE}}{V_A}\right) \tag{3.108}$$

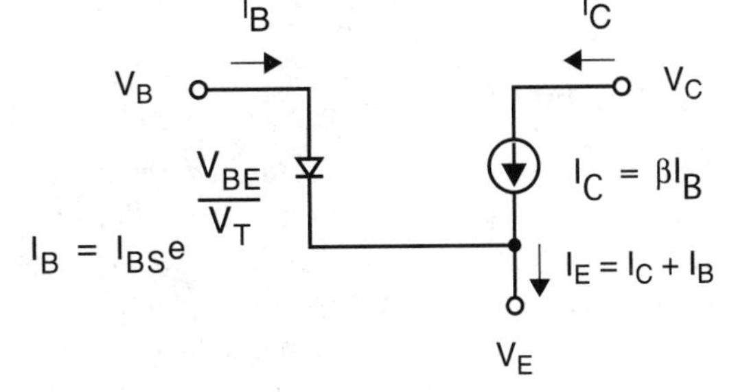

Figure 3.34 A large-signal model for a BJT in the active region.

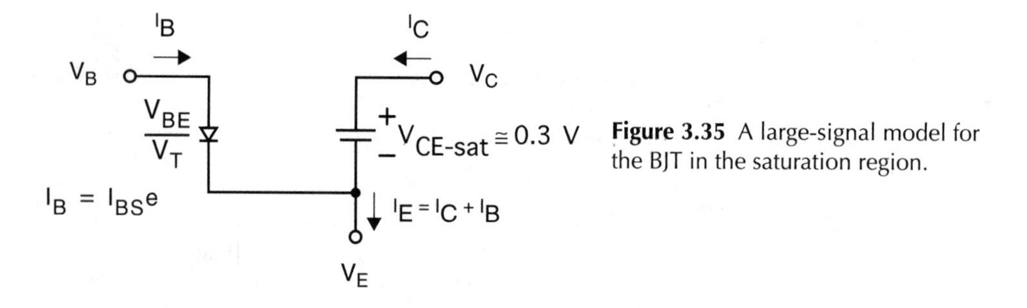

Figure 3.35 A large-signal model for the BJT in the saturation region.

where V_A is the Early-voltage constant. This additional modeling of the finite output impedance is normally not used in large-signal analysis without the use of a computer due to the extra complication involved.

As the collector-emitter voltage approaches $V_{CE\text{-}sat}$ (typically around 0.2 to 0.3 V), the base-collector junction becomes forward biased and holes from the base will begin to diffuse to the collector. A common model for this case, when the transistor is *saturated* or in the saturation region, is shown in Fig. 3.35. It should be noted that the value of $V_{CE\text{-}sat}$ decreases for smaller values of collector current.

BASE CHARGE STORAGE IN THE ACTIVE REGION

When a transistor is in the active region, many *minority carriers* are stored in the base region (n electrons are stored in the p base of an npn transistor). Recall that this *minority charge* is responsible for I_C, so this charge must be removed (through the base contact) before a transistor can turn off. As in a forward-biased diode, this charge can be modeled as a diffusion capacitance, C_d, between the base and emitter given by (see Appendix)

$$C_d = \tau_b \frac{I_C}{V_T} \tag{3.109}$$

where τ_b is the *base-transit-time constant.* Thus, we see that the diffusion capacitance is proportional to I_C. The total base-emitter capacitance, C_{be}, will include the base-emitter depletion capacitance, C_j, in parallel with C_d. Normally, however, C_j is much less than C_d, unless the transistor current is small, and can often be ignored.

BASE CHARGE STORAGE OF A SATURATED TRANSISTOR

When a transistor becomes *saturated*, the minority charge storage in the base, and, even more so, in the lightly doped region of the collector, increases drastically. The major component of this charge storage is due to holes diffusing from the base, through the collector junction, and continuing on through the lightly doped n^- epitaxial region of the collector to the n^+ collector region. The n^- epitaxial region is so named because it is epitaxially grown on a p region. Most of the charge storage occurs in this region. Additional charge storage occurs because electrons that diffused from the collector are stored in the base, but this charge is normally smaller. The magnitude of the additional charge stored by a transistor that is saturated is given by

$$Q_S = \tau_S\left(I_B - \frac{I_C}{\beta}\right) \tag{3.110}$$

where the *base overdrive current*, defined to be $I_B - I_C/\beta$, is approximately equal to the hole current from the base to the collector. Normally, in saturation, $I_B >> I_C/\beta$, and (3.110) can be approximated by

$$Q_S \cong \tau_S I_B \tag{3.111}$$

The constant τ_S is approximately equal to another constant called the *epitaxial-region transit time constant*, τ_E (ignoring the storage of electrons in the base that diffused from the collector). Since the epitaxial region is much wider than the base, the constant τ_S is normally much larger than the base transit time constant, τ_b, often by up to two orders of magnitude. The specific value of τ_S is usually found empirically for a given technology.

When a saturated transistor is being turned off, first the base current will reverse. *However, before the collector current will change, the saturation charge,* Q_S, *must be removed.* After Q_S is removed, the base minority charge, Q_b, will be removed. During this time, the collector current will decrease until the transistor shuts off. Typically, the time to remove Q_S greatly dominates the overall charge removal time.

If the time required to remove the base saturation charge, t_S, is much shorter than the epitaxial-region transit time, τ_E, then one can derive a simple expression for the time required to remove the saturation charge. If the reverse base current (when the saturation charge is being removed), denoted by I_{BR}, remains constant while Q_S is being removed, then we have (Hodges and Jackson, 1988)

$$t_S \cong \frac{Q_S}{I_{BR}} \cong \frac{\tau_S[I_B - (I_C/\beta)]}{I_{BR}} \cong \tau_S \frac{I_B}{I_{BR}} \tag{3.112}$$

where $\tau_S \approx \tau_E$.

Normally, the forward base current during saturation, I_B, will be much smaller than the reverse base current during saturation-charge removal, I_{BR}. If this were not the case, then our original assumption that $t_S << t_E \cong t_S$ would not be true. In this case, the turn-off time of the BJT would be so slow as to make the circuit unusable in most digital applications. Nevertheless, the turn-off time for this case, when t_S is not much less than τ_E, is given by (Hodges and Jackson, 1988)

$$t_S = \tau_S \ln\left(\frac{I_{BR} + I_B}{I_{BR} + \frac{I_C}{\beta}}\right) \tag{3.113}$$

The reader should verify that for $I_{BR} >> I_B$ and I_C / β, the expression in (3.113) is approximately equivalent to the much simpler one in (3.112) (Problem 3.22).

In both of the above cases, the time required to remove the storage charge of a saturated transistor is much larger than the time required to turn off a transistor in the active region.

In modern digital microcircuit designs using BJT transistors, transistors are never allowed to saturate to avoid the resulting large turn-off time.

Example 3.8

For $\tau_b = 0.2$ ns, $\tau_s = 100$ ns (a small value for τ_s), $I_B = 0.2$ mA, $I_C = 1$ mA, $\beta = 100$, and $I_{BR} = 1$ mA, calculate the time required to remove the base saturation charge using (3.112) and compare to the time obtained using the more accurate (3.113). Compare this to the time required to remove the base minority charge for the same I_{BR}.

Solution: Using (3.112), we have

$$t_s = \frac{10^{-7}(2 \times 10^{-4})}{10^{-3}} = 20 \text{ ns} \tag{3.114}$$

Using (3.113), we have

$$t_s = 10^{-7} \ln\left[\frac{10^{-3} + 2 \times 10^{-4}}{10^{-3} + \frac{10^{-3}}{100}}\right] = 17.2 \text{ ns} \tag{3.115}$$

which is fairly close. Next, the time required to remove Q_b is given by

$$t_A = \frac{Q_b}{I_{BR}} = \frac{\tau_b I_C}{I_{BR}} = 0.2 \text{ ns} \tag{3.116}$$

which is approximately 100 times smaller!

SMALL-SIGNAL MODELING

The small-signal modeling of BJTs is covered in the Appendix and is only briefly summarized here. The most commonly used small-signal model is the *hybrid-π model.* This is very similar to the small-signal model used for MOS transistors, as we shall see, except for the inclusion of a finite base-emitter impedance, r_π, and no emitter-to-bulk capacitance. The hybrid-π model is shown in Fig. 3.36. We will first discuss the d.c. components, that is, the transconductance, g_m, and the resistors. Next the parasitic capacitances will be discussed.

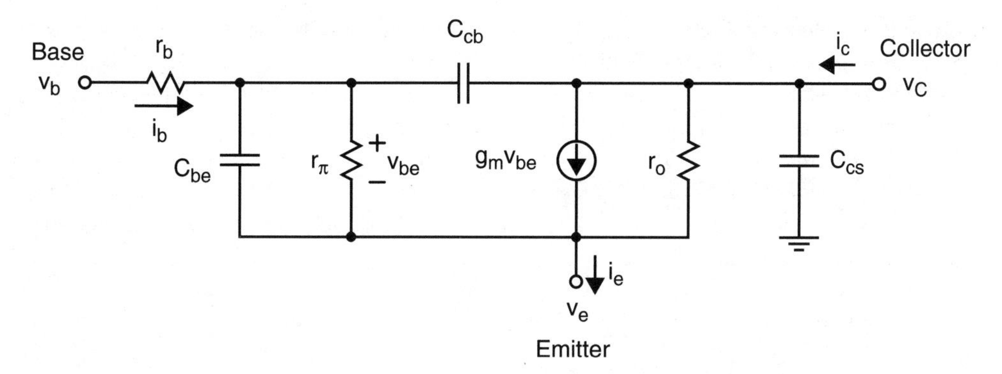

Figure 3.36 The small-signal model of a BJT.

The transistor transconductance, g_m, is perhaps the most important parameter of the small-signal model and is given by

$$g_m = \frac{I_C}{V_T} \tag{3.117}$$

where

$$V_T = \frac{kT}{q} \cong 26 \text{ mV at } T = 300°\text{K} \tag{3.118}$$

The presence of the resistor r_π reflects the fact that there is a finite base current. We have

$$r_\pi = \frac{V_T}{I_B} \tag{3.119}$$

or equivalently

$$r_\pi = \beta\frac{V_T}{I_C} = \frac{\beta}{g_m} \tag{3.120}$$

The resistor r_o models the dependence of the collector current on the collector-emitter voltage.

$$r_o = \frac{V_A}{I_C} \tag{3.121}$$

which is inversely proportional to collector current. Resistor r_o is of only minor importance in bipolar logic circuits. The resistor r_b models the resistance of the semiconductor material between the base contact and the effective base region, namely the moderately lightly doped base p material (see Fig. 3.29). This resistor, although small and typically 200 to 500 Ω, can be very important in limiting the speed of BJT logic gates.

The high-frequency operation of a BJT is limited by the capacitances of the small-signal model. We have already encountered one of these capacitances, C_{be}, in a previous section. Recapping, we have

$$C_{be} = C_j + C_d \tag{3.122}$$

where C_j is the depletion capacitance of the base-emitter junction. For a forward-biased junction, a rough approximation for C_j is

$$C_j \cong 2A_E C_{je0} \tag{3.123}$$

The diffusion capacitance, C_d, is given in (3.109) as

$$C_d = \tau_b \frac{I_C}{V_T} = g_m \tau_b \tag{3.124}$$

The capacitor C_{cb} models the depletion capacitance of the collector-base junction.

3.6 SPICE-Modeling Parameters

This section briefly describes some of the important model parameters for diodes, bipolar transistors, and MOS transistors, used during a SPICE simulation (Massobrio and Antognetti, 1993). Not all SPICE model parameters are described; however, enough are described to enable the reader to understand the relationship between the relative parameters and the corresponding constants used when doing hand analysis. This section is relevant to SPICE modeling where a physical relationship exists between the SPICE parameters and the actual transistor; some modern models that do not possess this feature (such as the Berkeley *BSIM* models) are beyond the scope of this text.

Diode Model Parameters

There are a number of important d.c. parameters. The constant I_S is specified using either the parameter IS or JS in SPICE. These two parameters are synonyms for each other and only one should be specified. A typical value specified might be between 10^{-18} and 10^{-15} A for small diodes in a microcircuit. Another important parameter is called the *emission coefficient*, n. This constant multiplies V_T in the exponential diode I–V relationship given by

$$I_D = I_S e^{V_{BE}/nV_T} \tag{3.125}$$

The SPICE parameter for n is N and is defaulted to 1 when not specified (which is a reasonable value for junctions in a microcircuit). A third important d.c. characteristic is the *series resistance,* which is specified in SPICE using RS. Note that some SPICE programs allow the user to specify the area of the diode while others expect absolute parameters that already take into

Table 3.2 Important SPICE Parameters for Modeling Diodes

SPICE parameter	Model constant	Brief description	Typical value
IS	I_S	Transport saturation current	10^{-17} A
RS	R_d	Series resistance	30 Ω
CJ	$C_{j\text{-}0}$	Capacitance at 0 V bias	0.01 pF
MJ	m_j	Diode grading coefficient exponent	0.5
PB	Φ_0	Built-in diode contact potential	0.9 V

account the effective area. The relevant manual should be consulted to determine the correct usage for the program being utilized.

The most important capacitance parameter specified is CJ (CJO is a synonym—one should never specify both CJ and CJO). This parameter specifies the capacitance at 0 V bias. Once again, it may be specified as absolute or relative to the area (i.e., F/m^2) depending on the version of SPICE used. Also, the area junction grading coefficient (MJ) might be specified. This parameter specifies the exponent used in the capacitance equation. It is typically 0.5 for abrupt junctions and 0.33 for graded junctions. In some SPICE versions, it might also be possible to specify the side-wall capacitance at 0 V bias and its grading junction coefficient as well. Finally, the built-in potential of the junction that is also used in calculating the capacitance can be specified using PB (PHI, VJ, or PHA are all synonyms).

Reasonably accurate diode simulations can usually be obtained by specifying IS, CJ, MJ, and PB only. However, most modern versions of SPICE, such as HSPICE, have many more parameters that can be specified if one wants accurate temperature and noise simulations. Users should consult their manuals for more information.

Shown in Table 3.2 is a brief summary of some of the more important SPICE diode parameters. This set of parameters constitutes a minimal set for reasonable simulation accuracy under ordinary conditions.

MOS Transistors Parameters

Modern MOS models are quite complicated, so only some of the more important MOS parameters used in SPICE simulations are described here. These parameters are used in what is normally called the *Level 2* or *Level 3* models. The model level can be chosen by setting the SPICE parameter LEVEL to either 2 or 3. The oxide thickness, t_{ox}, is specified using SPICE parameter TOX. If it is specified, then it is not necessary to specify the thin gate oxide capacitance (i.e., C_{ox} specified by parameter COX). The mobility, μ_n, can be specified using UO. If UO is specified, the intrinsic transistor conductance (i.e., $\mu_n C_{ox}$) will be calculated automatically, unless this automatic calculation is overridden by specifying either KP (or its synonym BETA). The transistor threshold voltage at $V_S = 0$ V, V_{tn}, is specified by VTO. The body effect parameter, γ, can be specified using GAMMA or it will be automatically calculated if the substrate doping, N_A, is specified using NSUB. Normally, one would not want SPICE to calculate γ since the effective substrate doping under the channel can be significantly different from the substrate doping in the bulk due to threshold-voltage-adjust implants. The output impedance constant, λ, can be specified using LAMBDA. Normally, LAMBDA should not be specified as it takes precedence over internal calculations and does not change the output impedance as a function of different transistor lengths or bias voltages, which should be the case. Indeed, modeling the transistor output impedance is one of weakest points in SPICE. If LAMBDA is not specified, it is calculated automatically. The surface inversion potential, $|2\phi_F|$, can be specified using PHI or it will be calculated automatically. Another parameter that is usually specified is the lateral diffusion of the junctions under the gate, L_D, which is specified by LD. For accurate simulations, one might also specify the resistances in series with the source and drain by specifying RS and RD (typically only the source resistance is important). In addition, there are many other parameters to model short-channel effects, subthreshold effects, channel-width effects, etc., which are outside the scope of this book.

The modeling of parasitic capacitances in SPICE is quite involved. Originally, this modeling was not very accurate as it did not include charge conservation for the gate charge. However, this modeling has greatly improved in recent commercial versions of SPICE. These more modern models might be invoked by setting a parameter such as CAPOP=9 in HSPICE. The capacitances under the junctions per unit area at 0 V bias (i.e., $C_{j\text{-}0}$) can be specified using CJ or can be calculated automatically from the specified substrate doping. The side-wall capacitances at 0 V, $C_{j\text{-}0\text{-}sw}$, are specified using CJSW. The bulk grading coefficient, specified by MJ, can usually be defaulted to 0.5. Similarly, the side-wall grading coefficient, specified by MJSW, can usually be defaulted to 0.33 (it assumes a graded junction). The built-in bulk-to-junction contact potential, Φ_0, can be specified using PB, or defaulted to 0.8 V (note that 0.9 or 1.0 V would typically be more accurate, but the resulting simulation differences are small). Sometimes, the gate-to-source or drain overlap capacitances can be specified using CGSO or CGDO, but normally these would be left to be calculated automatically using COX and LD.

Some of the more important parameters, that should result in reasonable simulations (except for modeling short-channel effects), are summarized in Table 3.3 for both n- and p-channel transistors. Included are reasonable parameters for a typical 0.6 μm technology.

Bipolar-Junction Transistor Parameters

For historical reasons, most parameters for modeling BJT transistors are specified absolutely. As well, rather than specifying the emitter area of a BJT in micrometers squared on the line where the individual transistor connections are specified, most SPICE versions have multiplication factors. These multiplication factors can be used to automatically multiply parameters when a transistor is composed of a number of transistors connected in parallel. This parameter would normally be called M.

The most important d.c. parameters are the transistor current gain, β, specified using the SPICE parameter BF, the transistor transport saturation current, I_{CS}, specified using the parameter IS, and the Early-voltage constant, specified by the parameter VAF. Typical values for these might be 100, 10^{-17} A, and 50 V, respectively. If it is desired to model the transistor in reverse mode (where the emitter voltage is higher than the collector voltage for an npn), then

Table 3.3 A Reasonable Set of MOS Parameters for a Typical 0.6 μm Technology

SPICE parameter	Model constant	Brief description	Typical value
VTO	V_{tn}/V_{tp}	Transistor threshold voltage (in V)	0.7 / −0.8
UO	μ_n/μ_p	Carrier mobility in bulk (in $cm^2/V{\cdot}s$)	500 / 175
TOX	t_{ox}	Thickness of gate oxide (in m)	1×10^{-8}
LD	L_D	Lateral diffusion of junction under gate (in m)	5×10^{-8}
GAMMA	γ	Body-effect parameter	0.6
NSUB	N_A/N_D	The substrate doping (in cm^{-3})	1.5×10^{17}
PHI	$\lvert 2\phi_F \rvert$	Surface inversion potential (in V)	0.7
PB	Φ_0	Built-in contact potential of junction to bulk (in V)	1.0
CJ	$C_{j\text{-}0}$	Junction depletion capacitance at 0 V bias (in F/m^2)	5×10^{-4}
CJSW	$C_{j\text{-}0\text{-}sw}$	Side-wall capacitance at 0 V bias (in F/m)	2.0×10^{-10}
MJ	m_j	Bulk-to-junction exponent (grading coefficient)	0.5
MJSW	$m_{j\text{-}sw}$	Side-wall-to-junction exponent (grading coefficient)	0.3

one might specify BR, ISS, and VAR, as well, which are the corresponding parameters to BF, IS, and VAF, in the reverse direction. Typically, this reverse-mode modeling is not important for most digital circuits. Other important d.c. parameters for accurate simulations are the base, emitter, and collector resistances, which are specified by RB, RE, and RC, respectively. It is especially important to specify RB (which might be 200 to 500 Ω).

The important capacitance parameters include the depletion capacitances at 0 V bias voltage, their grading coefficients, and their built-in voltages for base-emitter, base-collector, and collector-substrate junctions. The corresponding SPICE parameters to these parameters are (in the same respective order) CJE, CJC, CJS, MJE, MJC, MJS, VJE, VJC, and VJS. Again, the 0 V depletion capacitances should be specified in absolute values for a unit-size transistor. Normally the base-emitter and base-collector junctions are graded (i.e., MJE, MJC = 0.33), whereas the collector-substrate junction may be abrupt (MJS = 0.5) or graded (MJS = 0.33) depending on processing details. Typical built-in voltages might be 0.75 to 1.0 V. In addition, for accurate simulations, one should specify the forward-base transit time, τ_F specified by TF, and, if the transistor is to be operated in reverse mode or under saturated conditions, the reverse-base transit time, τ_R specified by TR should be given.

The more important of the model parameters just described are summarized in Table 3.4.

Once again, there are many other parameters that can be specified if accurate simulation is desired. Other parameters might include those to model β degradation under high or low current applications and parameters for accurate noise and temperature analysis. The reader should refer to their respective SPICE manuals for descriptions of these parameters.

Table 3.4 The Most Important SPICE Parameters for Modeling BJTs

SPICE parameter	Model constant	Brief description	Typical value
BF	β	Transistor current gain in forward direction	100
ISS	I_{CS}	Transport saturation current in forward direction	2×10^{-18} A
VAF	V_A	Early voltage in forward direction	50 V
RB	r_b	Series base resistance	500 Ω
RE	R_E	Series emitter resistance	30 Ω
CJE	$C_{je\text{-}0}$	Base-emitter depletion capacitance at 0 V	0.015 pF
CJC	$C_{jc\text{-}0}$	Base-collector depletion capacitance at 0 V	0.018 pF
CJS	$C_{js\text{-}0}$	Collector-substrate depletion capacitance at 0 V	0.040 pF
MJE	m_e	Base-emitter junction exponent (grading factor)	0.30
MJC	m_c	Base-collector junction exponent (grading factor)	0.35
MJS	m_s	Collector-substrate junction exponent (grading factor)	0.29
TF	τ_F	Base-forward transit time	12 ps
TR	τ_R	Base-reverse transit time	4 ns

3.7 Appendix

The purpose of this Appendix is to present many derivations for device equations that rely more heavily on device physics knowledge. Specifically, equations are derived for the exponential relationship and diffusion capacitance of diodes, the threshold voltage and triode relationship for MOS transistors, and the exponential relationship and base charge storage for bipolar transistors. In addition, small-signal models for diodes, MOS transistors, and bipolar transistors are given in greater detail. These are subjects that are not used often in digital IC design except at the most advanced level where the circuits are considered from an *analog* perspective.

Diode Exponential Relationship

The concentration of minority carriers in the bulk, far from the junction, is given by equations (3.8) and (3.9). Close to the junction for a forward bias, the minority carrier concentrations are much larger. Indeed, the concentration next to the junction increases exponentially with the external voltage, V_D, that is applied in the forward direction. The concentration of holes in excess of equilibrium in the n side next to the junction, p_n, is given by (Sze, 1981)

$$\Delta p_n = p_{n\text{-}0} e^{V_D/V_T} = \frac{n_i^2}{N_D} e^{V_D/V_T} \tag{3.126}$$

Similarly, the concentration of electrons in excess of equilibrium in the p side next to the junction is given by

$$\Delta n_p = n_{p\text{-}0} e^{V_D/V_T} = \frac{n_i^2}{N_A} e^{V_D/V_T} \tag{3.127}$$

As the carriers diffuse away from the junction, their concentration exponentially decreases. The relationship for holes in the n side is

$$p_n(x_n) = \Delta p_n(0) e^{-\frac{x_n}{L_p}} + p_{n\text{-}0} \tag{3.128}$$

where x_n is the distance from the junction into the n region and L_p is a constant known as the diffusion length for holes in the n side. Similarly, for electrons in the p side we have

$$n_p(x_p) = \Delta n_p(0) e^{-x_p/L_N} + n_{p\text{-}0} \tag{3.129}$$

where x_p is the distance from the junction into the p region and L_n is a constant known as the diffusion length of electrons in the p side. Note that Δp_n (0) and Δn_p (0) are given by (3.126) and (3.127), respectively. Note also that the constants L_n and L_p are dependent on the doping concentrations N_A and N_D, respectively.

The current density of diffusing carriers moving away from the junction is given by the well-known diffusion equations (Sze, 1981). For example, the current density of diffusing electrons is given by

$$J_{D\text{-}n} = -qD_n \frac{dn_p(x_p)}{dx_p} \tag{3.130}$$

where D_n is the diffusion constant of electrons in the p side of the junction. The negative sign is due to electrons having negative charge. Note that $D_n = (kT/q)\mu_n$ where μ_n is the mobility of electrons. Using (3.129), we have

$$\frac{dn_p(x_p)}{dx} = \frac{\Delta n_p(0)}{L_n} e^{-x_p/L_N} = -\frac{\Delta n_p(x_p)}{L_n} \tag{3.131}$$

Therefore

$$J_{D\text{-}n} = \frac{qD_n}{L_n} n_p(x_p) \tag{3.132}$$

Thus, the current density due to diffusion is proportional to the minority carrier concentration. Next to the junction, all the current flow is due to the diffusion of minority carriers. Further away from the junction, some of the current flow is due to diffusion and some of the current flow is due to majority carriers *drifting by* to replace carriers that recombined with minority carriers or diffused across the junction.

Continuing, next to the junction, using (3.127) and (3.132), we have the current density of electrons in the p side as

$$J_{D\text{-}n} = \frac{qD_n}{L_n}\Delta n_p(0) = \frac{qD_n}{L_n}\frac{n_i^2}{N_A}e^{V_D/V_T} \tag{3.133}$$

For the total current of electrons in the p side, one simply multiples (3.133) by the effective junction area, A_D. The total current remains constant as one moves away from the junction, for in the steady state, the minority carrier concentration at any particular location remains constant with time. Stated differently, if the current changed as one moved away from the junction, there would be a change of charge concentrations with time.

Using a similar derivation, we obtain the total current of holes in the n side, $I_{D\text{-}p}$, as

$$I_{D\text{-}p} = \frac{A_D q D_n n_i^2}{L_p N_D}e^{V_D/V_T} \tag{3.134}$$

where D_n is the diffusion constant of electrons in the p side of the junction, L_p is the diffusion length of holes in the n side, and N_D is the impurity concentration of donors in the n side. This current, consisting of positive carriers, flows in the opposite direction to the flow of minority electrons in the p side. But since electron carriers are negatively charged, the direction of the current flow is the same. Note also that if the p side is more heavily doped than the n side, most of the carriers will be holes, whereas if the n side is more heavily doped, most of the carriers will be electrons.

The total current is the sum of the minority currents at the junction edges

$$I_D = A_D q n_i^2\left(\frac{D_n}{L_n N_A} + \frac{D_p}{L_p N_D}\right)e^{V_D/V_T} \tag{3.135}$$

Equation (3.135) is often expressed as

$$I_D = I_S e^{V_D/V_T} \tag{3.136}$$

where

$$I_S = A_D q n_i^2\left(\frac{D_n}{L_n N_A} + \frac{D_p}{L_p N_D}\right) \tag{3.137}$$

tion (3.54) is the well-known exponential current–voltage relationship of forward-biased s.

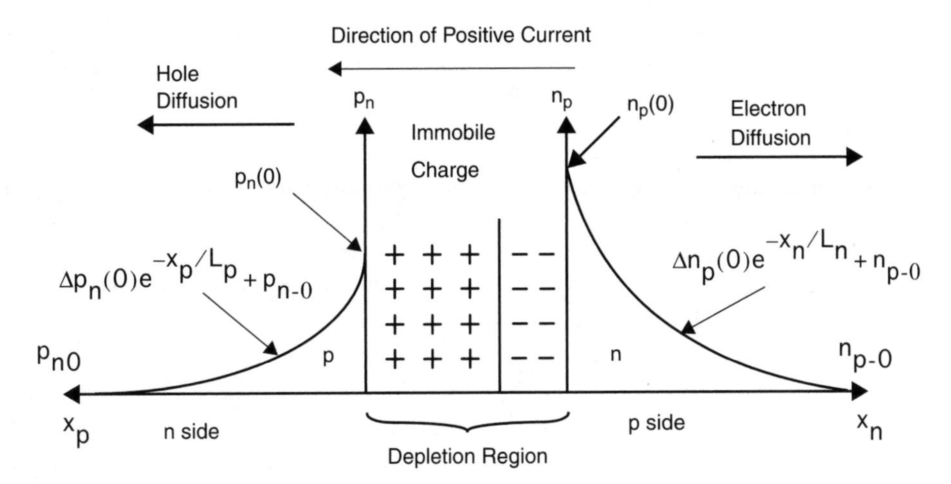

Figure 3.37 The concentration of minority carriers and the direction of diffusing carriers near a forward-biased junction.

The concentrations of minority carriers near the junction and the direction of current flow are shown graphically in Fig. 3.37.

Diode Diffusion Capacitance

To find the diffusion capacitance, C_d, we must first find the minority charge close to the junction, Q_d, and then differentiate it with respect to V_D. The minority charge close to the junction, Q_d, can be found by integrating either (3.128) or (3.129) over a few diffusion lengths. For example, if we assume $n_{p\text{-}0}$ (the minority electron concentration in the p side far from the junction) is much less than $\Delta n_p(0)$ (the excess electron concentration at the junction edge), we have through the use of (3.129)

$$
\begin{aligned}
Q_n &= qA_D\int_0^{\infty}\Delta n_p(x)dx \\
&= qA_D\int_0^{\infty}\Delta n_p(0)e^{-\frac{x}{L_n}}dx \\
&= qA_DL_n\Delta n_p(0)
\end{aligned}
\tag{3.138}
$$

Using (3.11) for $\Delta n_p(0)$ results in

$$
Q_n = \frac{qA_DL_nn_i^2}{N_A}e^{V_D/V_T}
\tag{3.139}
$$

In a similar manner, we also have

$$Q_p = \frac{qA_D L_n n_i^2}{N_D} e^{V_D/V_T} \tag{3.140}$$

For a typical junction, one side will be much more heavily doped than the other side, and therefore the minority charge storage in the heavily doped side can be ignored since it will be much less than that in the lightly doped side. Assuming the n side is heavily doped, we find the total charge, Q_d, to be approximately given by Q_n, the minority charge in the p side. Thus, the small-signal diffusion capacitance, C_d, is given by

$$C_d = \frac{dQ_d}{dV_D} \cong \frac{dQ_n}{dV_D} = \frac{qA_D L_n n_i^2}{N_A V_T} e^{V_D/V_T} \tag{3.141}$$

Using (3.135) and again noting that $N_D \gg N_A$, we have

$$C_d = \frac{L_n^2}{D_n} \frac{I_D}{V_T} \tag{3.142}$$

Equation (3.142) is often expressed as

$$C_d = \tau_T \frac{I_D}{V_T} \tag{3.143}$$

where τ_T is the transit time of the diode given by

$$\tau_T = \frac{L_n^2}{D_n} \tag{3.144}$$

for a single-sided diode with the n side being more heavily doped.

SMALL-SIGNAL MODEL OF A FORWARD-BIASED DIODE

A small-signal equivalent model for a forward-biased diode is shown in Fig. 3.38. The resistor, r_d, models the change in the diode voltage, V_D, when I_D changes. Using (3.54), we have

$$\frac{1}{r_d} = \frac{dI_D}{dV_D} = I_S \frac{e^{V_D/V_T}}{V_T} = \frac{I_D}{V_T} \tag{3.145}$$

This resistance is called the incremental resistance of the diode. For very accurate modeling, it is sometimes necessary to add the series resistance due to the bulk and also the resistance associated with the contacts. Typical values for the contact resistances (caused by the work-function difference between metal and silicon) might be 5 to 20 Ω.

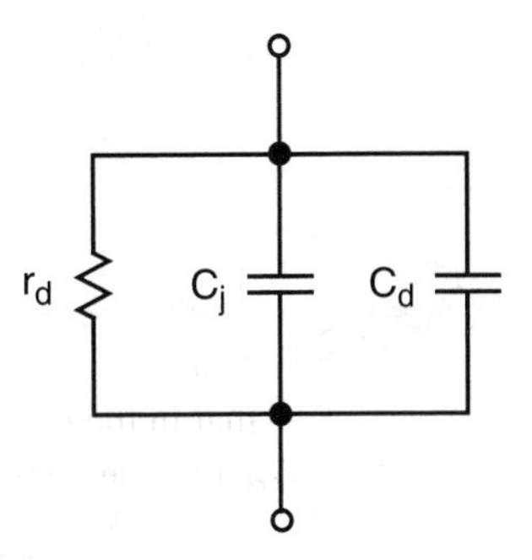

Figure 3.38 The small-signal model for a forward-biased junction.

By combining (3.56) and (3.145), we see that an alternative equation for the diffusion capacitance, C_d, is

$$C_d = \frac{\tau_T}{r_d} \tag{3.146}$$

Since for moderate forward-bias currents, $C_d \gg C_j$, the total small-signal capacitance $C_T \cong C_d$ and

$$r_d C_T \cong \tau_T \tag{3.147}$$

Thus, for charging or discharging a forward-biased junction with a current source having an impedance much larger than r_d, the time constant of the charging is approximately equal to the transit time of the diode, and is independent of the diode current. For smaller diode currents, where C_j becomes important, the charging or discharging time constant of the circuit becomes larger than τ_T.

Example 3.9

A given diode has a transit time of 100 ps and is biased at 1 mA. What are the values of its small-signal resistance and diffusion capacitance? Assume room temperature such that $V_T = kT/q = 26\ \text{mV}$.

Solution: We have

$$r_d = \frac{V_T}{I_D} = \frac{26\ \text{mV}}{1\ \text{mA}} = 26\ \Omega$$

and

$$C_d = \frac{\tau_T}{r_d} = \frac{100\ \text{ps}}{26\ \Omega} = 3.8\ \text{pF}$$

Note that this diffusion capacitance is over 100 times larger than the total depletion capacitance found in the previous examples.

MOS THRESHOLD VOLTAGE AND THE BODY EFFECT

There are many factors that affect the gate-source voltage at which the channel becomes conductive:

1. The work-function difference between the gate material and the substrate material.
2. The voltage drop between the channel and the substrate required for the channel to exist.
3. The voltage drop across the thin oxide required for the depletion region with its immobile charge to exist.
4. The voltage drop across the thin oxide due to unavoidable charge trapped in the thin oxide.
5. The voltage drop across the thin oxide due to implanted charge at the surface of the silicon. The amount of implanted charge is adjusted in order to realize the desired threshold voltage.

These factors will each be described in turn.

The first factor affecting the threshold is the built in *Fermi potential* due to the different materials and doping concentrations used for the gate material and the substrate material. If one refers these potentials to that of *intrinsic* silicon, we have (Tsividis, 1987)

$$\phi_{F\text{-Gate}} = \frac{kT}{q}\ln\left(\frac{N_D}{n_i}\right) \tag{3.148}$$

for a polysilicon gate with doping concentration N_D, and

$$\phi_{F\text{-Sub}} = \frac{kT}{q}\ln\left(\frac{n_i}{N_A}\right) \tag{3.149}$$

for a p substrate with doping concentration N_A. The work function difference is then given by

$$\begin{aligned}\phi_{MS} &= \phi_{F\text{-Gate}} - \phi_{F\text{-Sub}} \\ &= \frac{kT}{q}\ln\left(\frac{N_D N_A}{n_i^2}\right)\end{aligned} \tag{3.150}$$

The next factor in determining the transistor threshold voltage is the voltage drop from the channel to the substrate, which is required for the channel to exist. The question of exactly when the channel exists does not have a precise answer. Rather, the channel is *defined to exist when the concentration of electron carriers in the channel is equal to the concentration of holes in the substrate*. At this gate voltage, the channel is said to be *inverted*. As the gate voltage changes from a low value to the value at which the channel becomes inverted, the voltage drop in the silicon also changes, as does the depletion region between the channel and the bulk. After the channel becomes inverted, any additional increase in gate voltage is closely equal to the increase in voltage drop across the thin oxide. It has little effect on the voltage drop in the silicon or the depletion region between the channel and the substrate.

The electron concentration in the channel is equal to the hole concentration in the substrate when the voltage drop from the channel to the substrate is equal to two times the difference between the Fermi potential of the substrate and intrinsic silicon, ϕ_F, where

$$\phi_F = -\frac{kT}{q}\ln\left(\frac{N_A}{n_i}\right) \tag{3.151}$$

This equation is a factor in a number of equations used in modeling MOS transistors. For typical processes, it can usually be approximated by $|\phi_F| \cong 0.35$ V, for typical doping levels at room temperature.

The third factor in determining the threshold voltage is due to the immobile negative charge in the depletion region that is left behind after the p mobile carriers are repelled. This effect gives rise to a voltage drop across the thin oxide of $-Q_B/C_{ox}$, where

$$Q_B = -qN_A x_d \tag{3.152}$$

with x_d being the width of the depletion region. Since

$$x_d = \sqrt{\frac{2K_s\varepsilon_0|2\phi_F|}{qN_A}} \tag{3.153}$$

we have

$$Q_B = -\sqrt{2qN_A K_s \varepsilon_0 |2\phi_F|} \tag{3.154}$$

The fourth factor in determining V_{tn} is due to the unavoidable charge trapped in the thin oxide. Typical values for the effective ion density of this charge, N_{ox}, might be 2×10^{14} to 10^{15} ions/m^3. These ions are almost always positive. This effect gives rise to a voltage drop across the thin oxide, V_{ox}, given by

$$V_{ox} = \frac{-Q_{ox}}{C_{ox}} = \frac{-qN_{ox}}{C_{ox}} \tag{3.155}$$

Thus, the *native transistor threshold voltage* (i.e., without an ion implantation to adjust it to its desired value) is given by

$$V_{t\text{-native}} = \phi_{MS} - 2\phi_F - \frac{Q_B}{C_{ox}} - \frac{Q_{ox}}{C_{ox}} \tag{3.156}$$

A typical native threshold value might be around –0.1 V for n-channel transistors. This value is the threshold voltage that would occur naturally if one does not include a special ion implant used to adjust the threshold voltage. *It should be noted that transistors having native transistor threshold voltages are becoming more important in circuit design where they might be used as pass transistors in transmission gates and/or as source-follower buffers.*

The fifth factor is a charge implanted in the silicon under the gate to change the threshold voltage from that given by (3.156) to the desired value, which might be 0.7 V for an n-channel transistor.

In the case of the source-to-substrate voltage being increased, the effective threshold voltage is increased by what is known as the body effect. The reason for the body effect is that as the source-bulk voltage, V_{SB}, becomes larger, the depletion region between the channel and the substrate becomes wider, and therefore more immobile negative charge becomes *uncovered.* This increase in charge changes the third factor in determining the transistor threshold voltage. Specifically, instead of using (3.154) to determine Q_B, one should now use

$$Q_B = -\sqrt{2qN_AK_s\varepsilon_0(V_{SB} + |2\phi_F|)} \tag{3.157}$$

If the threshold voltage when $V_{SB} = 0$ is denoted $V_{tn\text{-}0}$, then using (3.156) and (3.157), it is simple to show that

$$\begin{aligned} V_{tn} &= V_{tn\text{-}0} + \Delta V_{tn} \\ &= V_{tn\text{-}0} + \frac{\sqrt{2qN_AK_s\varepsilon_0}}{C_{ox}}\left(\sqrt{V_{SB} + |2\phi_F|} - \sqrt{|2\phi_F|}\right) \\ &= V_{tn\text{-}0} + \gamma\left(\sqrt{V_{SB} + |2\phi_F|} - \sqrt{|2\phi_F|}\right) \end{aligned} \tag{3.158}$$

where

$$\gamma = \frac{\sqrt{2qN_AK_s\varepsilon_0}}{C_{ox}} \tag{3.159}$$

The factor γ is often called the *body-effect constant.*

MOS TRIODE RELATIONSHIP

The current flow in an MOS transistor is due to *drift current* rather than *diffusion current.* This type of current flow is the same mechanism that determines the current in a resistor. The current density, J, is proportional to the electrical field, E, where the constant of proportionality, σ, is called the electrical conductivity. Thus,

$$J = \sigma E \tag{3.160}$$

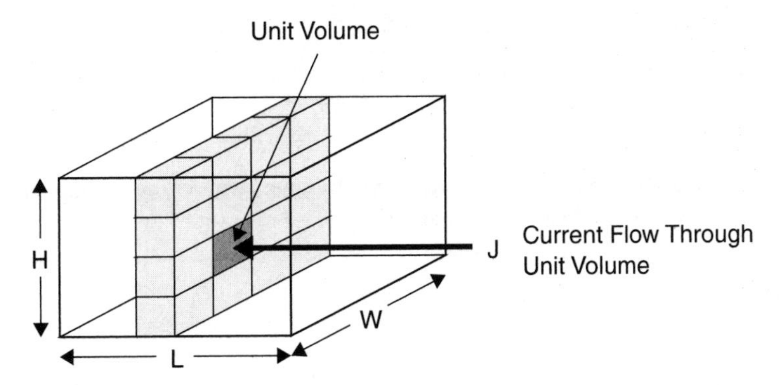

Figure 3.39 Current flowing through a unit volume.

This constant for an n-type material is given by

$$\sigma = qn\mu_n \tag{3.161}$$

where n is the concentration per unit volume of negative carriers and μ_n is the mobility of electrons. Thus, the current density is given by

$$J = qn\mu_n E \tag{3.162}$$

Next, consider the current flow through the volume shown in Fig. 3.39 where the volume has a height H and a width W. The current is flowing perpendicular to the plane $H \times W$ down the length of the volume, L. The current, I, everywhere along the length of the volume is given by

$$I = JWH \tag{3.163}$$

The voltage drop along the length of the volume in the direction of L for a distance dx is denoted dV and is given by

$$dV = E(x)dx \tag{3.164}$$

Combining (3.162), (3.163), and (3.164), we obtain

$$q\mu_n WH\,n(x)\,dV = Idx \tag{3.165}$$

where the carrier density $n(x)$ is now assumed to change along the length L, and is therefore a function of x.

As an aside, it is interesting to look at the case of a resistor where $n(x)$ is usually constant. A resistor of length L would therefore have a current given by

$$I = \frac{q\mu_n WH}{L}\Delta V \tag{3.166}$$

Thus, the resistance is given by

$$R = \frac{L}{q\mu_n WH} \tag{3.167}$$

Often this resistance is presented in a relative manner where the length and width are removed (since they can be design parameters) but the height remains included. In this case, the resulting expression is commonly referred to as the resistance per square and designated as $R/\square$ where

$$R/\square = q\mu_n H \tag{3.168}$$

The total resistance is then given by

$$R_{total} = R/\square \frac{L}{W} \tag{3.169}$$

This equation is important when calculating the resistance of interconnects used in integrated circuits.

In the case of an MOS transistor, the charge density is not constant down the channel. If instead of the carrier density per unit volume, one expresses $n(x)$ as a function of charge density per square area from the top looking down, we have

$$Q_n(x) = qHn(x) \tag{3.170}$$

Substituting (3.170) into (3.171) results in

$$\mu_n W Q_n(x)\, dV = I dx \tag{3.171}$$

Equation (3.171) applies to drift current through any structure having varying charge density in the direction of the current flow. It can also be applied to an MOS transistor in the triode region to derive its I–V relationship. Note that in this derivation, it is assumed the source voltage is the same as the substrate voltage.

Since the transistor is in the triode region, we have $V_{DG} < -V_{tn}$. This requirement is equivalent to $V_{DS} < V_{GS} - V_{tn} = V_{eff}$. It is assumed that the effective channel length is L. Assuming the voltage in the channel at distance x from the source is given by $V_{ch}(x)$, from Fig. 3.40, we have

$$Q_n(x) = C_{ox}[V_{GS} - V_{ch}(x) - V_{tn}] \tag{3.172}$$

Substituting (3.172) into (3.171) results in

$$\mu_n W C_{ox}[V_{GS} - V_{ch}(x) - V_{tn}] dV_{ch} = I_D dx \tag{3.173}$$

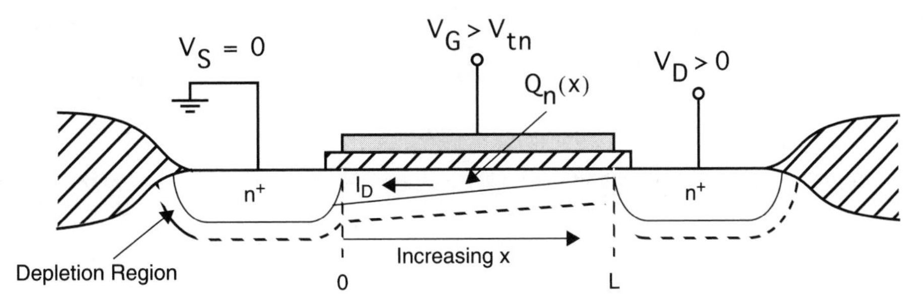

Figure 3.40 The transistor definitions used in developing its I–V relationship.

Integrating both sides of (3.173), and noting that the total voltage along the channel of length L is V_{DS}, we obtain

$$\int_0^{V_{DS}} \mu_n W C_{ox}[V_{GS} - V_{ch}(x) - V_{tn}]dV_{ch} = \int_0^L I_D dx \tag{3.174}$$

which results in

$$\mu_n W C_{ox}\left[(V_{GS} - V_{tn})V_{DS} - \frac{V_{DS}^2}{2}\right] = I_D L \tag{3.175}$$

Thus, solving for I_D results in the well-known triode relationship for an MOS transistor:

$$I_D = \mu_n \frac{W}{L} C_{ox}\left[(V_{GS} - V_{tn})V_{DS} - \frac{V_{DS}^2}{2}\right] \tag{3.176}$$

Small-Signal MOS Modeling in the Active Region

The most commonly used small-signal model for an MOS transistor operating in the active region is shown in Fig. 3.41. We will first consider the d.c. parameters where all the capacitors are ignored (i.e., replaced by open circuits). This leads to the low-frequency small-signal model shown in Fig. 3.42. The voltage-controlled current source, $g_m v_{gs}$, is the most important component of the model, with the transistor transconductance g_m defined as

$$g_m = \frac{\partial I_D}{\partial V_{GS}} \tag{3.177}$$

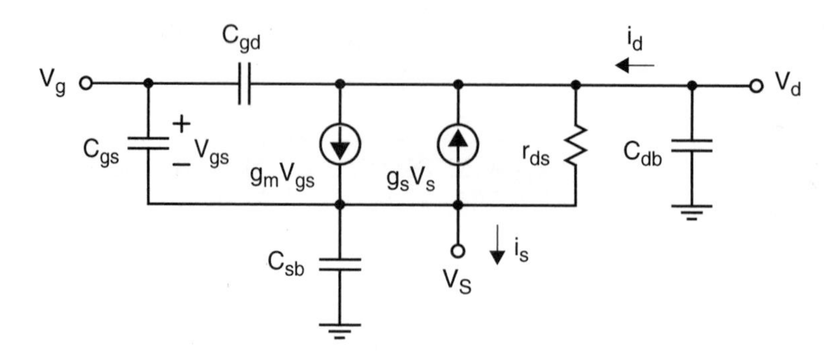

Figure 3.41 The small-signal model for an MOS transistor in the active region.

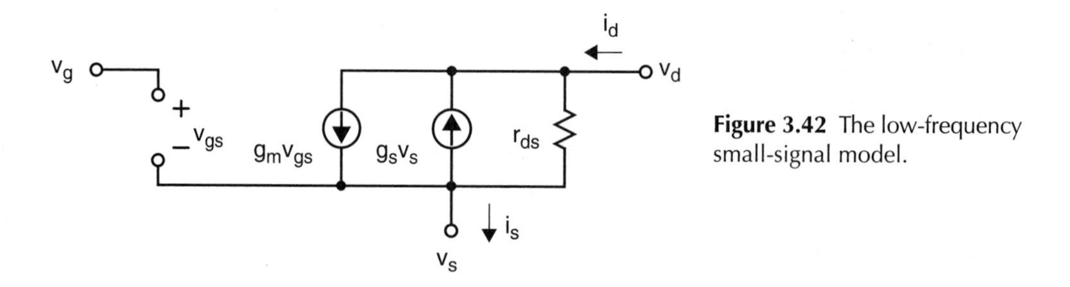

Figure 3.42 The low-frequency small-signal model.

In the active region, we make use of (3.71), which is repeated here for convenience,

$$I_D = \frac{\mu_n C_{ox}}{2}\left(\frac{W}{L}\right)(V_{GS} - V_{tn})^2 \tag{3.178}$$

Applying the derivative shown in (3.177) results in

$$g_m = \frac{\partial I_D}{\partial V_{GS}} = \mu_n C_{ox} \frac{W}{L}(V_{GS} - V_{tn}) \tag{3.179}$$

or equivalently,

$$g_m = \mu_n C_{ox} \frac{W}{L} V_{eff} \tag{3.180}$$

where the effective gate-source voltage, V_{eff}, is defined to be $V_{eff} \equiv V_{GS} - V_{tn}$. Thus, we see that the transconductance of an MOS transistor is directly proportional to V_{eff}.

Sometimes it is desirable to express g_m in terms of I_D rather than V_{GS}. From (3.178), we have

$$V_{GS} = V_{tn} + \sqrt{\frac{2I_D}{\mu_n C_{ox}(W/L)}} \tag{3.181}$$

The second term in (3.181) is the *effective gate-source voltage*, V_{eff}, where

$$V_{eff} = V_{GS} - V_{tn} = \sqrt{\frac{2I_D}{\mu_n C_{ox}(W/L)}} \tag{3.182}$$

Substituting (3.182) in (3.180) results in an alternate expression for g_m:

$$g_m = \sqrt{2\mu_n C_{ox}\frac{W}{L}I_D} \tag{3.183}$$

Thus, the transistor transconductance is proportional to $\sqrt{I_D}$ for an MOS transistor as contrasted to being proportional to I_C for a BJT.

A third alternative formula for g_m is found by rearranging (3.183) and then using (3.182) to obtain

$$g_m = \frac{2I_D}{V_{eff}} \tag{3.184}$$

Note that this third expression is independent of $\mu_n C_{ox}$ and W/L and relates the transconductance to the ratio of drain current to effective gate-source voltage. This simple relationship can be quite useful during an initial circuit design.

The second voltage-controlled current source in Fig. 3.42, shown as $g_s v_s$, models the body effect on the small-signal drain current, i_d. When the source is connected to signal ground, or does not have its voltage change appreciably, then this current source can be ignored. When the body effect cannot be ignored, we have

$$g_s = \frac{\partial I_D}{\partial V_{SB}} = \frac{\partial I_D}{\partial V_{tn}}\frac{\partial V_{tn}}{\partial V_{SB}} \tag{3.185}$$

From (3.178) we have

$$\begin{aligned}\frac{\partial I_D}{\partial V_{tn}} &= -\mu_n C_{ox}\frac{W}{L}(V_{GS} - V_{tn}) \\ &= -g_m\end{aligned} \tag{3.186}$$

Using (3.158), which gives V_{tn} as

$$V_{tn} = V_{tn0} + \gamma\left(\sqrt{V_{SB} + |2\phi_F|} - \sqrt{|2\phi_F|}\right) \tag{3.187}$$

we have

$$\frac{\partial V_{tn}}{\partial V_{SB}} = \frac{\gamma}{2\sqrt{V_{SB} + |2\phi_F|}} \tag{3.188}$$

The negative sign of (3.188) is eliminated by having the current $g_s v_s$ subtract from the major component of the drain current, $g_m v_{gs}$ as shown in Fig. 3.42. Thus, using (3.186) and (3.188), we have

$$g_s = \frac{\gamma g_m}{2\sqrt{V_{SB} + |2\phi_F|}} \tag{3.189}$$

Note that although g_s is nonzero for $V_{SB} = 0$, in the case where the source is connected to the bulk, ΔV_{SB} would be zero and so the effect of g_s does not need to be taken into account. However, if the source happens to be biased at the same potential as the bulk but is not directly connected to it, then the effect of g_s should be taken into account.

The resistor, r_{ds}, shown in Fig. 3.42 accounts for the finite output impedance (i.e., it models the *channel-length modulation* and its effect on the drain current due to changes in V_{DS}). Using (3.75), repeated here for convenience,

$$I_D = \frac{\mu_n C_{ox}}{2} \frac{W}{L} (V_{GS} - V_{tn})^2 [1 + \lambda(V_{DS} - V_{eff})] \tag{3.190}$$

we have

$$\frac{1}{r_{ds}} = g_{ds} = \frac{\partial I_D}{\partial V_{DS}} = \lambda\left(\frac{\mu_n C_{ox}}{2}\right)\left(\frac{W}{L}\right)(V_{GS} - V_{tn})^2 = \lambda I_{D\text{-sat}} \cong \lambda I_D \tag{3.191}$$

for λ small. Thus,

$$r_{ds} \cong \frac{1}{\lambda I_D} \tag{3.192}$$

where

$$\lambda = \frac{k}{2L\sqrt{V_{DS} - V_{eff} + \Phi_0}} \tag{3.193}$$

and

$$k = \sqrt{\frac{2K_s\varepsilon_0}{qN_A}} \tag{3.194}$$

but should be empirically adjusted to take into account second-order effects.

Example 3.10

Derive the low-frequency model parameters for an n-channel transistor having a substrate doping concentration of NA = $1.4 \times 10^{23}/\text{m}^3$ with $\mu_n C_{ox}$ = 188 μA/V², $|2\phi_F|$ = 0.7 V, W = 6 μm, L = 0.6 μm, V_{GS} = 1.2 V, V_{tn} = 0.8 V, $V_{DS} = V_{eff}$. Assume $\Upsilon = 0.62\ \sqrt{V}$ and V_{SB} = 0.5 V. What is the new value of r_{ds} if the drain-source voltage is increased by 0.5 V?

Solution: Since these parameters are the same as in Example 3.7, we have

$$g_m = \frac{2I_D}{V_{eff}} = \frac{2 \times 150\ \mu\text{A}}{0.4\ \text{V}} = 0.75\ \text{mA/V}$$

and from (3.189), we have

$$g_s = \frac{0.5 \times 0.75 \times 10^{-3}}{2\sqrt{0.5 + 0.7}} = 0.171\ \text{mA/V}$$

Note that this source-bulk transconductance value is about one-fourth that of the gate-source transconductance.

For r_{ds}, we make use of (3.192) to find

$$r_{ds} = \frac{1}{80.8 \times 10^{-3} \times 150 \times 10^{-6}} = 82.5\ \text{k}\Omega$$

At this point, it is interesting to calculate the gain $g_m r_{ds} = 61.9$, which is the largest voltage gain this single transistor can achieve for these operating bias conditions. As we shall see, this gain is much smaller than its corresponding value for a bipolar transistor.

If V_{DS} is increased by 0.5 to 0.9 V, the new value for λ is

$$\lambda = \frac{96.6 \times 10^{-9}}{2 \times 0.6 \times 10^{-6}\sqrt{0.5 + 0.99}} = 0.0659\ \text{V}^{-1}$$

resulting in a new value of r_{ds} given by

$$r_{ds} = \frac{1}{\lambda I_{D\text{-}2}} = \frac{1}{65.9 \times 10^{-3} \times 156 \times 10^{-6}} = 97.3\ \text{k}\Omega$$

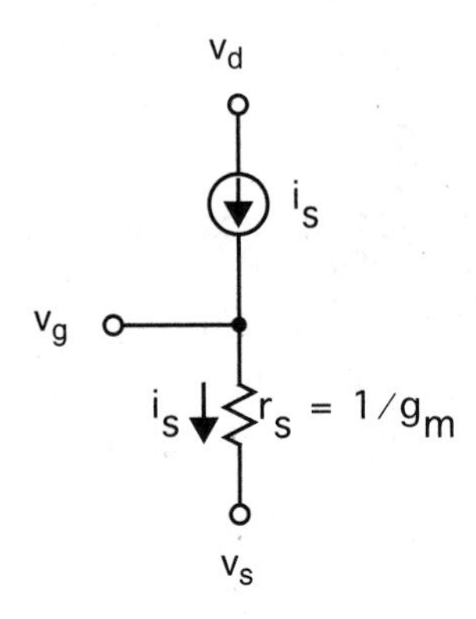

Figure 3.43 The small-signal low-frequency T-model that can be used when the body effect is not important.

An alternate low-frequency model often used when the body effect is not important is shown in Fig. 3.43. This *T-model* can often result in simpler equations and is often used by experienced designers for a quick analysis. At first glance, it might appear that it allows for finite gate current, but a quick check confirms that the drain current must always equal the source current, and, therefore, the gate current must always be zero. For this reason, when using the T-model, one assumes from the beginning that the gate current is zero.

Example 3.11

Find the T-model parameter, r_s, for the transistor of Example 3.10.

Solution: The value of r_s is simply the inverse of g_m, resulting in

$$r_s = \frac{1}{g_m} = \frac{1}{0.75 \times 10^{-3}} = 1.33\ \text{k}\Omega$$

This T-model can also account for transistor finite output impedance by placing r_{ds} between the drain and source nodes.

Most of the capacitors in the small-signal model are easily related to the physical transistor. Shown in Fig. 3.44 is a cross-sectional view of an MOS transistor where the parasitic capacitances have been drawn in where they exist. The largest capacitor is C_{gs}. This is primarily due to the change in channel charge as a result of a change in V_{GS}. It can be shown that C_{gs} is approximately given by (Tsividis, 1987)

$$C_{gs} \cong \frac{2}{3}WLC_{ox} \tag{3.195}$$

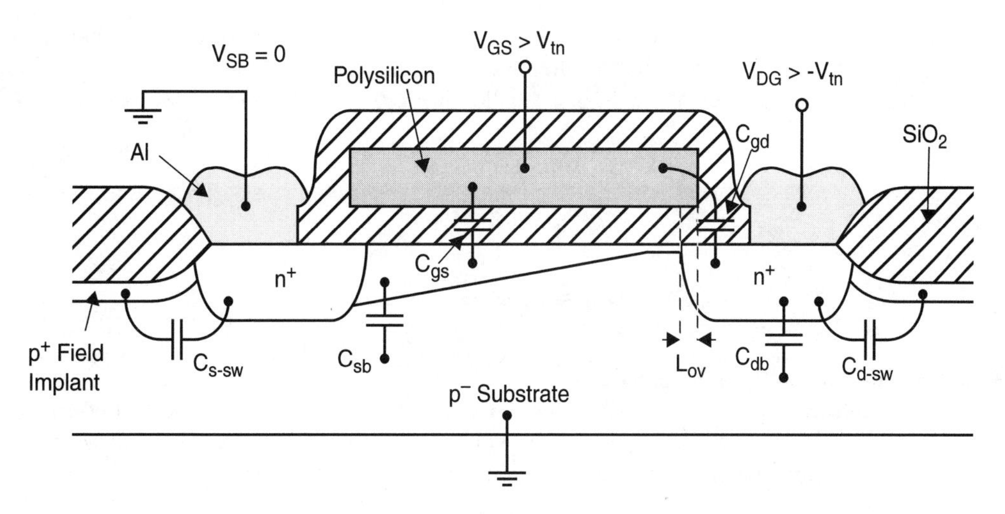

Figure 3.44 A cross section of an n-channel MOS transistor showing the small-signal capacitances.

When accuracy is important, an additional term should be added to (3.195) to take into account the overlap between the gate and the source junction and *fringing* capacitance. This additional component is given by

$$C_{ov} = WL_{ov}C_{ox} \tag{3.196}$$

where L_{ov} is the overlap distance and is usually empirically derived. Thus

$$C_{gs} = WC_{ox}\left(\frac{2}{3}L + L_{ov}\right) \tag{3.197}$$

when higher accuracy is needed.

The next largest capacitor is C_{sb}, the capacitor between the source and the substrate. This capacitor is due to the depletion capacitance of the reverse-biased source junction and includes the channel-to-bulk capacitance (assuming the transistor is *on*). Its size is given by

$$C_{sb} = (A_s + A_{ch})C_{js} \tag{3.198}$$

where A_s is the area of the source junction, A_{ch} is the area of the channel (i.e., WL), and C_{js} is the depletion capacitance of the source junction and is given by

$$C_{js} = \frac{C_{j\text{-}0}}{\sqrt{1 + V_{SB}/\Phi_0}} \tag{3.199}$$

Note that the total area of the effective source includes the original area of the junction (when no channel is present) together with the effective area of the channel.

The depletion capacitance of the drain is smaller due to the smaller area. Here, we have

$$C_{db} = A_d C_{jd} \tag{3.200}$$

where

$$C_{jd} = \frac{C_{j\text{-}0}}{\sqrt{1 + V_{DB}/\Phi_0}} \tag{3.201}$$

and A_D is the area of the drain junction.

The capacitance C_{gd}, which is sometimes called the *Miller capacitor*, is important when the transistor is being used in circuits with large voltage gain. It is primarily due to the overlap between the gate and the drain and due to fringing capacitance. Its value is given by

$$C_{gd} = C_{ox} W L_{ov} \tag{3.202}$$

where again L_{ov} is usually empirically derived.

There are two other capacitors that often are important in integrated circuits. These are the source and drain side-wall capacitances. These capacitances can be large because of some highly doped regions that have not been mentioned yet. These p^+ regions are under the thick field oxide. They are called field implants. The major reason for these regions is to make sure there is no leakage current between transistors. Because they are highly doped, and are adjacent to the highly doped source and drain junctions, they can result in appreciable additional capacitances that must be added to C_{sb} and C_{db}. They are particularly important in modern technologies as dimensions shrink. For the source-to-bulk capacitor C_{sb}, an additional term,

$$C_{s\text{-}sw} = P_s C_{j\text{-}sw} \tag{3.203}$$

must be added where P_s is the length of the perimeter of the source junction excluding the side adjacent to the channel,

$$C_{j\text{-}sw} = \frac{C_{j\text{-}sw\text{-}0}}{\sqrt{1 + V_{SB}/\Phi_{SW}}} \tag{3.204}$$

but it should be noted that $C_{j\text{-}sw\text{-}0}$, the side-wall capacitance per unit length at 0 V bias voltage, can be quite large because the field implants are heavily doped.

The situation is similar for the drain-to-bulk capacitance where an additional term, $C_{d\text{-}sw}$, must be added to C_{db}, where

$$C_{d\text{-}sw} = P_d C_{j\text{-}sw} \tag{3.205}$$

Again, P_d is the drain perimeter excluding the portion adjacent to the gate.

Example 3.12

An n-channel transistor is modeled as having the following capacitance parameters: $C_j = 5.4 \times 10^{-4}\ \text{pF}/(\mu\text{m})^2$, $C_{jsw} = 1.5 \times 10^{-4}\ \text{pF}/\mu\text{m}$, $C_{ox} = 3.4 \times 10^{-3}\ \text{pF}/(\mu\text{m})^2$, $C_{gs\text{-}ov} = C_{gd\text{-}ov} = 3.7 \times 10^{-4}\ \text{pF}/\mu\text{m}$ (overlap capacitance per unit width). Find the capacitances C_{gs}, C_{gd}, C_{db}, and C_{sb} for a transistor having $W = 100\ \mu\text{m}$ and $L = 0.6\ \mu\text{m}$. Assume the source and drain junctions extend $1.2\ \mu\text{m}$ beyond the gate resulting in source and drain areas of $A_s = A_d = 120\ \mu\text{m}^2$ and the perimeter of each junction being $P_s = P_d = 102.4\ \mu\text{m}$.

Solution: We calculate the various capacitances as follows:

$$C_{gs} = \left(\frac{2}{3}\right)WLC_{ox} + C_{gs\text{-}ov} \times W = 0.17\ \text{pF}$$

$$C_{gd} = C_{gd\text{-}ov} \times W = 0.04\ \text{pF}$$

$$C_{sb} = C_j(A_s + WL) + (C_{j\text{-}sw} \times P_s) = 0.11\ \text{pF}$$

$$C_{db} = (C_j \times A_d) + (C_{j\text{-}sw} \times P_d) = 0.08\ \text{pF}$$

Note here that the source-bulk and drain-bulk capacitance are very significant compared to the gate-source capacitance. *Thus, for high-speed circuits, it is important to keep the areas and perimeters of drain and source junctions as small as possible.*

Small-Signal MOS Modeling in the Triode and Cutoff Regions

The low-frequency small-signal model of an MOS transistor in the triode region (which is sometimes referred to as the linear region) is a resistor. Using (3.69), the large-signal equation for I_D in the triode region

$$I_D = \mu_n C_{ox}\frac{W}{L}\left[(V_{GS} - V_{tn})V_{DS} - \frac{V_{DS}^2}{2}\right] \tag{3.206}$$

results in

$$\frac{1}{r_{ds}} = g_{ds} = \frac{dI_D}{dV_{DS}} = \mu_n C_{ox}\frac{W}{L}(V_{GS} - V_{tn} - V_{DS}) \tag{3.207}$$

where r_{ds} is the small-signal drain-source resistance (and g_{ds} is the admittance). For the common case of V_{DS} near zero, we have

$$g_{ds} = \frac{1}{r_{ds}} = \mu_n C_{ox} \frac{W}{L}(V_{GS} - V_{tn}) = \mu_n C_{ox} \frac{W}{L} V_{eff} \tag{3.208}$$

which is similar to the I_D versus V_{DS} relationship given earlier [i.e., (3.64)].

Example 3.13

For the transistor of Example 3.10, find the triode model parameters when V_{DS} is near zero.

Solution: From (3.208), we have

$$g_{ds} = 188 \times 10^{-6} \times \left(\frac{6}{0.6}\right) \times 0.4 = 0.752\ \text{mA/V}$$

Note that this conductance value is the same as the transconductance of the transistor, g_m, in the active region. The resistance r_{ds} is simply $1/g_{ds}$ resulting in $r_{ds} = 1.33\ \text{k}\Omega$.

The accurate modeling of the high-frequency operation of a transistor in the triode region is nontrivial (even assuming the use of a computer simulation). A moderately accurate model is shown in Fig. 3.45 where the gate-to-channel capacitance and the channel-to-substrate capacitance are modeled as distributed elements. However, the I–V relationships of the distributed RC elements are highly nonlinear. This is because the junction capacitances of the source and drain are nonlinear depletion capacitances as is the channel-to-substrate capacitance. Also, if

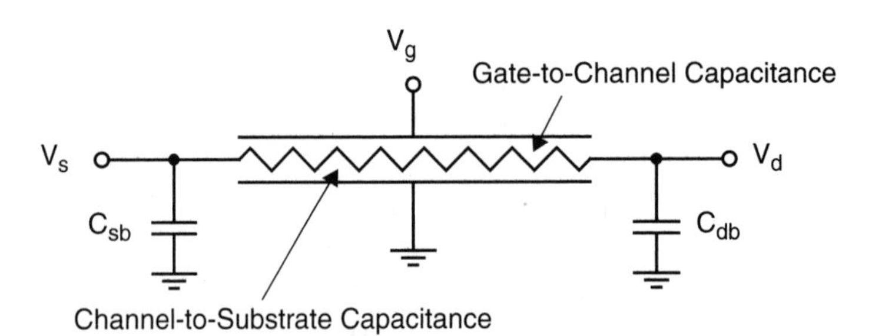

Figure 3.45 A distributed RC model for a transistor in the saturation region.

V_{DS} is not small, then the channel resistance per unit length should increase as one moves closer to the drain. This model is too complicated for use in hand analysis.

A simplified model often used for small V_{DS} is shown in Fig. 3.46 where the resistance r_{ds} is given by (3.208). Here, the gate-to-channel capacitance has been evenly divided between the source and drain nodes,

$$C_{gs} = C_{gd} = \frac{A_{ch}C_{ox}}{2} = \frac{WLC_{ox}}{2} \tag{3.209}$$

Note that this equation ignores the gate-to-junction overlap capacitances, as given by (3.196), which should be taken into account when accuracy is very important. The channel-to-substrate capacitance has also been divided into two and shared between the source and drain junctions. Each of these capacitors should be added to the junction-to-substrate capacitance and the junction side-wall capacitance at the appropriate node. Thus, we have

$$C_{sb\text{-}0} = C_{j\text{-}0}\left(A_s + \frac{A_{ch}}{2}\right) + C_{j\text{-}sw\text{-}0}P_s \tag{3.210}$$

and

$$C_{db\text{-}0} = C_{j\text{-}0}\left(A_d + \frac{A_{ch}}{2}\right) + C_{j\text{-}sw\text{-}0}P_d \tag{3.211}$$

Also,

$$C_{sb} = \frac{C_{sb\text{-}0}}{\sqrt{1 + \frac{V_{sb}}{\Phi_0}}} \tag{3.212}$$

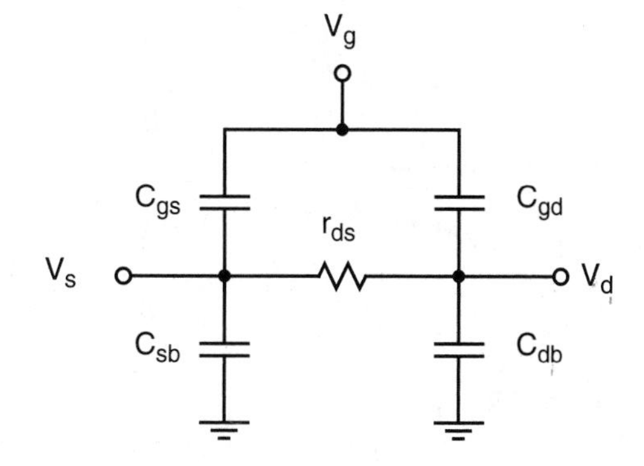

Figure 3.46 A simplified triode-region model valid for small V_{DS}.

and

$$C_{db} = \frac{C_{db\text{-}0}}{\sqrt{1 + \frac{V_{db}}{\Phi_0}}} \tag{3.213}$$

It might be noted that C_{sb} is often comparable in size to C_{gs} due to its larger area and the sidewall capacitance.

When the transistor turns *off*, the model changes considerably. A reasonable model is shown in Fig. 3.47. Perhaps the biggest difference is that r_{ds} is now infinite. Another major difference is that C_{gs} and C_{gd} are now much smaller. Since the channel has disappeared, they are now due only to overlap and fringing capacitance. Thus we have

$$C_{gs} = C_{gd} = WL_{ov}C_{ox} \tag{3.214}$$

However, the reduction of C_{gs} and C_{gd} does not mean that the total gate capacitance is necessarily smaller. We now have a *new* capacitor, C_{gb}, which is the gate-to-substrate capacitance. This capacitor is highly nonlinear and dependent on the gate voltage. If the gate voltage has been very negative for some time and the gate is *accumulated*, then we have

$$C_{gb} = A_{ch}C_{ox} = WLC_{ox} \tag{3.215}$$

If the gate-to-source voltage is around 0 V, then C_{gb} is C_{ox} in series with the channel-to-bulk depletion capacitance and is considerably smaller, especially when the substrate is lightly doped. Another case when C_{gb} is small is just after a transistor has been turned *off* before the channel has had time to *accumulate*. Because of the complicated nature of correctly modeling C_{gb} when the transistor is turned *off*, equation (3.215) is usually used for hand analysis as a worst-case estimate.

The capacitors C_{sb} and C_{db} are also smaller when the channel is not present. We now have

$$C_{sb\text{-}0} = A_sC_{j\text{-}0} \tag{3.216}$$

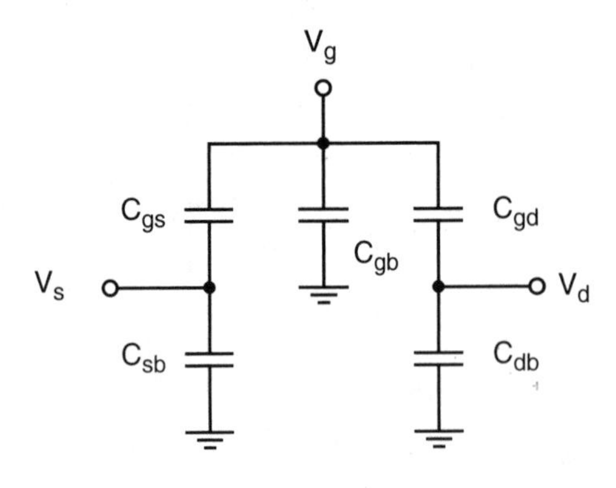

Figure 3.47 A small-signal model for a MOSFET that is turned off.

and

$$C_{db\text{-}0} = A_d C_{j\text{-}0} \tag{3.217}$$

BIPOLAR TRANSISTOR EXPONENTIAL RELATIONSHIP

The various components of the base, collector, and emitter currents are shown in Fig. 3.32. Shown in Fig. 3.48 are plots of the minority carrier concentrations in the emitter, base, and collector regions. The current flow of these minority carriers is due to *diffusion*. By calculating the gradient of the minority-carrier concentrations near the base-emitter junction, in a manner similar to what was done for diodes, it is possible to derive a relationship between the electron current and the hole current of Fig. 3.48.

The concentration of holes in the emitter, at the edge of the base-emitter depletion region, is denoted $p_e(0)$. This concentration decreases exponentially the farther one gets from the junction, in a manner similar to that described for diodes, until far from the junction the concentration, $p_{e\text{-}0}$, is given by

$$p_{e\text{-}0} = \frac{n_i^2}{N_D} \tag{3.218}$$

where N_D is the doping density of the n^+ emitter. At a distance x from the edge of the emitter-base depletion region, we have

$$\Delta p_e(x) \cong \Delta p_e(0) e^{-\frac{x}{L_p}} \tag{3.219}$$

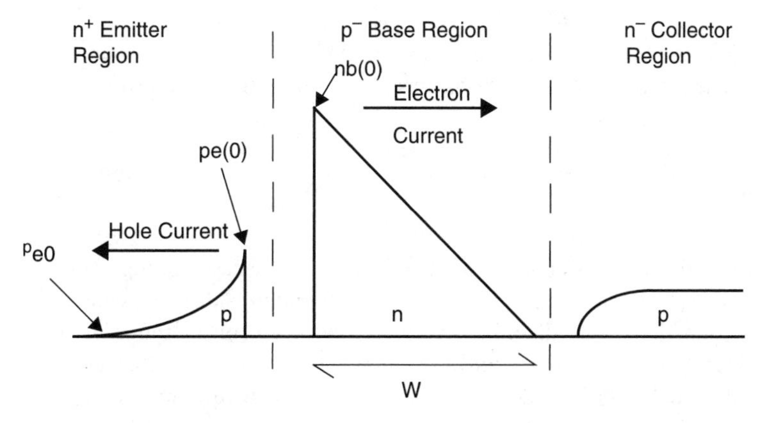

Figure 3.48 The concentrations of minority carriers in the emitter, base, and collector.

where

$$\Delta p_e(0) = p_{e\text{-}0}e^{V_{BE}/V_T}$$
$$= \frac{n_i^2}{N_D}e^{V_{BE}/V_T} \tag{3.220}$$

where V_{BE} is the forward-bias voltage of the base-emitter junction.

At the edge of the base-emitter depletion region, the gradient of the hole concentration in the emitter is easily found using (3.219) to be

$$\left.\frac{dp_e(x)}{dx}\right|_{x=0} = \frac{\Delta p_e(0)}{L_p} \tag{3.221}$$

which using (3.220) can be rewritten as

$$\left.\frac{dp_e(x)}{dx}\right|_{x=0} = \frac{n_i^2}{L_pN_D}e^{V_{BE}/V_T} \tag{3.222}$$

The hole current is now easily found using the diffusion equation

$$I_{pe} = A_EqD_p\left.\frac{dp_e(x)}{dx}\right|_{x=0} \tag{3.223}$$

where A_E is the effective area of the emitter. Recalling that the minority hole current in the emitter, I_{pe}, is closely equal to the base current, I_B, we obtain after combining (3.222) and (3.223)

$$I_B = \frac{A_EqD_pn_i^2}{L_pN_D}e^{V_{BE}/V_T} \tag{3.224}$$

The situation on the base side of the base-emitter junction is somewhat different. The concentration of the minority carriers, in this case electrons that diffused from the emitter, is given by a similar equation, that is

$$n_b(0) = \frac{n_i^2}{N_A}e^{V_{BE}/V_T} \tag{3.225}$$

However, the gradient of this concentration at the edge of the base-emitter depletion region is calculated differently. This difference in gradient concentration is due to the close proximity of the collector-base junction where the minority carrier (electron) concentration, $n_b(W)$, must be zero. This zero concentration at the collector-base junction is due to the fact that any electrons diffusing to the edge of the collector-base depletion region immediately *drift* across the junction to the collector as stated previously. If the base *width* is much less than the *diffusion length* of electrons in the base, L_n, then almost no electrons will recombine with base majority carriers (holes) before they diffuse to the collector-base junction. Given this fact, the decrease in electron or minority concentration from the base-emitter junction to the collector-base junction is a linear relationship decreasing from $n_b(0)$ at the emitter junction to 0 at the collector junction in distance W. This assumption ignores any *recombination* of electrons in the base as they travel to the collector, which is reasonable in modern transistors having very narrow bases. Thus, throughout the base region, the gradient of the minority carrier concentration is closely given by

$$\begin{aligned}\frac{dn_b(x)}{dx} &= -\frac{n_b(0)}{W} \\ &= -\frac{n_i^2}{WN_A}e^{V_{BE}/V_T}\end{aligned} \tag{3.226}$$

Combining (3.226) with the diffusion equation we obtain

$$\begin{aligned}I_{nb} &= -A_E q D_n \frac{dn_b(0)}{dx} \\ &= \frac{A_E q D_n n_i^2}{WN_A}e^{V_{BE}/V_T}\end{aligned} \tag{3.227}$$

Remembering that I_{nb} is closely equal to the collector current I_C, we have

$$I_C \cong I_{CS}e^{V_{BE}/V_T} \tag{3.228}$$

where

$$I_{CS} = \frac{A_E q D_n n_i^2}{WN_A} \tag{3.229}$$

The ratio of the collector current to the base current, commonly called the transistor common-emitter current gain and denoted β, is easily found using (3.228), (3.229), and (3.224). We have

$$\beta = \frac{I_C}{I_B} = \frac{D_n N_D L_p}{D_p N_A W} \cong 2.5\frac{N_D L_p}{N_A W} \tag{3.230}$$

which is a constant independent of voltage and current. Noting that $N_D >> N_A$, $L_p > W$, and $D_n \cong 2.5\ D_p$ results in $\beta >> 1$. A typical value might be between 50 and 200. The derivation of β just presented ignores many second-order effects that make β somewhat current and voltage dependent and are beyond the scope of this book. The interested reader is referred to Roulston, (1990) for more details. Irrespective, equation (3.230) does reflect the approximate relationships among β, doping levels, and base width. For example, equation (3.230) explains why heavily doped emitters are important to achieve large current gain.

BASE CHARGE STORAGE OF AN ACTIVE BIPOLAR-JUNCTION TRANSISTOR

With reference to Fig. 3.48, there is *minority carrier storage* in the base region, Q_b, given by

$$Q_b = A_E q \frac{n_b(0)W}{2} \tag{3.231}$$

Using (3.225) for $n_b(0)$, we have

$$Q_b = \frac{A_E q n_i^2 W}{2N_A} e^{V_{BE}/V_T} \tag{3.232}$$

This equation can be rewritten using (3.228) and (3.229) to obtain

$$Q_b = \frac{W^2}{2D_n} I_C = \tau_b I_C \tag{3.233}$$

where τ_b is called the base-transit time constant and is given approximately by

$$\tau_b = \frac{W^2}{2D_n} \tag{3.234}$$

ignoring second-order effects. Normally, the base-transit time constant is specified for a given technology. It will also take into account other charge-storage effects, not considered here, and is therefore often denoted τ_T, but since the base storage of electrons dominates the other effects, we have $\tau_T \cong \tau_b$ as approximately given by (3.234).

If the current in a BJT changes, the base charge storage must also change. This change can be modeled by a diffusion capacitance, C_d, between the base and the emitter terminals. Using (3.233), we have

$$C_d \cong \frac{dQ_b}{dV_{BE}} = \frac{d(\tau_b I_C)}{dV_{BE}} \tag{3.235}$$

Using (3.235) and $I_C = I_{CS} e^{V_{BE}/V_T}$ results in

$$C_d = \tau_b \frac{I_C}{V_T} \tag{3.236}$$

similar to the case of a diode.

SMALL-SIGNAL BIPOLAR MODELING

A good understanding of the small-signal model of BJTs is critical when designing analog BJT circuits. It is also becoming very important when designing digital *BJT* circuits as well, as these often operate with very small voltage swings where the model is valid. With the growing popularity of BiCMOS technology, where BJT transistors and MOS transistors can be realized in the same IC, this understanding is becoming even more important.

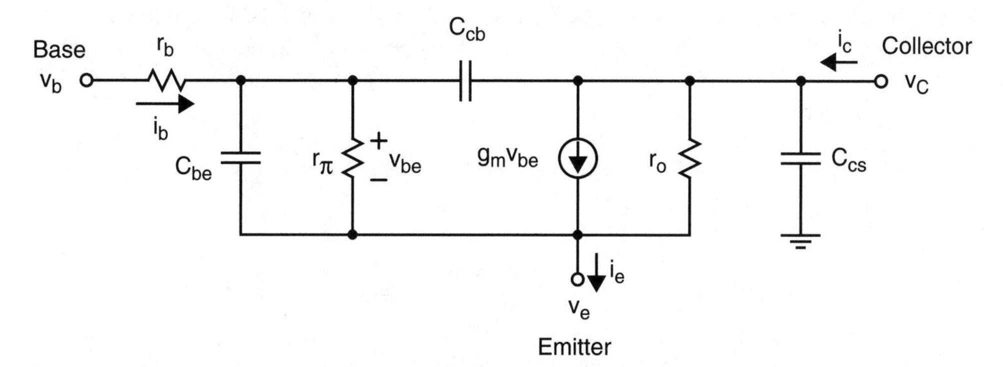

Figure 3.49 The small-signal model of a BJT.

The most commonly used small-signal model is the hybrid-π model. This is very similar to the small-signal model used for MOS transistors, as we shall see, except for the inclusion of a finite base-emitter impedance, r_π, and no emitter-to-bulk capacitance. The hybrid-π model was shown in Fig. 3.36 on page 117. We will first discuss the d.c. components, that is, the transconductance, g_m, and the resistors. Next the parasitic capacitances will be discussed.

The transistor transconductance, g_m, is perhaps the most important parameter of the small-signal model. The transconductance is the ratio of the small-signal collector current, i_c, to the small-signal base-emitter voltage, v_{be}. Thus, we have

$$g_m = \frac{i_c}{v_{be}} = \frac{dI_C}{dV_{BE}} \tag{3.237}$$

where the lower-case letters with lower-case subscripts designate small-signal voltages and currents and the upper-case letters designate large-signal voltages and currents. Since, from (3.101), we have

$$I_C = I_{CS} e^{V_{BE}/V_T} \tag{3.238}$$

then

$$g_m = \frac{dI_C}{dV_{BE}} = \frac{I_{CS}}{V_T} e^{V_{BE}/V_T} \tag{3.239}$$

Using (3.238), we obtain

$$g_m = \frac{I_C}{V_T} \tag{3.240}$$

where again

$$V_T = \frac{kT}{q} \cong 26 \text{ mV at } T = 300°\text{K} \tag{3.241}$$

Thus, the transconductance is proportional to the bias current of a BJT. In integrated-circuit design where it is important that the speeds remain temperature independent, then the bias currents are usually made proportional to absolute temperature to keep the transconductance constant with temperature.

The presence of the resistor r_π reflects the fact that there is a finite base current. We have

$$r_\pi = \frac{dV_{BE}}{dI_B} \tag{3.242}$$

Since from (3.104), we have

$$I_B = \frac{I_C}{\beta} = \frac{I_{CS}}{\beta} e^{V_{BE}/V_T} \tag{3.243}$$

we therefore have

$$\frac{1}{r_\pi} = \frac{dI_B}{dV_{BE}} = \frac{I_{CS}}{\beta V_T} e^{V_{BE}/V_T} \tag{3.244}$$

Using (3.243) again, we have

$$r_\pi = \frac{V_T}{I_B} \tag{3.245}$$

or equivalently

$$r_\pi = \beta \frac{V_T}{I_C} = \frac{\beta}{g_m} \tag{3.246}$$

Since

$$I_E = I_C + I_B \tag{3.247}$$

we also have

$$\begin{aligned} \frac{dI_E}{dV_{BE}} &= \frac{dI_C}{dV_{BE}} + \frac{dI_B}{dV_{BE}} \\ &= g_m + \frac{g_m}{\beta} \\ &= g_m\left(\frac{1+\beta}{\beta}\right) \\ &= \frac{g_m}{\alpha} \end{aligned} \tag{3.248}$$

Some alternative models, including the T-model (to be described shortly), use the emitter resistance, r_e, where

$$r_e = \frac{dV_{BE}}{dI_E} = \frac{\alpha}{g_m} \tag{3.249}$$

Continuing, we have

$$\frac{1}{r_o} = \frac{dI_C}{dV_{CE}} \tag{3.250}$$

The resistor r_o models the dependence of the collector current on the collector-emitter voltage. Repeating (3.103) here for convenience,

$$I_C = I_{CS} e^{V_{BE}/V_T}\left(1 + \frac{V_{CE}}{V_A}\right) \tag{3.251}$$

we have[13]

$$\frac{1}{r_o} = \frac{\partial I_C}{\partial V_{CE}} = \frac{I_{CS}}{V_A} e^{V_{BE}/V_T} \tag{3.252}$$

Thus,

$$r_o = \frac{V_A}{I_C} \tag{3.253}$$

which is inversely proportional to collector current. As an aside, note that $g_m r_o = V_A / V_T$, a constant. This constant, commonly called η and usually between 2000 and 8000 for an npn BJT, is an upper limit on the attainable voltage gain for a single-transistor BJT amplifier.

The resistor r_b models the resistance of the semiconductor material between the base contact and the effective base region, namely the moderately lightly doped base p material (see Fig. 3.29). This resistor, although small and typically 200 to 500 Ω, can be important in limiting the speed of high-frequency low-gain BJT circuits.

[13] The partial derivative symbols are used to indicate V_{BE} is considered to be constant. In fact, it would have been more correct to have used partial derivatives in deriving all of the preceding small-signal parameters.

Example 3.14

For $I_C = 1$ mA, $\beta = 100$, and $V_A = 100$ V, calculate g_m, r_π, r_e, r_o, and $g_m r_o$.

Solution: We have

$$g_m = \frac{I_C}{V_T} = \frac{10^{-3}\ \text{A}}{0.026\ \text{V}} = 38.5\ \text{mA/V} \tag{3.254}$$

$$r_\pi = \frac{\beta}{g_m} = 2.6\ \text{k}\Omega \tag{3.255}$$

$$r_e = \frac{\alpha}{g_m} = \frac{100}{101}26 = 25.7\ \Omega \tag{3.256}$$

$$r_o = \frac{V_A}{I_C} = \frac{100}{10^{-3}} = 100\ \text{k}\Omega \tag{3.257}$$

and $g_m r_o$, the maximum possible gain that one can achieve with a single-transistor amplifier, is given by

$$g_m r_o = \frac{V_A}{V_T} = 3846 \tag{3.258}$$

It is interesting to compare this gain with that of 61.9, which was found for a single MOS transistor in Example 3.10. Note that this BJT maximum gain is independent of the bias current. For MOS transistors, it can be shown that the maximum gain decreases with larger bias currents in a square-root relationship. This is one of the reasons why it is possible to realize a single transistor BJT amplifier with a much larger gain than would result if an MOS transistor were used, especially at high current levels (and therefore at high frequencies).

The high-frequency operation of a BJT is limited by the capacitances of the small-signal model. One of these capacitances, C_{be}, we have already encountered in a previous section. Recapping, we have

$$C_{be} = C_j + C_d \tag{3.259}$$

where C_j is the depletion capacitance of the base-emitter junction. For a forward-biased junction, a rough approximation for C_j is

$$C_j \cong 2A_E C_{je\text{-}0} \tag{3.260}$$

The diffusion capacitance, C_d, is given in (3.109) as

$$C_d = \tau_b \frac{I_C}{V_T} = g_m \tau_b \tag{3.261}$$

The capacitor C_{cb} models the depletion capacitance of the collector-base junction. Since this is a graded junction, we can approximate C_{cb} by

$$C_{cb} = \frac{A_C C_{jc\text{-}0}}{1 + (V_{CB}/\Phi_{c\text{-}0})^{1/3}} \tag{3.262}$$

where A_C is the effective area of the collector-base interface.

Due to the lower doping levels in the base and especially in the collector (i.e., perhaps 5×10^{22} acceptors/m^3 and 10^{21} donors/m^3, respectively), Φ_{c0}, the built-in potential for the collector-base junction, will be less than that of the base-emitter junction (perhaps 0.75 V as opposed to 0.9 V). It should be noted, however, that typically the cross-sectional area of the collector-base junction, A_C, is much larger than the effective area of the base-emitter junction, A_E, as is easily seen from Fig. 3.29. This normally results in $A_C\, C_{jc\text{-}0}$ being larger than $A_E\, C_{je\text{-}0}$, the base-emitter junction capacitance at 0 V bias, despite the lower doping levels.

An even larger capacitor is the depletion capacitance of the collector-to-substrate junction, C_{CS}. Since this area is quite large, the depletion capacitance due to it, C_{CS}, will be much larger than either C_{cb}, or the depletion capacitance component of C_{be}, that is, C_j. Its value can be calculated using

$$C_{cs} = \frac{A_T C_{js\text{-}0}}{[1 + (V_{CS}/\Phi_{s\text{-}0})]^{1/2}} \tag{3.263}$$

where A_T is the effective transistor area and $C_{js\text{-}0}$ is the collector-to-substrate capacitance per unit area at 0 V bias voltage.

A common figure of merit for the *speed* of a BJT is the frequency where its current gain drops to unity, when its collector is connected to a small-signal ground. This frequency is denoted f_t and is called the transistor unity-gain frequency. This frequency is easily related to the transistor model parameters by analyzing the small-signal circuit of Fig. 3.50. The resistor r_b has been ignored because it has no effect on i_b since the circuit is being driven by a perfect current source. We have

$$V_{be} = i_b\left(r_\pi \,\Big\|\, \frac{1}{sC_{be}} \,\Big\|\, \frac{1}{sC_{cb}}\right) \tag{3.264}$$

and

$$i_c = g_m V_{be} \tag{3.265}$$

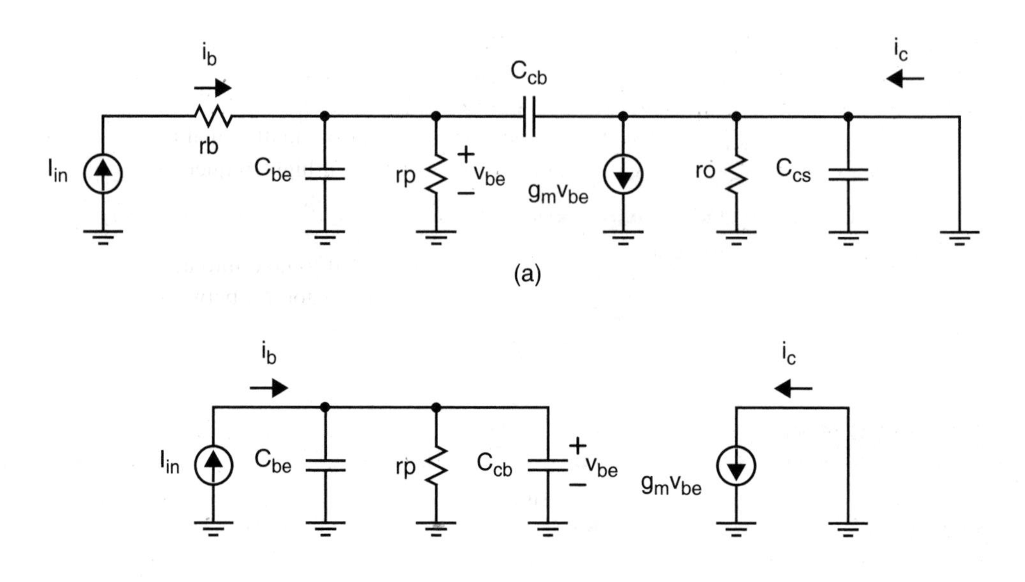

Figure 3.50 (a) A small-signal model that can be used to find f_t and (b) an equivalent simplified model.

Solving for i_c / i_b, gives

$$\frac{i_c}{i_b} = \frac{g_m r_\pi}{1 + s(C_{be} + C_{cb})r_\pi} \tag{3.266}$$

At low frequencies, the current gain is $g_m r_\pi$, which equals β [from (3.120)], as expected. At frequencies where the current gain is approximately unity, i_c / i_b is approximately given by

$$\left|\frac{i_c}{i_b}(\omega)\right| \cong \frac{g_m r_\pi}{\omega(C_{be} + C_{cb})r_\pi} = \frac{g_m}{\omega(C_{be} + C_{cb})} \tag{3.267}$$

Setting $|(i_c/i_b)(\omega_t)| = 1$ and solving for ω_t results in

$$\omega_t = \frac{g_m}{C_{be} + C_{cb}} \tag{3.268}$$

or

$$f_t = \frac{g_m}{2\pi(C_{be} + C_{cb})} \tag{3.269}$$

Often, either f_t, ω_t, or $\tau_t = 1 / \omega_t$ will be specified for a transistor at a particular bias current. These values indicate an upper limit on the maximum frequency at which the transistor can be effectively used.

The hybrid-π model just presented is only one of a number of small-signal models that can be used. An alternative T-model that is approximately correct at high frequencies is shown in Fig. 3.51. This model is accurate for $f_t / \beta << f << f_t$ in circuits where the gain is not too large. This T-model often results in a simplified analysis as compared to the use of the hybrid-π model and thus is useful for hand analysis. It is easily extended to take into account the finite output impedance of a BJT transistor by simply including a resistor, r_o, between the collector and emitter.

3.8 SPICE Simulations

The circuit diagrams, netlist files, and output data associated with SPICE simulations of selected examples are presented in this section. The device models, unless explicitly present in the netlist file, are assumed to be as given in Section 2.4, where SPICE-modeling parameters are discussed.

Simulation of Example 3.6

The circuit diagram of Example 3.6 is shown in Fig. 3.52. Node labels correspond to those used in the input netlist below. The reverse-biased diode is modeled with a junction potential of

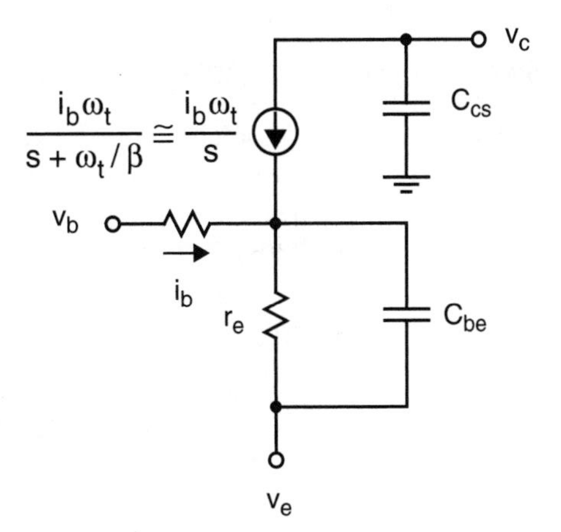

Figure 3.51 An alternative, high-frequency small-signal model for a BJT.

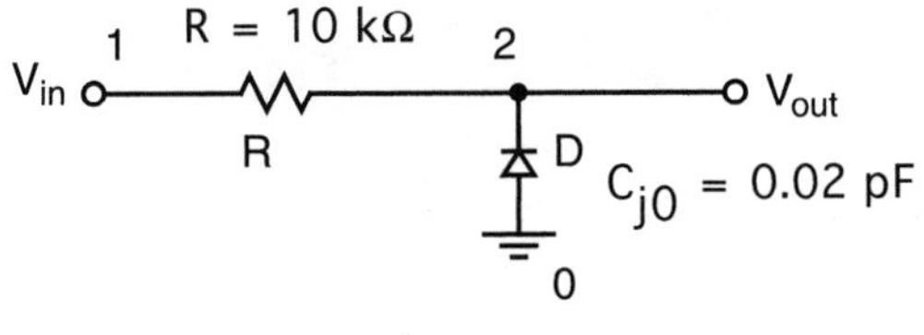

Figure 3.52 The reverse-biased diode of Example 3.6.

0.9 V and a total junction capacitance of 0.002 pF for a 0 V applied voltage. The remaining diode parameters assume their default values (refer to the *SPICE* manuals for default parameter values specific to the release of SPICE in use).

```
The transient response of charging and discharging a diode
*
R 1 2 10k
D 0 2 DMOD
*
VIN 1 0 PULSE (0 3.3 0p 10p 10p 0.99n 2.0n)
*
.MODEL DMOD D (CJ0=0.02e-12, PB=0.9)
*
.OPTION NOMOD POST INGOLD=2 NUMDGT=6 BRIEF
.TRAN 0.01n 2.0n
.PRINT TRAN V(1) V(2)
.END
```

The transient response of the simulation is plotted in Fig. 3.53. The respective rise and fall times are $t_{+70\%} = 0.16$ ns and $t_{-70\%} = 0.13$ ns, which compare favorably to the previous calculations, which yielded 0.17 and 0.13 ns, respectively.

Simulation of Example 3.7

The circuit required to obtain a solution to Example 3.7 is shown in Fig. 3.54, where the voltage source V_{DS} is varied. The SPICE netlist used for the simulation is shown below.

```
The n-channel enhancement MOSFET
*
M 2 1 0 0 nch W=6u L=0.6u

*
VGS 1 0 d.c. 1.2V
VDS 2 0 d.c. 0.4V
*
.LIB '../mod_06' typical
*
```

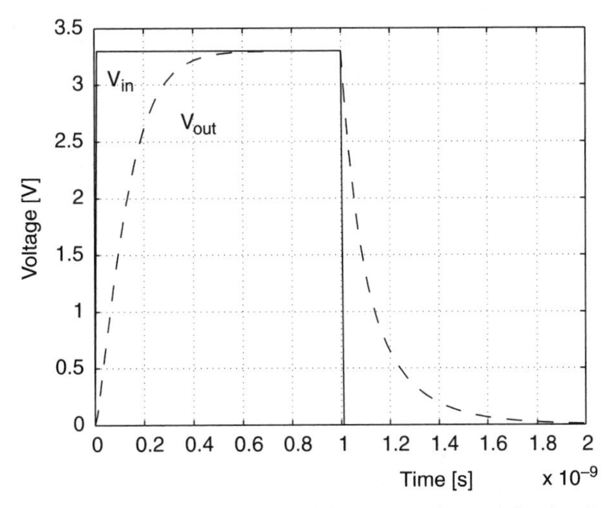

Figure 3.53 The transient response of the reverse-biased diode of Example 3.6.

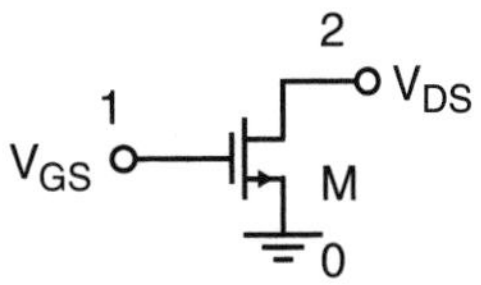

Figure 3.54 The MOSFET of Example 3.8.

```
.OPTION NOMOD POST INGOLD=2 NUMDGT=6 BRIEF
.DC VDS 0V 1.0V 0.001V
.PRINT d.c. I1(M)
.END
```

The result of the simulation is plotted in Fig. 3.55. For a drain-source voltage of 0.4 V, the drain current is 164 μA while a drain-source voltage of 0.9 V yields a drain current of 175 μA. The values obtained by hand calculation are 150 and 156 μA, respectively. The discrepancy can be attributed to two short-channel effects that are considered in the SPICE simulation, but ignored in the hand calculations. The threshold voltage V_{tn} is reduced by drain-induced barrier lowering. Also affecting the current–voltage characteristics, but to a lesser extent, is electron mobility reduction.

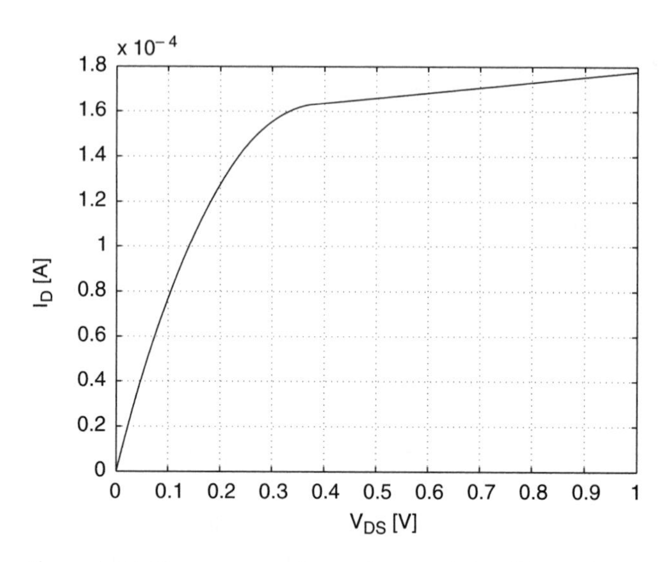

Figure 3.55 The current–voltage characteristic of Example 3.7.

3.9 Bibliography

R. Geiger, P. Allen, and N. Strader, *VLSI, Design Techniques for Analog and Digital Circuits*, McGraw-Hill, 1990.

P. Gray and R.G. Hodges, *Analog Integrated Circuits,* 3rd ed., John Wiley & Sons, 1993.

D. Hodges and H. Jackson, *Analysis and Design of Digital Integrated Circuits,* 2nd ed., McGraw-Hill, 1988.

G. Massobrio and P. Antognetti, *Semiconductor Device Modelling with SPICE,* 2nd ed., McGraw-Hill, 1993.

D. Roulston, *Semiconductor Devices,* McGraw-Hill, 1990.

S.M. Sze, *Physics of Semiconductor Devices,* Wiley Interscience, 1981.

Y. Tsividis, *Operation and Modelling of the MOS Transistor*, McGraw-Hill, 1987.

J. Uyemura, *Fundamentals of MOS Digital Integrated Circuits,* Addison-Wesley, 1988.

S. Wolf, *Silicon Processing for the VLSI Era—Volume 3: The Submicron MOSFET,* Lattice Press, 1995.

3.10 Problems

For the problems in this chapter and all future chapters, assume the following transistor parameters:

- npn bipolar transistors:

  $\beta = 100$

  $V_A = 80\text{ V}$

$\tau_b = 13\ \text{ps}$
$\tau_s = 4\ \text{ns}$
$r_b = 330\ \Omega$

- n-channel MOS transistors:

  $\mu_n C_{ox} = 190\ \mu A/V^2$
  $V_{tn} = 0.7\ V$
  $\gamma = 0.6\ V^{1/2}$
  $r_{ds}\ (\Omega) = 5000L\ (\mu m)/I_D\ (mA)$ in active region
  $C_j = 5 \times 10^{-4}\ pF/(\mu m)^2$
  $C_{j\text{-}sw} = 2.0 \times 10^{-4}\ pF/\mu m$
  $C_{ox} = 3.4 \times 10^{-3}\ pF/(\mu m)^2$
  $C_{gs(overlap)} = C_{gd(overlap)} = 2.0 \times 10^{-4}\ pF/\mu m$

- p-channel MOS transistors:

  $\mu_p C_{ox} = 50\ \mu A/V^2$
  $V_{tp} = -0.8\ V$
  $\gamma = 0.7\ V^{1/2}$
  $r_{ds}\ (\Omega) = 6000L\ (\mu m)/I_D\ (mA)$ in active region
  $C_j = 6 \times 10^{-4}\ pF/(\mu m)^2$
  $C_{j\text{-}sw} = 2.5 \times 10^{-4}\ pF/\mu m$
  $C_{ox} = 3.4 \times 10^{-3}\ pF/(\mu m)^2$
  $C_{gs(overlap)} = C_{gd(overlap)} = 2.0 \times 10^{-4}\ pF/\mu m$

3.1 Estimate the hole and electron concentration in silicon doped with arsenic at a concentration of 10^{25} atoms/m^3 at a temperature 22°C above room temperature. Is the resulting material n type or p type?

3.2 For the pn junction of Example 3.2, does the built-in potential, Φ_0, increase or decrease when the temperature is increased 11°C above room temperature?

3.3 Calculate the amount of charge per μm^2 in each of the n and p regions of the pn junction of Example 3.2 for a 5-V reverse-bias voltage. How much charge on each side would be present in a $10 \times 10\ \mu m$ diode?

3.4 A silicon diode has $\tau_t = 12$ ps and $C_{je\text{-}0} = 15$ fF. It is biased by a 43-kΩ resistor connected between the *cathode* of the diode and the input signal, as shown in Fig. P3.4. Initially, the input is 5 V, and then at time 0 it changes to 0 V. Estimate the time it

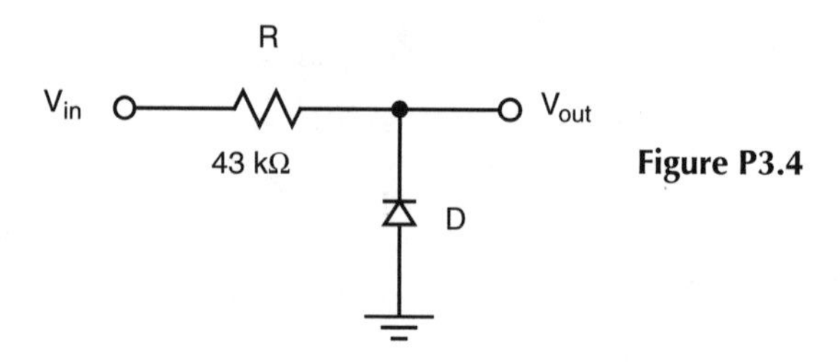

Figure P3.4

takes for the output voltage to change from 5 to 1.5 V (i.e., the $\Delta t_{-70\%}$ time). Repeat for an input voltage change of 0 to 5 V and an output voltage change of 0 to 3.5 V.

3.5 Compare your answers for Problem 3.4 to those obtained using a SPICE simulation.

3.6 A silicon diode has $\tau_t = 12$ ps and $C_{je\text{-}0} = 15$ fF. It is biased by a 43-kΩ resistor connected between the anode of the diode and the input signal, as shown in Fig. P3.6. Initially, the input is 5 V, and then at time 0 it changes to 0 V. Estimate the time it takes for the diode to stop conducting. Estimate the time for a Schottky diode having the same $C_{je\text{-}0}$. (*Note:* a Schottky diode has no capacitance due to minority charge storage when forward biased—all of the capacitance in the small-signal model is just due to depletion capacitance.)

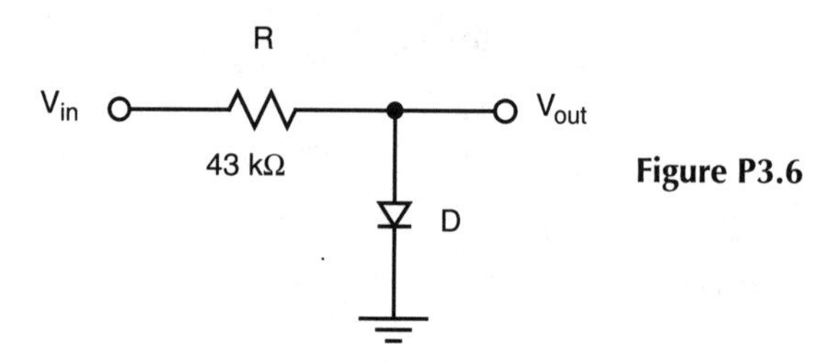

Figure P3.6

3.7 Compare your answers for Problem 3.6 to those obtained using SPICE.

3.8 Verify that when $V_{DS} = V_{eff}$ is used in the triode equation for an MOS transistor, the current equals that of the active region equation given in (3.71).

3.9 Find I_D for an n-channel transistor having doping concentrations of $N_D = 10^{25}\,/\,m^3$ and $N_A = 10^{22}\,/\,m^3$ with $W = 50\ \mu m$, $L = 1.5\ \mu m$, $V_{GS} = 1.1$ V, and $V_{DS} = V_{eff}$. Assuming λ remains constant, estimate the new value of I_D if V_{DS} is increased by 0.3 V.

3.10 An MOS transistor in the active region is measured to have a drain current of 20 μA when $V_{DS} = V_{eff}$. When V_{DS} is increased by 0.5 V, I_D increases to 23 μA. Estimate the output impedance, r_{ds}, and the output impedance constant, λ.

3.11 Assume all dimensions are scaled by S, but the voltages and doping levels are only scaled $\sqrt{S}$. Fill in all other entries of Table 3.1.

3.12 Derive the low-frequency model parameters for an n-channel transistor having doping concentrations of $N_D = 10^{25} / m^3$ and $N_A = 10^{22} / m^3$ with W = μm, L = 1.2 μm, $V_{GS} = 1.1$ V, $V_{DS} = V_{eff}$. Assume $V_{SB} = 1.0$ V.

3.13 Find the capacitances C_{gs}, C_{gd}, C_{db}, C_{sb}, for a transistor having W = 50 μm and L = 1.2 μm. Assume the source and drain junctions extend 4 μm beyond the gate resulting in source and drain areas of $A_s = A_d = 200\ (\mu m)^2$ and the perimeter of each being $P_s = P_d = 58$ μm.

3.14 Consider the circuit shown in Fig. P3.14 where V_{in} is a d.c. signal of 1 V. Taking into account only the channel charge storage, what is the final value of V_{out} when the transistor is turned off assuming half the channel charge goes to C_L.

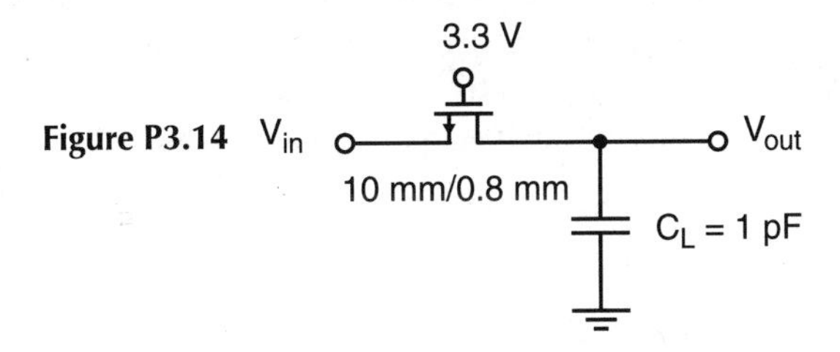

Figure P3.14

3.15 For the same circuit as in Problem 3.14, the input voltage has a step voltage change at time 0 from 1 to 1.2 V (the gate voltage remains at 5 V). Find its 99% settling time (i.e., the time it takes to settle to within 1% of the total voltage change). You may ignore the body effect and all capacitances except C_L. Also assume $V_{tn} = V_{tn\text{-}0}$. Repeat the question for V_{in} changing from 3 to 3.1 V.

3.16 Repeat Problem 3.14, but now take into account the body effect on V_{tn}.

3.17 Assume a CMOS inverter has $V_{in} = V_{out} = V_{TH}$ and that both transistors are in the active region. Find V_{TH} for an n-channel transistor having a size of (W/L = 5 μm/0.8 μm) and a p-channel transistor of size (W/L = 7.5 μm/0.8 μm). Repeat this problem for a p-channel transistor of size (W/L = 5 μm/0.8 μm) and a p-channel transistor of size (W/L = 10 μm/0.8 μm).

3.18 Using the small-signal model of Fig. 3.42, and assuming a CMOS inverter has an n-channel transistor of size (W/L = 5 μm/0.8 μm) and a p-channel transistor of size (W/L = 7.5 μm/0.8 μm), find an equation for the gain of the inverter when $V_{in} = V_{TH}$. Note that using the small-signal model of Fig. 3.42 implicitly assumes both transistors are in the active region.

3.19 For an npn transistor having $I_C = 0.1$ mA, calculate g_m, r_p, r_e, r_o, and $g_m r_o$.

3.20 A bipolar junction transistor has the following SPICE parameters (note that the SPICE name for the parameter is included in parentheses).

I_S(IS)=2×10^{-18}A B_F(BF)=100 B_R(BR)=1 V_A(VA)=50 V

τ_F(TF)=12×10^{-12} s τ_R(TR)=4×10^{-9} s C_{je0}(CJE)=15×10^{-15} F Φe(VJE)=0.9 V

m_e(MJE)=0.27 C_{jc0}(CJC)=18×10^{-15} F Φc(VJC)=0.7 V m_c(MJC)=0.37

C_{js0}(CJS)=40×10^{-15} F Φs(VJS)=0.64 V m_s(MJS)=0.29 R_e(RE)=30 Ω

R_b(RB)=500 Ω R_c(RC)=90 Ω

Initially, the circuit shown in Fig. P3.20 has a 0 V input. At time 0, its input changes to 5 V. Estimate the time it takes its output voltage to saturate, using the concepts of average capacitance and first-order transient solutions for each node. The individual time constants of each node can be added to arrive at an overall time constant for the approximate first-order transient response of the circuit. Next, assume the input changes from 5 to 0 V at time 0. How long does it take the output voltage to change to 3.5 V?

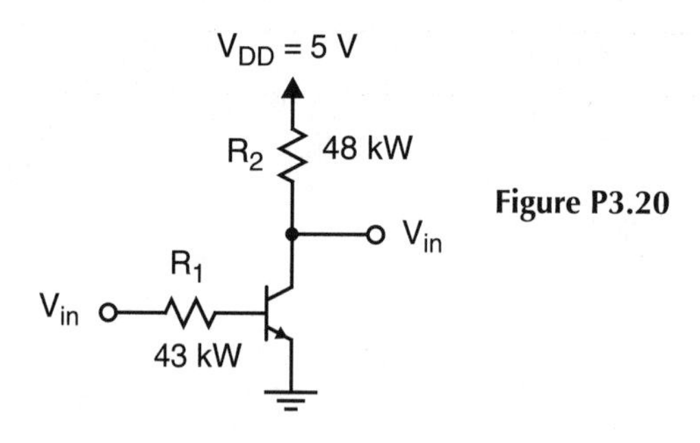

Figure P3.20

3.21 Compare your answers to Problem 3.20 to those obtained using SPICE.

3.22 Verify that for $I_{BR} >> I_B$ and I_C / β, equation (3.113) simplifies to equation (3.112).

3.11 Device Model Summary

As a useful aid, all of the equations for the large-signal and small-signal modeling of diodes, MOS transistors, and bipolar transistors, along with values for the various constants, are listed in the next few pages.

Constants

$q = 1.602 \times 10^{-19}$ C	$k = 1.38 \times 10^{-23}$ J K^{-1}
$n_i = 1.1 \times 10^{16}$ carriers/m^3 at $T = 300°K$	$\varepsilon_0 = 8.854 \times 10^{-12}$ F/m
$K_{ox} \cong 3.9$	$K_s \cong 11.8$
$\mu_n = 0.05\ m^2/V \cdot s$	$\mu_p = 0.02\ m^2/V \cdot s$

Diode Equations

Reverse-Biased Diode (Abrupt Junction)

$C_j = \frac{C_{j\text{-}0}}{\sqrt{1 + (V_R/\Phi_0)}}$	$Q = 2C_{j0}\Phi_0\sqrt{1 + \frac{V_R}{\Phi_0}}$
$C_{j0} = \sqrt{\frac{qK_s\varepsilon_0}{2\Phi_0}\frac{N_D N_A}{N_A + N_D}}$	$C_{j0} = \sqrt{\frac{qK_s\varepsilon_0 N_D}{2\Phi_0}}$ if $N_A >> N_D$
$\Phi_0 = \frac{kT}{q}\ln\left(\frac{N_A N_D}{n_i^2}\right)$	

Forward-Biased Diode

$I_D = I_S e^{V_D/V_T}$	$I_S = A_D q n_i^2\left(\frac{D_n}{L_n N_A} + \frac{D_p}{L_p N_D}\right)$
$V_T = \frac{kT}{q} \cong 26$ mV at 300°K	

Small-Signal Model of Forward-Biased Diode

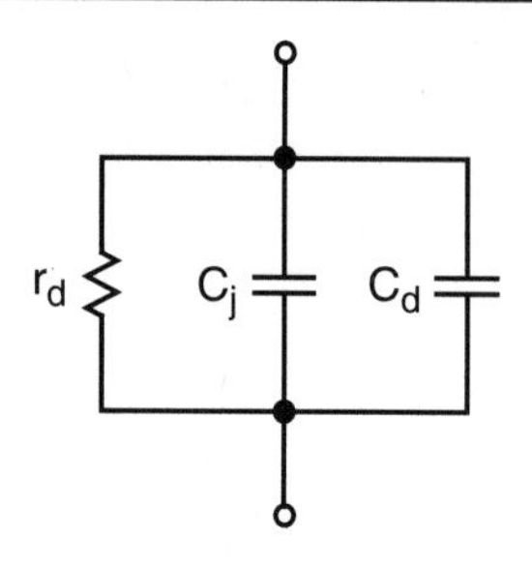

$r_d = \frac{V_T}{I_D}$	$C_T = C_d + C_j$
$C_d = \tau_T \frac{I_D}{V_T}$	$C_j \cong 2C_{j0}$
$\tau_T = \frac{L_n^2}{D_n}$	

MOS Transistor Equations

The following equations are for n-channel devices; for p-channel devices, put negative signs in front of all voltages. These equations do not account for short-channel effects (i.e., $L < 2L_{min}$).

Triode Region ($V_{GS} > V_{tn}$, $V_{DS} \le V_{eff}$)

$I_D = \mu_n C_{ox}\left(\frac{W}{L}\right)\left[(V_{GS} - V_{tn})V_{DS} - \frac{V_{DS}^2}{2}\right]$	
$V_{eff} = V_{GS} - V_{tn}$	$V_{tn} = V_{tn\text{-}0} + \gamma(\sqrt{V_{SB} + 2\phi_F} - \sqrt{2\phi_F})$
$\phi_F = \frac{kT}{q}\ln\left(\frac{N_A}{n_i}\right)$	$\gamma = \frac{\sqrt{2qK_{si}\varepsilon_0 N_A}}{C_{ox}}$
$C_{ox} = \frac{K_{ox}\varepsilon_0}{t_{ox}}$	

Small-Signal Model in Triode Region (for $V_{DS} \ll V_{eff}$)

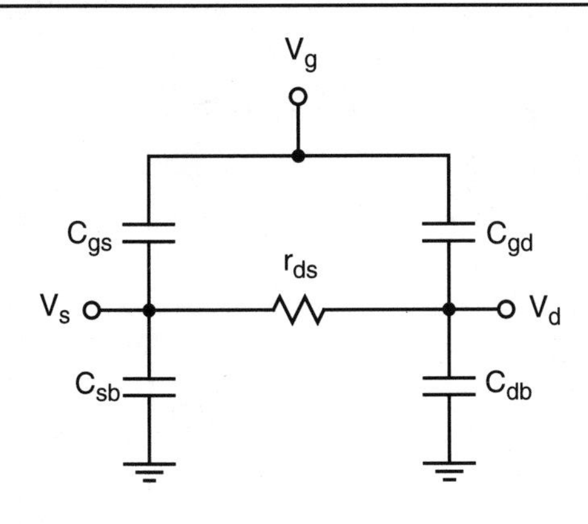

$r_{ds} = \frac{1}{\mu_n C_{ox}(W/L)V_{eff}}$	
$C_{gd} = C_{gs} \cong \frac{1}{2}WLC_{ox} + WL_{ov}C_{ox}$	$C_{sb} = C_{db} = \frac{C_{j\text{-}0}(A_s + WL/2)}{\sqrt{1 + (V_{sb}/\Phi_0)}}$

Active (or Pinch-Off) Region ($V_{GS} > V_{tn}$, $V_{DS} \ge V_{eff}$)

$I_D = \frac{\mu_n C_{ox}}{2}\frac{W}{L}(V_{GS} - V_{tn})^2[1 + \lambda(V_{DS} - V_{eff})]$	
$\lambda \propto \frac{1}{L\sqrt{V_{DS} - V_{eff} + \Phi_0}}$	$V_{tn} = V_{tn\text{-}0} + \gamma(\sqrt{V_{SB} + 2\phi_F} - \sqrt{2\phi_F})$
$V_{eff} = V_{GS} - V_{tn} = \sqrt{\frac{2I_D}{\mu_n C_{ox} W/L}}$	

Small-Signal Model (Active Region)

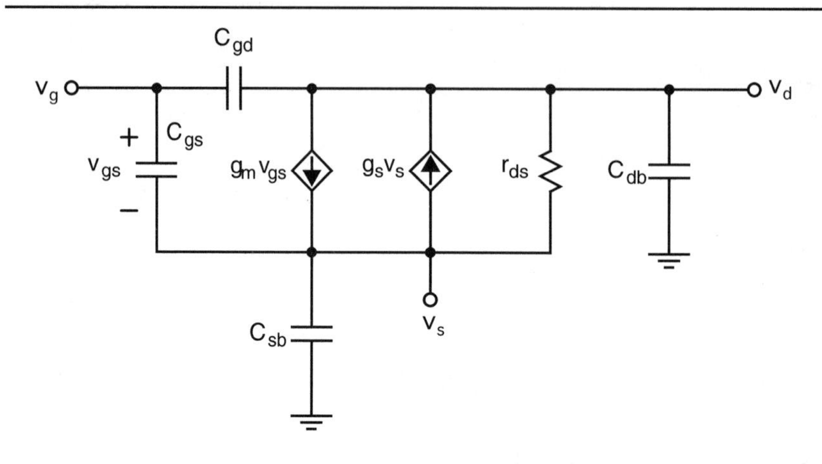

$g_m = \mu_n C_{ox}\left(\frac{W}{L}\right)V_{eff}$	$g_m = \sqrt{2\mu_n C_{ox}(W/L)I_D}$
$g_m = \frac{2I_D}{V_{eff}}$	$g_s = \frac{\gamma g_m}{2\sqrt{V_{SB} + \lvert 2\phi_F \rvert}}$
$r_{ds} = \frac{1}{\lambda I_D}$	$g_s \cong 0.2 g_m$
$\lambda = \frac{k_{rds}}{2L\sqrt{V_{DS} - V_{eff} + \Phi_0}}$	$k_{rds} = \sqrt{\frac{2K_s \varepsilon_0}{qN_A}}$
$C_{gs} = \frac{2}{3}WLC_{ox} + WL_{ov}C_{ox}$	$C_{gd} = WL_{ov}C_{ox}$
$C_{sb} = (A_s + WL)C_{js} + P_s C_{j\text{-}sw}$	$C_{js} = \frac{C_{j\text{-}0}}{\sqrt{1 + V_{SB}/\Phi_0}}$
$C_{db} = A_d C_{jd} + P_d C_{j\text{-}sw}$	$C_{jd} = \frac{C_{j\text{-}0}}{\sqrt{1 + V_{DB}/\Phi_0}}$

Typical Values for a 0.5-μm Process

$V_{tn} = 0.7\ \text{V}$	$V_{tp} = -0.8\ \text{V}$
$\mu_n C_{ox} = 170\ \mu\text{A/V}^2$	$\mu_p C_{ox} = 60\ \mu\text{A/V}^2$
$C_{ox} = 3.4 \times 10^{-3}\ \text{pF}/(\mu\text{m})^2$	$C_j = 5.4 \times 10^{-4}\ \text{pF}/(\mu\text{m})^2$
$C_{j\text{-}sw} = 1.5 \times 10^{-4}\ \text{pF}/\mu\text{m}$	$C_{gs(overlap)} = 1.7 \times 10^{-4}\ \text{pF}/\mu\text{m}$
$\phi_F = 0.42\ \text{V}$	$\Phi_0 = 0.99\ \text{V}$
$\gamma = 0.6\ \text{V}^{1/2}$	$t_{ox} = 0.01\ \mu\text{m}$
$N_B = 1.3 \times 10^{23}\ \text{impurities/m}^3$	

Bipolar-Junction Transistors

Active Transistor

$I_C = I_{CS} e^{V_{BE}/V_T}$	$V_T = \frac{kT}{q} \cong 26\ \text{mV at } 300°\text{K}$
For more accuracy, $I_C = I_{CS} e^{V_{BE}/V_T}\left(1 + \frac{V_{CE}}{V_A}\right)$	
$I_{CS} = \frac{A_E q D_n n_i^2}{W N_A}$	$I_B = \frac{I_C}{\beta}$
$I_E = \left(1 + \frac{1}{\beta}\right) I_C = \frac{1}{\alpha} I_C = (\beta + 1) I_B$	$\beta = \frac{I_C}{I_B} = \frac{D_n N_D L_p}{D_p N_A W} \cong 2.5 \frac{N_D L_p}{N_A W}$
$\alpha = \frac{\beta}{1 + \beta}$	

Small-Signal Model of an Active BJT

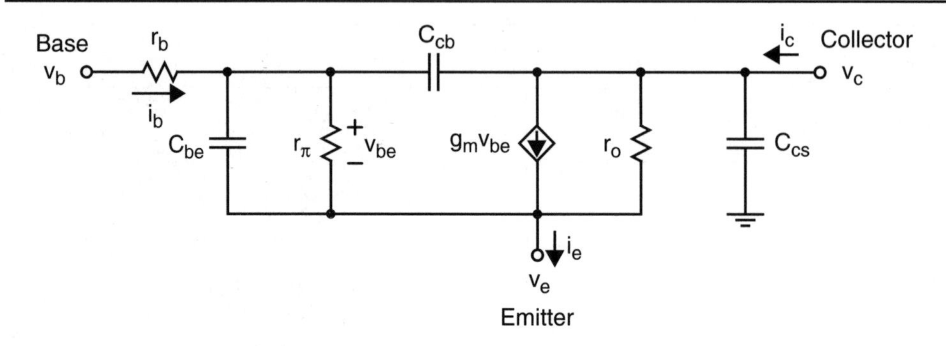

$g_m = \frac{I_C}{V_T}$	$r_\pi = \frac{V_T}{I_B} = \frac{\beta}{g_m}$
$r_e = \frac{\alpha}{g_m}$	$r_o = \frac{V_A}{I_C}$
$g_m r_o = \frac{V_A}{V_T}$	$C_d = \tau_b \frac{I_C}{V_T} = g_m \tau_b$
$C_{be} = C_j + C_d$	$C_{cs} = \frac{A_T C_{js\text{-}0}}{[1 + (V_{CS}/\Phi_{s\text{-}0})]^{1/2}}$
$C_{cb} = \frac{A_C C_{jc\text{-}0}}{[1 + (V_{CB}/\Phi_{c\text{-}0})]^{1/3}}$	

CHAPTER 4

Traditional MOS Design

MOS logic gates are by far the most popular choice for realizing digital integrated circuits at the present time. This is primarily due to the large number of MOS transistors that can be fabricated in a single integrated circuit. MOS logic gates were introduced in Chapter 1, where the transistors were modeled very simply as switches. In this chapter, we shall analyze the MOS gate circuits in greater detail, now that a more complete understanding of MOS transistors and the MOS IC fabrication process has been gained.

Although, the logic family called CMOS logic is by far the most popular due to its low power dissipation, a different logic family called pseudo-NMOS technology will be discussed first. This logic family is realized primarily using n-channel transistors with a single p-channel transistor used as the load for each logic gate. The design of these logic gates is very similar to what was used previously for the NMOS logic family, which was the first technology that allowed for the manufacture of very large-scale integrated (VLSI) circuits. Pseudo-NMOS gates are not used very often in actual ICs (except perhaps for programmable logic arrays, which are described in Chapter 10) because they dissipate power even when their outputs are not changing. However, they are simple, result in dense circuitry, are moderately fast when capacitive loading is small, and are designed in a manner similar to many other logic families. An example of this is a logic family that is realized using the very high-speed gallium arsenide (GaAs) technology. More importantly, pseudo-NMOS logic gates are relatively simple examples that can be used to explain how n-channel driver networks are designed. These are required in most MOS logic families. After an understanding of NMOS logic design is gained, it is a relatively minor extension to understand traditional CMOS logic design as well.

4.1 Pseudo-NMOS Logic

The original idea behind the pseudo-NMOS inverter is to realize a gate as a common-source amplifier with a current source as a load, as shown in Fig. 4.1. If the gate input voltage is less

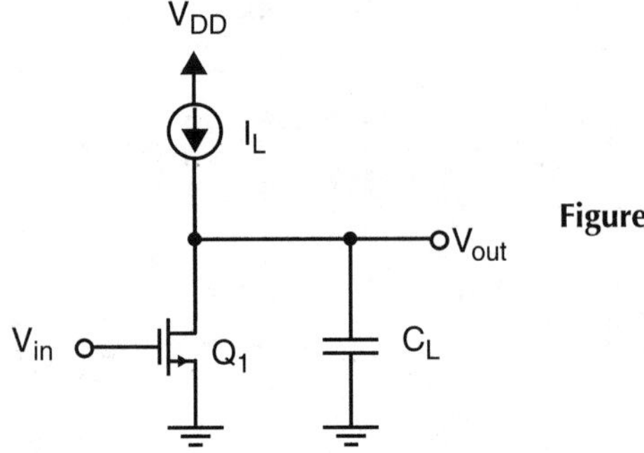

Figure 4.1 A current-source load inverter.

than the threshold voltage[1] of Q_1, then Q_1 will be cut off and I_L will charge the load capacitances to a high voltage, ideally V_{DD}. If, on the other hand, the input signal changes to a high voltage, then initially Q_1 will conduct substantially more current than I_L and the load capacitance will be discharged to a low voltage. When this happens, the voltage across Q_1, that is $V_{DS\text{-}1}$, will be less than the effective gate-source voltage of Q_1, and Q_1 will enter the triode region. This will cause its current to decrease to be equal to I_L. When this occurs, the output voltage will be close to 0 V, assuming the transistor widths have been reasonably chosen.

Normally, to guarantee that the output low voltage, V_{OL}, is close enough to 0 V, it is necessary to make Q_1 sufficiently wide so that its current is many times greater than the value of I_L for the case in which the output is in its transition region around the gate threshold V_{th} while changing from a high to a low voltage. This, however, results in the rise time being considerably larger than the fall time. Thus, most *current-load* gates have poor drive capability for positive-going output transitions, unless they are followed by buffers. Also, they have a power dissipation equal to $I_L V_{DD}$ when the output is low. When the output is high, there is no power dissipation. Thus, approximately half the time they dissipate power. This d.c. power dissipation means that a modern IC containing millions of gates could not be realized using just pseudo-NMOS gates. Nevertheless, these circuits are simple, take up little area, and usually introduce only small loads on preceding gates.

Using a p-Channel Transistor to Realize a Current Source

The basic idea behind pseudo-NMOS logic is to use a p-channel transistor to realize a current-source load as shown in Fig. 4.2. To understand this, consider the equation describing the current of a p-channel MOS transistor in the active or saturation region. For this case, we have

---

[1]Note that the threshold voltage of an MOS transistor, V_t, is different than the threshold voltage of a logic gate V_{th} as was discussed in Chapter 1.

Figure 4.2 Using a p-channel transistor to approximate a current source.

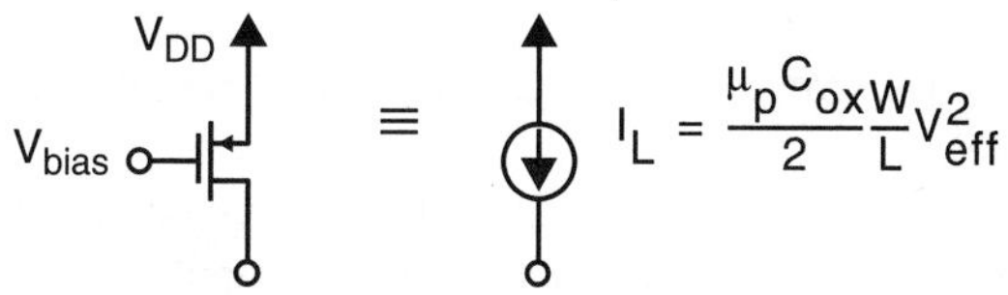

$$I_D = \frac{\mu_p C_{ox}}{2}\frac{W}{L}(V_{SG} + V_{tp})^2 = \frac{\mu_p C_{ox}}{2}\frac{W}{L}V_{eff}^2 \tag{4.1}$$

where $V_{eff} = V_{SG} + V_{tp}$ is the effective gate-source voltage of the transistor. This equation ignores the finite output impedance, r_{ds}, of the transistor. For equation (4.1) to apply, it is necessary that the drain voltage is not greater than the gate voltage by more than $|V_{tp}|$ which is usually about 0.8 to 0.9 V. If we assume that $V_{DD} = 3.3$ V, and that the bias voltage V_{bias} is about half way between ground and V_{DD}, or at 1.65 V, then (4.1) applies as long as the drain voltage is less than 2.45 V, approximately. We now have $V_{SG} = V_{DD} - V_{bias} = 1.65$ V and $V_{eff} = V_{SG} + V_{tp} \approx 0.75$ V in equation (4.1). Therefore, $I_D = I_L$ is independent of the voltage across the transistor as long as this voltage is greater than 0.75 V, or so. Thus, the p-channel transistor can be used to realize an approximate current source as long as the output voltage of the logic gate is less than 2.45 V, or so. This is the case most of the time during an output logic change or transition. When the output voltage becomes greater than approximately 2.45 V, then the load current starts to decrease and a more accurate model would be a current source in parallel with a resistor.

A PSEUDO-NMOS INVERTER

Figure 4.3 shows a pseudo-NMOS inverter for which the W/L of the drive transistor is one half the W/L of the load device; that is

$$(W/L)_1 = \frac{(W/L)_2}{2} \tag{4.2}$$

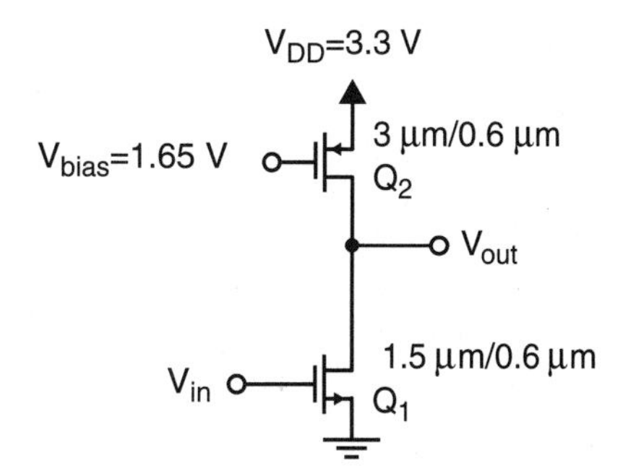

Figure 4.3 A typical pseudo-NMOS inverter in a 0.6-µm technology.

This would be a typical choice for a pseudo-NMOS inverter having a p-channel load with a 1.65-V gate-bias voltage. A simple circuit capable of realizing this bias voltage will be described shortly.

In this section, we will analyze the pseudo-NMOS inverter for its approximate gate threshold voltage, gain at threshold, typical output high and low voltages, and approximate transient responses. To make these analyses tractable, a number of approximations will be made. The errors arising from these approximations vary somewhat, but in all cases are typically less than the errors caused by the inability to accurately predict the processing and transistor parameters, prior to manufacture. Also, by making reasonable approximations, the essential operation of the gate becomes more obvious without miring the reader in complex calculations that are of limited use in designing practical digital integrated circuits. Irrespective of the use of approximations, this section relies heavily on understanding the equations that describe MOS transistors (which were presented in Chapter 3). Thus, not only will this section give greater insight into pseudo-NMOS logic, but it will also serve as the first in-depth example of the application of MOS equations to actual circuits.

Inverter Threshold Voltage (V_{TH})

As was mentioned in Chapter 1, the threshold voltage of the pseudo-NMOS inverter is defined as the input voltage that gives an identical output voltage. Normally, for a properly designed MOS inverter, this voltage would be approximately half the power-supply voltage. To calculate this voltage, it is first necessary to determine the region in which a transistor is operating and which equations apply. That is, is the transistor in the triode region or the saturation region (which we typically call the active region)? First, it is safe to say that the enhancement n-channel drive transistor (Q_1) is definitely in the active or saturation region. This is true because for $V_{in} = V_{out}$, the gate-drain voltage of Q_1 is zero. Any enhancement transistor with $V_{DG} = 0$ is in the active region. It will also be assumed that the p-channel load transistor is also in the active region. This will be the case as long as the gate-threshold voltage happens to be less than 2.45 V, which is almost certainly true for a properly designed inverter. If this were not the case, then the device sizes would have been poorly chosen and the gate would be impractical, in which case it would not be described in this text. Under these assumptions, we may write the following equations. For Q_2 we have

$$I_{D\text{-}2} = \frac{\mu_p C_{ox}}{2}\left(\frac{W}{L}\right)_2 \left(\frac{V_{DD}}{2} + V_{tp}\right)^2 \tag{4.3}$$

where it is assumed that $V_{SG\text{-}2} = V_{DD}/2$. For Q_1, we have $V_{GS\text{-}1} = V_{in} = V_{th}$ and

$$I_{D\text{-}1} = \frac{\mu_n C_{ox}}{2}\left(\frac{W}{L}\right)_1 (V_{th} - V_{tn})^2 \tag{4.4}$$

Since it is always assumed that MOS gates drive only capacitive loads, we have $I_{D-1} = I_{D-2}$ and we may equate (4.3) and (4.4) and solve for V_{th} to obtain

$$V_{th} = V_{tn} + \sqrt{\frac{\mu_p (W/L)_2}{\mu_n (W/L)_1}}\left(\frac{V_{DD}}{2} + V_{tp}\right) \tag{4.5}$$

Example 4.1

For $\mu_n/\mu_p = 4.2$, $V_{tn} = 0.8$ V and $V_{tp} = -0.9$ V, find V_{th} for the inverter of Fig. 4.3.

Solution: For $(W/L)_2/(W/L)_1 = 2$, we have using equation (4.5) $V_{th} = 1.32$ V.

TYPICAL OUTPUT HIGH VOLTAGE V_{OH}

The output voltage is high when the input voltage is a "0". Assuming the typical input "0" voltage is less than V_{tn}, then Q_1 will be completely cut off. Under this condition, the source-drain voltage of Q_2 will be very small. For this case, we have $V_{SD2} \ll V_{eff\text{-}2}$ and Q_2 is *hard in the triode region.* A transistor that is hard in the triode region has its current given by

$$I_{D2} \cong \mu_p C_{ox}\left(\frac{W}{L}\right)_2 V_{eff\text{-}2} V_{SD\text{-}2} \tag{4.6}$$

Thus, its current is approximately proportional to the voltage across it. Since $V_{eff\text{-}2} = V_{SG\text{-}2} + V_{tp} = V_{DD}/2 + V_{tp} = 0.75$ V, we can approximate Q_2 by a resistor of size

$$r_{ds\text{-}2} = \frac{1}{\mu_p C_{ox}(W/L)_2 V_{eff\text{-}2}} \tag{4.7}$$

Since Q_1 is off, there will be no current through Q_2, and according to equation (4.6), $V_{SD\text{-}2} = 0$, which implies $V_{out} = V_{OH} = V_{DD}$. Thus, the typical output high voltage is V_{DD}.

It should be emphasized that the approximation of equation (4.7) is valid only for $V_{SD\text{-}2}$ close to 0 V or equivalently for V_{out} close to V_{DD}. It is not valid when the output is in the transition region around V_{th} when changing from a "0" to a "1". Rather, it is valid only when the output voltage has completely finished its change during a transition.

Example 4.2

Assuming $\mu_n C_{ox} = 188\ \mu A/V^2$, $\mu_p C_{ox} = 44.5\ A/V^2$, and that the load capacitance is 1 pF, how long would it take the output voltage to change from 3.0 to 3.2 V during the end of a "0" to "1" transition. Assume $(W/L)_2 = (3\ \mu m / 0.6\ \mu m)$.

Solution: Using equation (4.7), we have

$$r_{ds\text{-}2} = \frac{1}{44.5 \times 10^{-6}(3/0.6)0.75} = 5.99\ k\Omega \tag{4.8}$$

Using equation (1.14) from Chapter 1 we have

$$\Delta t = \tau \ln\left[\frac{V_{out}(\infty) - V_{out}(t_1)}{V_{out}(\infty) - V_{out}(t_2)}\right] \tag{4.9}$$

where $\tau = r_{ds\text{-}2} C_L = 5.99$ ns, $V_{out}(\infty) = 3.3$ V, $V_{out}(t_1) = 3.0$ V, and $V_{out}(t_2) = 3.2$ V. Plugging these values into equation (4.9) gives $\Delta t = 2.4$ ns.

Typical Output Low Voltage V_{OL}

When the input is a typical "1", from the previous section we have $V_{GS\text{-}1} = V_{DD} = 3.3$ V. Thus, Q_1 will be *hard on*. Assuming we have chosen the device sizes correctly, the output will be a low voltage. Thus, it is safe to say that Q_2 will be in the active region and acts like a current source of size

$$I_L = I_{D\text{-}2} \cong \frac{\mu_p C_{ox}}{2}\left(\frac{W}{L}\right)_2 V_{eff\text{-}2}^2 \tag{4.10}$$

Also, since V_{out} is low, we have $V_{DS\text{-}1}$ very small (i.e., much less than its effective gate voltage, which is $V_{GS\text{-}1} - V_{tn}$ or 2.5 V) and Q_1 is hard in the triode region. Thus, Q_1 can be approximated by a resistor of size

$$r_{ds\text{-}1} = \frac{1}{\mu_n C_{ox}(W/L)_1(V_{DD} - V_{tn})} \tag{4.11}$$

The output low voltage is then simply given by

$$V_{OL} = I_{D\text{-}2} r_{ds\text{-}1} = \frac{1}{2} \frac{\mu_p}{\mu_n} \frac{(V_{DD}/2 + V_{tp})^2}{(V_{DD} - V_{tn})} \frac{(W/L)_2}{(W/L)_1} \tag{4.12}$$

Example 4.3

Using the same parameters as for the previous examples, calculate V_{OL} and the noise margins of the pseudo-NMOS inverter.

Solution: Using equation (4.12), we have

$$V_{OL} = \frac{1}{2}\left(\frac{44.5}{188}\right)\frac{0.75^2}{(3.3 - 0.8)^2} 2 = 0.053 \text{ V} \tag{4.13}$$

From Example 4.1, we have $V_{th} = 1.32$ V, and from the previous section we have $V_{OH} = 3.3$ V. Thus, $NM_H = V_{OH} - V_{th} = 1.98$ V and $NM_L = V_{th} - V_{OL} = 1.27$ V. The threshold, output-high and output-low voltages obtained from SPICE simulations are $V_{th} = 1.43$ V, $V_{OH} = 3.3$ V, and $V_{OL} = 0.13$ V. The discrepancies are primarily due to second-order effects not considered in modeling equations used for *hand analysis.* The noise margins are not symmetric but are still quite large.

GAIN AT THE GATE THRESHOLD VOLTAGE

One of the commonly used figures of merit for an inverter is the small-signal gain of the inverter when it is at the operating point $V_{in} = V_{out} = V_{th}$. To find this gain, the transistors should be replaced by their small-signal models described in Chapter 3. The small-signal model of an MOS transistor at low frequencies, where the capacitors have been removed, is shown in Fig. 4.4. This model was described in Chapter 3, but for convenience the equations for the various parameters will be repeated here.

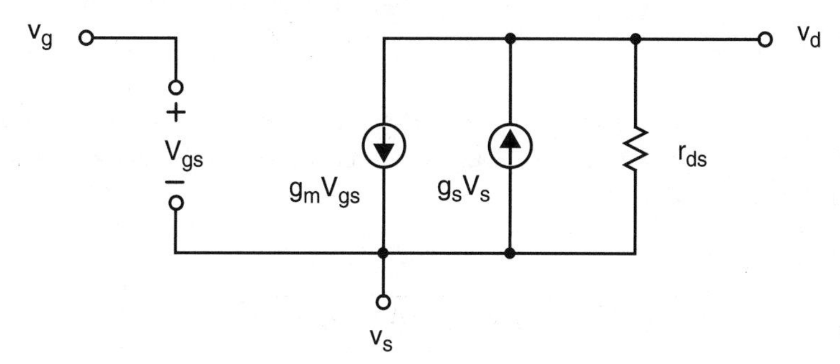

Figure 4.4 The small-signal model of an MOS transistor at low frequencies.

The voltage-controlled current source, $g_m v_{gs}$, is responsible for the transistor operation. In an ideal transistor, this would be the only component in the model. The transconductance g_m is given by

$$g_m = \mu_n C_{ox}\frac{W}{L}(V_{GS} - V_{tn}) = \mu_n C_{ox}\frac{W}{L}V_{eff} = \sqrt{2\mu_n C_{ox}\frac{W}{L}I_D} \quad (4.14)$$

The second voltage-controlled current source, $g_s v_s$, models the body effect. It is responsible for the decrease in current as the source voltage, v_s, increases. This is the reason its direction is opposite that of $g_m v_{gs}$. In many introductory texts, this current source is not included in MOS small-signal models. For the pseudo-NMOS inverter, the sources of both the drive transistor and the p-channel load transistor are connected to ground and V_{DD}, respectively, which are both *small-signal grounds*. Therefore, for both transistors, $v_s = 0$, and the current source modeling the body effect can be ignored. The final parameter of the model is the output resistance of the transistor, r_{ds}. For transistors that do not have short-channel effects, this models the reduction of the channel length due to the increase in the length of the pinch-off region at the drain end of the channel when the drain voltage increases. For this case it is approximately given by

$$r_{ds} \cong \frac{L\alpha\sqrt{V_{DS} + V_t}}{I_D} \quad (4.15)$$

where α is a process-dependent constant on the order of $5\sqrt{V/\mu m}$ (i.e., L in micrometers). For transistors having short-channel effects, the output resistance will be much smaller. For this case we have

$$r_{ds} = \frac{1}{\lambda I_D} \quad (4.16)$$

where λ must now be empirically derived from transistor measurements. This must be done for every different channel length that is to be used.[2]

In deriving the small signal-model of the inverter, V_{DD} is replaced by ground as it is assumed to be a constant voltage. This leads to the small-signal equivalent circuit shown in Fig. 4.5a. Note that $v_{g\text{-}2} = v_{s\text{-}2} = 0$ implies both of the current sources $g_{m\text{-}2}v_{gs\text{-}2}$ and $g_{s\text{-}2}v_{s\text{-}2}$ are zero. Also, since $v_{s\text{-}1} = 0$, $g_{s\text{-}1}v_{s\text{-}1} = 0$, as well. This allows the circuit to be simplified to that shown in Fig. 4.5b. The gain is now readily calculated to be

$$\frac{v_{out}}{v_{in}} = \frac{-g_{m\text{-}1}}{1/r_{ds\text{-}1} + 1/r_{ds\text{-}2}} \quad (4.17)$$

---

[2]In the MOS models used in SPICE, λ is specified by the LAMBDA parameter. If this is specified, the output impedance is calculated using (4.16) independent of the transistor's L. Thus, a different LAMBDA and therefore model statement should be used for every different transistor length used.

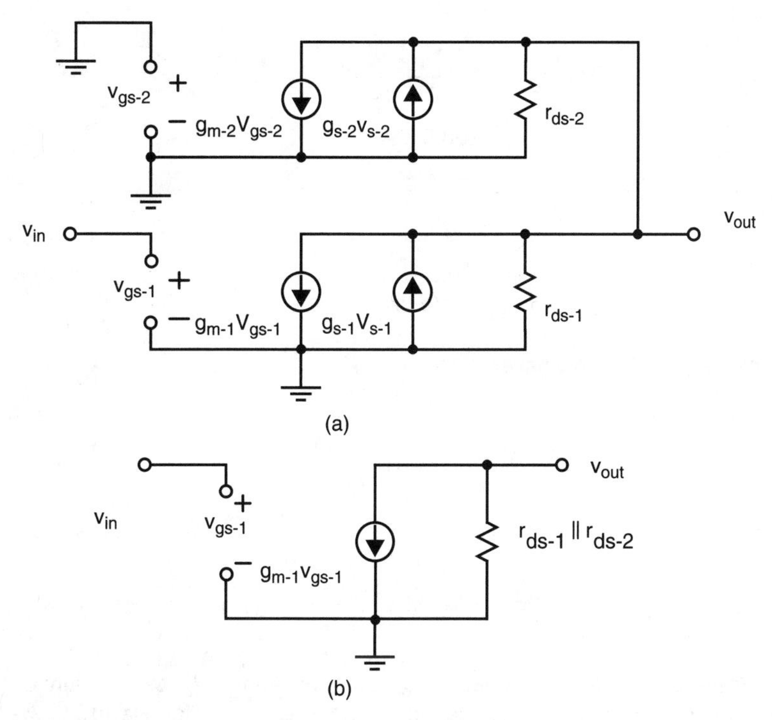

Figure 4.5 (a) The small-signal equivalent circuit of a pseudo-NMOS inverter and (b) a simplified small-signal equivalent circuit of the inverter.

This gives a gain at the gate threshold voltage on the order of –15 to –50 depending on processing parameters for a particular technology.

Example 4.4

Calculate the gain of a pseudo-NMOS inverter at the threshold voltage without ignoring the transistor output impedances. Assume $\lambda_n = 0.060$ and $\lambda_p = 0.065$.

Solution: From Example 4.1, $V_{th} = 1.32$ V. From equation (4.14), using $\mu_n C_{ox} = 188\ \mu A/V^2$, we have

$$g_{m-1} = \mu_n C_{ox}\left(\frac{W}{L}\right)_1 (V_{th} - V_{tn}) = 0.244\ \text{mA/V} \tag{4.18}$$

Also,

$$I_{D-1} = \frac{\mu_n C_{ox}}{2}\left(\frac{W}{L}\right)(V_{th} - V_{tn})^2 \tag{4.19}$$
$$= 63.5\ \mu A$$

Thus,

$$r_{ds-1} = \frac{1}{\lambda_n I_{D-1}} = 262\ k\Omega \tag{4.20}$$

Also, since $I_{D-2} = I_{D-1}$, we have

$$r_{ds-2} = \frac{1}{\lambda_p I_{D1}} = 242\ k\Omega \tag{4.21}$$

as well. Finally, using (4.17), we get

$$\frac{v_{out}}{v_{in}} = \frac{-0.244}{1/262 + 1/242} = -31.0 \tag{4.22}$$

The transfer curve, which is a plot of V_{out} as a function of V_{in}, was obtained using SPICE and is shown in Fig. 4.6. The gain at the threshold voltage is –10.9, which is significantly different from that obtained through hand analysis. Estimating circuit gain is an area in which SPICE is especially inaccurate. Luckily, accurate gain estimates are not important when realizing digital circuits.

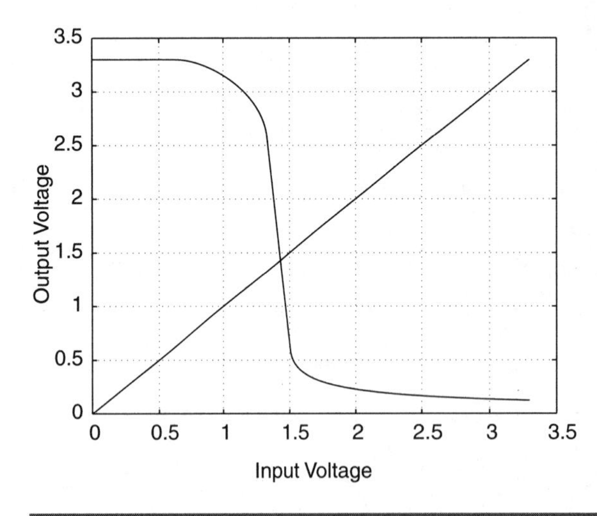

Figure 4.6 The transfer curve of the pseudo-NMOS inverter of Fig. 4.3 obtained using SPICE.

TRANSIENT RESPONSE

The speed with which the output of a gate can change state is limited because the transistors can supply only a finite amount of current to charge or discharge the parasitic capacitances that exist at every node. The parasitic capacitance is composed of three components: the input capacitance of the gates being driven by the output, the capacitance due to the interconnect wiring, and the capacitance due to reverse-biased junctions at the output node of the gate. There is often additional delay terms due to the requirement that internal nodes of a gate must change state before the output can change state. For the pseudo-NMOS inverter, these additional terms are not present, since there is only one node, the output node.

Often, when a gate is driving only a couple of other gates, the junction capacitance at the output node can be the dominant, if not the major, component of the load capacitance. As explained in Chapter 2, this capacitance is highly nonlinear. Its magnitude is also difficult to predict at design time because the area of the junctions is not known before the layout has been done and can only be estimated. For these reasons, *it is very important that the designer guarantees the ICs are functional irrespective of accurate knowledge of gate delays.* Also, the designer should supervise the layout of the circuit to ensure that the junction areas of the critical nodes are minimized.

The exact analytical derivation of the transient response of an NMOS inverter (Taub and Schilling, 1977; Hodges and Jackson, 1988) is tedious and complicated due to the highly nonlinear nature of the transistors and the capacitances (both junction capacitances and gate input capacitances). Given that the load capacitances are never known accurately at the time of design, it is deemed that the exact calculation of the transient delays is not of much use and would never be utilized by a digital designer working under usual time constraints. Rather, approximate methods that give the correct order of magnitude of the response times, and can be used to quickly determine critical nodes, must be developed.

In analyzing the pseudo-NMOS inverter for its approximate delay times, it will be assumed that the load capacitance is ideal and known. In the examples, it will be assumed to be 0.2 pF. Later in Chapter 6, we will discuss how to estimate this capacitance in greater detail.

RISE TIME

First, we will calculate the time from the point at which the input undergoes a step change from a "1" to a "0" to when the output has undergone a 70% change to approximately 2.3 V. As mentioned in Chapter 1, the reason for using the 70% rise time as a figure of merit is that *if a large number of inverters are cascaded, then the sum of the 70% rise and fall times is approximately equal to the total delay through the chain of inverters.* During the 70% rise time, Q_1 will be off and thus will be ignored. Initially, Q_2 is in the active region and has a current of approximately

$$I_{D\text{-}2} = \frac{\mu_p C_{ox}}{2}\left(\frac{W}{L}\right)_2 V_{eff\text{-}2}^2 = \frac{\mu_p C_{ox}}{2}\left(\frac{W}{L}\right)_2\left(\frac{V_{DD}}{2} + V_{tp}\right)^2 \tag{4.23}$$

This remains true until the voltage across Q_2 decreases to $V_{eff-2} = 0.75$ V or equivalently when $V_{out} = 2.55$ V. Since the output must reach 2.3 V to undergo a 70% change, the transistor Q_2 will remain in the active region for the duration of the 70% rise time. Therefore, I_{D-2} remains constant at the value given by (4.23). We have

$$t_{+70\%} = \frac{C_L}{I_{D-2}}\Delta V_{out} = \frac{2C_L 2.3}{\mu_p C_{ox}(W/L)_2[(V_{DD}/2) + V_{tp}]^2} \tag{4.24}$$

where C_L is the total load capacitance in Farads.

Example 4.5

What is the approximate rise time of the pseudo-NMOS inverter of Fig. 4.3, for $C_L = 0.2$ pF. Assume $\mu_p C_{ox} = 44.5\ \mu A/V^2$.

Solution: Using equation (4.24) with $(W/L)_2 = 3\ \mu m/0.6\ \mu m$, $\mu_p C_{ox} = 44.5\ \mu A/V^2$, $V_{tp} = -0.9$ V, and $V_{DD} = 3.3$ V, we get $t_{+70\%} = 5.7$ ns. This is a little faster than the rise time found from SPICE simulation and shown in Fig. 4.7. The rise time is seen to be 5.9 ns. The major reason for the difference is that SPICE adds device parasitic capacitances to the specified load capacitance of 0.2 pF.

Fall Time

When calculating the fall time, it is assumed that at time 0, the input undergoes a step change to V_{DD}, which turns Q_1 on *hard*. The fall time is defined as the time it takes the output to decrease from 3.3 V through a 70% change to 1.0 V. During most of this time, the current through Q_1 is much larger than the current through Q_2 and thus Q_2 will be ignored. This will make our estimate somewhat optimistic, but helps to simplify the analysis.

Initially Q_1 is in the active region until $V_{DG-1} < -V_{tn}$ or equivalently $V_{DS-1} < V_{eff-1} = V_{GS-1} - V_{tn}$. This occurs when the output is discharged to 2.5 V (for $V_{tn} = 0.8$ V). Next, Q_1 enters the triode region during most of the fall time. The analytic derivation of the fall time is possible and has been given (Hodges and Jackson, 1988), but is of little use during design. Instead, it is desired to find a formula for an approximately equivalent resistor that would give a similar fall time.

As mentioned, the current through Q_1 is highly nonlinear as the output voltage changes from 3.3 to 1.0 V. Figure 4.7 shows a plot of this current as a function of the output voltage. If we could find a value for a resistor that gives approximately the same current as the actual transistor, the analysis would be greatly simplified. To find a resistor that is approximately equivalent to Q_1, it is necessary to find an I–V curve that is straight, passes through the origin, and results in the same fall time as obtained with Q_1. One approximation that has been

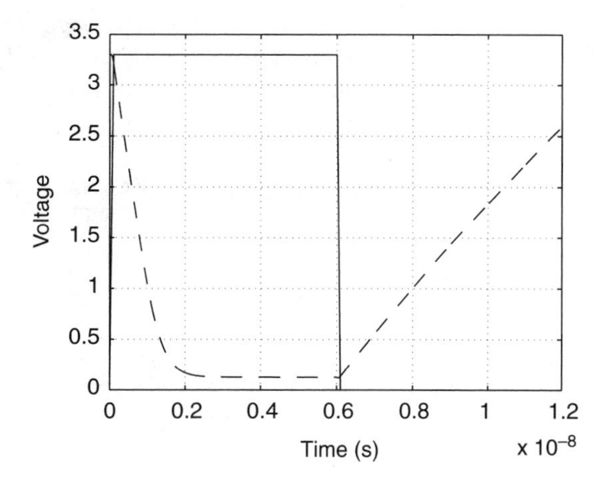

Figure 4.7 The transient response of the inverter of Fig. 4.3 for $C_L = 0.2$ pF.

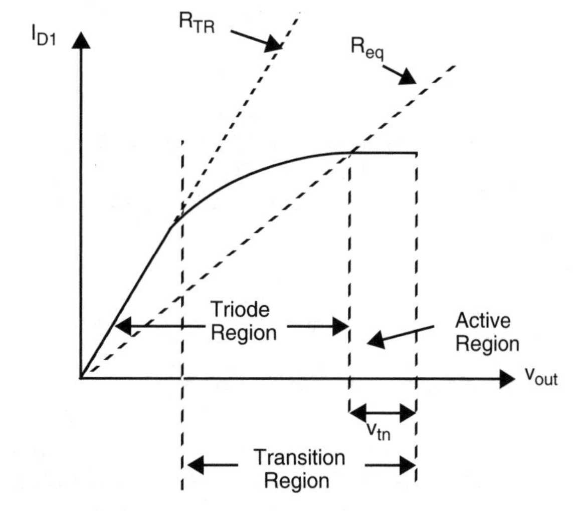

Figure 4.8 The current through Q_1 as a function of the output voltage.

suggested previously is to use a resistor that is equal to the resistance of Q_1 when V_{DS-1} is small (i.e., Q_1 is hard in the triode region). The equation for this resistance (denoted R_{TR}) is

$$R_{TR} = \frac{1}{\mu_n C_{ox}(W/L)_1(V_{DD} - V_{tn})} \tag{4.25}$$

However, as can easily be seen from Fig. 4.8, this approximation would result in a resistor that always has more current in the transition region than Q_1 has. A better approximation is a resistor whose I–V curve intersects the I–V curve of Q_1 at the edge of the triode region. This approximation is shown in Fig. 4.8 as the straight line labeled R_{eq}. It can be seen that for the initial interval of the fall time, the current through R_{eq} will be larger than that of Q_1, and for the second part of the transition, the current through R_{eq} will be less than that through Q_1.

The voltage at the intersection point of the I–V curves of R_{eq} and Q_1 is $V_{DD} - V_{tn}$. The current at the intersection point is the current of Q_1 in the active region, that is

$$I_{D\text{-}1} = \frac{\mu_n C_{ox}}{2}\left(\frac{W}{L}\right)_1 (V_{DD} - V_{tn})^2 \tag{4.26}$$

The value for R_{eq} is the ratio of the voltage to the current, that is

$$R_{eq} = \frac{V_{DD} - V_{tn}}{\frac{\mu_n C_{ox}}{2}(W/L)_1 (V_{DD} - V_{tn})^2} = \frac{2}{\mu_n C_{ox}(W/L)_1 (V_{DD} - V_{tn})} \tag{4.27}$$

Thus, the approximately equivalent resistance to Q_1 is twice R_{TR}, the triode resistance of Q_1. It turns out that (4.27) results in transient times that are a little over 20% too low. Therefore, some designers modify (4.27) to be somewhat larger according to

$$R_{eq} = \frac{2.5}{\mu_n C_{ox}(W/L)_1 (V_{DD} - V_{tn})} \tag{4.28}$$

Others will simply use (4.27) and then add on 20% additional time after the estimate. Either approach is valid.

Example 4.6

Using the equivalent resistance approximation, calculate the fall time of the pseudo-NMOS inverter assuming the output goes through a 70% change and the load capacitance is 0.2 pF.

Solution: Using equation (4.27) and assuming $\mu_n C_{ox} = 188\ \mu A/V^2$ and $(W/L)_1 = 2.5$ we have $R_{eq} = 1.7\ k\Omega$. Thus, $\tau = R_{eq} C_L = 0.34$ ns. Using equation (1.14) of Chapter 1

$$t_F = \tau \ln\left[\frac{v_{out}(\infty) - v_{out}(t_1)}{v_{out}(\infty) - v_{out}(t_2)}\right] \tag{4.29}$$

along with $v_{out}(t_1) = 3.3$ V, $v_{out}(t_2) = 1.0$ V, and $v_{out}(\infty) = 0.05$ V gives $t_{70\%} = 0.42$ ns. The fall time obtained from SPICE simulation is 0.94 ns. The major reason for the discrepancy is that the current through the load transistor Q_2 was ignored.

4.2 Pseudo-NMOS Logic Gates

In Chapter 1, a number of NMOS logic gates were introduced. In this section, *pseudo-NMOS* logic gates will be revisited, in a somewhat more formal manner, where extensive use will be made of the equivalent-resistor approximation. This section will also deal with how to choose the device sizes for NMOS logic gates.

An example of NMOS logic is the straightforward realization of the *exclusive-or* function. The *exclusive-or* is defined by

$$\begin{aligned} y &= x_1 \oplus x_2 = x_1\overline{x}_2 + \overline{x}_1 x_2 \\ &= \overline{x_1 x_2 + \overline{x}_1\overline{x}_2} \\ &= \overline{x_1 x_2 + \overline{x_1 + x_2}} \end{aligned}$$

A logic circuit that realizes this function is shown in Fig. 4.9a. A pseudo-NMOS realization of this logic circuit is shown in Fig. 4.9b. The *nor* gate *a* is realized by transistors Q_1, Q_2, and Q_3. The second NMOS gate, composed of transistors Q_4 to Q_7, is a compound gate; it realizes both the *nor* gate *c* and the *and* gate *b* by a series connection of drive transistors. Notice, how the series connected transistors (Q_5 and Q_6) are chosen to be wider than the parallel transistor Q_4.

The general form of a typical NMOS logic gate is shown in Fig. 4.10. It consists of a single p-channel load transistor, between V_{DD} and V_{out}, that has its gate connected to a bias voltage.

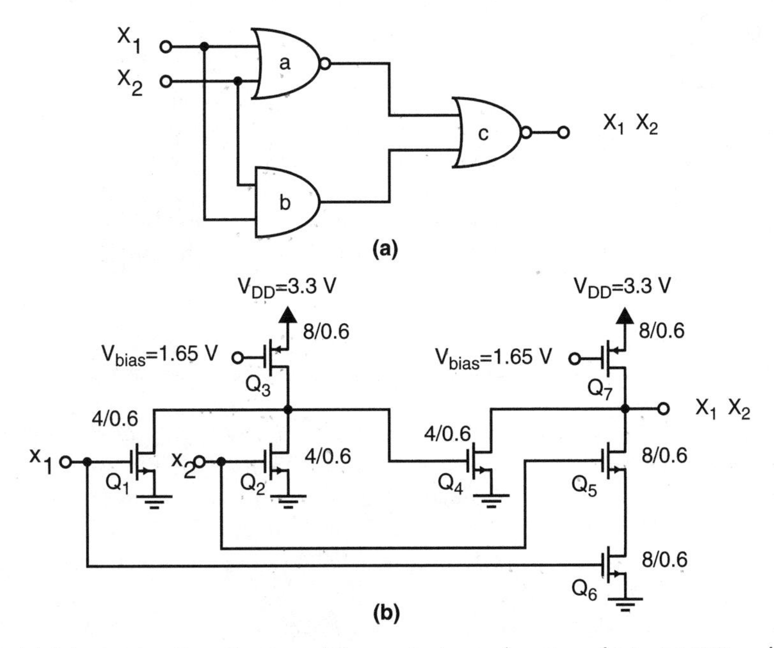

Figure 4.9 (a) A logic circuit realization of the *exclusive-or* function. (b) An NMOS realization of the *exclusive-or* function.

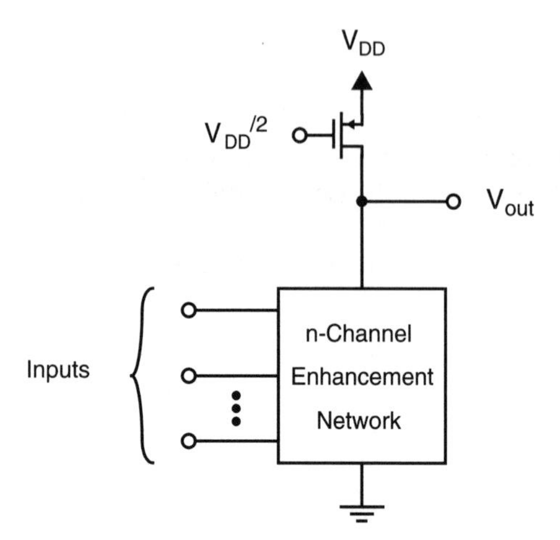

Figure 4.10 A generalized pseudo-NMOS logic gate.

In addition, there is a network of n-channel-enhancement *drive transistors* between V_{out} and ground. Each of the drive transistors in this network has its gate connected to an input. Depending on the logic values of the inputs, the network of drive transistors will either provide an infinite impedance between V_{out} and ground, in which case the output will be V_{DD}, or will provide a low enough impedance so that V_{out} will be pulled low to a "0".

In the event the network of drive transistors is low impedance, such impedance must not be larger than that of a single enhancement transistor, with its gate connected to V_{DD}, and having its W/L at least one-half the size of the load transistor. This constrains the minimum W/Ls of the drive transistors.

The maximum W/Ls of the drive transistors are constrained by the desire for speed. Making them larger than necessary makes the logic slower than it need be for two reasons; first, it increases the parasitic capacitances of the internal nodes of the gate; second, it increases the output load capacitance of preceding gates. Taking the W/Ls too large does not cause the gate to function incorrectly.

Before describing how the W/Ls of the n-channel network should be chosen, it is necessary to state some important theorems regarding approximate transistor equivalency that will be used often in this text.

4.3 Transistor Equivalency

1. *Two transistors are equivalent if the ratio of their W/Ls is equal independent of the absolute values of* W *or* L. This theorem is depicted pictorially in Fig. 4.11. This statement is obvious since the large signal I–V equations are only functions of W/L. In reality, there are second-order effects that make this statement only approximate. A transistor with a smaller W tends to have slightly less current than that predicted by (4.1) due to the cross section of the channel, as viewed along the channel being rounded rather than square. Also, a transistor with a shorter length tends to inaccuracies due to short-channel effects. Because of these second-order effects, transistors having very large Ws (50 to 1000 μm) are usually realized by connect-

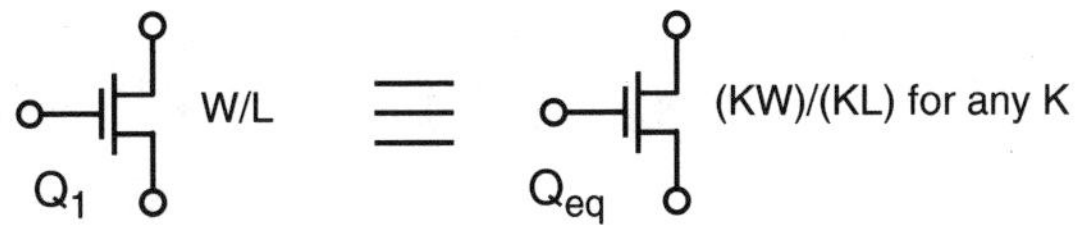

Figure 4.11 Scaling both W and L of a transistor has no effect on a first-order approximation.

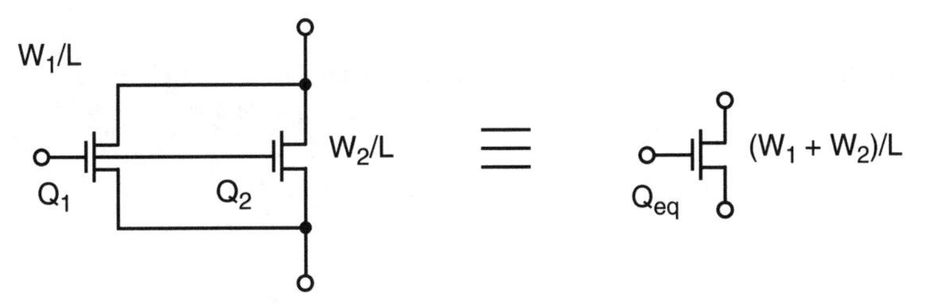

Figure 4.12 Two transistors in parallel (Q_1 and Q_2) having equal Ls, are equivalent to a single transistor having a width equal to the sum of the widths of Q_1 and Q_2.

ing a number of smaller transistors (with Ws on the order of 10 to 25 μm) in parallel. As we will be doing only approximate analyses by hand, these second-order effects will be ignored.

2. *Two transistors, with the same length, connected in parallel, are equivalent to a single transistor having a width equal to the sum of the widths of the two transistors.* This theorem is depicted graphically in Fig. 4.12. The proof is simple and straightforward. Assume there are two transistors, Q_1 and Q_2, having identical lengths L and widths W_1 and W_2, respectively. Furthermore, assume their gates, sources, and drains are all connected together, and the transistors are in the active region. Then the total current, I_T, is given by

$$I_T = I_{D\text{-}1} + I_{D\text{-}2} = \frac{\mu C_{ox}}{2} \frac{(W_1 + W_2)}{L} (V_{GS} - V_t)^2 \tag{4.30}$$

But this is the I–V equation of a single transistor having a width $W_1 + W_2$. A similar proof can be given for transistors in the triode region.

If the lengths of the transistors are not equal, then one of the transistors can have both its W and L scaled (according to Theorem (1)) to make the lengths equal and Theorem (2) can then be applied to simplify the network.

3. *Two transistors having equal widths, connected in series, and having their gates connected together, are equivalent to a single transistor with a length equal to the sum of the individual lengths.* This theorem is shown pictorially in Fig. 4.13. The proof of this theorem, although straightforward, does require a little algebra. Q_1 will certainly be operating in the triode region, since its drain-source voltage is small, and has a current given by

$$I_{D\text{-}1} = \mu C_{ox} \frac{W}{L_1} \left[(V_{GS\text{-}1} - V_t) V_{DS\text{-}1} - \frac{V_{DS\text{-}1}^2}{2} \right] \tag{4.31}$$

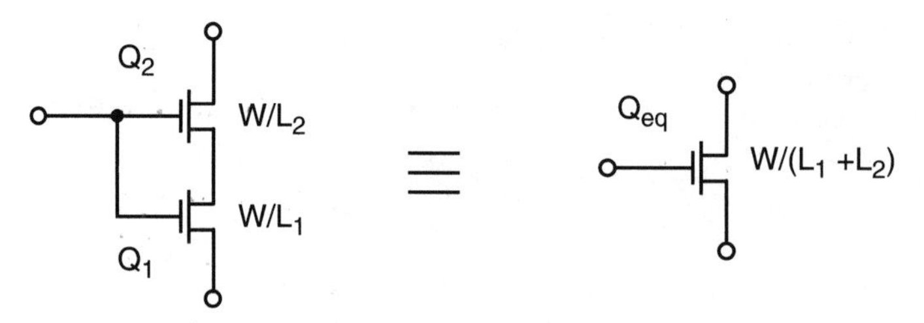

Figure 4.13 Two transistors (Q_1 and Q_2), having equal Ws and connected in series, with their gates connected together are equivalent to a single transistor having a length equal to the sum of the lengths of Q_1 and Q_2.

Assuming Q_2 is in the active region, its gate-source voltage is given by

$$V_{GS\text{-}2} = V_{GS\text{-}1} - V_{DS\text{-}1} \tag{4.32}$$

Then, the drain current of Q_2 is given by

$$\begin{aligned} I_{D\text{-}2} &= \frac{\mu C_{ox}}{2}\frac{W}{L_2}(V_{GS\text{-}1} - V_{DS\text{-}1} - V_t)^2 \\ &= \frac{\mu C_{ox}}{2}\frac{W}{L_2}[(V_{GS\text{-}1} - V_t) - V_{DS\text{-}1}]^2 \end{aligned} \tag{4.33}$$

Expanding (4.33), we have

$$I_{D\text{-}2} = \frac{\mu C_{ox}}{2}\frac{W}{L_2}[(V_{GS\text{-}1} - V_t)^2 - 2(V_{GS\text{-}1} - V_t)V_{DS\text{-}1} + V_{DS\text{-}1}^2] \tag{4.34}$$

Since $I_{D\text{-}1} = I_{D\text{-}2}$, we can equate (4.31) and (4.34) and rearrange to get

$$\left(\frac{W}{L_1} + \frac{W}{L_2}\right)(V_{GS\text{-}1} - V_t)V_{DS\text{-}1} - \left(\frac{W}{L_1} + \frac{W}{L_2}\right)\frac{V_{DS\text{-}1}^2}{2} = \frac{1}{2}\frac{W}{L_2}(V_{GS\text{-}1} - V_t)^2 \tag{4.35}$$

$$\Rightarrow (V_{GS\text{-}1} - V_t)V_{DS\text{-}1} - \frac{V_{DS\text{-}1}^2}{2} = \frac{1}{2}\frac{L_1}{L_1 + L_2}(V_{GS\text{-}1} - V_t)^2 \tag{4.36}$$

Substituting (4.36) into (4.31) we get

$$I_{D1} = \frac{\mu C_{ox}}{2}\frac{W}{L_1 + L_2}(V_{GS\text{-}1} - V_t)^2 \tag{4.37}$$

But this is the I–V equation of a single transistor having length $L_1 + L_2$ and the proof is completed. If Q_2 had been in the triode region, a similar proof is possible. This is left as an exercise for the reader (Problem 4.7).

As before, if the widths of Q_1 and Q_2 are not equal, then one of the transistors can be scaled and the circuit can then be simplified.

It is now possible to describe a procedure for simplifying the network of drive transistors of an NMOS gate to either an open circuit, or a single equivalent transistor:

1. All transistors that have "0" inputs at their gates are off and can be eliminated from the circuit.
2. Any transistors in series with transistors that are off can be eliminated from the circuit.
3. All the remaining transistors have equal gate voltages. Theorems 1 and 3 can now be repeatedly applied to replace parallel and series connections by equivalent transistors until eventually a single transistor (or an open circuit) results. This procedure is best illustrated by an example.

Example 4.7

Replace the drive-transistor network of Fig. 4.14 by a single equivalent transistor.

Solution: The first thing that should be noticed is that Q_3 and Q_6 are off and can be ignored. Next Q_5 in parallel with Q_7 can be replaced by a single transistor Q_1' of size 24/2. Next, Q_2 in series with Q_1' can be replaced by a single transistor, but first one of them must be scaled so they have equal width. Q_2 can be scaled to Q_2', a

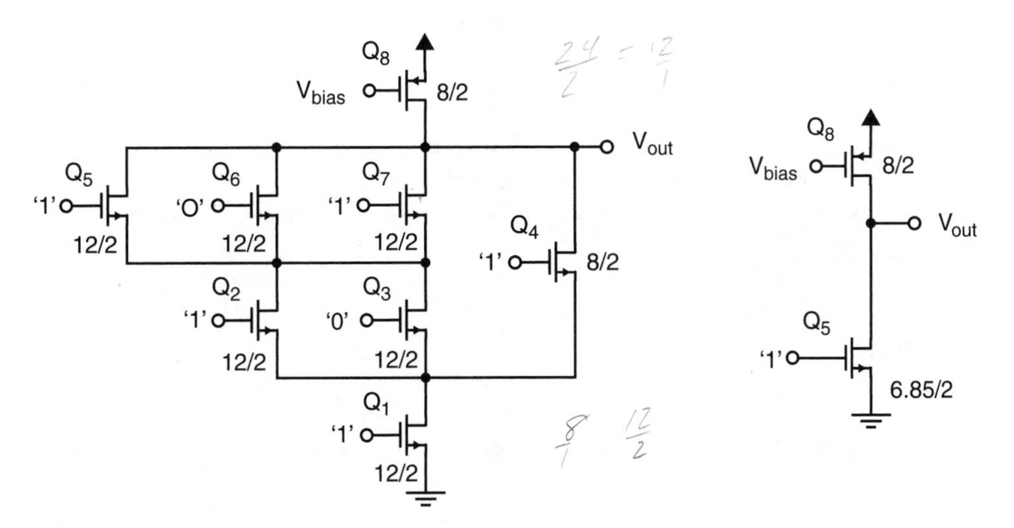

Figure 4.14 A complex pseudo-NMOS gate used to illustrate the principles of transistor equivalence.

transistor of size 24/4. Now we can combine Q_2' and Q_1' to get a single transistor Q_3' of size 24/6, which is equivalent to a transistor of size 8/2. Q_3' can then be combined with Q_4 to get Q_4' with a size of 16/2. Next, scaling Q_1 to 16/2.67, we can combine it with Q_3' to get Q_4' with a size 16/4.67, which can be scaled to a size of 6.85/2. Notice that the equivalent transistor has a W/L greater than its minimum size [i.e., $(W/L)_8/2 = 4/2$].

RESISTOR EQUIVALENCY

The procedure above can be simplified somewhat by using the principle of resistor equivalence introduced in Section 3.1, where it was stated that an n-channel drive transistor Q_1 that has a "1" input is approximately equivalent to a resistor of size

$$R_i = \frac{2.5}{\mu_n C_{ox}(W/L)_i(V_{DD} - V_{tn})} \tag{4.38}$$

If one works with scaled admittances

$$G_i = \frac{1}{R_i}\frac{2.5}{\mu_n C_{ox}(V_{DD} - V_{tn})} = \left(\frac{W}{L}\right)_i \tag{4.39}$$

during the simplification, then each transistor, Q_i, can be considered to be equivalent to a scaled admittance equal to its relative size [i.e., $(W/L)_i$]. Now, when simplifying the network, one simply combines parallel and series admittances. After all the admittances are combined into a single admittance, G_{eq}, the scaling is undone by multiplying it by $\mu_n C_{ox}(V_{DD} - V_{tn})/2.5$. Again, this is best illustrated by an example.

Example 4.8

Using the concept of equivalent resistances (or admittances), simplify the n-channel driver network of Fig. 4.14 to a single equivalent resistance. Second, estimate the output low voltage. Finally, estimate the 70% fall time.

Solution: As in Example 4.7, Q_3 and Q_6 will be ignored as they are off. Next Q_5 in parallel with Q_7 is equal to a scaled admittance G_1 where

$$G_1 = \frac{12}{2} + \frac{12}{2} = 12 \tag{4.40}$$

G_1 in series with Q_2 is equal to a scaled resistance R_2 where

$$R_2 = \frac{1}{12} + \frac{2}{12} = \frac{1}{4} \tag{4.41}$$

R_2 in parallel with Q_4 is equal to a scaled admittance G_3 where

$$G_3 = \frac{8}{2} + 4 = 8 \tag{4.42}$$

G_3 in series with Q_1 is equal to a scaled resistance R_4 where

$$R_4 = \frac{2}{12} + \frac{1}{8} = \frac{7}{24} = 0.292 \tag{4.43}$$

Note that $1/R_4$ is equal to the W/L of the equivalent transistor Q_4' of Example 4.7. Continuing, the equivalent unscaled resistor is now given by

$$\begin{aligned} R_{eq} &= 0.292\frac{2.5}{\mu_n C_{ox}(V_{DD} - V_{tn})} \\ &= 1.55 \text{ k}\Omega \end{aligned} \tag{4.44}$$

assuming $\mu_n C_{ox} = 188\ \mu A/V^2$ and $V_{tn} = 0.8$ V. Now the current of the p-channel load is given by

$$I_{D\text{-}8} = \frac{\mu_p C_{ox}}{2}\left(\frac{W}{L}\right)_8 V_{eff\text{-}2}^2 = 50.1\ \mu A \tag{4.45}$$

assuming $\mu_p C_{ox} = 44.5\ \mu A/V^2$ and $V_{tp} = -0.9$ V. One might then assume that $V_{out\text{-}0}$ would be simply given by $I_{D\text{-}8} R_{eq}$ where R_{eq} is given by (4.44). This would be incorrect, because R_{eq} as given by (4.44) is an average resistance to be used for calculating transient fall times. The resistance to be used to calculate $V_{out\text{-}0}$ should be the triode-region resistance, that is the drain-source resistance when V_{DS} is very small. This is a factor of 2.5 smaller. Thus,

$$V_{out\text{-}0} = I_{D\text{-}8}\frac{R_{eq}}{2.5} = 31 \text{ mV} \tag{4.46}$$

For calculating the fall time

$$\tau = C_L R_{eq} = 0.2 \times 10^{-12} \times 1.55 \times 10^3 = 0.31 \text{ ns} \tag{4.47}$$

and

$$t_{\text{-}70\%} = \tau \ln\left(\frac{0.07 + 3.3}{0.07 + 1.0}\right) = 0.37 \text{ ns} \tag{4.48}$$

This is expected to be somewhat too small because the current of the p-channel load has been ignored.

EVALUATING THE LOGIC FUNCTION OF AN NMOS GATE

It is quite simple to determine the logic function being implemented by a pseudo-NMOS gate, even when the gates are quite complex. To do this, one analyzes the n-channel driver network to determine when it is low impedance. Assuming that the widths of the drivers have been chosen large enough, then whenever the n-channel driver network is a low impedance, the output will be a zero. To determine when the driver network is low impedance, one need only remember that a series connection of subnetworks corresponds to the *and* function, whereas a parallel connection corresponds to the *or* function. The procedure for determining the logic function is simply

1. Start with any transistor connected to ground and write the name of its input. Let us assume the transistor is Q_1 with input x_1.
2. If Q_1 has one or more transistors in parallel with it, their inputs should be *or*ed with x_1. Similarly, if Q_1 has one or more transistors in series with it, their inputs should be *and*ed with x_1. This procedure should be done separately with all other groups of series or parallel transistors. Each of these groups now represents a subnetwork with its respective logic term.
3. Recursively starting at ground combine each group of parallel or series subnetworks into larger subnetworks by *or*ing or *and*ing their individual logic terms depending on whether they are in parallel or series, respectively.
4. After a single logic term has been found, the complement of the logic function being implemented has been found.

Although this procedure may sound complicated, with only a little practice it may be applied immediately by inspection, even for quite complicated NMOS gates.

Example 4.9

As an example, consider the gate of Fig. 4.14, which has been repeated for clarity in Fig. 4.15, and assume that transistor Q_i has x_i as an input. Q_1 is a subnetwork in series with a subnetwork consisting of Q_4 in parallel with a subnetwork consisting of two series subnetworks. The first has Q_2 in parallel with Q_3. The second has Q_5 in parallel with Q_6 in parallel with Q_7. Thus by inspection

$$\overline{V_{out}} = [x_1(x_4 + (x_2 + x_3)(x_5 + x_6 + x_7))] \quad (4.49)$$

and

$$V_{out} = \overline{x_1(x_4 + (x_2 + x_3)(x_5 + x_6 + x_7))} \quad (4.50)$$

An equivalent logic circuit is shown in Fig. 4.16.

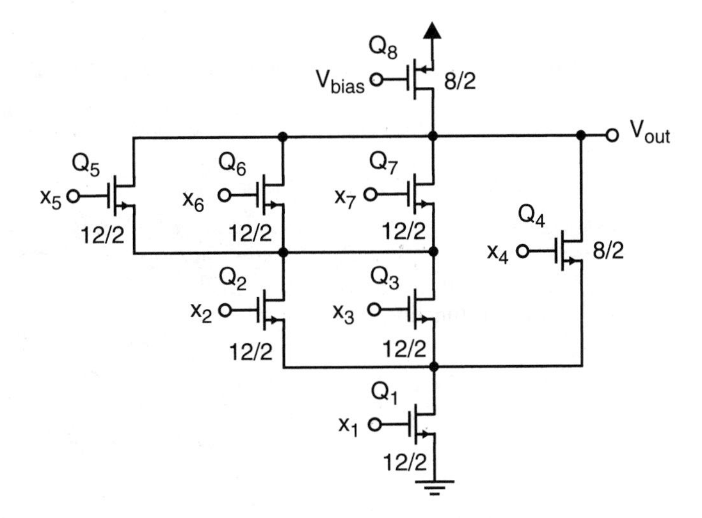

Figure 4.15 The circuit of Fig. 4.14 that is being analyzed to determine its logic function.

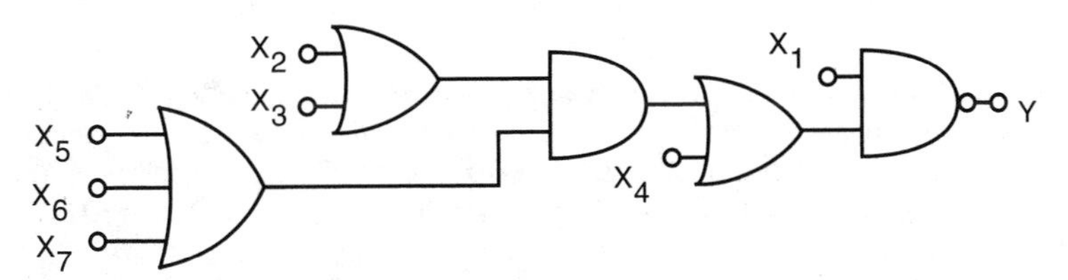

Figure 4.16 An equivalent logic diagram to the NMOS gate of Fig. 4.15.

REALIZING COMPLEX PSEUDO-NMOS GATES

As has been seen, quite complex logic functions can be realized by single NMOS gates. In general, any combination of noninverting *and* and *or* gates feeding into a single inverting *nand* or *nor* gate can be realized by a single pseudo-NMOS gate. The procedure for doing this (synthesis) is the reverse of the procedure for finding the logic function of an NMOS gate (analysis). The final inversion is realized because the n-channel drive network pulls the output low. Therefore, starting at the output, realize the inversion by the combination of the n-channel drive network along with the load transistor. The final *nand* or *nor* gate should now be considered an *and* or *or* gate, respectively. Now, working from the output to the input, recursively realize the gate by subnetworks. An n-input *and* gate is realized by n series subnetworks; an n-input *or* gate is realized by *n* parallel subnetworks. Each of these logic functions in turn is decomposed. If the input to an *and* or *or* gate does not come from another logic gate, but is an input to the overall circuit from the *outside world*, then the recursion is terminated by an enhancement drive transistor. As before, this procedure is most readily understood with the use of some simple examples.

Example 4.10

Realize the logic circuit of Fig. 4.17 by an NMOS gate.

Solution: The step-by-step procedure is illustrated in Fig. 4.18.

1. The inversion of *or* gate *a* is realized by the p-channel transistor load in combination with the complete n-channel driver network. Since gate *a* is now considered a three input or gate, it translates to three parallel subnetworks between the output and ground. Since one of the inputs of gate *a* is from the outside world (i.e., x_8), one of the subnetworks is simply a transistor (Q_8). This is illustrated in Fig. 4.18a.

2. Next, subnetwork 1 is decomposed. It corresponds to gates *b* and c in Fig. 4.17. *And* gate *b* corresponds to two series subnetworks between the output and ground, one of which is a transistor Q_6 having input x_6. The other subnetwork, corresponding to *or* gate *c* translates to two parallel transistors Q_1 and Q_2 having inputs x_1 and x_2. This step of the realization is illustrated in Fig. 4.18b.

3. Finally, a similar procedure is used for realizing subnetwork 2. The complete circuit is shown in Fig. 4.18c.

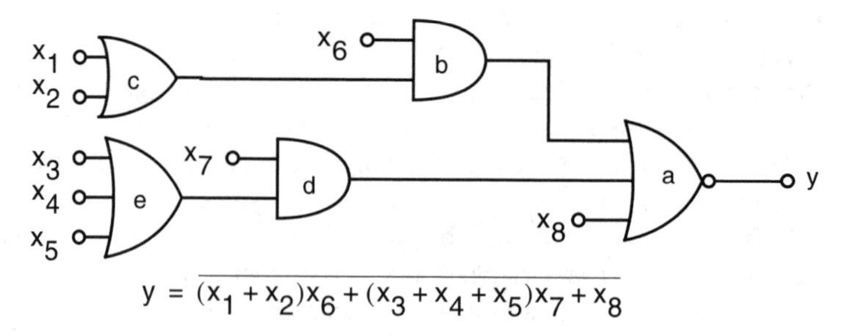

$$y = \overline{(x_1 + x_2)x_6 + (x_3 + x_4 + x_5)x_7 + x_8}$$

Figure 4.17 A logic circuit used in Example 4.10.

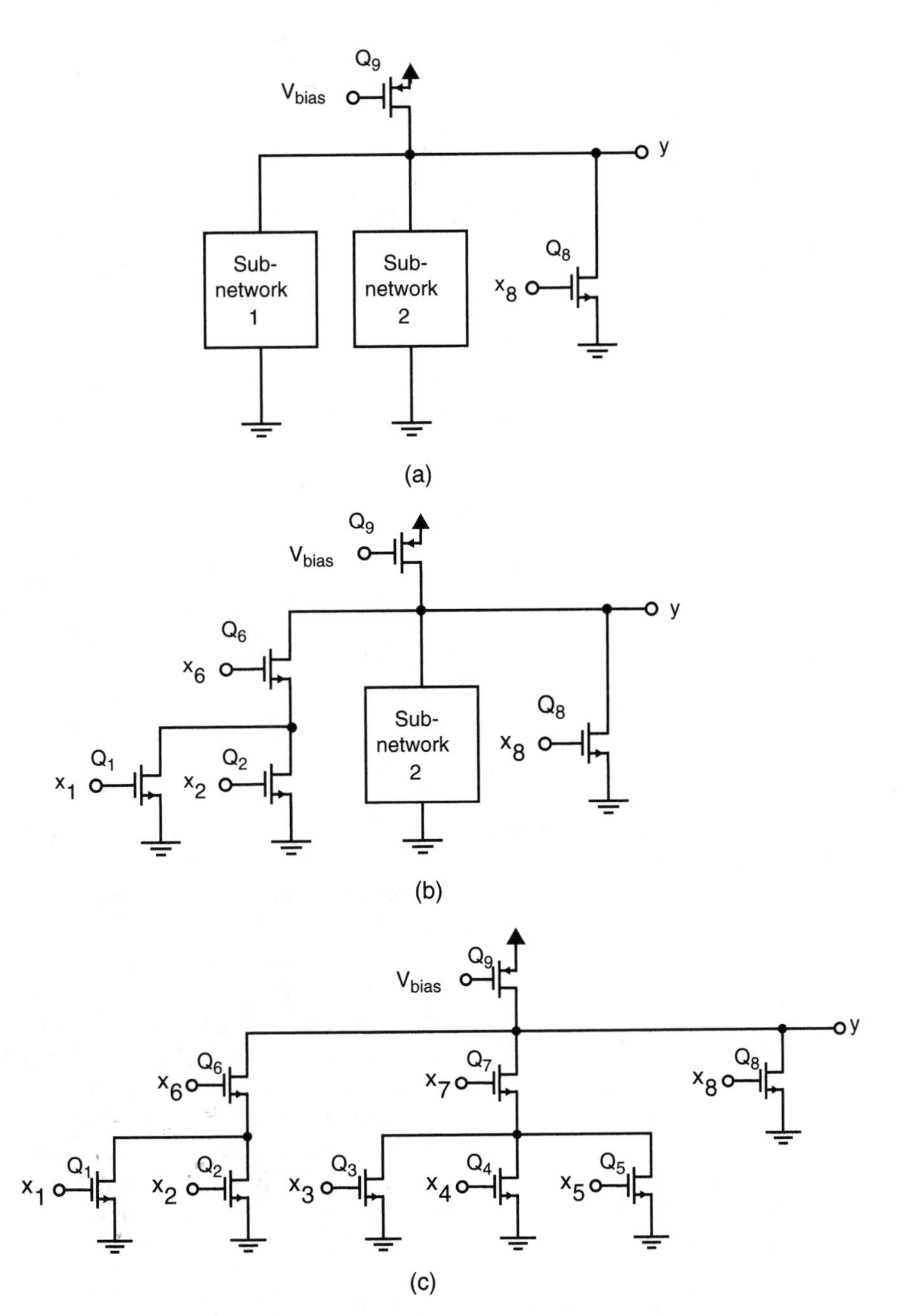

Figure 4.18 The various stages of realizing the logic function of Fig. 4.17.

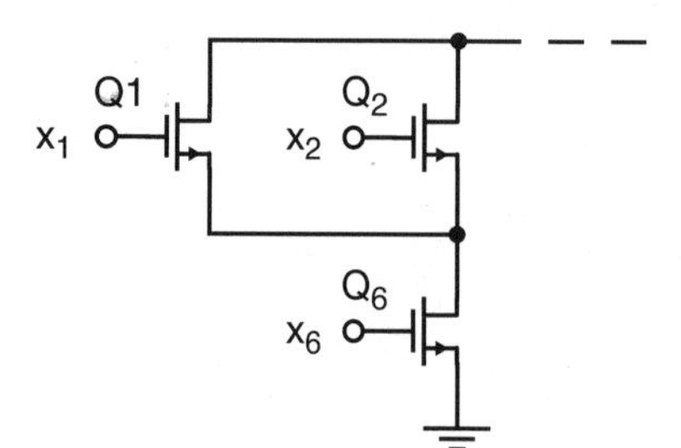

Figure 4.19 An alternative (and inferior) realization of subnetwork 1 of Fig. 4.18a.

It should be noted that the realization of subnetwork 1 in the above example is not unique. It would have been possible to have Q_6 connected to ground and Q_1 and Q_2 connected to the output, as shown in Fig. 4.19. As will be shown later in this chapter, this would result in a slower gate, as it would increase the capacitance at the output node. The following rule is used for minimum delay: *when realizing and functions by a series connection of subnetworks, place the more complicated subnetworks closer to ground.* This keeps the parasitic capacitances introduced by the transistor junctions as close to ground as possible, which in turn means they are being discharged by the smallest equivalent resistance possible. This principle is particularly important for dynamic gates, which will be described in a later chapter. As a final note, never realize NMOS gates that can have more than four drive transistors in series as this would seriously degrade the speed of the gate.

Choosing Transistor Sizes

As mentioned previously, the basic rule in choosing transistor sizes is that for any possible input combination that results in a finite impedance between the output and ground, the equivalent transistor to the n-channel driver network should have a W/L at least one-half that of the load, assuming the load is a p-channel transistor with its gate connected to $V_{DD}/2$. Given this constraint, the W/Ls of the driver network should be as small as possible.

The relative size of the p-channel load transistor (i.e., its W/L) is chosen as a compromise between speed and size versus power dissipation. The larger the W/L of the load transistor, the faster the gate will be, particularly when driving many other gates or a bus. Unfortunately, this increases the power dissipation and the area of the driver network. A typical W/L might be around 5 $\mu m/L_{min}$ or 10 $\mu m/L_{min}$ where L_{min} is the minimum channel length of the process.

Once the size of the load transistor has been chosen, then a simple procedure can be used to choose the W/Ls of the drive transistor. Although it is not an optimum procedure for maximizing speed, the difference in speed between the resulting gate and the optimum gate is usually small. The procedure is as follows:

1. Let $(W/L)_{eq}$ be equal to one-half the W/L of the p-channel load transistor.
2. For each transistor Q_i, determine the maximum number of drive transistors it will be in series with, for all possible inputs. Denote this number n_i. This can almost always be determined by a very cursory examination.
3. Take $(W/L)_i = n_i\,(W/L)_{eq}$.

Example 4.11

Choose appropriate sizes for the pseudo-NMOS logic gate shown in Fig. 4.20. Since the size of the p-channel load transistor $(W/L)_8$ is 5 μm/0.8 μm, we have $(W/L)_{eq}$ = (5 μm/0.8 μm)/2 = 3.125. A 0.8 μm technology is assumed since the length of the load transistor is 0.8 μm; thus, the smallest allowable dimension for any device is 0.8 μm. Normally, the gate lengths of the drive transistors are always taken at their minimum to minimize parasitic capacitances and size and to maximize speed. Therefore, the length of Q_1 will be taken equal to 0.8 μm. The width of Q_1 is chosen equal to $L_{min}(W/L)_{eq}$ = 2.5 μm, as it is not in series with any other transistors. Next, considering Q_2, the worst possible case is when all drive transistors are off except Q_2, Q_6, and Q_7. This means the size of Q_2 should be taken equal to $3(W/L)_{eq}$ = 7.5 μm/0.8 μm. Similarly, the worst case for Q_3 is when all transistors are off except Q_3, Q_4, and Q_5, and thus the size of Q_3 should also be 7.5 μm/0.8 μm. Next, considering Q_4, the worst case is when all transistors are off except Q_4, Q_5, Q_6, and Q_7. Therefore, the size of Q_4 should be taken equal to $4(W/L)_{eq}$ = 10 μm/0.8 μm. Similar arguments hold for Q_5, Q_6, and Q_7, which should all have sizes equal to 10 μm/0.8 μm. All of the selected sizes are listed in the following table.

Again, these sizes are not optimum; Q_2 and Q_3 could have been chosen a little smaller, but the difference in gate speed would be small.

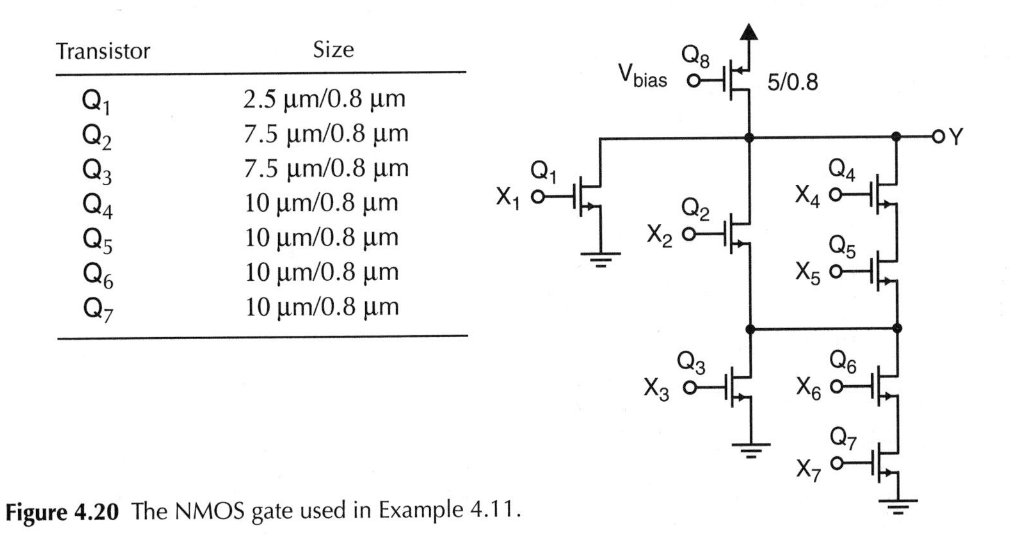

Transistor	Size
Q_1	2.5 μm/0.8 μm
Q_2	7.5 μm/0.8 μm
Q_3	7.5 μm/0.8 μm
Q_4	10 μm/0.8 μm
Q_5	10 μm/0.8 μm
Q_6	10 μm/0.8 μm
Q_7	10 μm/0.8 μm

Figure 4.20 The NMOS gate used in Example 4.11.

Power Dissipation

A pseudo-NMOS logic gate having a "1" output has no d.c. power dissipation. A pseudo-NMOS logic gate having a "0" output has a d.c. power dissipation equal to the current of the p-channel load transistor multiplied by the power supply voltage. Thus, the power dissipation of a gate with a "0" output is given by

$$P_d = \frac{\mu_p C_{ox}}{2}\left(\frac{W}{L}\right)_p V_{eff\text{-}p}^2 V_{DD} \tag{4.51}$$

where $(W/L)_p$ is the size of the p-channel load transistor. Assuming that a pseudo-NMOS logic gate will have a "1" output for half the time and a "0" output for half the time, the average power dissipation will be

$$P_d = \frac{\mu_p C_{ox}}{4}\left(\frac{W}{L}\right)_p V_{eff\text{-}p}^2 V_{DD} \tag{4.52}$$

In addition to this d.c. power, there is a.c. power dissipated whenever a capacitor is charged or discharged. Also, most ICs require buffers for driving output pins and internal buses. These buffers often dissipate an order of magnitude more power than a typical gate. Altogether, this implies that pseudo-NMOS ICs with tens of thousands of gates (which is not a large IC by modern standards) dissipate a very large amount of power, typically too much for an IC package to dissipate. For this reason, pseudo-NMOS logic is seldom used for a complete circuit; rather, it might be used only in select locations in which the fan-out is small and speed is critical. *However, the design procedures used for the n-channel drive network are identical to the design procedures for the n-type drive networks of other logic, families such as traditional CMOS logic, dynamic CMOS logic, and GaAs logic.*

Example 4.12

What is the d.c. power dissipation of a 20,000 gate pseudo-NMOS gate-array package if the p-channel transistor loads have a size of $(W/L)_p = 4/2$ and the output buffers are ignored.

Solution: Using equation (4.52), we have

$$P_{total} = 20,000\frac{44.5\times10^{-6}}{4}\frac{4}{2}(0.75)^2 3.3 = 0.83\text{ W} \tag{4.53}$$

assuming $\mu_p C_{ox} = 44.5\ \mu\text{A/V}^2$ and $V_{eff\text{-}p} = 0.75$ V. This is very high for an IC containing only 20,000 gates; modern ICs may contain millions of gates.

Other Pseudo-NMOS Circuits

The bias voltage for pseudo-NMOS circuits can be easily obtained using a reference circuit such as that shown in Fig. 4.21. The ratio of $(W/L)_1$ to $(W/L)_2$ should be the same as the ratio of $(W/L)_{eq}$ to $(W/L)_p$; $(W/L)_{eq} = (1/2)(W/L)_p$ is a reasonable choice as shown. The transistor Q_3 has been included to realize capacitive loading, which helps minimize noise being injected onto V_{bias}. Its exact size is dependent on how many gates the reference circuit is connected to and can be determined using SPICE-level simulation.

Alternative pseudo-NMOS circuits can be realized that do not require a reference (or bias) circuit by simply connecting the gate of the p-channel load to ground as is shown in Fig. 4.22. Notice that the relative size of the n-channel drive transistor relative to the p-channel load transistor is now different. To obtain reasonable gate threshold voltages and output-low voltages, the typical choice is to now take $(W/L)_{eq} = 2(W/L)_p$. The transfer curve of the inverter of Fig. 4.22, obtained using SPICE, is shown in Fig. 4.23. It is seen that the gate threshold voltage is 1.58 V and the output low voltage is 0.17 V; both are reasonable values. Also note that the gain at the gate threshold voltage is –6.5, which is less than the pseudo-NMOS gate when the p-channel load is biased using a bias voltage. The reason for the lower gain is that when the gate output voltage is at the threshold voltage, the p-channel load transistor is in the triode region and has a smaller drain-source impedance (i.e., r_{ds}) than when it is in the active or saturation region. The analysis for the gain at the gate threshold voltage is left as an exercise for the interested reader (see Problem 4.12).

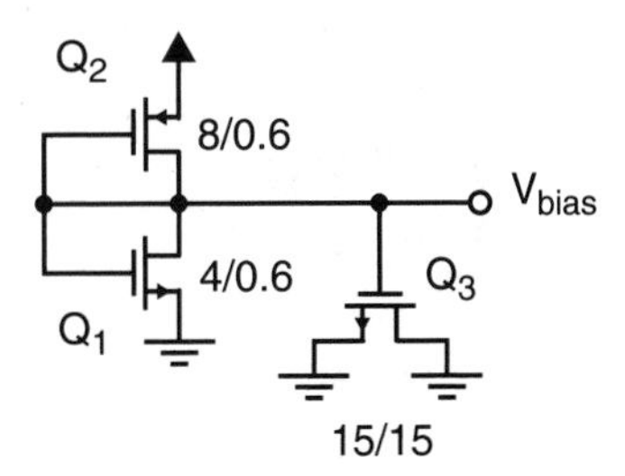

Figure 4.21 A circuit that can be used to generate the bias voltage for pseudo-NMOS gates.

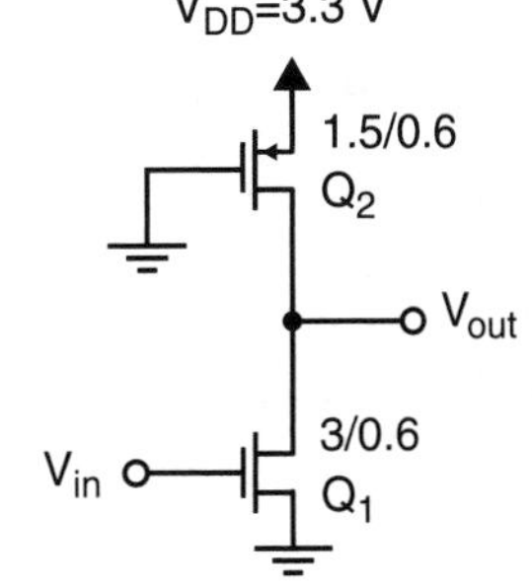

Figure 4.22 An alternative typical pseudo-NMOS inverter in a 0.6 μm technology.

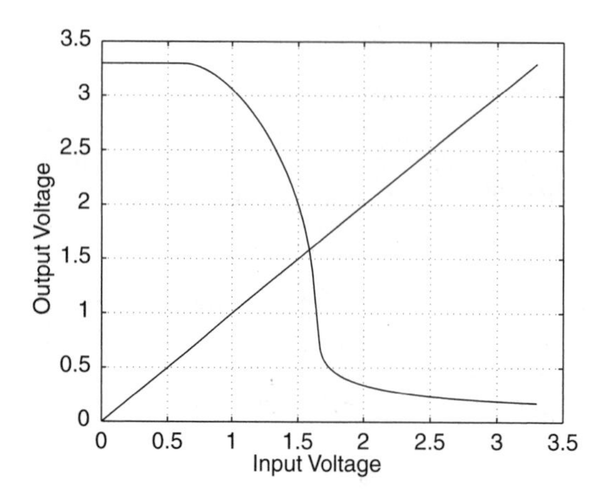

Figure 4.23 The transfer function of the pseudo-NMOS inverter of Fig. 4.22 obtained using SPICE.

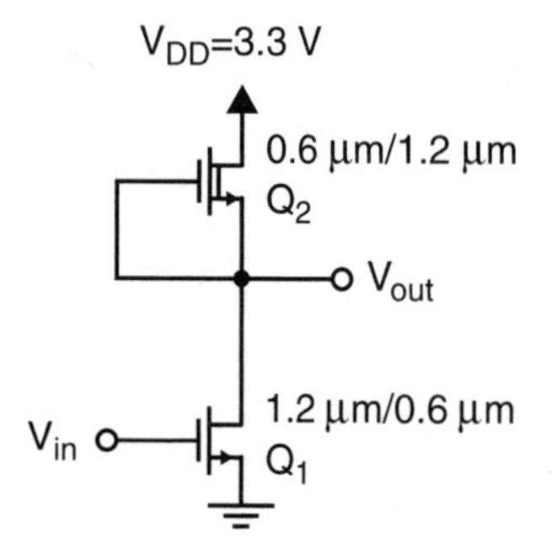

Figure 4.24 A typical NMOS depletion-load inverter in a 0.6 μm technology.

NMOS Logic with Depletion-Load Transistors

The pseudo-NMOS gates we have just described were actually based on an earlier type of logic where n-channel depletion transistors were used as the loads. This logic family was called NMOS logic and was prevalent in the early days of 8-bit microprocessors; now it is seldom used. NMOS logic is very similar to pseudo-NMOS logic; although historically it came second, pseudo-NMOS logic was described first because it is currently used sometimes for critical circuits in an IC, whereas depletion-load NMOS logic is almost never used now. Also, the analysis of pseudo-NMOS logic for the gain at the gate threshold voltage is simpler than for traditional NMOS logic where a second-order effect called the body effect becomes dominant. This section should be considered optional for a first-level course; it has been included because of its historical importance and as an example of using depletion transistors.

An example NMOS depletion-load inverter is shown in Fig. 4.24. The load is realized by a depletion transistor having its gate connected to its source. Since for a depletion transistor, we have $V_{td} < 0$, where V_{td} is the threshold voltage of the transistor, a channel is present for $V_{GS} = 0$. The depletion transistor will be in the active region for $V_{DS} > V_{eff\text{-}d} = -V_{td}$, which might be on the order of 2 V or a little larger. Assuming this to be the case, we have the drain current of the depletion load transistor given by

$$I_{D-2} = \frac{\mu_n C_{ox}}{2}\left(\frac{W}{L}\right)_2 V_{td}^2 \tag{4.54}$$

Thus, the depletion transistor with its gate connected to its source approximates a d.c. current source ignoring second-order effects; as such it is an extremely efficient realization. In reality, it approximates only a moderate-quality current source. A major source of error is the fact that as the source-to-substrate voltage changes, the depletion-region width between the channel and the substrate changes, which in turn causes the threshold voltage to change according to the equation

$$V_{td} = V_{td\text{-}0} + \gamma(\sqrt{V_{SB} + |2\phi_F|} - \sqrt{|2\phi_F|}) \tag{4.55}$$

This effect is called the *body effect*. When the inverter output voltage is around the gate threshold voltage, this causes the output impedance of the inverter to be finite, perhaps on the order of only eight times greater than $1/g_{m\text{-}2}$ where $g_{m\text{-}2}$ is the transconductance of the depletion transistor. In addition, the finite output impedance of the depletion and driver transistors also reduce the gain of the inverter somewhat. Together, these two effects cause the gain of the inverter of Fig. 4.24 at its threshold voltage to be given by

$$\frac{V_{out}}{V_{in}} = A_{inv} = \frac{-g_{m\text{-}1}}{g_{s\text{-}2} + g_{ds\text{-}1} + g_{ds\text{-}2}} \tag{4.56}$$

when the output voltage is around the gate threshold voltage. Note that $g_{s\text{-}2}$ is the body-effect transconductance parameter and that $g_{ds\text{-}1}$ and $g_{ds\text{-}2}$ are the drain-source admittances of Q_1 and Q_2, respectively. The derivation (4.56) is left as an exercise for the interested reader (see Problem 4.13). Thus, the gain of the inverter at the gate-threshold voltage might be between –6 and –9. However, this somewhat low gain has little impact on digital circuits. A typical transfer curve of an NMOS inverter obtained using SPICE is shown in Fig. 4.25. A value of $V_{td} = -2$ V was used to obtain this curve. Also, the W/L of the drive transistor was taken to be four times the W/L of the drive transistor as is shown in Fig. 4.25. This ratio of sizes was a typical ratio often chosen by NMOS circuit designers. The gate-threshold voltage is seen to be 1.44 V and the gain at the threshold voltage is –8.3. The output low voltage is 0.14 V. This transfer curve is acceptable.

NMOS logic dominated integrated circuit design during the late 1970s. Indeed, it was NMOS technology that was a key factor in the development of 8- and 16-bit microprocessors that enabled the home-computer revolution. In the early 1980s, CMOS logic, which will be described next, began to become more popular because its power dissipation was much less, despite the fact that the circuits took more area, the yield was less, and the processing was more complicated. In the late 1990s, NMOS logic has practically disappeared due to its power dissipation. However, as was mentioned previously, similar logic families with similar design methodologies are still quite popular.

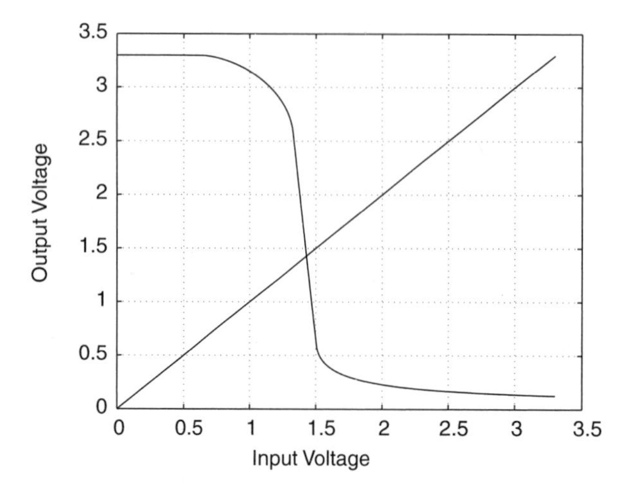

Figure 4.25 The transfer curve of the NMOS inverter of Fig. 4.24.

4.4 CMOS Logic

CMOS Logic, first developed for commercial use in the early 1970s, was originally thought to be too complicated, expensive, and slow to compete with NMOS technology. It was also originally prone to a failure mechanism called *latch-up* (which was described in Chapter 3). However, with improvements in technology, the increasing complexity of NMOS processing, and the increasing importance of power dissipation as ICs got larger, CMOS technology has almost completely supplanted NMOS technology at present. Not only does the availability of p-channel transistors allow for much lower power dissipation, but it also gives a much better drive capability for positive-going signals. As ICs become faster, this becomes more critical, especially for output buffers that need to drive large capacitive loads.

Normally, a CMOS IC has only n-channel and p-channel *enhancement* transistors; it is possible to realize *depletion* transistors as well, but typically this is not done because the few times they would be useful does not justify the extra mask required.

As was done for pseudo-NMOS logic, the CMOS inverter will first be revisited in greater detail, and then CMOS logic will be studied more fully and in greater detail than was done in the introduction in Chapter 1.

CMOS Inverter

A typical CMOS inverter is shown in Fig. 4.26. Note that the size of the p-channel transistor is wider than that of the n-channel transistor. This width difference is not needed for functionally correct operation. Rather, it somewhat compensates for the difference in the mobilities of n-channel and p-channel transistors. The effective mobility of n-channel transistors is between two and four times that of p-channel transistors. Making the p-channel device wider by a ratio equal to the inverse of the corresponding mobility ratio gives a gate threshold voltage close to $V_{DD}/2$ and more nearly equal rise and fall times than would otherwise be the case.

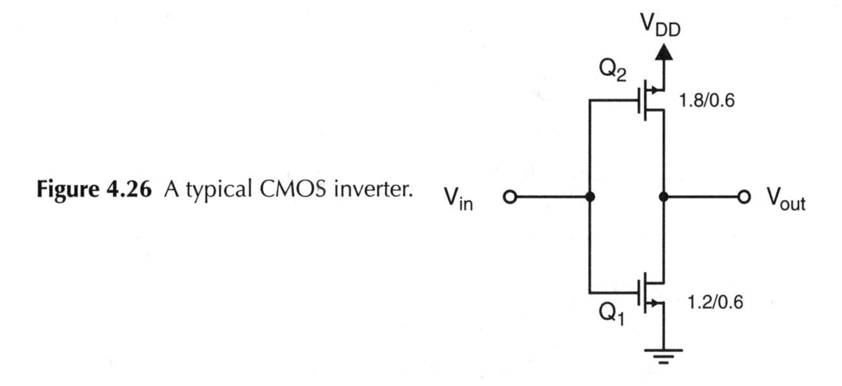

Figure 4.26 A typical CMOS inverter.

Alternatively, taking the p-channel width more closely equal to the width of the n-channel saves on area and helps minimize loads on preceding gates. Irrespective of the relative sizes chosen, traditional CMOS gates will function correctly. For this reason, *traditional CMOS logic is called a ratioless logic family.* Typical choices for the width of p-channel transistors might be between 1 and 2.5 times the width of the n-channel devices. Unless large capacitive loads are being driven, a width 1.5 times that of the n-channel transistor is a reasonable choice most of the time; the exception is when a large number of p-channel transistors are in series, in which case they should be taken correspondingly wider.

THRESHOLD VOLTAGE

When analyzing the CMOS inverter for its threshold voltage (V_{th}), it is known a priori that both the p-channel and n-channel transistors are in their active or saturation regions. This is true because since V_{in} and V_{out} are equal (when $V_{in} = V_{th}$), the drain-gate voltages of both the p- and the n-channel transistors are zero. Since they are both enhancement transistors, this implies they must be in the active region.[3] Thus for the n-channel transistor Q_1, we have

$$I_{D\text{-}1} = \frac{\mu_n C_{ox}}{2}\left(\frac{W}{L}\right)_1 (V_{th} - V_{tn})^2 \tag{4.57}$$

and for the p-channel transistor Q_2, we also have[4]

$$I_{D\text{-}2} = \frac{\mu_p C_{ox}}{2}\left(\frac{W}{L}\right)_2 (V_{DD} - V_{th} + V_{tp})^2 \tag{4.58}$$

[3]For the n-channel transistor, we have $V_{DS} = 0 > -V_{tn}$ and for the p-channel transistor we have $V_{GD} = 0 > V_{tp}$, which are the necessary conditions for the transistors to be in the active region.

[4]Notice that for p-channel transistors, a negative sign has been inserted in front of every voltage variable of the large-signal I–V equations that were used for n-channel transistors.

where we recall that V_{tp} is negative. Setting (4.57) equal to (4.58), and solving for V_{th}, gives

$$V_{th} = \frac{V_{tn} + (V_{DD} + V_{tp})\sqrt{\mu_p (W/L)_2 / \mu_n (W/L)_1}}{1 + \sqrt{\mu_p (W/L)_2 / \mu_n (W/L)_1}} \tag{4.59}$$

Example 4.13

Find the threshold voltage of the inverter of Fig. 4.26. Assume $\mu_n = 545\ \text{cm}^2/\text{V} \cdot \text{s}$ and $\mu_p = 130\ \text{cm}^2/\text{V} \cdot \text{s}$, $V_{tn} = 0.8$ V, $V_{tp} = -0.9$ V, and $V_{DD} = 3.3$ V.

Solution: Substituting the given values in equation (4.59) yields

$$V_{th} = \frac{0.8 + (3.3 - 0.9)\sqrt{(130/545)1.5}}{1 + \sqrt{(130/545)1.5}} = 1.40\ \text{V} \tag{4.60}$$

This should be compared with the threshold voltage obtained using SPICE, which is 1.45 V. Since, for CMOS gates, V_{OL} is 0 V and V_{OH} is 3.3 V, this would give noise margins of $NM_L = 1.45$ V and $NM_H = 1.85$ V for the device sizes chosen. If we had chosen $(W/L)_2 = (W/L)_1$, we would have had $V_{th} = 1.32$ V for the same parameters, whereas if we had chosen $(W/L)_2 = (\mu_n/\mu_p)(W/L)_1 = 4.2(W/L)_1$, we would have $V_{th} = V_{DD}/2 = 1.65$ V.

INVERTER GAIN AT $V_{IN} = V_{TH}$

In analyzing the CMOS inverter for its gain at the threshold voltage, the procedure is the same as for the pseudo-NMOS inverter. The small-signal equivalent circuit for the inverter is used, which is then analyzed for its gain. The small-signal equivalent circuit for the CMOS inverter is shown in Fig. 4.27.

Notice that as in all small-signal circuits, the d.c. power supplies have been set to zero. There are no current sources to model the body effect, as the sources of both transistors are connected to small-signal grounds. Note also that the small-signal circuit for the p-channel transistor is identical to the small-signal circuit for the n-channel transistor.[5] Thus, there are two small-signal models for transistors in parallel. This allows Fig. 4.27 to be redrawn as shown in Fig. 4.28.

---

[5]The change of the sign of the voltage variables for the large-signal equations of p-channel transistors is cancelled by the fact that the large-signal current of p-channel transistors is from the source to the drain (as opposed to being from the drain to the source for n-channel transistors).

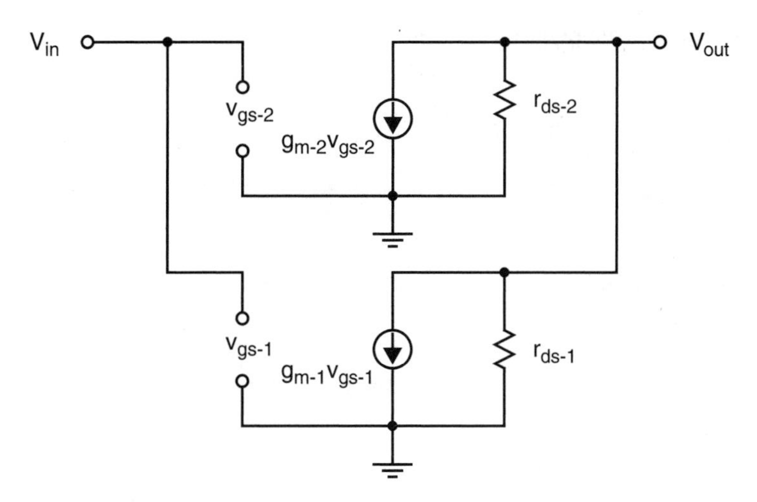

Figure 4.27 The equivalent small-signal model of a CMOS inverter.

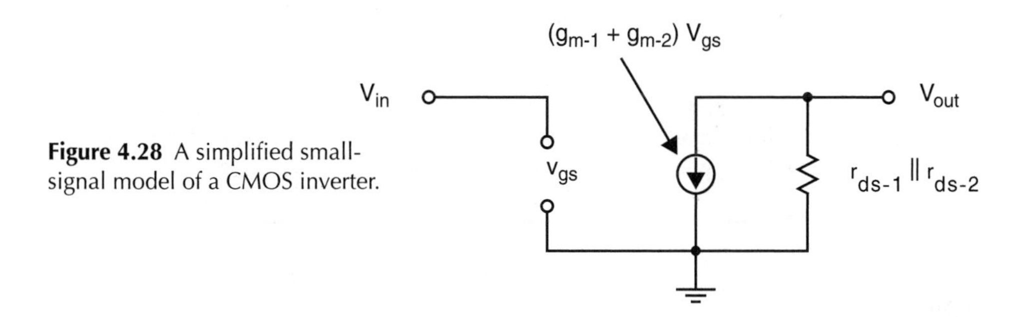

Figure 4.28 A simplified small-signal model of a CMOS inverter.

The gain is now immediately found to be

$$\frac{v_{out}}{v_{in}} = -(g_{m-1} + g_{m-2})(r_{ds-1} \parallel r_{ds-2}) \tag{4.61}$$

Example 4.14

For $\mu_n C_{ox} = 188\ \mu A/V^2$, $\mu_p C_{ox} = 44.5\ \mu A/V^2$, $\lambda_n = 0.06$ for n-channel transistors and $\lambda_p = 0.07$ for p-channel transistors, and all other parameters taken equal to those of Example 4.13, find the gain at the threshold voltage.

Solution: From Example 4.13, we have

$$I_{D-1} = \frac{\mu_n C_{ox}}{2}\left(\frac{W}{L}\right)_1 (V_{th} - V_{tn})^2 = 67.7\ \mu A \tag{4.62}$$

Thus,

$$r_{ds-n} = \frac{1}{\lambda_n I_{D1}} = 246\ k\Omega \tag{4.63}$$

and

$$r_{ds-p} = \frac{1}{\lambda_p I_{D1}} = 211\ k\Omega \tag{4.64}$$

Also, using

$$g_{m-1} = \sqrt{2\mu_i C_{ox}\left(\frac{W}{L}\right)_i I_{Di}} \tag{4.65}$$

gives $g_{m-1} = 226\ \mu A/V$ and $g_{m2} = 134\ \mu A/V$. Therefore, using (4.61) we find the gain to equal –40.1. SPICE calculated the gain to be –20.1. Again, SPICE is often inaccurate when calculating gains, but this is seldom significant for digital circuits.

Transient Response

The calculation of the transient response for both the rise and the fall times is almost identical to the calculation of the fall time of the NMOS inverter. For example, with respect to the fall time of the CMOS inverter, the p-channel transistor is off and can be ignored, and the n-channel transistor can be approximated by an equivalent resistor of size

$$R_{eq-1} = \frac{2.5}{\mu_n C_{ox}(W/L)_1 (V_{DD} - V_{tn})} \tag{4.66}$$

Similarly, during the rise time the p-channel transistor can be approximated by

$$R_{eq-2} = \frac{2.5}{\mu_p C_{ox}(W/L)_2 (V_{DD} + V_{tp})} \tag{4.67}$$

Using these approximations, the rise or fall times can now be approximated using the solution to a first-order RC circuit:

$$t_f \cong R_{eq\text{-}1} C_L \ln\left[\frac{V_{out}(\infty) - V_{out}(t_1)}{V_{out}(\infty) - V_{out}(t_2)}\right] \tag{4.68}$$

For a –70% voltage change, this yields

$$t_{-70\%} \cong 1.2 R_{eq\text{-}1} C_L \tag{4.69}$$

Similarly, for the rise time we have

$$t_{+70\%} \cong 1.2 R_{eq\text{-}2} C_L \tag{4.70}$$

Example 4.15

For the CMOS inverter of Fig. 4.26, find approximate values for the 70% rise and fall times assuming a load capacitance of 0.2 pF.

Solution: Using equations (4.66) and (4.67), we have $R_{eq\text{-}1} = 2.66\ \text{k}\Omega$ and $R_{eq\text{-}2} = 7.80\ \text{k}\Omega$. Substituting these values into (4.69) gives

$$t_{-70\%} = 0.64\ \text{ns} \tag{4.71}$$

and from (4.70)

$$t_{+70\%} = 1.87\ \text{ns} \tag{4.72}$$

These should be compared to transient times from SPICE of 1.07 and 1.67 ns, respectively. The differences are primarily due to junction capacitances and short-channel effects that we ignored in our hand analysis.

THE EFFECT OF TRANSISTOR SIZES ON THE TRANSIENT RESPONSES

In NMOS logic, the correct choice of the sizes of the transistors was necessary for the gates to work. This is not necessary for typical CMOS logic to operate functionally correctly. However, as was mentioned, the ratio of the W/Ls of the p-channel transistors to the W/Ls of the n-channel transistors does affect the gate threshold voltages and, more importantly, the transient response.

There is no exact optimum ratio for the sizes of p-channel transistors to n-channel transistors that can be specified independent of the application. However, there are two situations in which

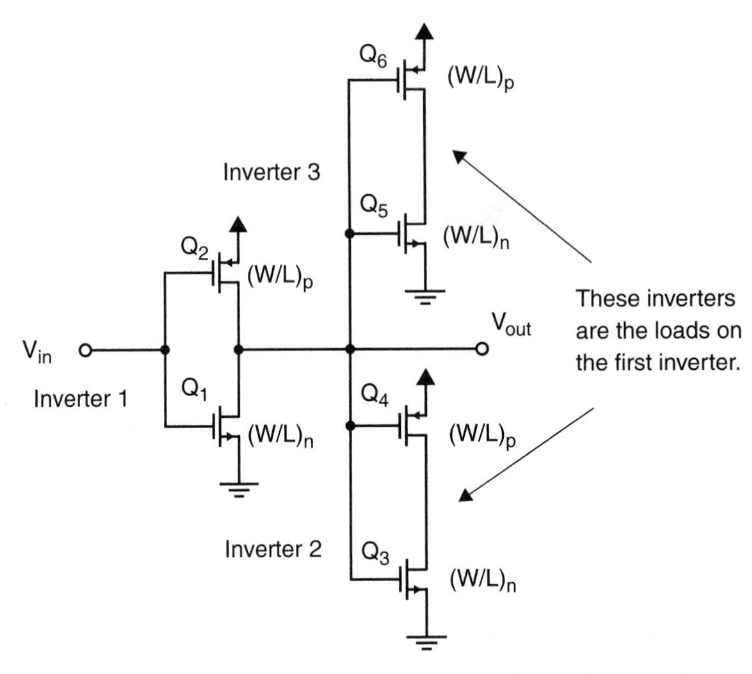

Figure 4.29 A CMOS inverter driving two identical inverters.

something can be said about the relative W/Ls. The first case is when a single gate is driving a number of identical gates.

Consider the situation shown in Fig. 4.29 in which a CMOS inverter is loaded by the input capacitance of two identical inverters. This example is to be analyzed for the transient response times of the first inverter. It is assumed that all n-channel transistors have identical W/Ls. Also, all p-channel transistors are assumed to have the same W/Ls, but these may be different from those of the n-channel transistors. This situation is representative of many typical situations encountered in CMOS design where the fan-out is small.

When considering the transient response of the first inverter, it is first necessary to approximate the capacitive load. There are two components to this load. The first is the junction capacitance of the drains of Q_1 and Q_2. These capacitances are roughly proportional to the widths of Q_1 and Q_2, assuming Q_1 and Q_2 are not too large[6]. In addition, there are the capacitances due to driving the inverters 2 and 3. The input capacitances of these inverters consist principally of the gate capacitances of two n-channel transistors and two p-channel transistors. The exact

[6]When the widths are large, the transistors are realized by a multiple-finger structure that minimizes the junction capacitance (as described in Chapter 3).

determination of these capacitances is a difficult and highly nonlinear problem, but fortunately conservative estimates are easily arrived at using (3.61) of Chapter 3, which is repeated here. When a transistor is *hard* in the triode region, its gate-channel capacitance is approximately given by

$$C_{gs} = WLC_{ox} \tag{4.73}$$

Using equation (4.73), we find the capacitive loading of the two inverters is approximately given by

$$C_L = 2C_{ox}L(W_n + W_p) \tag{4.74}$$

assuming the n- and p-channel transistors have equal lengths, which are also chosen equal to the minimum length. Thus, these input capacitances are also proportional to the widths of the transistors.

Assuming all transistors have equal minimum lengths, then scaling the widths of all the transistors equally, both n-channel and p-channel, has little effect on the transient delays. This is because as the widths go up, the capacitive loads go up proportionally, but the equivalent resistances go down inverse proportionally and the rise and fall time constants are roughly unchanged. This is not exactly true for very small widths in which the junction side-wall capacitances become more important causing larger delays. It also ignores wiring capacitance and external capacitances. When these become important, wider transistors result in smaller delays. However, for the simple case being considered, the independence of the delay on the scaling of all transistor widths is a reasonable approximation. Thus, it is necessary to optimize the delay only as a function of the ratio of the W/L of the p-channel transistors to the W/L of the n-channel transistors.

If a number of gates are in series, with no memory or local feedback (i.e., acyclic logic), then the average of the rise and the fall times determines the overall gate delay. This might not be the case for flip-flops or pipelined logic, but will be the case considered now. Also, only the load capacitance due to the gates being driven will be considered. If the junction capacitance was also considered, the conclusions would change very little, but the analysis would be more complicated. This extra complication is not felt to be merited by any substantial increase in insight.

The average of the rise and the fall times of the first inverter of Fig. 4.29 is

$$t_{AV} = 1.2C_L\frac{R_{eq\text{-}1} + R_{eq\text{-}2}}{2} \tag{4.75}$$

where C_L is given by (4.74) and $R_{eq\text{-}1}$ and $R_{eq\text{-}2}$ are given by (4.66) and (4.67), respectively. Therefore, we have

$$t_{AV} = 1.2 \times 2C_{ox}L(W_n + W_p)\frac{1}{2}\left[\frac{2.5}{\mu_n C_{ox}\frac{W_n}{L}(V_{DD} - V_{tn})} + \frac{2.5}{\mu_p C_{ox}\frac{W_p}{L}(V_{DD} - V_{tp})}\right] \tag{4.76}$$

Assuming $V_{DD} - V_{tn} = V_{DD} + V_{tp}$ and rearranging, we have

$$t_{AV} = \frac{3L^2}{(V_{DD} - V_{tn})\mu_n}\left(1 + \frac{W_p}{W_n}\right)\left(1 + \frac{\mu_n W_n}{\mu_p W_p}\right) \tag{4.77}$$

To find the optimum W_p/W_n to minimize the average of the rise and fall times, we can differentiate (4.77) with respect to W_p/W_n, set the result equal to zero, and solve for W_p/W_n. Continuing, after some algebra, we have

$$\frac{\partial t_{AV}}{\partial(W_p/W_n)} = \frac{3L^2}{(V_{DD} - V_{tn})\mu_n}\left[1 - \frac{\mu_n}{\mu_p}\left(\frac{W_n}{W_p}\right)^2\right] \tag{4.78}$$

Setting (4.78) equal to zero gives the optimum (W_p / W_n) as

$$\left(\frac{W_p}{W_n}\right)_{opt} = \sqrt{\frac{\mu_n}{\mu_p}} \tag{4.79}$$

For example, if $\mu_n/\mu_p = 2.5$ the optimum ratio for W_p/W_n is 1.58. This was the reason for the earlier statement in Section 4 that a ratio of W_p/W_n equal to 1.5 is usually a reasonable choice. For this choice we have

$$t_{AV} = \frac{3L^2}{(V_{DD} - V_{tn})\mu_n}\left[1 + \sqrt{\frac{\mu_n}{\mu_p}}\right]^2 \tag{4.80}$$

It is interesting to see how the average rise and fall time increases for the special cases $W_p/W_n = 1$ and $W_p/W_n = \mu_n/\mu_p$. For $W_p/W_n = 1$ and using (4.77), we have

$$t_{AV} = \frac{6L^2}{(V_{DD} - V_{tn})\mu_n}\left(1 + \frac{\mu_n}{\mu_p}\right) \tag{4.81}$$

Also, for $W_p / W_n = \mu_n/\mu_p$, we have

$$t_{AV} = \frac{6L^2}{(V_{DD} - V_{tn})\mu_n}\left(1 + \frac{\mu_n}{\mu_p}\right) \tag{4.82}$$

which is identical to (4.81). For example, if $\mu_n/\mu_p = 2.5$, a typical value, using (4.80), (4.81), and (4.82), we see the increase in the average of the rise and fall times is 5%. This very small increase indicates that though an optimum exists, taking $W_p/W_n = 1$ would cause very little penalty and would save on space. This choice is often used except in more critical designs.

It should be emphasized that the preceding analysis ignored any wiring capacitance or external capacitances that do not scale with the transistor widths. When these load capacitances dominate, taking the p-channel wider definitely improves the rise time and therefore the average of the rise and fall times. In this extreme, designers often take $W_p/W_n = 2$ or 3. As a compromise, for when the load of the gate is not known, a reasonable choice again might be $W_p/W_n = 1.5$.

The constant term in equation (4.80), that is the term

$$T_{proc} = \frac{L^2}{(V_{DD} - V_{tn})\mu_n} \tag{4.83}$$

is independent of geometry and dependent only on the technology. It is a good figure of merit when comparing two different technologies. Also, it makes it obvious how important short-channel lengths are for high-speed operation.

Example 4.16

For $\mu_n = 545\ \text{cm}^2/\text{V}\cdot\text{s}$, $\mu_p = 130\ \text{cm}^2/\text{V}\cdot\text{s}$, $L_{min} = 0.6\ \mu\text{m}$, and $V_{DD} - V_{tn} = 2.5\ \text{V}$, find the average rise and fall times for the cases $W_p/W_n = 1$, $\sqrt{\mu_n/\mu_p}$, and equal to μ_n / μ_p.

Solution: Using (4.77), we have for $W_p/W_n = 1$, $W_p/W_n = \sqrt{\mu_n/\mu_p}$, and $W_p/W_n = (\mu_n/\mu_p)$, $t_{AV} = 82.3$, 73.6, and 82.3 ps, respectively. In reality, the actual times would be larger due to the junction capacitances that we ignored, but the relative differences would be about the same.

POWER DISSIPATION

As has been mentioned many times previously," the major reason for the popularity of CMOS logic is that traditional gates have no d.c. power dissipation when their outputs are not changing. However, whenever the output of a CMOS gate changes, there is power dissipated in the gate transistors. The primary reason for this power dissipation is the movement of charge required to charge or discharge external load capacitances and internal parasitic capacitances. In addition, for input signals with finite rise and fall times, it is possible that d.c. paths exist between the power supply and ground temporarily while the output is changing. This is especially true for large inverters used as buffers.

The power dissipated by an inverter charging or discharging a load capacitor is particularly easy to calculate. Consider the inverter of Fig. 4.30 that has a square-wave input of frequency f_{clk} and a load capacitance C_L. Each time the output changes from a "0" to a "1", the load capacitor is charged from 0 V to V_{DD} by the n-channel transistor. The energy dissipated in the n-channel transistor during this time is equal to

$$E_n = \frac{C_L V_{DD}^2}{2} \tag{4.84}$$

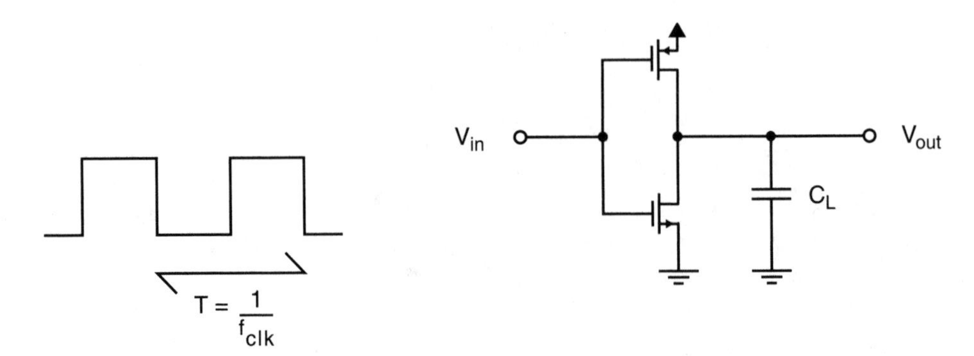

Figure 4.30 A CMOS inverter with a square-wave input (at a frequency f_{clk}).

Similarly, the energy dissipated in the n-channel transistor when the inverter output changes from a "1" to a "0" is

$$E_p = \frac{C_L V_{DD}^2}{2} \tag{4.85}$$

Since the output changes from a "0" to a "1" and back to a "0" each period (T) of the input, the total energy dissipated each period is equal to

$$E_T = C_L V_{DD}^2 \tag{4.86}$$

The average power dissipated during a period is the total energy dissipated divided by T. Thus, the average power dissipation due to the dynamic charging and discharging of the load capacitance is

$$P_{dyn\text{-}avg} = \frac{C_L V_{DD}^2}{T} = C_L V_{DD}^2 f_{clk} \tag{4.87}$$

Thus, the average power dissipation is proportional to the clock frequency. At higher clock frequencies, more power is proportionally dissipated. Also, the power dissipated is very sensitive to the power-supply voltage. Decreasing the power-supply voltage from 5 to 3.3 V decreases the power dissipation by approximately one-half.

A CMOS gate continuously changing at maximum speed dissipates a similar amount of power as an NMOS gate running at a similar speed. This would seem to contradict the stated superiority of CMOS logic with respect to power dissipation. *Luckily, most modern chips have tens of thousands or even millions of gates that do not change in any given clock cycle.* For example, in a computer chip, most of the memory and registers are unchanging in any given clock cycle. The gates that do not change do not dissipate power and the overall dissipation is

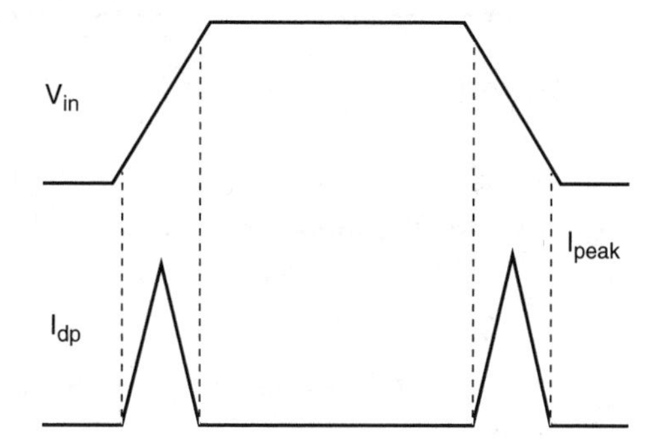

Figure 4.31 The input voltage waveform and the direct-path current of a CMOS inverter when the input voltage has finite rise and fall times.

much less. However, for the particular case of a CMOS gate changing continuously at full speed, the designer might consider logic circuits that have d.c. power dissipation, if smaller size or greater speed results. The overall difference of the power dissipation of the total chip will be small for this case.

There is additional power dissipated over and above that given by (4.87). This additional component is due to that fact that during transitions there is some d.c. current going through both the p-channel and n-channel transistors at the same time. This current is often called *direct-path current.* This additional power is usually less than 20% of that due to charging and discharging parasitic capacitances (Veendrick, 1984), but can be substantial, particularly if the input changes slowly.

This power dissipation can be estimated assuming the area under a plot of the direct-path current of an inverter can be approximated by a triangular wave (Rabaey, 1996). For example, consider the plots of the input voltage and direct-path current of a CMOS inverter shown in Fig. 4.31. Once the inverter threshold voltage, V_{th}, has been found, then the peak direct-path current is found from

$$I_{peak} = \frac{\mu_n C_{ox}}{2}\left(\frac{W}{L}\right)_1 (V_{th} - V_{tn})^2 \tag{4.88}$$

where it is assumed the n-channel drive transistor is Q_1. The energy dissipated each period is then given by

$$E_{dp} = V_{DD}\left(\frac{I_{peak}t_r}{2} + \frac{I_{peak}t_f}{2}\right) = V_{DD}I_{peak}\left(\frac{t_r + t_f}{2}\right) \tag{4.89}$$

The average power dissipated is the energy dissipated per period divided by the time of the period. Thus, we have

$$P_{dp\text{-}avg} = \frac{1}{T}V_{DD}I_{peak}\left(\frac{t_r + t_f}{2}\right) = V_{DD}I_{peak}\left(\frac{t_r + t_f}{2}\right)f_{clk} \tag{4.90}$$

Example 4.17

For the CMOS inverter of Fig. 4.26, find the average power dissipated due to dynamic power and due to the direct-path current. Assume the rise and fall times are as found in Example 4.15 and frequency of the input signal is 100 MHz.

Solution: From Example 4.15, we have $C_L = 0.2$ pF. Using (4.87) and $V_{DD} = 3.3$ V gives

$$P_{dyn\text{-}avg} = 5\times10^{-13}(3.3^2)1\times10^8 = 0.54 \text{ mW} \tag{4.91}$$

Next, assuming $V_{th} = 1.4$ V from Example 4.13, we have

$$I_{peak} = \frac{188\times10^{-6}}{2}\left(\frac{1.2}{0.6}\right)(1.4-0.8)^2 = 68\ \mu\text{A} \tag{4.92}$$

Continuing, using $t_r = 0.64$ ns and $t_f = 1.87$ ns from Example 4.15, and using (4.90), we have

$$P_{dp\text{-}avg} = 3.3(6.8\times10^{-5})(0.64\times10^{-9} + 1.87\times10^{-9})1\times10^8 = 56\ \mu\text{W} \tag{4.93}$$

Note the average direct-path power dissipation is about 10% of the dynamic power dissipation for this example.

4.5 CMOS Gate Design

There are many different design methodologies for realizing logic functions using CMOS transistors. In Chapter 1, we were introduced to traditional CMOS gates. These will be dealt with in greater detail in this section. Different types of CMOS logic families are now gaining in popularity. These newer families tend to minimize the number of p-channel transistors required and often use dynamic techniques such as precharging outputs to a high voltage before evaluation. In addition, they often operate on fully differential signals. That is, every signal is propagated along with its complement. These advanced techniques will be covered in Chapter 9.

TRADITIONAL LOGIC DESIGN

The traditional approach to CMOS logic design is to realize an n-channel driver circuit the same as one would for pseudo-NMOS logic, although the transistor sizing might be done differently. The load is then realized as a *complementary* p-channel network.

As an example of a moderately complicated CMOS logic circuit, consider the realization of a full adder. A full adder is one of the most often used building blocks in arithmetic circuits. These circuits are often in the critical path of microcomputers and digital signal-processing circuits; thus, a lot of effort has been spent on optimizing them. A full adder has three inputs (A, B, and C) and two outputs (Sum and Carry). The functions realized by a full adder are

$$\mathsf{Sum} = \mathsf{A} \oplus \mathsf{B} \oplus \mathsf{C} = \mathsf{ABC} + \mathsf{A}\overline{\mathsf{B}}\overline{\mathsf{C}} + \overline{\mathsf{A}}\overline{\mathsf{B}}\mathsf{C} + \overline{\mathsf{A}}\mathsf{B}\overline{\mathsf{C}} \tag{4.94}$$

and

$$\mathsf{Carry} = \mathsf{AB} + \mathsf{AC} + \mathsf{BC} = \mathsf{AB} + \mathsf{C}(\mathsf{A} + \mathsf{B}) \tag{4.95}$$

In words, the *sum* function realizes the three-input *exclusive-or* function; its output is a "1" if an odd number of inputs are "1"s. The *carry* is a "1" if two or more inputs are "1"s. A logic diagram that realizes both functions is shown in Fig. 4.32. Since there are only four inverting gates in the diagram, the function can be realized by four CMOS gates; two of these gates are simple inverters.

The first complex gate can realize the same function as logic gates *a, b, c,* and *d.* The n-channel drive network, which is arrived at using the same procedures as described previously for pseudo-NMOS logic gates, is shown in Fig. 4.33. The complementary p-channel network,

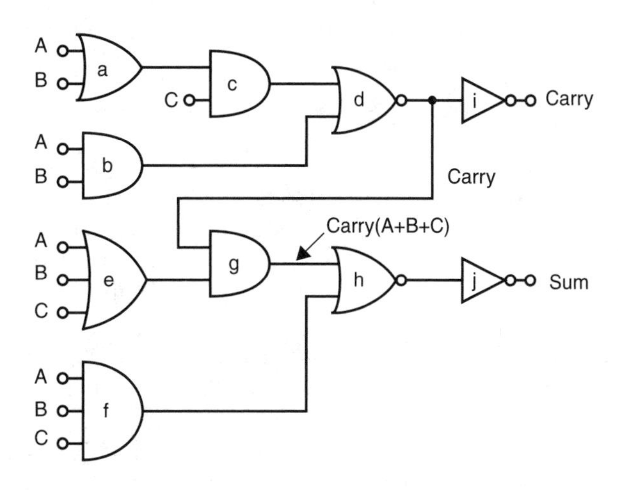

Figure 4.32 A logic diagram that realizes the full adder.

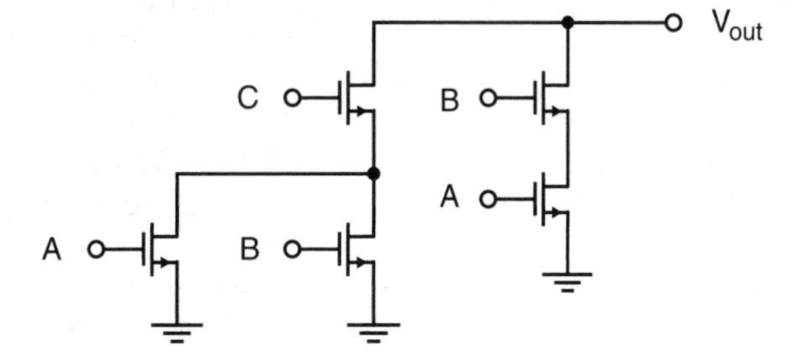

Figure 4.33 The n-channel drive network for realizing the gates *a, b, c,* and *d* of Fig. 4.32.

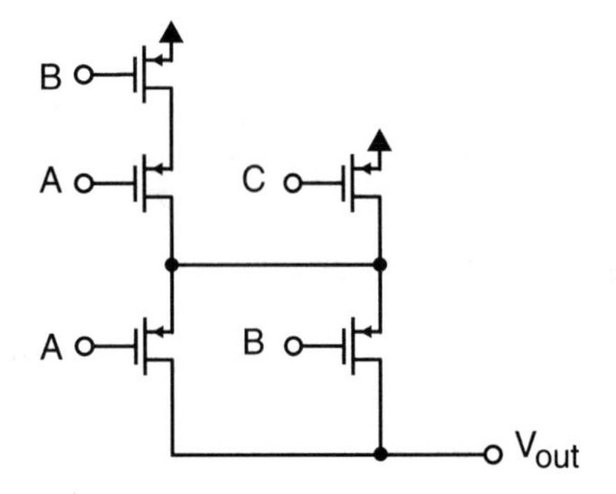

Figure 4.34 The complementary p-channel network to the n-channel network of Fig. 4.33.

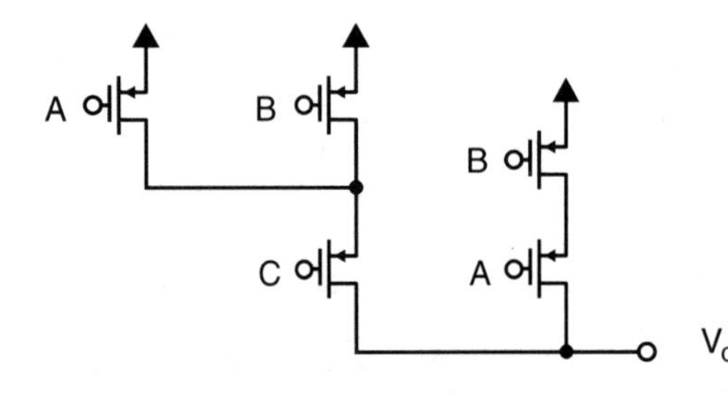

Figure 4.35 An equivalent p-channel network to the p-channel network of Fig. 4.34.

derived using the techniques of Section 1.2, is shown in Fig. 4.34. Figure 4.34 could be simplified somewhat by noting that the p-channel network of Fig. 4.35 realizes the same function and only has two transistors in series worst-case. This and similar simplifications are often possible; unfortunately, a formal procedure to identify and implement these simplifications has not been found.

In a similar manner, the CMOS realizations of logic gates *e, f, g,* and *h* can be found and simplified. The realization of inverters *i* and *j* are obvious. The complete CMOS realization of a full-adder cell is shown in Fig. 4.36.

There is no *exact* optimum procedure for choosing the W/Ls of transistors used to realize traditional CMOS logic gates. The gates will always function correctly independent of the sizes chosen. However, there are a few guidelines that are often followed. These guidelines mimic the procedures used for choosing the W/Ls of transistors used in NMOS gates. Basically, the transistors are taken wider if it is possible they might be in series with a larger number of transistors. However, in traditional CMOS circuits, the Ws are seldom taken larger than 10 to 20 μm to conserve space. Also, the W/Ls of p-channel networks are sometimes taken larger than the W/Ls of n-channel networks to give more equal rise and fall times, given the unequal mobilities of electrons and holes.

When the designer has the choice, *nand* gates are preferable in traditional CMOS design compared to *nor* gates. This is because *nand* gates have the n-channel transistors in series and

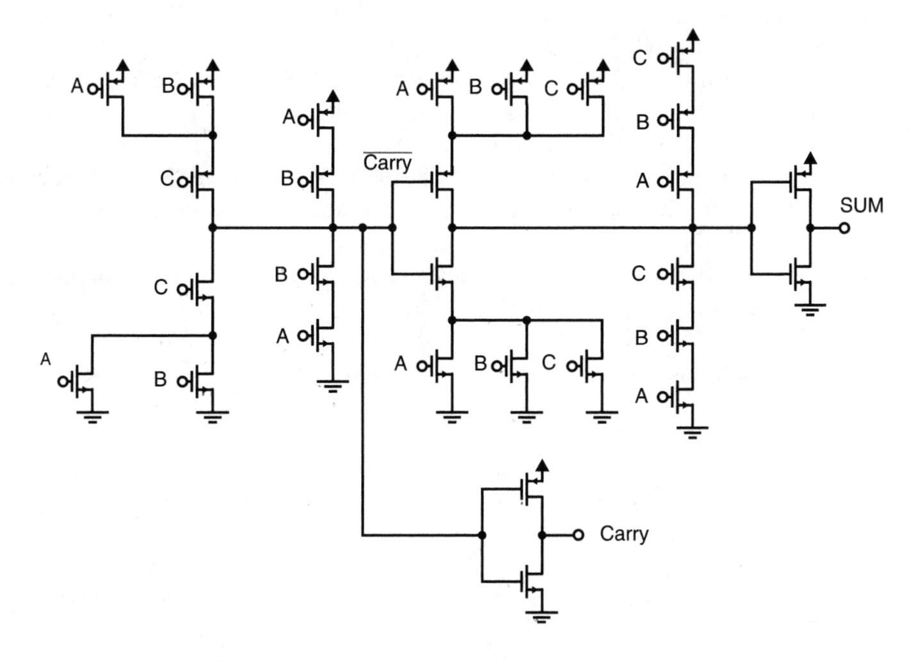

Figure 4.36 A CMOS realization of the full-adder function.

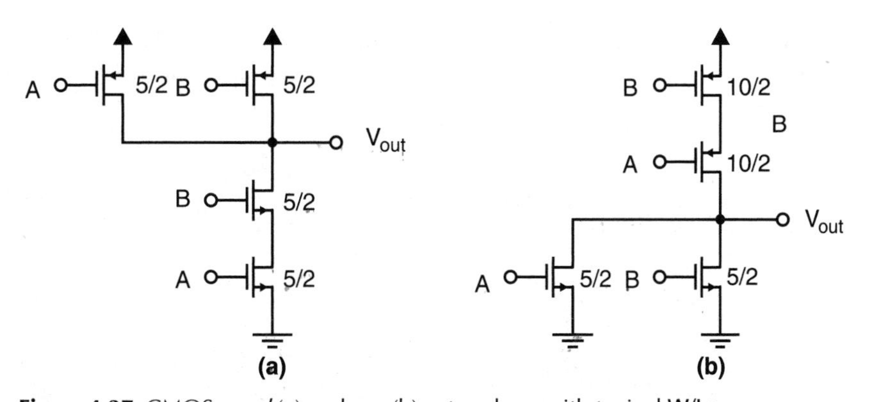

Figure 4.37 CMOS *nand* (a) and *nor* (b) gates along with typical W/Ls.

the p-channel transistors in parallel, where the opposite is true for *nor* gates. When designing for almost-equal rise and fall times, *nand* gates have more equally sized transistors; in *nor* gates the W/Ls of the p-channel transistors get quite large. For example, Fig. 4.37a and b shows typical realizations of two input *nand* and *nor* gates with reasonable sizes. Notice the smaller W/Ls of the *nand* gate.

Despite the more reasonable W/Ls, the worst case rise time of the *nand* gate is the same as that of the *nor* gate when driving large capacitive loads.

Given the preference for the *and* function when formal logic design techniques are used, functions that result in *and-or* functions are usually preferable to those that result in *or-and* functions, as the former is traditionally realized by two levels of *nand* gates, whereas the latter is normally realized by two levels of *nor* gates. This means grouping the "1"s is often preferable to grouping the "0"s when using Karnaugh map realization techniques for logic minimization (Mano, 1984).

This generalization only holds when simple *nand* or *nor* gates are used. When complex gates are used, one should check all possibilities.

Example 4.18

Find traditional CMOS realizations of the following truth table using Karnaugh map techniques. Find realizations based on grouping the "0"s as well as grouping the "1"s. Also, give reasonable sizes. You may assume the inputs and their complements are available. Note that "d" designates a "don't care" situation.

A	B	C	D	F
0	0	0	0	0
0	0	0	1	0
0	0	1	0	0
0	0	1	1	1
0	1	0	0	d
0	1	0	1	1
0	1	1	0	0
0	1	1	1	1
1	0	0	0	0
1	0	0	1	1
1	0	1	0	1
1	0	1	1	1
1	1	0	0	0
1	1	0	1	1
1	1	1	0	d
1	1	1	1	d

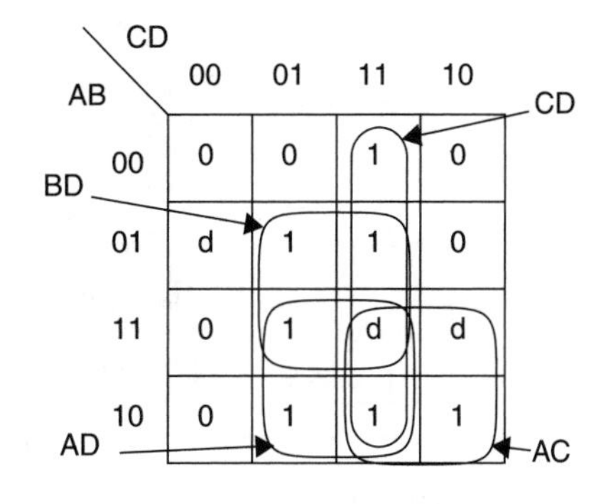

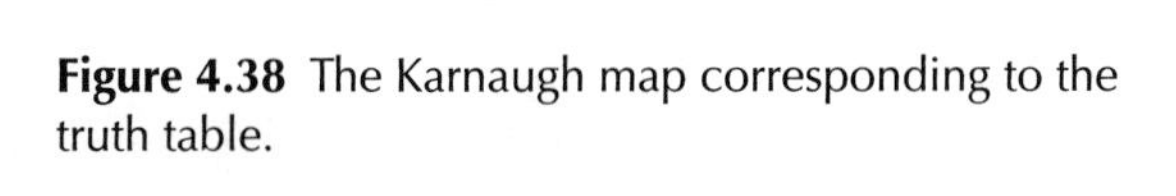
Figure 4.38 The Karnaugh map corresponding to the truth table.

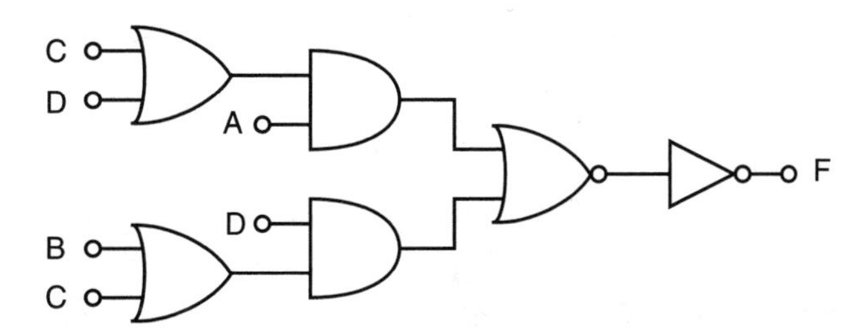

Figure 4.39 A logic-gate realization of the truth table of Example 4.18.

Solution: The Karnaugh map with the grouping of the "1"s is shown in Fig. 4.38. Using standard Karnaugh map techniques, this is realized by the function

$$F = AC + AD + BD + CD = A(C + D) + D(B + C) \tag{4.96}$$

A logic diagram realization of this is shown in Fig. 4.39. One possible CMOS realization of this logic function is shown in Fig. 4.40. Notice how only a single complex gate and an inverter are required to realize a function that if standard Karnaugh map procedures had been followed would have required five *nand* gates (requiring 28 transistors as opposed to 14 transistors required for the realization shown in Fig. 4.40). Also, the sizes of the transistors in the inverter have been taken somewhat larger so it can drive a larger load. Alternatively, (4.96) could be written [using De Morgan's theorem (Mano, 1984)] as

$$\begin{aligned} F &= \overline{\overline{A(C + D) + D(B + C)}} \\ &= \overline{\overline{A(C + D)}\,\overline{D(B + C)}} \\ &= \overline{(\bar{A} + \overline{(C + D)})(\bar{D} + \overline{(B + C)})} \\ &= \overline{(\bar{A} + \bar{D}\bar{C})(\bar{D} + \bar{B}\bar{C})} \end{aligned} \tag{4.97}$$

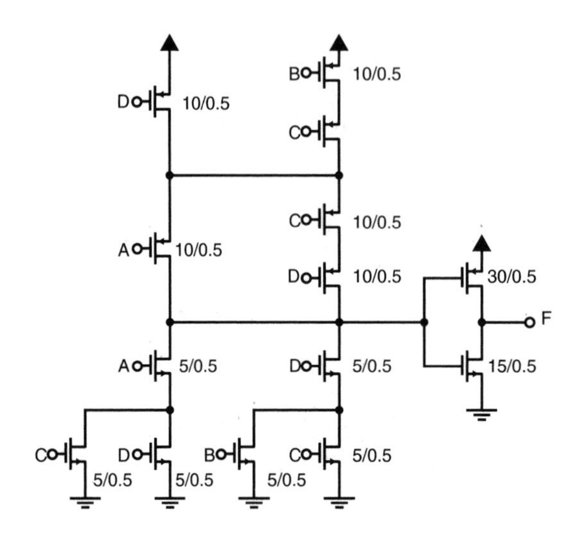

Figure 4.40 A CMOS realization of the logic circuit of Fig. 4.39.

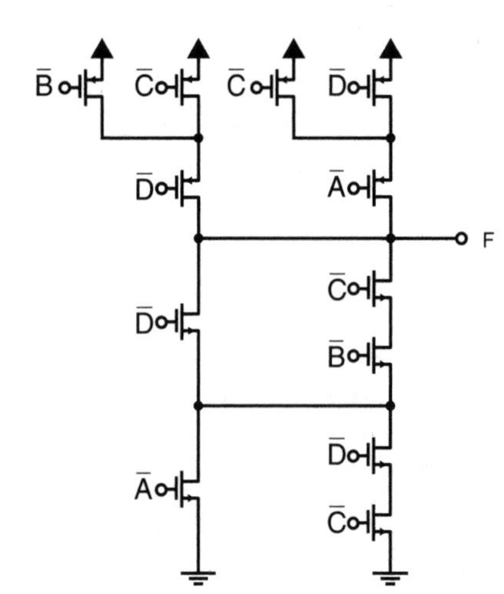

Figure 4.41 A complementary realization of the same logic function as the circuit of Fig. 4.40.

This function could be realized by the single CMOS gate shown in Fig. 4.41. Notice that this gate is complementary to the CMOS circuit of Fig. 4.40. That is, the n-channel and p-channel networks are interchanged, all inputs are complemented, and the output is inverted. In general, this complementary transformation can be applied to any traditional CMOS gate without changing the function being realized. *This technique is sometimes useful to eliminate inversions from arithmetic circuits.* Alternatively, grouping the zeros results in the Karnaugh map shown in Fig. 4.42. This map realizes the function

$$\bar{F} = \bar{A}\bar{B}\bar{C} + \bar{A}\bar{D} + \bar{C}\bar{D} \tag{4.98}$$

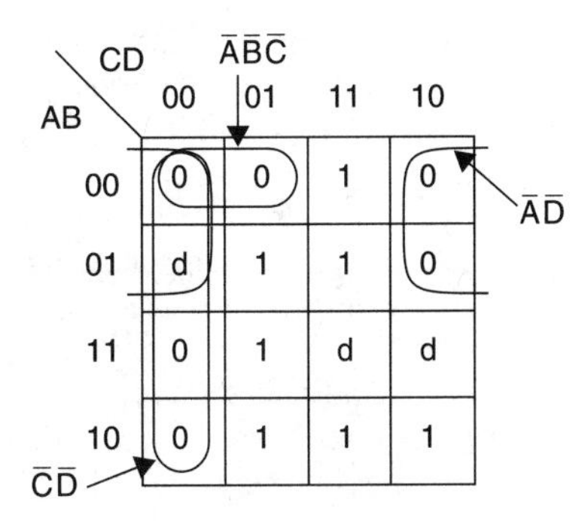

Figure 4.42 The Karnaugh map with the "0"s grouped.

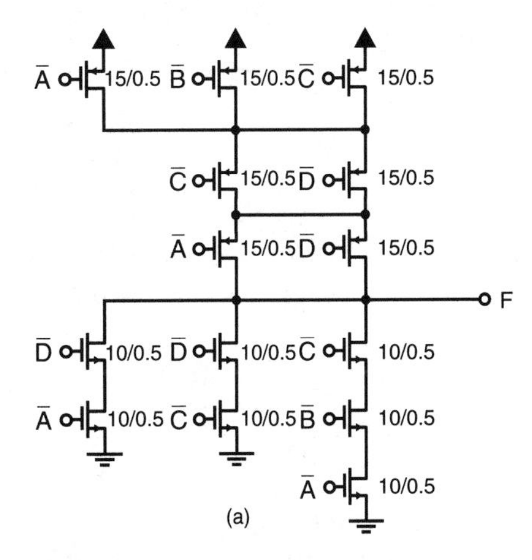

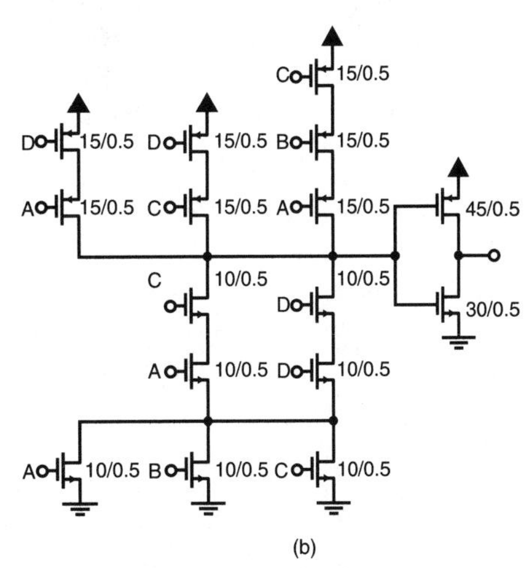

Figure 4.43 CMOS realizations of the logic circuit of Fig. 4.39 obtained by grouping the "0"s in a Karnaugh map.

or equivalently

$$F = (A + B + C)(A + D)(C + D) \tag{4.99}$$

This function could be realized by four *nor* gates. Alternatively, equations (4.98) and (4.99) could be realized by either of the complex CMOS circuits shown in Fig.4.43a and b. Notice again that the circuits are complementary. Even more alternatives can be realized by interchanging the n-channel networks of Fig. 4.40 and Fig. 4.43b, or alternatively Fig. 4.41 and Fig. 4.43a. Which alternative is preferable depends on the application and considerations such as the availability of inverted inputs and the load being driven. Probably, the circuit of Fig 4.43b would be reasonable in most general-purpose applications as it does not require complementary inputs. Also, the inverter buffer at the output would allow much larger capacitive loads to be driven more quickly, particularly given the large number of series transistors of all realization possibilities. If inverted inputs are available (which might be the case if the signals come from latches having differential outputs), then the circuit of Fig. 4.41 has the advantage of fewer transistors and only two p-channel transistors in series.

It might have been noticed that traditional CMOS logic design often requires more transistors than NMOS logic design. Although, this historically was the case, modern design techniques for CMOS logic are reducing the differences. One example of this, transmission-gate-based logic design, is described in the next chapter.

4.6 SPICE Simulations

This section presents the circuit diagrams, netlist, and output data for SPICE simulations of selected examples from the preceding sections.

Simulation of Examples 4.1, 4.3, 4.5, and 4.6

The pseudo-NMOS inverter of Fig. 4.3 is illustrated in Fig. 4.44 with all nodes enumerated according to the following netlist.

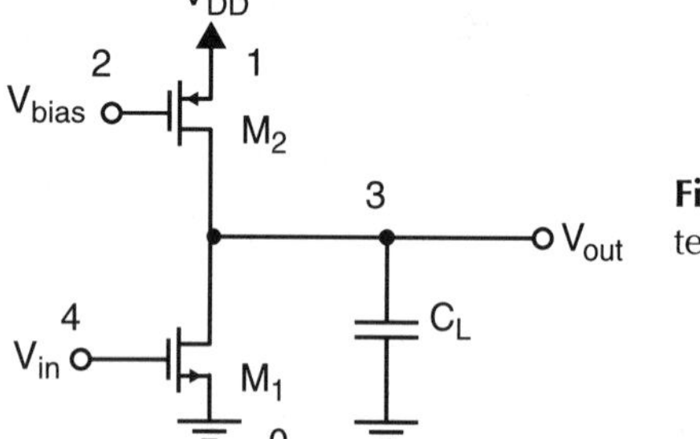

Figure 4.44 A pseudo-NMOS inverter in a 0.6-μm technology.

```
A Pseudo-NMOS Inverter
*
M1 3 4 0 0 nch W=1.5u L=0.6u AS=2.7p PS=5.1u AD=2.7p PD=5.1u
M2 3 2 1 1 pch W=3.0u L=0.6u AS=5.4p PS=6.6u AD=5.4p PD=6.6u
CL 3 0 0.2pF
*
VDD 1 0 3.3
VBIAS 2 0 1.65
VIN 4 0 DC 0 PULSE (0 3.3 0ns 100ps 100ps 5.9ns 12ns)
*
.LIB '../mod_06' typical
*
.OPTION NOMOD POST INGOLD=2 NUMDGT=6 BRIEF
.TRAN 0.01n 12n
.PRINT TRAN V(4) V(3)
.DC VIN 0V 3.3V 0.001V
.PRINT DC V(3)
.END
```

The results of the d.c. sweep simulation are plotted in Fig. 4.45. From the SPICE simulation, we have V_{th} = 1.43 V, V_{OH} = 3.3 V, and V_{OL} = 0.13 V. These values compare favorably with those calculated by hand, which were V_{th} = 1.32 V, V_{OH} = 3.3 V, and V_{OL} = 0.054 V. The discrepancy in the output-low voltage is caused by a discontinuity in the drain current dependence on drain voltage.

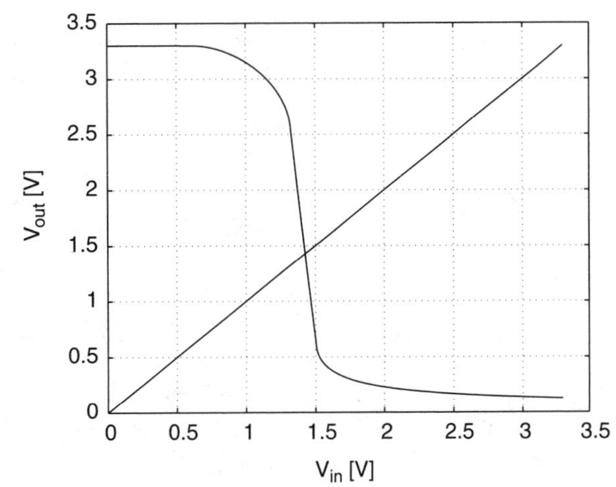

Figure 4.45 The transfer curve of the pseudo-NMOS inverter of Fig. 4.44.

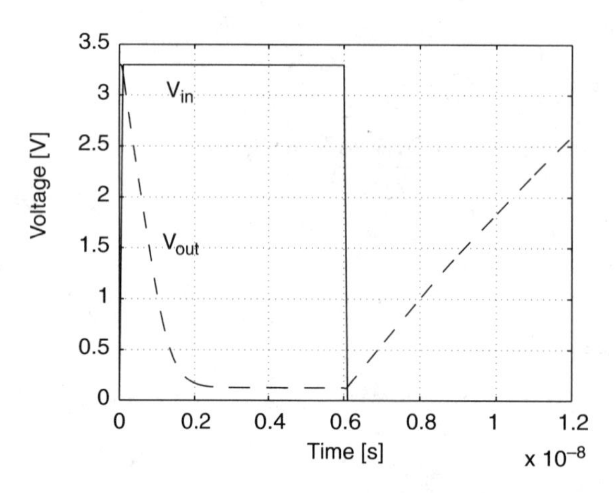

Figure 4.46 The transient response of the pseudo-NMOS inverter of Fig. 4.44.

The simulation results of the transient analysis are plotted in Fig. 4.46. The SPICE simulation gives rise and fall times of 5.90 and 0.94 ns. The values from hand calculation are 5.65 and 0.41 ns, respectively. The discrepancy is due to short-channel effects (including drain-induced barrier lowering and mobility reduction) as well as parasitic device capacitances.

SIMULATION OF EXAMPLE 4.4

To determine the gain of the pseudo-NMOS inverter of Fig. 4.3, the netlist below was used. The circuit description used is the same as that derived from Fig. 4.44. The hand calculated value of the gain was –30.1. The simulation result is –10.9. The discrepancy is due to both short-channel effects and insufficiently accurate modeling of the output impedance by SPICE.

```
A Pseudo-NMOS Inverter
*
M1 3 4 0 0 nch W=1.5u L=0.6u AS=2.7p PS=5.1u AD=2.7p PD=5.1u
M2 3 2 1 1 pch W=3.0u L=0.6u AS=5.4p PS=6.6u AD=5.4p PD=6.6u
CL 3 0 0.2pF
*
VDD 1 0 3.3
VBIAS 2 0 1.65
VIN 4 0 1.43
*
.LIB '../mod_06' typical
*
.OPTION NOMOD POST INGOLD=2 NUMDGT=6 BRIEF
.TF V(3) VIN
.END
```

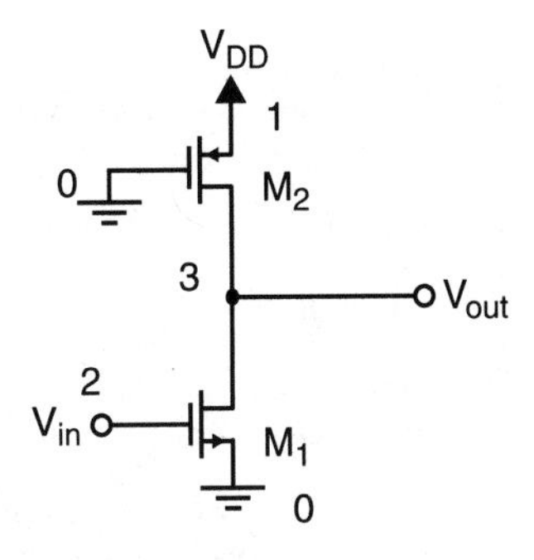

Figure 4.47 An alternative pseudo-NMOS inverter in a 0.6 μm technology.

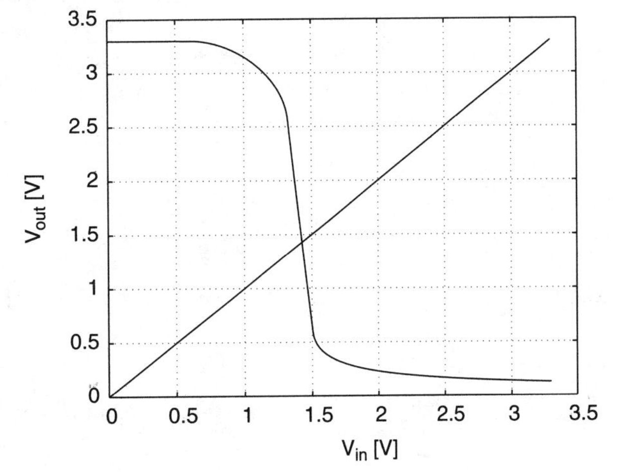

Figure 4.48 The transfer curve of the pseudo-NMOS inverter of Fig. 4.47.

Simulation of the Circuit of Fig. 4.22

The circuit of Fig. 4.22, an alternative implementation of the pseudo-NMOS inverter, is shown in Fig. 4.47 with all nodes labeled. To obtain a transfer curve from a SPICE simulation, the netlist below is used. The transfer curve is illustrated in Fig. 4.48. The threshold voltage is 1.58 V and the output low voltage is 0.17 V.

```
An Alternative Pseudo-NMOS Inverter
*
M1 3 2 0 0 nch W=3.0u L=0.6u AS=5.4p PS=6.6u AD=5.4p PD=6.6u
M2 3 0 1 1 pch W=1.5u L=0.6u AS=2.7p PS=5.1u AD=2.7p PD=5.1u
*
VDD 1 0 3.3
VIN 2 0 DC 0
*
.LIB '../mod_06' typical
```

```
*
.OPTION NOMOD POST INGOLD=2 NUMDGT=6 BRIEF
.DC VIN 0V 3.3V 0.01V
.PRINT DC V(3)
.END
```

SIMULATION OF THE CIRCUIT OF FIG. 4.24

The depletion-load NMOS inverter of Fig. 4.24 is illustrated in Fig. 4.49 with appropriate node labeling. The SPICE netlist to generate the transfer curve is shown below, where the transistor models are now for the n-channel enhancement and depletion transistors. In the models used, it was assumed that all parameters remained unchanged except the threshold voltage, which became VTO = –2.0. It must be noted that there are differences in other parameters between n-channel enhancement and n-channel depletion transistors, however they are assumed to be negligible for the purposes of the simulation.

```
An NMOS Inverter
*
M1 3 2 0 0 enh W=1.2u L=0.6u AS=2.2p PS=4.8u AD=2.2p PD=4.8u
M2 1 3 3 0 dep W=0.6u L=1.2u AS=2.2p PS=7.8u AD=2.2p PD=7.8u
*
VDD 1 0 3.3
VIN 2 0 DC 0 PULSE (0 3.3 0ns 100ps 100ps 0.4ns 1ns)
*
.LIB '../nmos' typical
*
.OPTION NOMOD POST INGOLD=2 NUMDGT=6 BRIEF
.DC VIN 0V 3.3V 0.001V
.PRINT DC V(3)
.END
```

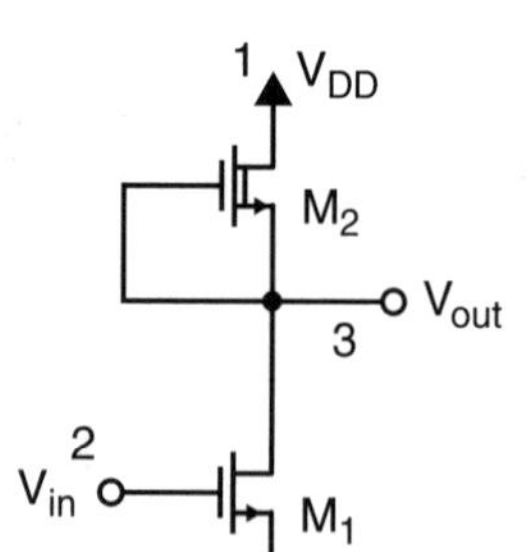

Figure 4.49 A depletion-load NMOS inverter in a 0.6-μm technology.

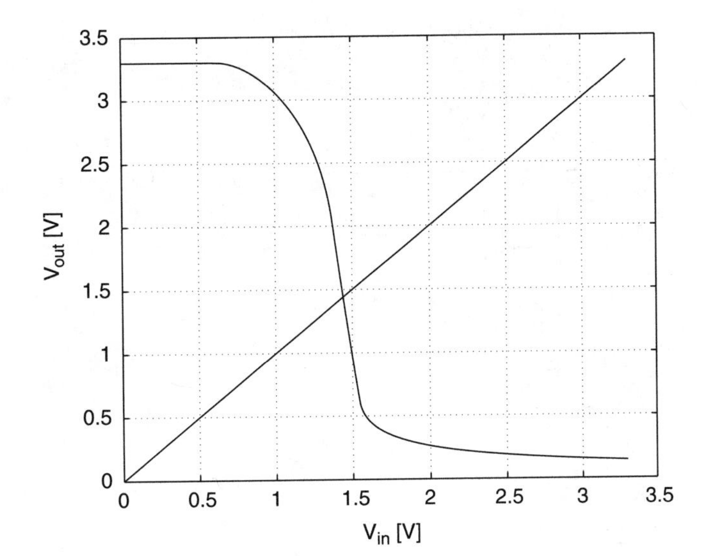

Figure 4.50 The transfer curve of the NMOS inverter of Fig. 4.49.

From the simulation, a transfer curve as shown in Fig. 4.50 is obtained. The output-low voltage and threshold voltage are 0.14 and 1.44 V, respectively. A similar netlist, shown below, may be used to perform a simulation and determine the gain at threshold, yielding a value of –8.3.

```
An NMOS Inverter
*
M1 3 2 0 0 enh W=1.2u L=0.6u AS=2.2p PS=4.8u AD=2.2p PD=4.8u
M2 1 3 3 0 dep W=0.6u L=1.2u AS=2.2p PS=7.8u AD=2.2p PD=7.8u
CL 2 0 0.2pF
*
VDD 1 0 3.3
VIN 2 0 DC 1.44
*
.LIB '../nmos' typical
*
.OPTION NOMOD POST INGOLD=2 NUMDGT=6 BRIEF
.TF V(3) VIN
.END
```

Simulation of Examples 4.13 and 4.15

The CMOS inverter of Fig. 4.26 is illustrated in Fig. 4.51 with all nodes enumerated for the purpose of simulation. To produce a transfer curve as in Fig. 4.52 and transient response data as

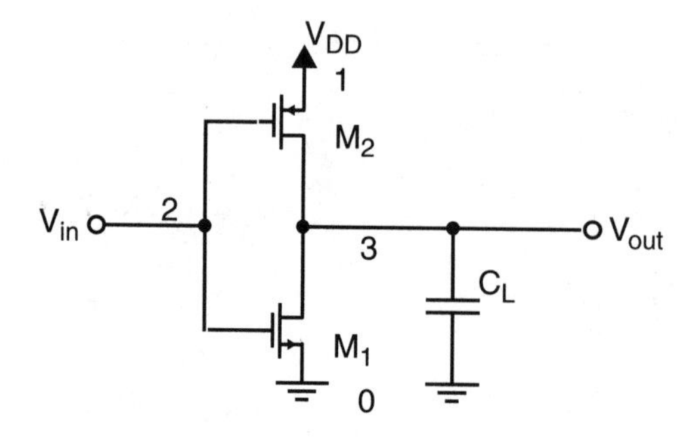

Figure 4.51 A CMOS inverter in a 0.6-μm technology, as in Fig. 4.26.

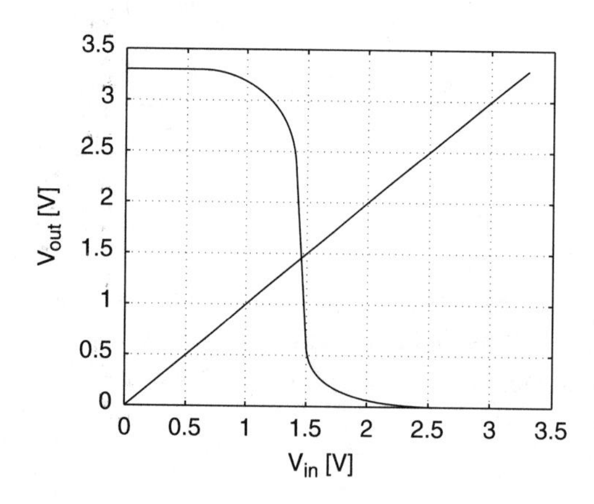

Figure 4.52 The transfer curve of the CMOS inverter of Fig. 4.51.

in Fig 4.53, the netlist below is used. From the transfer curve, it may be seen that V_{th} = 1.45 V, thus NM_L = 1.45 V and NM_H = 1.85 V. The hand-calculated threshold voltage of 1.40 V compares favorably with the simulation result. The rise and fall times of the inverter are determined from the simulation to be 1.67 and 1.07 ns, respectively. These values are similar to the hand-calculated values of 1.87 and 0.64 ns, respectively, where discrepancies are due to short-channel effects and parasitic device capacitances.

```
A CMOS Inverter
*
M1 3 2 0 0 nch W=1.2u L=0.6u AS=2.16p PS=4.8u AD=2.16p PD=4.8u
M2 3 2 1 1 pch W=1.8u L=0.6u AS=3.24p PS=5.4u AD=3.24p PD=5.4u
CL 3 0 0.2pF
*
VDD 1 0 3.3
VIN 2 0 DC 0 PULSE (0 3.3 0ns 100ps 100ps 2.4ns 5ns)
*
```

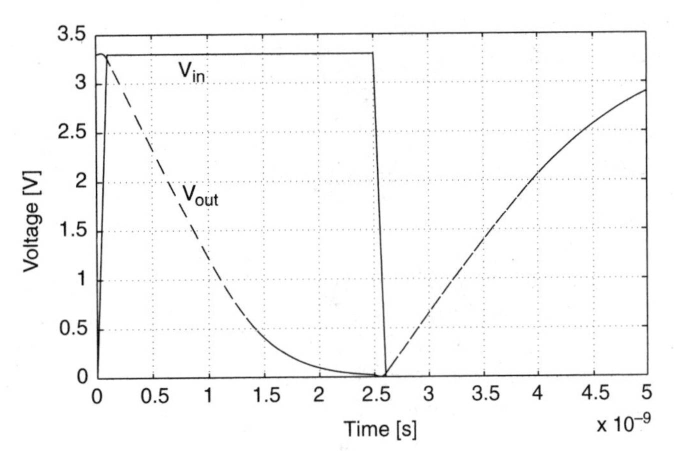

Figure 4.53 The transient response of the CMOS inverter of Fig. 4.51.

```
.LIB '../mod_06' typical
*
.OPTION NOMOD POST INGOLD=2 NUMDGT=6 BRIEF
.DC VIN 0V 3.3V 0.001V
.PRINT DC V(3)
.TRAN 0.001n 5n
.PRINT TRAN V(2) V(3)
.END
```

SIMULATION OF EXAMPLE 4.14

The SPICE netlist below was used to find the transistor output impedances, transistor transconductances, and inverter gain at the threshold voltage of the CMOS inverter illustrated in Fig. 4.51.

```
A CMOS Inverter
*
M1 3 2 0 0 nch W=1.2u L=0.6u AS=2.16p PS=4.8u AD=2.16p PD=4.8u
M2 3 2 1 1 pch W=1.8u L=0.6u AS=3.24p PS=5.4u AD=3.24p PD=5.4u
CL 3 0 0.2pF
*
VDD 1 0 3.3
VIN 2 0 DC 1.45
*
.LIB '../mod_06' typical
```

```
*
.OPTION NOMOD POST INGOLD=2 NUMDGT=6 BRIEF
.OP
.TF V(3) VIN
.END
```

The simulation yielded transistor transconductances of $g_{m\text{-}1}$ = 0.156 mA/V and $g_{m\text{-}2}$ = 0.117 mA/V as well as output impedances of $r_{DS\text{-}1}$ = 157 kΩ and $r_{DS\text{-}2}$ = 139 kΩ. The hand-calculated transconductances are 0.226 and 0.134 mA/V, respectively. SPICE simulation is not sufficiently accurate for modeling output impedance, while the hand calculations do not take short-channel effects, such as mobility degradation and drain-induced barrier lowering, into account. The output impedances were calculated by hand to be 246 and 211 kΩ, respectively. The discrepancies are again caused by insufficient accuracy of the SPICE simulations for output impedances and the lack of consideration of short-channel effects within the hand calculations. The inverter gain at threshold was simulated to be –20.1 and calculated by hand to be –40.1.

SIMULATION OF THE CIRCUIT OF FIG. 4.26

The netlist below was used to run a simulation of the total power consumed by the CMOS inverter of Fig. 4.26 and Fig. 4.51 during its transient response. A plot of the simulation results is illustrated in Fig. 4.54.

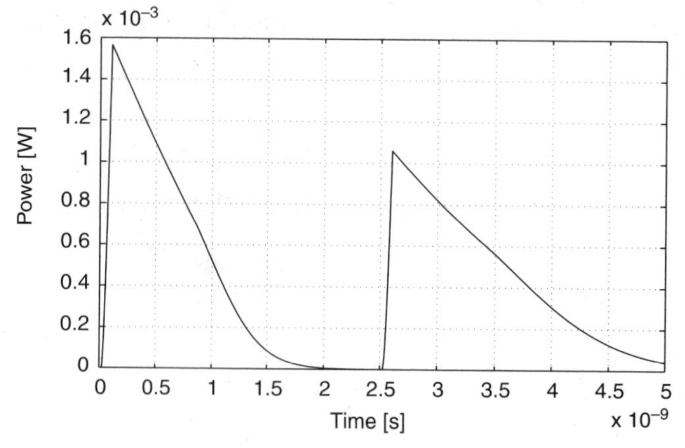

Figure 4.54 The transient power response of the CMOS inverter of Fig. 4.51.

```
A CMOS Inverter
*
M1 3 2 0 0 nch W=1.2u L=0.6u AS=2.16p PS=4.8u AD=2.16p PD=4.8u
M2 3 2 1 1 pch W=1.8u L=0.6u AS=3.24p PS=5.4u AD=3.24p PD=5.4u
CL 3 0 0.2pF
*
VDD 1 0 3.3
VIN 2 0 DC 0 PULSE (0 3.3 0ns 100ps 100ps 2.4ns 5ns)
*
.LIB '../mod_06' typical
*
.OPTION NOMOD POST INGOLD=2 NUMDGT=6 BRIEF
.TRAN 0.001n 5n
.PRINT POWER+
.END
```

4.7 Bibliography

M. Annaratone, *Digital CMOS Circuit Design*, Kluwer, 1986.

M. Elmasry, ed., *Digital MOS Integrated Circuits, II,* IEEE Press, 1991.

L. Glasser and D. Dopperpuhl, *The Design and Analysis of VLSI Circuits*, Addison-Wesley, 1985.

D. Hodges and J. Jackson, *Analysis and Design of Digital Integrated Circuits*, McGraw-Hill, 1988.

M. Mano, *Digital Design,* Prentice Hall, 1984.

M. Mano and C. Kime, *Logic and Computer Design Fundamentals,* Prentice Hall, 1997.

C. Mead and L. Conway, *Introduction to VLSI Systems*, Addison-Wesley, 1980.

J. Rabaey, *Digital Integrated Circuits, A Design Perspective*, Prentice Hall, 1996.

C. Roth, *Fundamentals of Logic Design*, West, 1985.

H. Taub and D. Schilling, *Digital Integrated Electronics*, McGraw-Hill, 1977.

H. Veendrick, "Short-Circuit Dissipation of Static CMOS Circuitry and Its Impact on the Design of Buffer Circuits," *IEEE Journal of Solid-State Circuits*, SC-19 (4), 468–473, 1984.

N. Weste and K. Eshragian, *Principles of CMOS VLSI Design: A Systems Perspective*, Addison-Wesley, 1983.

4.8 Problems

For the problems in this chapter, assume the following transistor parameters:

- npn bipolar transistors:

  $\beta = 100$

  $V_A = 80$ V

$\tau_b = 13$ ps
$\tau_s = 4$ ns
$r_b = 330\ \Omega$

- n-channel MOS transistors:
  $\mu_n C_{ox} = 190\ \mu A/V^2$
  $V_{tn} = 0.7$ V
  $\gamma = 0.6\ V^{1/2}$
  $r_{ds}\ (\Omega) = 5000L\ (\mu m)/I_D\ (mA)$ in active region
  $C_j = 5 \times 10^{-4}\ pF/(\mu m)^2$
  $C_{j\text{-}sw} = 2.0 \times 10^{-4}\ pF/\mu m$
  $C_{ox} = 3.4 \times 10^{-3}\ pF/(\mu m)^2$
  $C_{gs(overlap)} = C_{gd(overlap)} = 2.0 \times 10^{-4}\ pF/\mu m$

- p-channel MOS transistors:
  $\mu_p C_{ox} = 50\ \mu A/V^2$
  $V_{tp} = -0.8$ V
  $\gamma = 0.7\ V^{1/2}$
  $r_{ds}\ (\Omega) = 6000L\ (\mu m)/I_D\ (mA)$ in active region
  $C_j = 6 \times 10^{-4}\ pF/(\mu m)^2$
  $C_{j\text{-}sw} = 2.5 \times 10^{-4}\ pF/\mu m$
  $C_{ox} = 3.4 \times 10^{-3}\ pF/(\mu m)^2$
  $C_{gs(overlap)} = C_{gd(overlap)} = 2.0 \times 10^{-4}\ pF/\mu m$

4.1 For the pseudo-NMOS inverter shown in Fig. P4.1,calculate V_{th}.

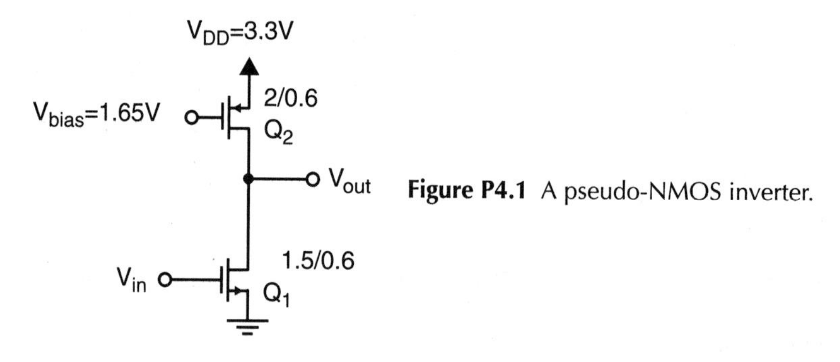

Figure P4.1 A pseudo-NMOS inverter.

4.2 For the pseudo-NMOS inverter shown in Fig. P4.1, find V_{OL}.

4.3 For the pseudo-NMOS inverter shown in Fig. P4.1, find the gain at $V_{in} = V_{th}$ assuming V_{th} is as found in Problem 4.1.

4.4 For the pseudo-NMOS inverter shown in Fig. P4.1, estimate the rise time. Assume $C_L = 75$ fF.

4.5 For the pseudo-NMOS inverter shown in Fig. P4.1, estimate the fall time. Assume $C_L = 75$ fF. Try to approximately account for the load current in Q_2 at this time.

4.6 Design a pseudo-NMOS logic gate that realizes the function $out = x_1(\overline{x_2} + x_2x_3)$. Give reasonable dimensions assuming $L_{min} = 0.5\ \mu m$.

4.7 Prove that two series transistors connected as shown in Fig. 4.13 and having the same widths but different lengths are equivalent to a single transistor having a length equal to the sum of the lengths. Assume Q_2 is in the triode region and ignore the body effect.

4.8 Using the concept of equivalent resistances (or admittances), simplify the n-channel driver network of Fig. P4.8 to a single equivalent resistance.

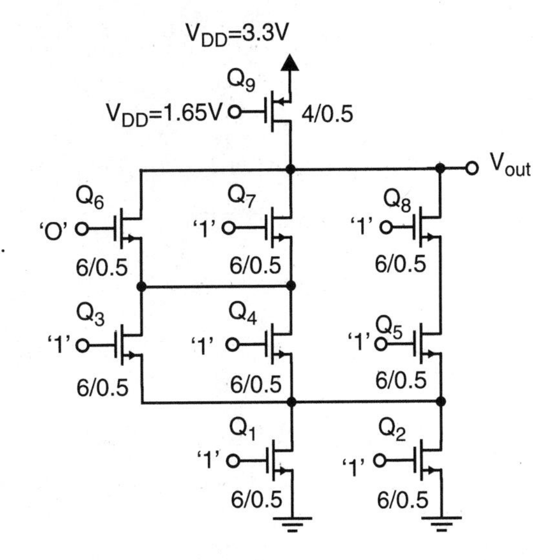

Figure P4.8 A complex pseudo-NMOS gate.

4.9 Estimate the output-low voltage. Also, estimate the 70% fall time assuming a load capacitance of 80 fF.

4.10 What is the logic function realized by the gate of Fig. P4.8 assuming transistor Q_i has x_i as an input.

4.11 For the alternative pseudo-NMOS inverter shown in Fig. P4.11, find V_{th} and V_{OL}.

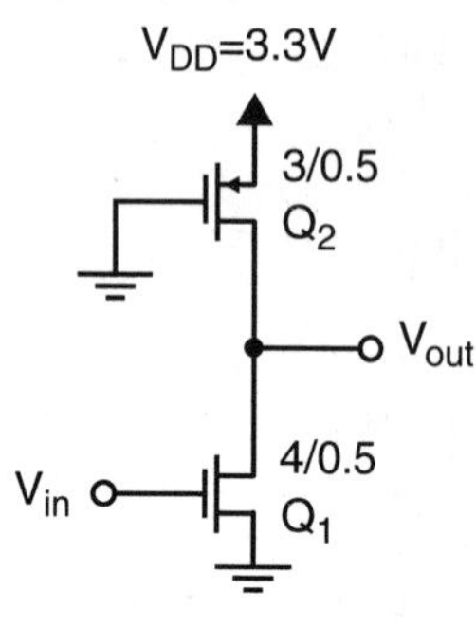

Figure P4.11 An alternative typical pseudo-NMOS inverter.

4.12 For the alternative pseudo-NMOS inverter shown in Fig. P4.11, estimate the 70% rise and fall times assuming $C_L = 60$ fF.

4.13 For the alternative pseudo-NMOS inverter shown in Fig. P4.11, calculate the gain of the inverter at the gate threshold voltage.

4.14 Derive equation (4.56) using small-signal analysis.

4.15 Give a pseudo-NMOS realization of the logic function shown in Fig. P4.15.

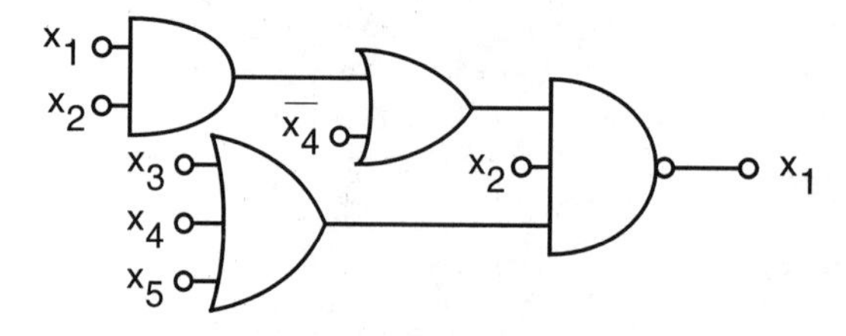

Figure P4.15 A logic circuit used Problem 4.15.

4.16 What logic function does the circuit of Fig. P4.16 realize?

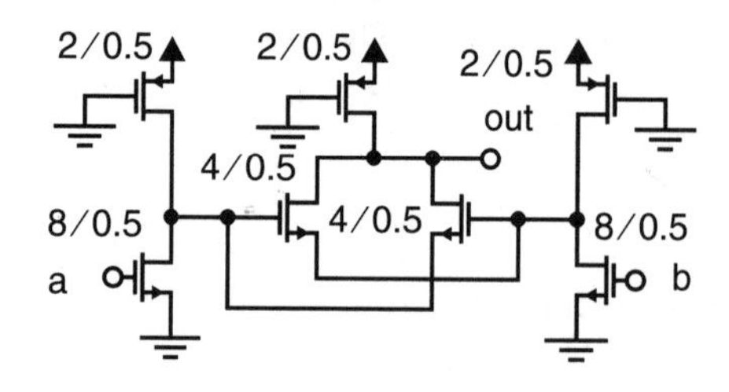

Figure P4.16 A logic circuit used in Problem 4.16.

4.17 For the CMOS inverter shown in Fig. P4.17, calculate V_{th}.

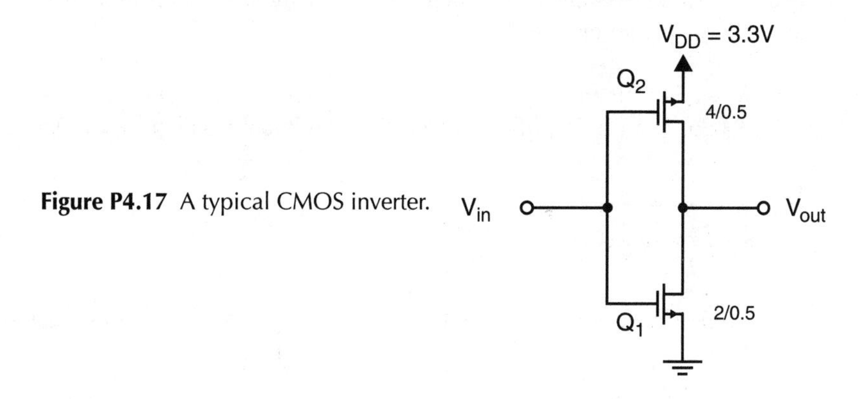

Figure P4.17 A typical CMOS inverter.

4.18 For the CMOS inverter shown in Fig. P4.17, find V_{OL}.

4.19 For the CMOS inverter shown in Fig. P4.17, find the gain at $V_{in} = V_{th}$ assuming V_{th} is as found in Problem 4.17.

4.20 For the CMOS inverter shown in Fig. P4.17, estimate the rise time. Assume $C_L = 125$ fF.

4.21 For the CMOS inverter shown in Fig. P4.17, estimate the fall time. Assume $C_L = 125$ fF.

4.22 For the CMOS inverter shown in Fig. P4.17, estimate the average dynamic power dissipation. Assume $C_L = 125$ fF and the input frequency is 125 MHz.

4.23 For the CMOS inverter shown in Fig. P4.17, estimate the average direct-path power dissipation. Assume $C_L = 125$ fF and the input frequency is 125 MHz. Use the rise and fall times found in Problems 4.20 and 4.21, respectively.

4.24 Give a complementary p-channel network to the n-channel network shown Fig. P4.24. What is the logic function realized? For the signals a,b,c,d = "0110" and assuming the

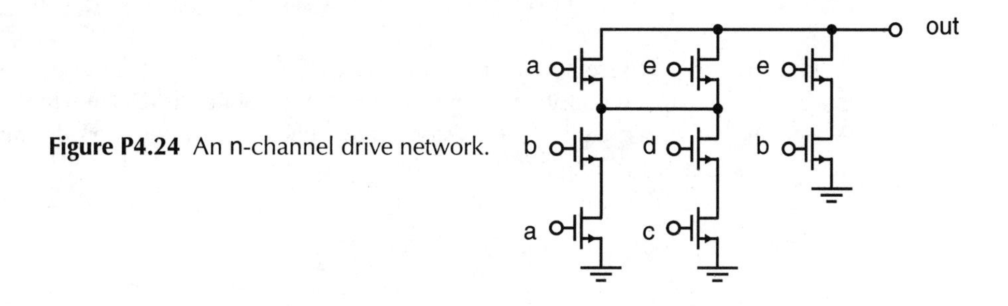

Figure P4.24 An n-channel drive network.

input signal *e* changes from a "1" to a "0", what is an equivalent pull-up resistor during the transition. Assume all the p-channel transistors have sizes of 5 μm/0.5 μm and the power-supply voltage is 3.3 V.

4.25 (a) Show a complementary p-channel network to the n-channel drive network shown in Fig. P4.25. Show a simplified p-channel network that would have a faster rise time.

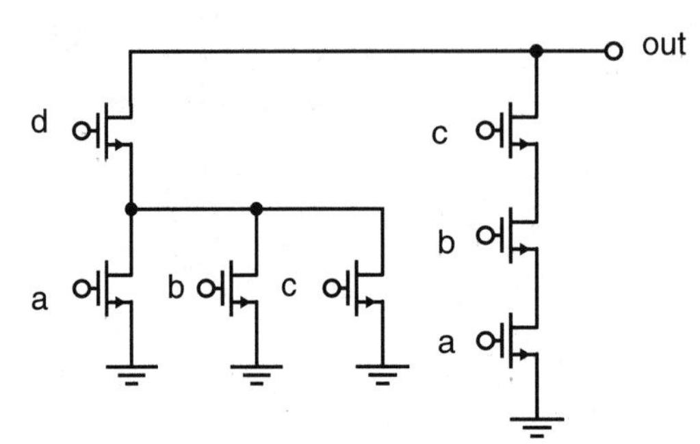

Figure P4.25 An n-channel drive network.

4.26 Design a two-input CMOS *exclusive-nor* gate. Its output should be "1" if an even number of inputs are "1"s. Assume the inputs and their complements are available. Your design should be a single gate.

4.27 Design a three-input CMOS *exclusive-or* gate. Its output should be "1" if an odd number of inputs are "1"s. Assume the inputs and their complements are available. Your design should be as simple as possible assuming traditional CMOS logic is used.

4.28 Design a CMOS circuit that realizes a 32-input *nor* function. Your solution may consist of multiple gates. Try to optimize speed.

4.29 Design a CMOS gate that realizes the function $\text{out} = \overline{x_1(x_2 + \overline{x}_2 x_3)}$. If possible, try to simplify the gate.

4.30 Design a CMOS gate that realizes the function $\text{out} = \overline{x_1 + \overline{x}_2(x_3 + x_2)}$. Try to simplify as much as possible. Give reasonable sizes assuming $L_{min} = 0.35\ \mu m$.

4.31 Realize the following truth table using traditional CMOS logic by grouping the "1"s in a Karnaugh map minimization. Assume all the p-channel transistors are 6 μm/0.6 μm. What is the worst case equivalent resistor of the n-channel drive network during a "1" to "0" transition of the output?

a	b	c	d	Out
0	0	0	0	0
0	0	0	1	1
0	0	1	0	1
0	0	1	1	0
0	1	0	0	1
0	1	0	1	d
0	1	1	0	0
0	1	1	1	0
1	0	0	0	0
1	0	0	1	1
1	0	1	0	1
1	0	1	1	0
1	1	0	0	0
1	1	0	1	d
1	1	1	0	1
1	1	1	1	1

4.32 Repeat Problem 4.31, but base your design on grouping the "0"s.

4.33 Realize the following truth table using traditional CMOS logic by grouping the "0"s in a Karnaugh map minimization. Assume all the p-channel transistors are 6 μm/0.6 μm. What is the worst case equivalent resistor during a "0" to "1" transition of the output?

a	b	c	d	Out
0	0	0	0	0
0	0	0	1	d
0	0	1	0	1
0	0	1	1	0
0	1	0	0	1
0	1	0	1	1
0	1	1	0	0

a	b	c	d	Out
0	1	1	1	d
1	0	0	0	0
1	0	0	1	0
1	0	1	0	1
1	0	1	1	0
1	1	0	0	1
1	1	0	1	d
1	1	1	0	0
1	1	1	1	0

4.34 Repeat Problem 4.33, but base your design on grouping the “1”s.

4.35 Check the accuracy of your answers to Problem 4.1 using SPICE simulation.

4.36 Check the accuracy of your answers to Problem 4.2 using SPICE simulation.

4.37 Check the accuracy of your answers to Problem 4.3 using SPICE simulation.

4.38 Check the accuracy of your answers to Problem 4.4 using SPICE simulation.

4.39 Check the accuracy of your answers to Problem 4.5 using SPICE simulation.

4.40 Check the accuracy of your answers to Problem 4.11 using SPICE simulation.

4.41 Check the accuracy of your answers to Problem 4.12 using SPICE simulation.

4.42 Check the accuracy of your answers to Problem 4.17 using SPICE simulation.

4.43 Check the accuracy of your answers to Problem 4.18 using SPICE simulation.

4.44 Check the accuracy of your answers to Problem 4.19 using SPICE simulation.

4.45 Check the accuracy of your answers to Problem 4.20 using SPICE simulation.

4.46 Check the accuracy of your answers to Problem 4.22 using SPICE simulation.

4.47 Check the accuracy of your answers to Problem 4.23 using SPICE simulation.

CHAPTER 5

Transmission-Gate and Fully Differential CMOS Logic

This chapter describes two alternative design methodologies for CMOS gates. The first of these uses *transmission gates* to minimize the number of transistors required. *Transmission gates* are very simple switches composed of one or two transistors. They function very much like *relays,* which were perhaps the first components ever used to realize electronic digital circuits. Many logic circuits, such as adders and D flip-flops, are often realized using transmission gates.

The second logic family introduced in this chapter, and covered in more detail in Chapter 9, is fully differential logic design. This family uses two wires to represent every digital signal; the voltage difference between the lines determines the logic value. This design family is rapidly gaining in popularity and therefore might often be described at an introductory level in a first-level course.

5.1 Transmission-Gate Logic Design

Until now, we have seen only traditional techniques for MOS logic circuit design. An alternative technique often used, particularly for multiplexors and circuits that require *exclusive-or* functions, makes use of transmission gates. A transmission gate operates much like a voltage-controlled switch or a relay. The simplest example consists of a single n-channel transistor as is shown in Fig. 5.1a. This type of transmission gate, which is often called a *pass transistor,* was used extensively in NMOS design. It is also often used in CMOS design, except for

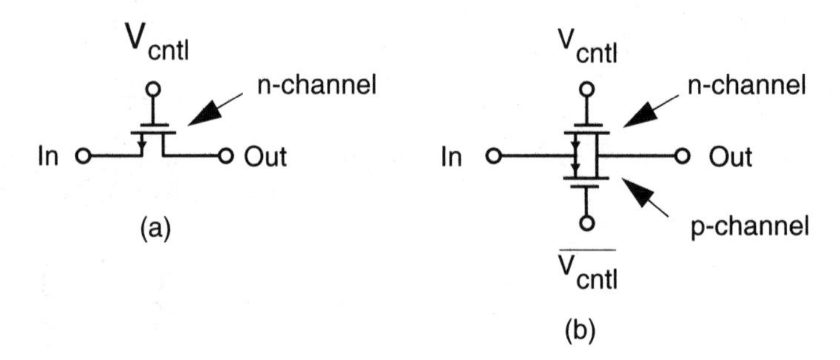

Figure 5.1 (a) A simple n-channel transistor transmission gate (or pass transistor) and (b) a CMOS transmission gate.

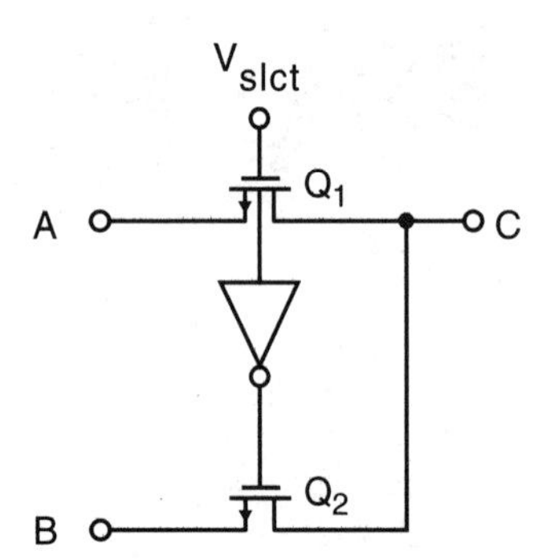

Figure 5.2 A 2-to-1 multiplexor realized using transmission gates. Note the connection to the gate of Q_1, not to be mistaken for a substrate connection.

p-well processes.[1] For p-well processes, the body effect is quite large and, as we shall see, this excessively reduces the noise margins. Alternatively, one can use transmission gates composed of n-channel and p-channel transistors in parallel with complementary signals controlling their gates voltages, as is shown in Fig. 5.1b. Normally, the two transistors have the same widths when used in digital circuits.

One of the simplest examples of using pass transistors is to realize a 2-to-1 multiplexor, as is shown in Fig. 5.2. In this example, if V_{slct} is high, Q_1 passes its input signal and Q_2 is off. In this case, the logic value of C is equal to the logic value of A. Alternatively, if V_{slct} is low, Q_2 passes its signal, Q_1 is off, and the logic value of C is equal to the logic value of B.

This circuit can be slightly augmented to implement an *exclusive-or* function, as is shown in Fig. 5.3. In this case, if B is a "1", C will be $\overline{A}$, otherwise C will be A, resulting in the truth table shown in Table 5.1. This is the truth table of the *exclusive-or* function. If standard NMOS logic were used to realize the inverters, this *exclusive-or* realization would require six transistors, as opposed to the seven transistors required for the pseudo-NMOS *exclusive-or* of Fig. 4.9b.[2]

[1] p-well processes used to be popular in the early 1980s but are seldom seen now.

[2] Although fewer transistors are required for a pass-transistor realization, the layout is somewhat more complicated and the area is not necessarily smaller.

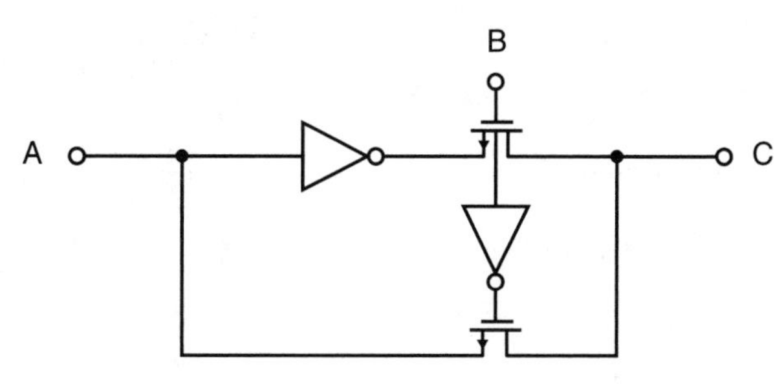

Figure 5.3 A transmission-gate-based realization of the *exclusive-or* function.

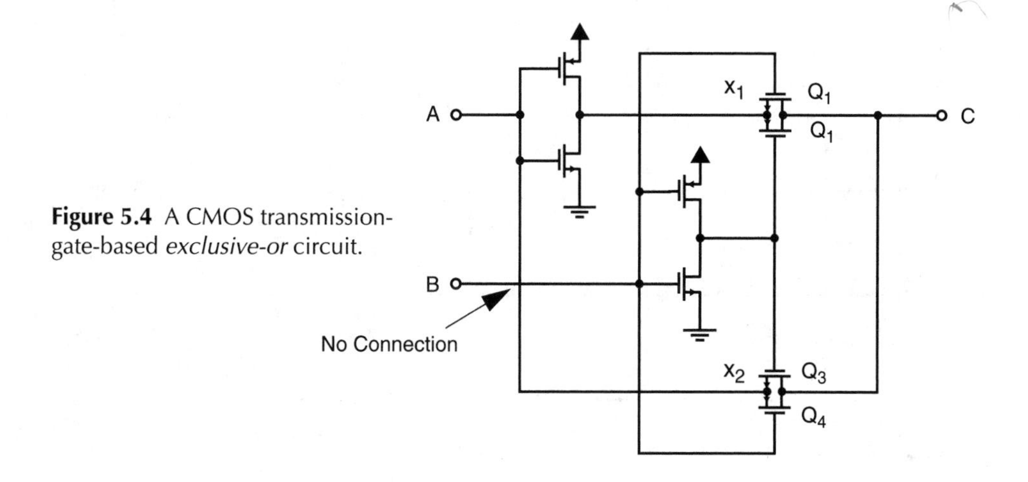

Figure 5.4 A CMOS transmission-gate-based *exclusive-or* circuit.

Table 5.1 The Truth Table of the *exclusive-or* Function

A	B	C
0	0	0
0	1	1
1	0	1
1	1	0

As another example, a CMOS transmission-gate-based *exclusive-or* function is shown in Fig. 5.4. This *exclusive-or* uses the transmission gate of Fig. 5.1b. For this circuit, if B is a "1", n-channel transistor Q_1 and p-channel transistor Q_2 are both on, which guarantees that transmission gate (X gate) X_1 is on. At the same time, Q_3 and Q_4 are both off, guaranteeing that X_2 is off. Thus, $C = \bar{A}$. Similarly, if B is a "0", X_1 is off, X_2 is on, and C = A. Thus, the *exclusive-or* function is realized using eight transistors, as opposed to 10 transistors required for a traditional CMOS design.

Example 5.1

Design an 8-to-1 multiplexor, using pass-transistor logic.

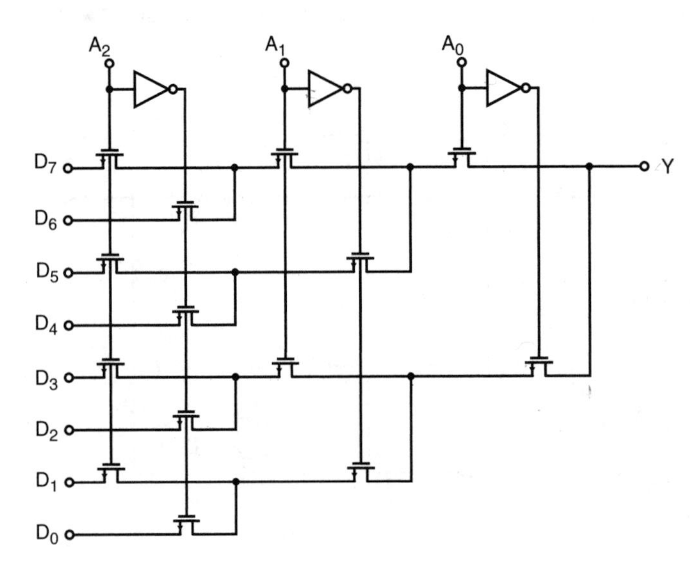

Figure 5.5 An X-gate-based 8-to-1 multiplexor.

Solution: An 8-to-1 multiplexor has eight data inputs, three address inputs, and one data output. For each value of the address inputs, the output will be equal to one of the data inputs. A possible realization is shown in Fig. 5.5. This is one of the most efficient multiplexors possible. Normally, the largest fan-in X-gate multiplexor (MUX) one would build is an 8-to-1 gate. If larger MUXs are required, then a number of 8-to-1 MUXs are used with each one followed by an inverter or a buffer. The reason for not using larger than 8-to-1 MUXs is, as we shall see in the next chapter, the delay is proportional to the number of X gates in series squared. This implies that the logic becomes excessively slow if four or more X gates are in series.

Another popular circuit based on transmission-gate logic is the clocked latch of Fig. 5.6. This circuit is often used for digital storage in registers or pipelined circuits. Its operation is fairly straightforward. When Clk is high, transmission gate X_1 is on and transmission gate X_2

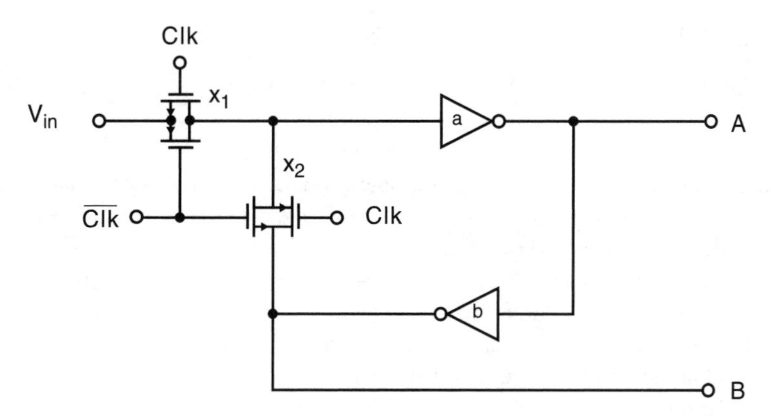

Figure 5.6 A transmission-gate-based clocked latch.

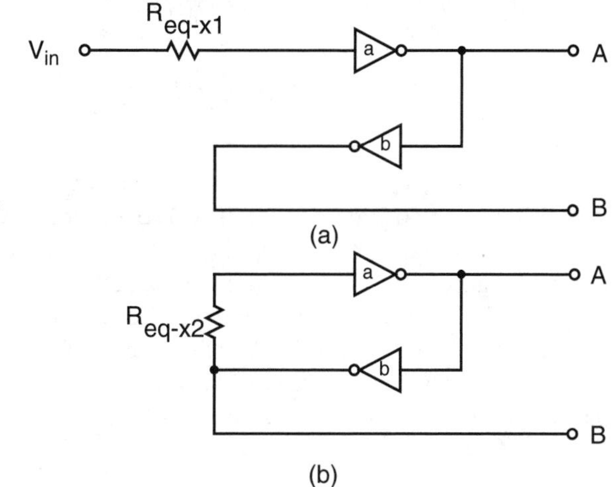

Figure 5.7 An equivalent circuit to the clocked latch of Fig. 5.6. when (a) Clk is high and (b) Clk is low.

is off. An equivalent circuit is shown in Fig. 5.7a. During this phase, the latch is being loaded. A is equal to $\overline{V_{in}}$ and B is equal to V_{in}. Next, when Clk goes low, transmission-gate X_1 goes off and transmission gate X_2 now comes on. The equivalent circuit is shown in Fig. 5.7b. Inverters *a* and *b* are now connected in a positive-feedback loop. Since this is a positive-feedback loop composed of two inverters, it has two possible stable states: A high and B low or vice versa. The state it is in is determined by its initial conditions at the instance Clk goes low. Once it is in a stable state, it will stay there indefinitely, barring any external influences such as excessive noise spikes. Thus, if V_{in} is low when Clk goes low, A will be a "1" and B will be a "0", indefinitely. If V_{in} is a "1" when Clk goes low, A will be a "0" and B will be a "1". This state will be stored in the latch indefinitely until Clk next goes high and the latch returns to the *track mode* of Fig. 5.7a. In Chapter 7, it will be shown how two clocked latches can be used to realize a master–slave D flip-flop.

VOLTAGE DROP OF n-CHANNEL X GATES

Although originally *only* n-channel transistors were used for X gates, CMOS X gates are used more often now. One of the reasons for this is that n-channel X gates have an output voltage much less than V_{DD} when they are on and their input voltage is V_{DD}. To understand this, consider the circuit of Fig. 5.8. As we shall see later, a number of these circuits cascaded, with adjacent cells having opposite-phase clocks for the X gates, realizes a simple delay line.

The idea of the circuit is this: if ϕ_{clk} is a "1", then the voltage at the gate of Q_2 ($V_{G\text{-}2}$) will be the same as the input voltage V_{in}. Next, when ϕ_{clk} goes low, the X gate Q_1 will turn off. From this time on, the voltage $V_{G\text{-}2}$ should not change, as it is stored on the parasitic capacitor C_p (which is primarily due to the gate-source capacitance of Q_2). The output voltage at this time is the inverse of what the input voltage was just before ϕ_{clk} went low.

Unfortunately, the circuit does have some problems. One of these occurs when V_{in} is at V_{DD} and ϕ_{clk} is a "1", also equal to V_{DD}. To understand this problem, first assume that originally V_{in} is at 0 V, when ϕ_{clk} goes high. At this time, the transmission gate is hard in the triode region and is equivalent to a resistor of size R_{TR} where

$$R_{TR} = \frac{1}{\mu_n C_{ox}(W/L)_1(V_{DD} - V_{tn})} \tag{5.1}$$

During this time, $V_{G\text{-}2}$ will be exactly equal to V_{in}, i.e., 0 V. Next, assume V_{in} goes to V_{DD} (i.e., 3.3 V) while ϕ_{clk} remains at V_{DD}. The V_{in} side of Q_1 then becomes the drain, because it is at a higher voltage than the Q_2 side (i.e., 3.3 versus 0 V). Q_1 is now in the active region as it is an enhancement transistor and both its gate and drain are at the same voltage, 3.3 V. At this time, Q_1 is operating as a source-follower (or common-drain) buffer. This situation is shown in Fig. 5.9. This quickly begins to charge C_p and thus raise $V_{G\text{-}2}$ to $V_{DD} - V_{tn}$. However, as V_{G2} gets larger, the effective gate voltage of Q_1 decreases causing $V_{G\text{-}2}$ to change more and more slowly particularly toward the end of its transition. Eventually, the effective

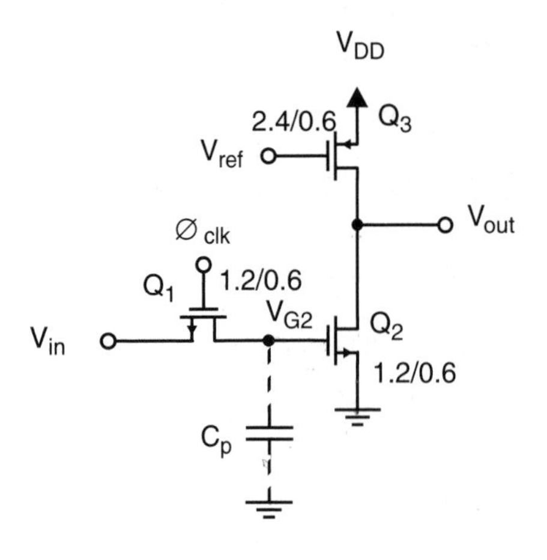

Figure 5.8 A pseudo-NMOS, X-gate-based delay-line cell.

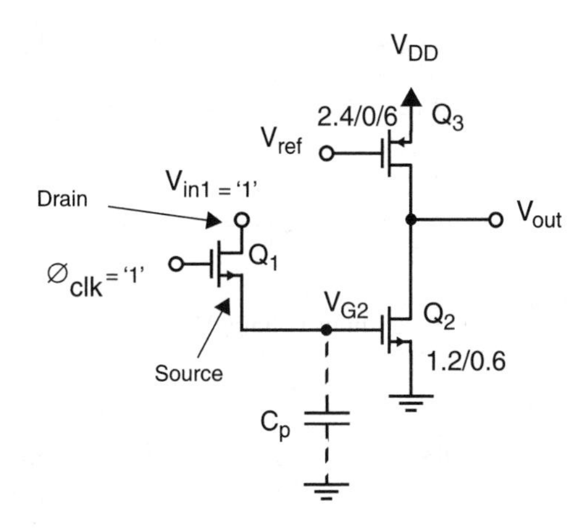

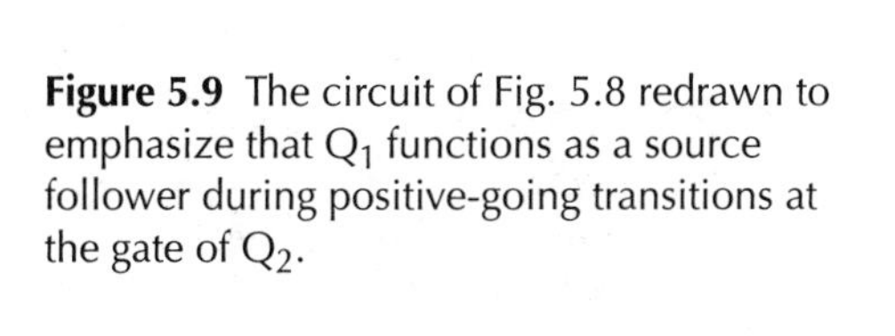

Figure 5.9 The circuit of Fig. 5.8 redrawn to emphasize that Q_1 functions as a source follower during positive-going transitions at the gate of Q_2.

gate voltage of Q_1 becomes zero (i.e., $V_{GS\text{-}1} - V_{tn} = 0$), Q_1 turns off, and $V_{G\text{-}2}$ no longer changes. This occurs when $V_{GS\text{-}1} = V_{tn}$ or equivalently $V_{G\text{-}2} = V_{DD} - V_{tn}$. Thus, $V_{G\text{-}2}$ never goes all the way to V_{DD} but ends up a threshold-voltage drop below it. As a result, V_{G2} can be substantially less than V_{DD}. The situation is more problematic when the body effect is taken into account; the body effect causes the value for V_{tn} to increase by roughly 0.5 V as compared to its value when the source voltage of Q_1 is zero. This is even more problematic for a p-well process where the n-channel transistor is formed in a well that is more heavily doped than the substrate. This results in a much larger body effect and accordingly a larger γ; a typical value might be 1.2 $V^{1/2}$ versus a γ value of 0.5 to 0.6 $V^{1/2}$ for a transistor in the bulk or substrate.

Example 5.2

What is the "1" voltage level of $V_{G\text{-}2}$ in Fig. 5.9 for $V_{tn\text{-}0} = 0.8$ V for the cases $\gamma = 0.5\ V^{1/2}$ and $\gamma = 1.2\ V^{1/2}$.

Solution: From equation (3.158), we have

$$V_{tn} = V_{tn\text{-}0} + \gamma\left(\sqrt{V_{SB\text{-}1} + |2\phi_F|} - \sqrt{|2\phi_F|}\right) \tag{5.2}$$

Also, with reference to Fig. 5.9, we have

$$V_{SB\text{-}1} = V_{G\text{-}2} = V_{DD} - V_{tn} \tag{5.3}$$

Substituting this into (5.2) gives

$$V_{tn} = V_{tn0} + \gamma\left(\sqrt{V_{DD} - V_{tn} + |2\phi_F|} - \sqrt{|2\phi_F|}\right) \tag{5.4}$$

Using $V_{tn\text{-}0} = 0.8$ V, $|2\phi_F| = 0.7$ V, and $\gamma = 0.5\ V^{1/2}$, we can initially guess that $V_{tn} = 1.3$ V. Using this in (5.4) gives $V_{tn} = 1.2$ V. Iterating once more, we get $V_{tn} = 1.22$ V and $V_{G2} = 2.08$ V. Repeating for $\gamma = 1.2\ V^{1/2}$ gives $V_{tn} = 1.64$ V and $V_{G\text{-}2} = 1.66$ V. Such a low input to a typical NMOS gate would seriously degrade the "1"-noise margin.

The lower noise margins, caused by the voltage drop across the n*-channel* X *gate, can be alleviated somewhat if the threshold voltage of the inverter following it is lowered.* This is achieved by taking the ratio of the W/L of the drive transistor to the W/L of the load transistor larger; a typical value might be twice as large as its normal size assuming a γ of 0.5 to 0.6 $V^{1/2}$. *Not only does this improve noise margins but it significantly decreases the time required for "0" to "1" transitions at the gate of* Q_2 *as the required voltage change is smaller and most of the transition occurs before the effective gate-source voltage of* Q_1 *becomes too small.* If the body-effect parameter γ is above 1 $V^{1/2}$, the n-channel only X gate is normally not used. Rather, in this case, the complementary CMOS X gate is utilized.

Example 5.3

Assuming the inverter of Fig. 5.9 has a ratio of $(W/L)_2$ to $(W/L)_3$ of 1, what would the noise margins be for $\gamma = 0.5\ V^{1/2}$. Assume $V_{tn\text{-}0} = 0.8$ V, $V_{tp\text{-}0} = -0.9$ V.

Solution: Repeating (4.5), we have,

$$V_{th} = V_{tn\text{-}0} + \sqrt{\frac{\mu_p (W/L)_3}{\mu_n (W/L)_2}}\left(\frac{V_{DD}}{2} + V_{tp\text{-}0}\right) = 1.2 \text{ V} \tag{5.5}$$

Since V_{OL} is still about 0.1 V, and from Example 5.2 we have $V_{G\text{-}2} = 2.1$ V, we have

$$NM_H = 2.1 - 1.2 = 0.9 \text{ V} \tag{5.6}$$

and

$$NM_L = 1.2 - 0.1 = 1.1 \text{ V} \tag{5.7}$$

Sometimes a "1" is passed through a number of series transistors before it reaches the input of a traditional logic gate. For example, in Fig. 5.5 the input signals would travel through three transistors before reaching the output. The voltage drop across a series of n-channel transistors,

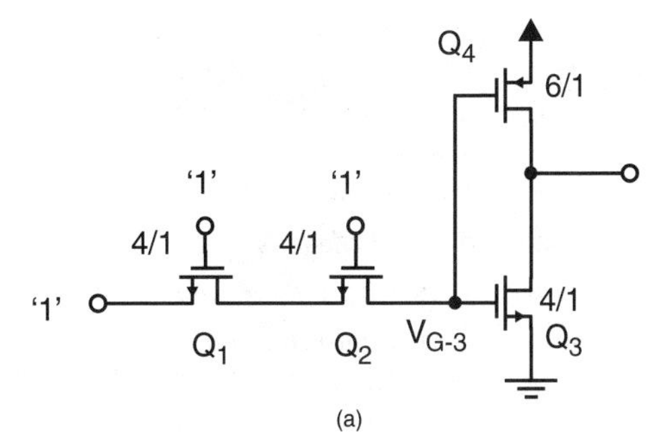

Figure 5.10 (a) Two series n-channel transmission gates and (b) its equivalent circuit.

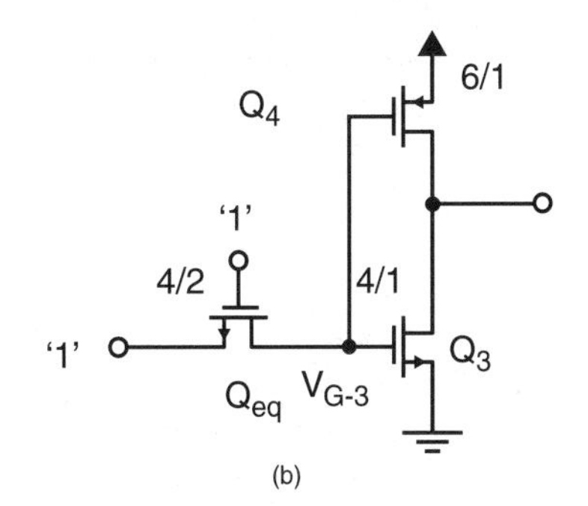

when a "1" is being transmitted, is easily found using the transistor equivalency theorems of Section 4.3. Consider, for example, the case shown in Fig. 5.10a where two transmission gates are in series and a "1" is being transmitted. Since the gates of series-connected Q_1 and Q_2 are both at the same voltage, Q_1 and Q_2 can be replaced by a single equivalent transistor. Since $(W/L)_1 = (W/L)_2 = 4\ \mu m/1\ \mu m$, the equivalent transistor will have a $(W/L)_{eq}$ given by

$$\left(\frac{W}{L}\right)_{eq} = \frac{W_1}{L_1 + L_2} = \frac{1}{2}\left(\frac{W}{L}\right)_1 = \frac{4}{2} \tag{5.8}$$

The equivalent circuit is shown in Fig. 5.10b. Thus, the effect of having two series transmission gates is the same (at low frequencies) as having a single transistor with twice the length. The gate voltage of Q_3 will be $V_{DD} - V_{tn}$, the same as if there were only one transmission gate. Thus, the voltage levels transmitted through a number of transmission gates connected in series is the same as the levels transmitted through a single transmission gate, but the circuit will be slower due to the extra distributed parasitic junction capacitances and larger equivalent resistances.

n-CHANNEL PASS TRANSISTORS VERSUS CMOS TRANSMISSION GATES

When designing transmission-gate logic, there is always a choice whether to simply use n-channel pass transistors only to realize the transmission gates or to use CMOS transmission gates consisting of parallel n-channel and p-channel transistors. When designing NMOS logic, where only n-channel transistors are available, then n-channel pass transistors must be used. However, now, with CMOS technology the norm, the decision is not so obvious. In CMOS logic, the practice is to use a CMOS X gate for p-well processes (where the n-channel transistor has large body effects).

For n-well processes, CMOS X gates are still most often used, except where space is very important, in which case n-channel only transmission gates (i.e., pass transistors) are used. There are two reasons for the popularity of the CMOS X gates. First, there never is a voltage drop across them; when the input voltage is low, the n-channel transistor is hard in the triode region and conducting the input signal (the p-channel is off); when the input signal is high, the n-channel transistor is off but now the p-channel transistor is hard in the triode region and transmits the input signal. Second, CMOS X gates have faster "0" to "1" transitions, particularly near the end of their transitions, again due to the parallel p-channel which gives a much lower resistance when the input is at a high voltage. CMOS transmission gates are particularly popular when the gate control signals and their complements are available, as is the case for the 8-to-1 multiplexor shown in Fig. 5.5.

A major disadvantage of using n-channel pass transistors is that because of the voltage drop when passing a "1", p-channel transistors in succeeding gates will not completely turn off. This can significantly increase the standby power dissipation, but is avoidable as discussed in the next section.

When speed is most important, then networks based on using n-channel pass transistors can be faster if the succeeding logic has its threshold voltages adjusted lower. CMOS transmission-gate impedance is smaller due to the p-channel transistor but the junction capacitance goes up by more than the impedance decrease due to the p-channel junction capacitance and the necessity to have contacts in all junctions. This can be significant. In large transmission-gate networks, where the junction capacitance dominates, pass-transistor realizations can be significantly faster than CMOS X-gate realizations; when the load capacitance dominates, the opposite can be true.

When power-supply voltages of 3.3 V or lower are used, which is presently the norm, n-channel pass transistors become impractical due to their voltage drops unless transistors having very small threshold voltages are available. Luckily, these low-threshold-voltage transistors, often called natural or native devices, are now becoming readily available.

FULL-SWING n-CHANNEL X-GATE LOGIC

It is possible to modify n-channel X gates so that full-level logic swings are realized. The basic principle is illustrated in Fig. 5.11. The method is applicable when an inverter follows a pass-transistor network. This is often the case. For this situation, an additional p-channel transistor can be added with its gate connected to the output of the inverter as shown. This additional p-channel transistor has its drain connected to the output of the pass-transistor network. When a "1" is being transmitted through the pass transistor, the output of the inverter will change to a "0". This will cause the additional p-channel transistor to turn on, which *pulls* the output of the pass transistor all the way to V_{DD}, eliminating the voltage drop through the network.

The p-channel transistor must be sized small enough (i.e., its W/L) so that when the output of the pass-transistor network is changing from a "1" to a "0", the impedance of the n-channel

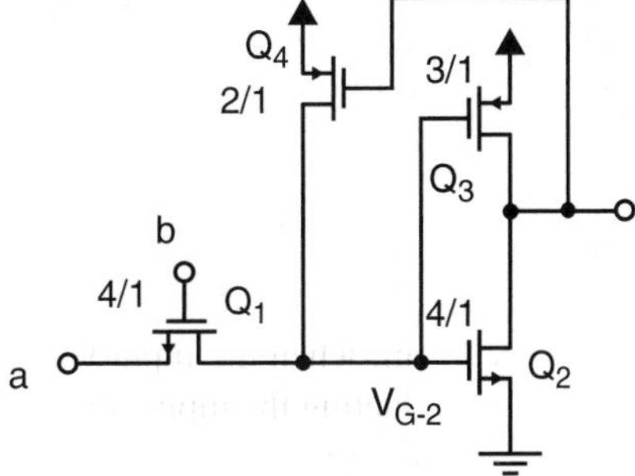

Figure 5.11 Modifying n-channel X-gate logic so full-level swings are realized.

pass-transistor network is less than the impedance of the p-channel transistor and the input to the inverter is *pulled* to a voltage less than the threshold voltage of the inverter. Otherwise "0" outputs of the pass-transistor network would not be propagated. This will be the case if sizing rules for the pseudo-NMOS gates described in the last chapter are followed.

An important advantage of adding this additional p-channel transistor is that it guarantees that the p-channel load transistor is completely turned off when a "1" is present at the output of the pass transistor. Otherwise, the p-channel load of the inverter would never turn off and the inverter would dissipate d.c. power when its output is low.

Adding the additional p-channel transistor (Q_4 in Fig. 5.11) has another advantage. It adds hysteresis to the inverter, which makes it less likely to have *glitches* when signals are noisy. This occurs because the additional transistor makes the threshold voltage of the inverter smaller when the pass transistor's output is changing from a "1" to a "0". The threshold voltage for "0" to "1" transitions is not affected. Since the threshold voltages are spread apart for the two different transitions, hysteresis results.

Leakage Currents

Although transmission-gate logic is widely used, designers should be wary of many of its possible dangers if they wish their circuits to be error free. These sources of error are especially troublesome in cases in which the output of the transmission gate is left as a high impedance node. An example of this was seen in Fig. 5.9 where during the time ϕ_{clk} is low the node $V_{G\text{-}2}$ is at high impedance. Ideally, $V_{G\text{-}2}$ will remain at the value it had when ϕ_{clk} went low. The trouble occurs in the case when V_{in} was high. At this time, the leakage current of the junctions of Q_1 can slowly cause $V_{G\text{-}2}$ to discharge. To see this, consider the circuit of Fig. 5.12 where diodes have been added to the circuit of Fig. 5.9 to model the reverse-biased junctions. These reverse-biased junctions have leakage currents that are a function of the process, the size of the junction, and temperature. The leakage current approximately doubles for every 10°C increase in temperature. Thus, it is especially troublesome at higher temperatures. If the circuit of Fig. 5.12 is left for a long enough time with ϕ_{clk} low, when $V_{G\text{-}2}$ is a high voltage, C_p will eventually discharge until $V_{G\text{-}2}$ becomes a "0" and V_{out} becomes incorrect. For this reason, no node should be left at high impedance for longer than about 200 μs or so.

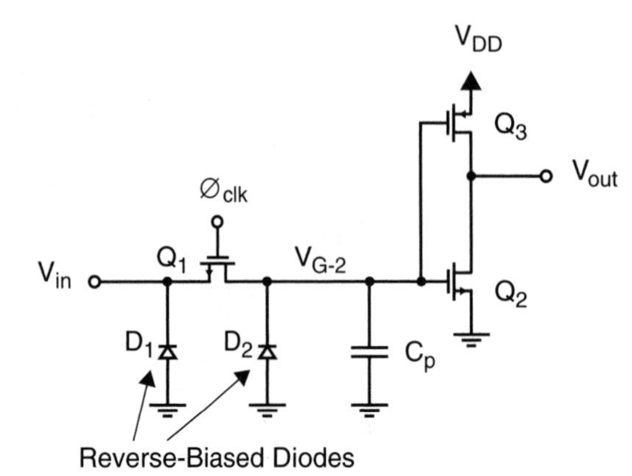

Figure 5.12 A CMOS circuit similar to the pseudo-NMOS circuit of Fig. 5.9 with diodes added to model the junctions of Q_1.

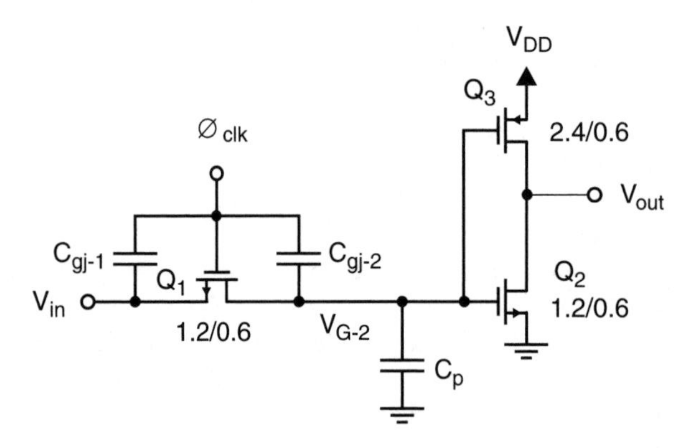

Figure 5.13 The circuit of Fig. 5.12 with the clock feedthrough capacitances (C_{gj}) of Q_1 shown.

Clock Feedthrough[3]

Another problem with transmission gates is the clock feedthrough that results when Q_1 turns off. This clock feedthrough can be modeled by gate-to-junction capacitances ($C_{gj\text{-}i}$) shown in Fig. 5.13. These capacitances are nonlinear and difficult to characterize accurately. They are composed of two components, the gate-to-channel capacitance (C_{ch}) and the overlap capacitance (C_{ov}). The first component models the effects of the channel charge that must flow out of the junctions when the transistor turns off. The second component models the capacitive coupling between the gate and the junctions due to the fact that the regions overlap a little and also due to electrical fringing fields.

[3]The material in this section is intended for advanced courses and professional engineers. It is not intended that it be covered in a first-level course.

The change in voltage due to the channel charge can be estimated if one assumes the transistor Q_1 turns off quickly in which case the channel charge flows approximately equally out of both junctions. This approximation represents a conservative worst-case estimate. From Chapter 3 (3.59), we have the total channel charge given by

$$Q_{ch\text{-}1} = -WLC_{ox}(V_{GS\text{-}1} - V_{tn\text{-}1}) = -WLC_{ox}V_{eff\text{-}1} \tag{5.9}$$

and since this charge is negative as the channel carriers are electrons, we have the voltage change at $V_{G\text{-}2}$ due to the channel charge given by

$$\Delta V_{G\text{-}2} = \frac{Q_{ch\text{-}1}/2}{C_p} = -\frac{WLC_{ox}V_{eff\text{-}1}}{2C_p} \tag{5.10}$$

The overlap capacitance is approximately constant and given by

$$C_{ov} = W_1 L_d C_{ox} \tag{5.11}$$

where L_d is the lateral diffusion of the junction under the gate. In practice, a slightly larger value should be used for L_d to account for electric-field fringing effects. *When there is no channel present before Q_1 turns off, which is the case when $V_{eff\text{-}1}$ is less than or equal to zero, then the clock feedthrough is just due to overlap capacitance.* When a channel is present before Q_1 turns off, then the clock feedthrough due to the channel charge and the clock feedthrough due to the overlap capacitance can be calculated separately and the individual voltage changes can be added when estimating the total clock feedthrough voltage change. This methodology is based on the principle of linear superposition. In this case, the clock feedthrough due to the channel charge, which is given by (5.10), usually dominates.

To calculate the clock feedthrough due to the overlap capacitance, one can make use of the voltage divider rule for capacitors. This rule states that when two series capacitors (C_1 and C_2) are connected to V_1, which has a voltage change ΔV_1, then the change in the voltage of the internal node, ΔV_2, is given by

$$\Delta V_2 = \frac{\Delta V_1 C_1}{C_1 + C_2} \tag{5.12}$$

as is illustrated in Fig. 5.14 (see Problem 5.7). When this formula is applied to the circuit of Fig. 5.13, the gate voltage of Q_1 corresponds to V_1, the gate voltage of Q_2 corresponds to V_2, C_{ov} corresponds to C_1, and the capacitance C_p corresponds to C_2. The change in the gate voltage, $V_{G\text{-}1}$, is equal to $-V_{DD}$. Therefore, the change in $V_{G\text{-}2}$ due to the overlap capacitance (only) is given by

$$\Delta V_{G\text{-}2} = \frac{-V_{DD}C_{ov}}{C_{ov} + C_p} \tag{5.13}$$

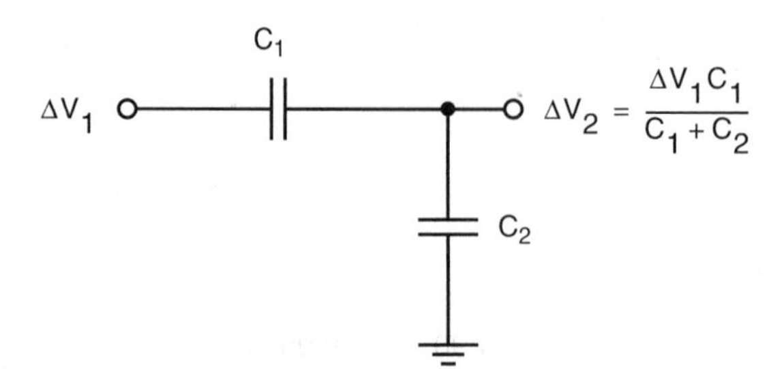

Figure 5.14 The voltage divider rule for capacitors.

This may be added to the voltage change caused by the channel charge in order to estimate the total voltage change caused by clock feedthrough.

When a "0" is being transmitted, then a channel exists before Q_1 turns off and the total change in the gate voltage of Q_2 is due to both channel charge and overlap capacitance. When a "1" is being transmitted, there is no channel present when Q_1 turns off as $V_{eff\text{-}1} = 0$ at this time, and, therefore, the clock feedthrough is due to overlap capacitance only. In both cases, the change in the gate voltage of Q_2 is negative and substantially greater in magnitude in the former case. However, since in the former case when a "0" is being transmitted, the clock feedthrough causes the gate voltage to change further away from the threshold voltage of the inverter consisting of Q_2 and Q_3, the noise margin is actually increased. In the latter case, when a "1" is being transmitted, the negative clock feedthrough decreases the noise margin, but since the clock feedthrough is due to overlap capacitance only, it hopefully does not decrease the noise margins too much. If the inverter consisting of Q_2 and Q_3 is designed to have a low gate threshold voltage, then errors due to the clock feedthrough in this case are made even less likely.

Example 5.4

In Fig. 5.13, assuming that C_p is equal to the sum of the gate capacitance of Q_2 and the junction capacitance of Q_1, calculate the voltage $V_{G\text{-}2}$ after ϕ_{clk} goes low for the cases $V_{in} = V_{DD}$ and separately for $V_{in} = 0$ V. Assume $C_{ox} = 3.4 \times 10^{-3}$ pF/mm^2, $V_{DD} = 3.3$ V, and $L_d = 0.04$ μm. Also assume the junction capacitance of Q_1 is 1.60 fF when a "0" is being transmitted and 0.94 fF when a "1" is being transmitted (in which case the reverse-bias voltage across the junctions is larger, that is 2.08 V from Example 5.2).

Solution: Assume that the gate capacitance of Q_2 is equal to $(WL)_2 C_{ox}$. This is not accurate for the 0 V case when the gate capacitance decreases (since $V_{GS\text{-}2} = 0$), but without this approximation the problem is intractable for hand analysis. Thus, for $V_{in} = 0$ V, we have the gate capacitance of Q_2 given by

$$C_{g\text{-}2} = C_{ox}(WL)_2 = 2.45 \text{ fF} \tag{5.14}$$

Also, assume the gate capacitance of Q_3 is equal to $(WL)_3C_{ox}$, where the inaccuracy arises for the 3.3 V case. We thus have

$$C_{g\text{-}3} = C_{ox}(WL)_3 = 4.90 \text{ fF} \tag{5.15}$$

and the total capacitance at the gate of Q_2, that is C_p, is given by

$$C_p = C_{j1} + C_{g2} + C_{g3} = 1.60 \text{ fF} + 2.45 \text{ fF} + 4.90 \text{ fF} = 8.95 \text{ fF} \tag{5.16}$$

When the input voltage is 0 V, we have $V_{eff\text{-}1} = 2.5$ V, and the clock feedthrough due to the channel charge is given by (5.10) as

$$\Delta V_{G\text{-}2} = -\frac{WLC_{ox}V_{eff\text{-}1}}{2C_p} = -\frac{(1.2\times 0.6)3.4\times 10^{-15}(2.5)}{2\times 8.95\times 10^{-15}} = -0.34\text{V} \tag{5.17}$$

The overlap capacitance, C_{ov}, is given by

$$C_{ov} = W_1L_dC_{ox} = (1.2\times 0.04)3.4\times 10^{-3} = 0.163 \text{ fF} \tag{5.18}$$

The clock feedthrough due to the overlap capacitance is found using (5.13) to be

$$\Delta V_{G\text{-}2} = -\frac{V_{DD}C_{ov}}{C_{ov}+C_p} = -\frac{3.3(0.163)}{0.163+8.95} = -0.059 \text{ V} \tag{5.19}$$

Therefore, the total clock feedthrough when a "0" is being transmitted is given by

$$\Delta V_{G\text{-}2} = -0.34 - 0.059 = -0.40 \text{ V} \tag{5.20}$$

When a "1" is being transmitted, the value for C_p is now given by

$$C_p = C_{j\text{-}1} + C_{g\text{-}3} + C_{g\text{-}2} = 0.94 \text{ fF} + 2.45 \text{ fF} + 4.90 \text{ fF} = 8.29 \text{ fF} \tag{5.21}$$

For this case, no channel is present just before Q_1 turns off and the clock feedthrough is due to overlap capacitance only. We have using (5.13)

$$\Delta V_{G\text{-}2} = -\frac{V_{DD}C_{ov}}{C_{ov}+C_p} = -\frac{3.3(0.163)}{0.163+8.29} = -0.064 \text{ V} \tag{5.22}$$

These numbers are fairly representative for the relative widths given for the transistors and do not change greatly as newer technologies with shorter channel lengths are used.

The clock feedthrough of CMOS transmission gates must take into account both the n-channel and the p-channel transistors. In most transmission gates, these transistors are sized to have equal widths, so the clock feedthrough of the two transistors cancel somewhat. In reality, assuming fast clock waveforms occurring at the same time, the clock feedthrough of CMOS transmission gates actually increases the noise margins somewhat.

To see this, consider the case in which the input is 0 V just before the transmission gate turns off. The clock feedthrough of the n-channel transistor is identical to that previously calculated for the NMOS case. At this time, the p-channel transistor has no channel and its clock feedthrough is due to overlap capacitance only but with a positive sign. The total clock feedthrough is the sum of the two. Since the clock feedthrough of the n-channel transistor is much larger than that of the p-channel transistor, the net effect will be a negative clock feedthrough that increases NM_L.

For the case in which the input is at V_{DD}, the p-channel transistor now has a channel, whereas the n-channel transistor is cut off. Thus, the clock feedthrough of the p-channel transistor dominates and the net effect is a positive voltage change, which increases NM_H. Thus, in both cases the noise margin increases. This is one of the reasons why CMOS transmission gates are often preferred.

Finally, it should be mentioned that for both types of transmission gates, there will also be clock feedthrough when the gates turn on that decreases the noise margins temporarily until the node can be charged or discharged through the transmission gate. This effect results in a *glitch*, which normally does not cause problems unless it is very large and goes to a clock input of a flip-flop.

5.2 Differential CMOS Circuits

A family of digital circuits that is gaining in importance is based on using differential circuits and signals. This class of circuits is covered in detail in Chapter 9, which deals with advanced CMOS design techniques. It is introduced here for first-level courses, where Chapter 9 may not be covered, since this is becoming a popular approach.

In differential logic design, each input signal is represented by the voltage difference between two wires. One wire will be denoted the positive wire and the other will be denoted the negative wire; if the voltage on the first wire is greater than the voltage on the second wire, the signal is considered to be a "1", otherwise it is considered to be a "0". Furthermore, each logic gate will have two output wires and the logic value will again be represented by the voltage difference between them. *One of the repercussions of this methodology is that logic inversions are trivially obtained by simply interchanging wires without incurring a time delay.* This fact often allows inverters to be eliminated in many arithmetic circuits such as the carry-generate circuits in full adders. This is one of the major advantages of differential circuits. A second advantage is that the load networks will often consist of two cross-coupled p-channel transistors only, as we shall see. This minimizes both the area and the number of series p-channel transistors compared to traditional CMOS logic designs. A further advantage is that differential circuits are less sensitive to noise corruption, particularly from power-supply

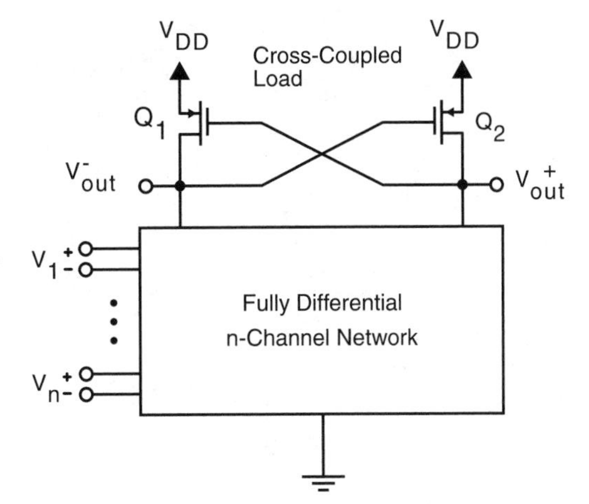

Figure 5.15 A fully differential logic circuit.

or ground noise. A disadvantage of differential logic gates is that since two wires must be used to represent every signal, the interconnect area can be significantly greater. In applications in which only a few close gates are being driven, this disadvantage is often not as significant as the advantages. For this reason, differential logic circuits are often a preferable consideration.

One simple and popular (but not the only) approach for realizing differential logic circuits is shown in Fig. 5.15, where the n-channel drive network is not shown explicitly. The inputs to the drive network come in pairs, a single-ended signal and its inverse, as was just mentioned. Depending on the values of these inputs, there will be a low-impedance path between one output of the n-channel network and ground; between the other output of the n-channel network and ground will (effectively) be infinite impedance. Therefore, one output of the logic gate will be pulled to a low voltage. When this occurs, the p-channel load at the other output is turned on, which pulls the other output to a high voltage. This causes the p-channel load at the first output to turn off, which implies the gate has no d.c. power dissipation, once it has finished its transition and settled. For these gates to work, the n-channel transistors must be taken wide enough so the impedance through the low-impedance path of the n-channel network is significantly less than the impedance of a p-channel load transistor when it is on; otherwise the gate can never change states.

There are many different approaches for designing the n-channel drive networks, and numerous examples are described in Chapter 9; a simple and general, but not necessarily optimum, approach is described here. The n-channel network can be divided into two separate networks, one between the inverting output and ground, and a complementary network between the noninverting output and ground. The first network can be designed using techniques described in Section 4.2 for designing drive networks for pseudo-NMOS gates. The second network can then be realized by a complementary network; parallel connections in the first network become series connections in the second network, and, correspondingly, series connections in the first network become parallel connections in the second network. This procedure is similar to that used in deriving p-channel networks in traditional CMOS

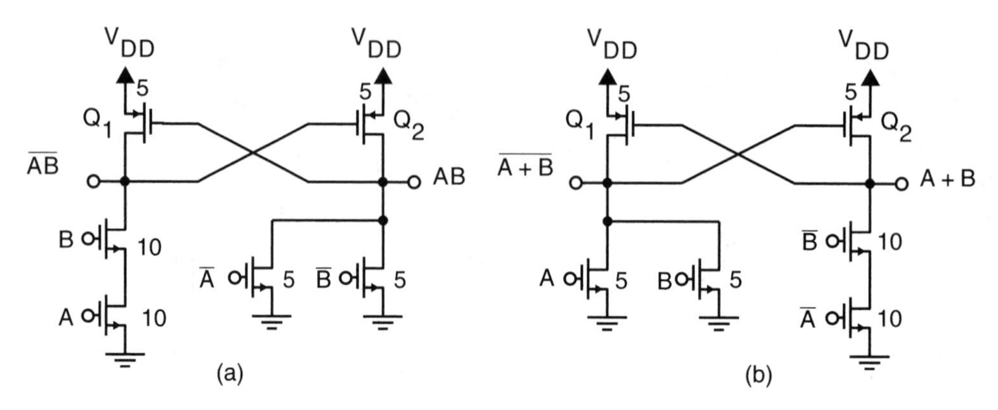

Figure 5.16 Examples of differential CMOS realizations of (a) *and* and (b) *or* functions.

logic gates. In addition, the single-ended inputs to the second network are taken as the complements of the single-ended inputs to the first network. Examples of this are shown in Fig. 5.16a and b for two-input fully differential *and* and *or* gates, respectively. Reasonable W/Ls for the various transistors that result in correct functional operation are also given. *Notice that the gates are identical in structure; the latter can be obtained from the former by simply interchanging inputs and outputs.* In the former case, a two-input *nand* gate can be realized by simply interchanging outputs only, and in the latter case, a two-input *nor* gate can be obtained by interchanging outputs only. Thus, a single structure suffices for all popular two-input gates. This can be used to simplify automatic synthesis programs in which the functionality can be determined by connections only. The extension of the gate to the three-input case is straightforward.

Example 5.5

Consider realizing the function

$$V_{out} = (A + \overline{B})C + \overline{A}E \tag{5.23}$$

Solution: A possible realization of this circuit is shown in Fig. 5.17. Note how the network connected to the noninverting output is the complement of the network connected to the inverting output.

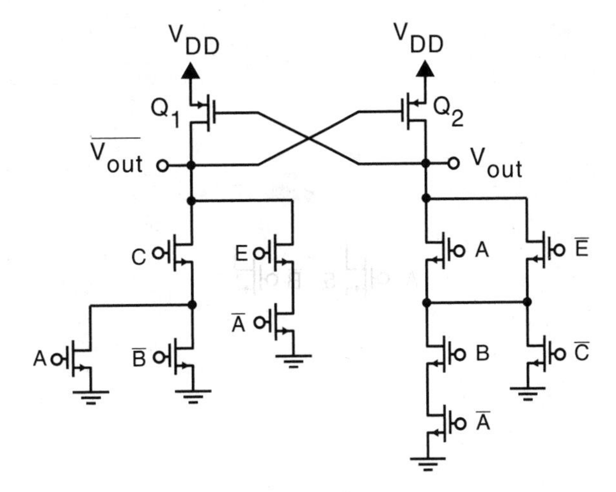

Figure 5.17 A fully differential realization of the function $V_{out} = (A + \bar{B})C + \bar{A}E$.

Although the above procedure is general, it does not always result in the simplest gates. As a final example, consider the fully differential full adder shown in Fig. 5.18, which also shows reasonable values for W/Ls. The sizes have been chosen not just to guarantee functionality, but to maximize the speed of carry-propagate adders, which are discussed in Chapter 10. It is left

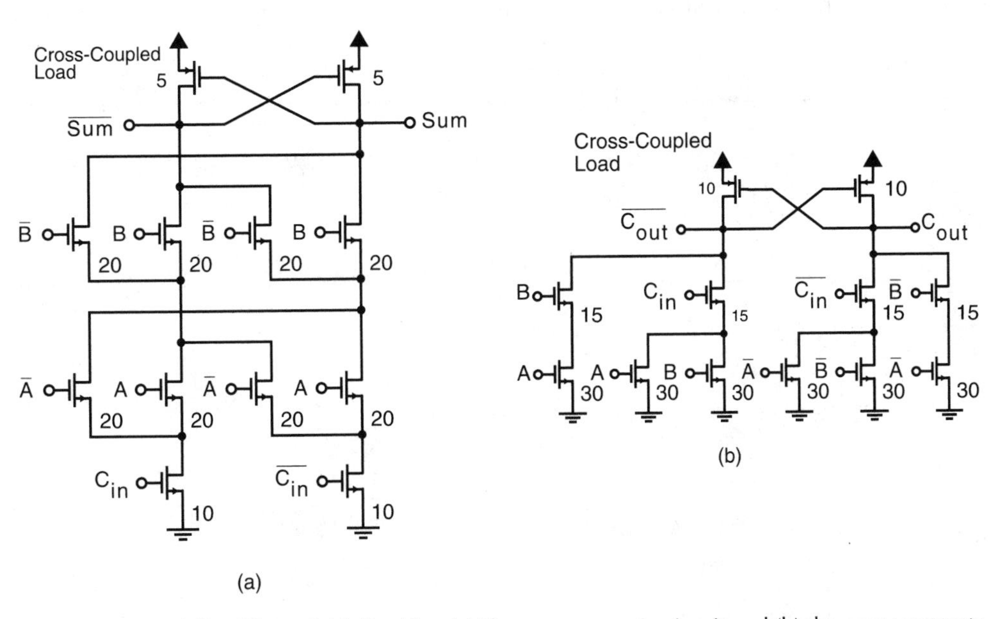

Figure 5.18 A fully differential full adder. (a) The sum-generate circuit and (b) the carry-generate circuit.

as an exercise for the reader to verify that the circuits Fig. 5.18a and b realize the sum-generate and carry-generate functions, respectively.

5.3 Bibliography

M. Annaratone, *Digital CMOS Circuit Design*, Kluwer, 1986.

K. Chu and D. Pulfrey, "Design Procedures for Differential Cascade Logic," *IEEE Journal of Solid-State Circuits*, SC-21 (6), 1082–1087, December 1986.

M. Elmasry, ed., *Digital MOS Integrated Circuits, II,* IEEE Press, 1991.

L. Glasser and D. Dopperpuhl, *The Design and Analysis of VLSI Circuits*, Addison-Wesley, 1985.

L. Heller et al., "Cascade Voltage Switch Logic: A Differential CMOS Logic Family," *Proceedings of the IEEE ISSCC Conference*, 16–17, February 1984.

D. Hodges and J. Jackson, *Analysis and Design of Digital Integrated Circuits*, McGraw-Hill, 1988.

M. Mano, *Digital Design,* Prentice Hall, 1984.

C. Mead and L. Conway, *Introduction to VLSI Systems,* Addison-Wesley, 1980.

J. Rabaey, *Digital Integrated Circuits, A Design Perspective*, Prentice Hall, 1996.

D. Radhakrishnan, S. Whittaker, and G. Maki, "Formal Design Procedures for Pass-Transistor Switching Circuits," *IEEE Journal of Solid-State Circuits,* SC-20 (2), 531–536, April 1985.

T. Sakurai and K. Tamaru, "Simple Formulas for Two- and Three-Dimensional Capacitances, *IEEE Transactions on Electron Devices*, ED-30 (2), 183–185, February 1983.

J. Uyemura, *Circuit Design for CMOS VLSI*, Kluwer, 1992.

N. Weste and K. Eshragian, *Principles of CMOS VLSI Design: A Systems Perspective*, Addison-Wesley, 1983.

5.4 Problems

For the problems in this chapter and all future chapters, assume the following transistor parameters:

- npn bipolar transistors:

  $\beta = 100$

  $V_A = 80$ V

  $\tau_b = 13$ ps

  $\tau_s = 4$ ns

  $r_b = 330\ \Omega$

- n-channel MOS transistors:

  $\mu_n C_{ox} = 190\ \mu A/V^2$

  $V_{tn} = 0.7$ V

$\gamma = 0.6\ V^{1/2}$
$r_{ds}\ (\Omega) = 5000L\ (\mu m)/I_D\ (mA)$ in active region
$C_j = 5 \times 10^{-4}\ pF/(\mu m)^2$
$C_{j\text{-}sw} = 2.0 \times 10^{-4}\ pF/\mu m$
$C_{ox} = 3.4 \times 10^{-3}\ pF/(\mu m)^2$
$C_{gs(overlap)} = C_{gd(overlap)} = 2.0 \times 10^{-4}\ pF/\mu m$

- p-channel MOS transistors:
  $\mu_p C_{ox} = 50\ \mu A/V^2$
  $V_{tp} = -0.8\ V$
  $\gamma = 0.7\ V^{1/2}$
  $r_{ds}\ (\Omega) = 6000L\ (\mu m)/I_D\ (mA)$ in active region
  $C_j = 6 \times 10^{-4}\ pF/(\mu m)^2$
  $C_{j\text{-}sw} = 2.5 \times 10^{-4}\ pF/\mu m$
  $C_{ox} = 3.4 \times 10^{-3}\ pF/(\mu m)^2$
  $C_{gs(overlap)} = C_{gd(overlap)} = 2.0 \times 10^{-4}\ pF/\mu m$

5.1 Design a transmission-gate-based *exclusive-nor* circuit using CMOS transmission gates.

5.2 Design a 4-to-1 multiplexor using CMOS X gates.

5.3 Design a 32-to-1 multiplexor using four 8-to-1 multiplexors and one 4-to-1 multiplexor.

5.4 Design a carry-propagate circuit using transmission gates. A carry-propagate circuit realizes the function $C_{out} = C_{in}\ (A + B)$. Minimize the delay through the path from C_{in} to C_{out}.

5.5 Repeat Example 5.2, but assume $V_{tn\text{-}0} = 0.7$ V and $\gamma = 0.65\ V^{1/2}$.

5.6 Repeat Example 5.3, but assume $V_{tn\text{-}0} = 0.7$ V, $V_{tp\text{-}0} = -8$ V, and $\gamma = 0.65\ V^{1/2}$.

5.7 Prove equation (5.12).

5.8 The CMOS *transmission gate* shown has an input voltage of 1.5 V when it turns off. The W/L of the n-channel is 4 μm/0.6 μm and the W/L of the p-channel is 8 μm/0.6 μm. Estimate the change in output voltage due to clock feedthrough. You may assume the total parasitic capacitance between the output node and ground is 50 fF, that V_{DD} = 3.3 V, and that the clock signals change very fast. Ignore changes due to overlap capacitance. Assume $V_{tn\text{-}0}$ = 1.1 V and $V_{tp\text{-}0}$ = −1.2 V (taking into account body effect). What will the final output voltage be?

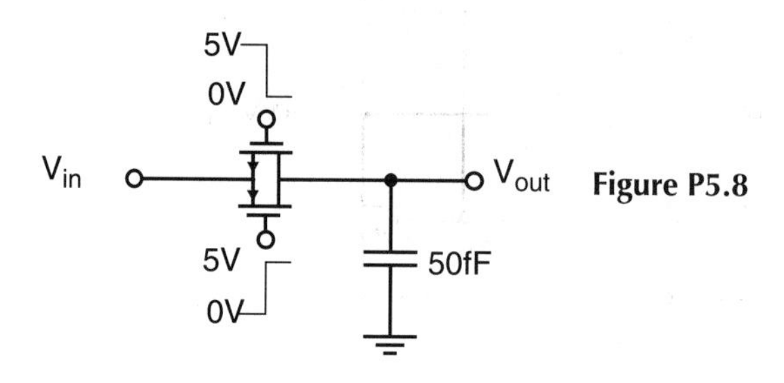

Figure P5.8

5.9 A CMOS transmission gate is transmitting a "1" (i.e., 3.3 V) when it turns off. The W/Ls of both the n-channel and p-channel transistors are 4 μm/0.4 μm. Estimate the change in output voltage due to clock feedthrough. You may assume the total parasitic capacitance between the output node and ground (or V_{DD}) is 25 fF, and that the clock signals change very fast. You may also ignore changes due to overlap capacitance. What will the final output voltage be? Repeat for when a "0" is being transmitted.

5.10 Repeat Problem 5.8 but calculate the change in output voltage considering overlap capacitance only.

5.11 Repeat Problem 5.9 but calculate the change in output voltage considering overlap capacitance only.

5.12 Design a fully differential CMOS gate with cross-coupled loads that realizes the function $\text{out} = x_1\overline{x_2}$. Give reasonable device sizes assuming L_{min} = 0.5 μm.

5.13 Design a fully differential CMOS gate with cross-coupled loads that realizes the function $\text{out} = x_1\overline{x_2}x_3$. Give reasonable device sizes assuming L_{min} = 0.5 μm.

5.14 Design a fully differential CMOS gate with cross-coupled loads that realizes the carry-propagate function $C_{out} = C_{in} + AB$. Give reasonable device sizes assuming L_{min} = 0.5 μm. Try to optimize the delay through the critical path from C_{in} to C_{out}.

5.15 What function does the following circuit realize? Does it have some limitations?

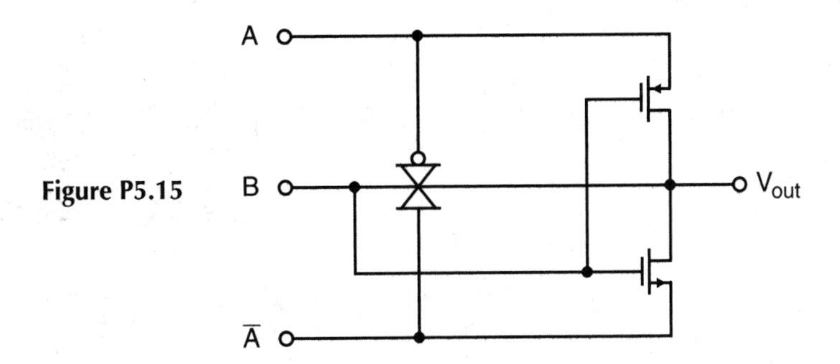

Figure P5.15

5.16 Using the circuit of Problem 5.15 as a building block, realize a transmission-gate-based full adder.

5.17 Using a combination of differential logic and X-gate-based logic, and some ingenuity, realize a two-input *and* function.

5.18 Using a combination of differential logic and X-gate-based logic realize a two-input *exclusive-or* function.

5.19 Use four n-channel transistors and a cross-coupled p-channel load to realize a fully differential two-input *and* gate.

5.20 Use four n-channel transistors and a cross-coupled p-channel load to realize a fully differential two-input *exclusive-or* gate.

CHAPTER 6

CMOS Timing and I/O Considerations

6.1 Delay of MOS Circuits

As we have seen previously, the transient delay of a CMOS inverter is often approximated using an RC approximation. Similarly, when analyzing more complicated gates for their transient responses, the gates are modeled using RC circuits. Often, the resulting RC circuit will be a ladder structure with resistors interconnected between nodes and capacitors connected between nodes and ground. The resistor values are found using the approximately equivalent resistor concept presented in Section 4.1. The calculation of the parasitic capacitances has been previously discussed, but will be described in more detail in this chapter.

As has been mentioned, there are typically three different components that are primarily responsible for the parasitic and load capacitances of MOS circuits. These are gate capacitances, junction capacitances, and interconnect capacitances. Each will be discussed in turn. Much of this section is repeated from the section on layout from Chapter 3.

Gate Capacitances

The transistor gate capacitances often dominate in situations in which only a number of close logic gates are being driven. Typical examples of this include arithmetic circuits such as adders, multipliers, controllers, and arithmetic logic units in microprocessors.

The exact calculation of the capacitances due to gate connections is complicated and highly nonlinear; a transistor's gate capacitance changes substantially depending on the node voltages of the transistor. For example, if the gate has a channel present and is in the *active region*, its gate capacitance is approximately given by

$$C_{gs} = \frac{2}{3}WLC_{ox} + 2WL_{ov}C_{ox} \quad (6.1)$$

The first term is the intrinsic gate capacitance, which usually dominates, and the second term is due to the overlap capacitance. However, when the transistor is *hard* in the *triode region*, its intrinsic capacitance is larger and given by

$$C_{gs} \approx WLC_{ox} \quad (6.2)$$

The gate-overlap capacitance should be added to this. When the transistor turns off, the situation gets even more complicated. When the gate-source voltage, while undergoing a negative transition, is less than the transistor threshold voltage but is greater than 0 V, the area directly under the gate is depleted. Also, the gate capacitance is formed from the series combination of the capacitance due to the gate oxide and the capacitance due to the depletion region. Because this is a series connection, the total capacitance is significantly less than that given by (6.2). But, as the gate voltage becomes more negative, the silicon region directly under the gate becomes *slowly* accumulated (that is heavily doped and the same type as the substrate, p^+ for an n-channel transistor) and the intrinsic gate capacitance is again approximately given by (6.2). In both cases, the gate-overlap capacitance should be added. The exact calculation of the gate capacitance is not possible during a quick simplified analysis by a human and can be approximated only using a computer. Luckily, if one simply assumes the gate capacitance is as given by (6.2), the errors are not large. The smaller gate capacitance during some of the states is roughly canceled by ignoring the overlap capacitance. The slightly pessimistic approximation given by (6.2) is about the best one can normally hope for during simplified hand analysis; the alternative, a very accurate and therefore complicated hand analysis is too time consuming and not merited by the benefits when designing digital integrated circuits.

JUNCTION CAPACITANCES

In addition to the gate capacitances, the parasitic capacitances due to drain and source transistor junctions are significant, particularly as technology dimensions become smaller with modern IC processes. The calculation of the junction capacitances is also complicated, if the desire is to be exact. As seen in Chapter 3, junction capacitances are highly nonlinear and are normally approximated by an average capacitance (per unit area) given by[1]

$$C_j = 0.63C_{j\text{-}0} \quad (6.3)$$

where $C_{j\text{-}0}$ is the junction capacitance at 0 V bias. The 0.63 constant is an approximation and assumes the voltage change across the junction is 3.3 V. This junction capacitance must be multiplied by the junction area. In addition, there are sidewall capacitances between the junction and the field implants under the field oxide surrounding the junction interfaces *that are not*

[1]Previously, we have used the more pessimistic and less accurate approximation $Cj = 2/3\ C_{j\text{-}0}$. Either approximation is acceptable.

next to a gate. Again, the nonlinear nature of this capacitance can be approximated by taking the sidewall capacitance (per unit length) as

$$C_{j\text{-sw}} = 0.63C_{jsw\text{-}0} \tag{6.4}$$

The total sidewall capacitance is found by multiplying C_{jsw} by the periphery of the junction excluding the periphery next to the transistor gate. There is sidewall capacitance at the junction interface adjacent to the gate; however, this capacitance is significantly less per unit length than at the interface next to the field oxide and field implants.

One of the most difficult steps in calculating the junction capacitance is estimating the junction areas and peripheries. This is difficult before the layout has been done; after the layout has been done it is typically done automatically using the computer and a circuit extraction program. However, before layout has been done, it is often necessary to estimate these dimensions; otherwise the timing accuracy when simulating the circuit would be substantially degraded. Accurately estimating the junction dimensions is often possible for small transistors if one considers the three different situations that are typically encountered: nonshared junctions with contacts, shared junctions without contacts, and shared junctions with contacts. Each case will be dealt with in turn.

A simplified layout of a transistor having nonshared junctions with contacts is shown in Fig. 6.1. It is easily seen that the junction area is given by

$$A_j = (5/2)LW \tag{6.5}$$

and the junction periphery not in contact with the gate is given by

$$P_j = W + 5L \tag{6.6}$$

The situation in which there is a shared junction without a contact is illustrated in Fig. 6.2. The distance between the gates is normally equal to L; for the shared junction we have

$$A_j = LW \tag{6.7}$$

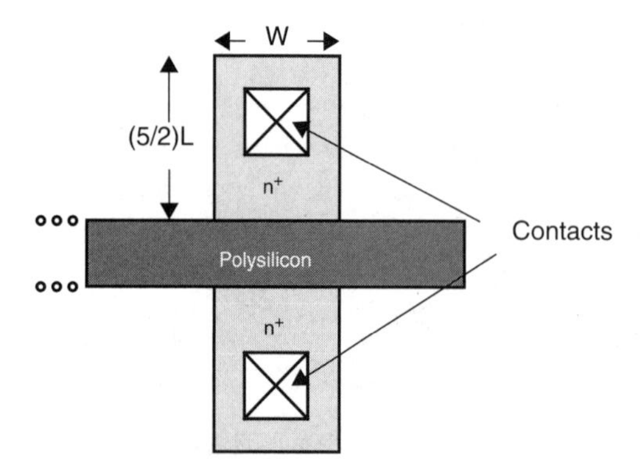

Figure 6.1 A typical layout of a transistor having nonshared junctions with contacts.

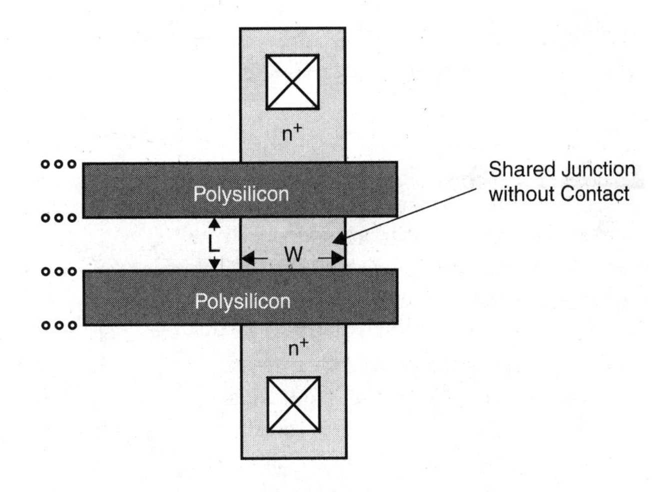

Figure 6.2 A typical layout of a transistor having a shared junction without a contact.

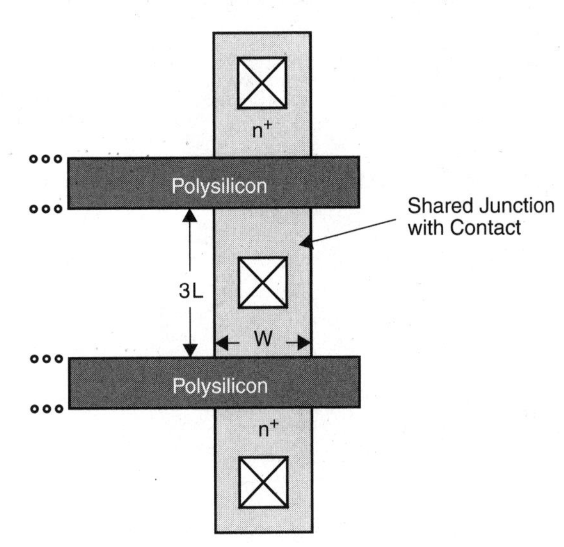

Figure 6.3 A typical layout of a transistor having a shared junction with a contact.

and

$$P_j = 2L \tag{6.8}$$

Notice that in this case the junction area and particularly the sidewall periphery are very small compared to the previous case. The last case, a shared junction with a contact, is illustrated in Fig. 6.3. The contact to the shared junction must normally be at least a distance of L away from the polysilicon gates. This means the minimum distance between the gate stripes is 3L and therefore

$$A_j = 3LW \tag{6.9}$$

Furthermore,

$$P_j = 6L \tag{6.10}$$

Obviously, whenever possible *and* particularly for *critical nodes*, shared junctions should be used as these have smaller areas and in particular smaller peripheries. **Also, a word of caution is in order; when simulating a circuit using SPICE, and when the junction areas and peripheries are specified, a shared junction should not be specified for each of the connected transistors.** Rather, it should be specified for only one transistor, or, alternatively, the area and periphery should be divided by two and then specified for both transistors (this is normally the safer of the two alternatives).[2]

Once an estimate of the junction capacitance at 0 V bias has been found, its value should be modified to reflect the fact that for large voltage changes, a junction capacitance is nonconstant. As a rough approximation, an *average* capacitance value can be used. Remembering from Chapter 3 that the charge stored on a single-sided junction is given by

$$Q_j(V_{jb}) = 2C_{j\text{-}0}\Phi_0\sqrt{1 + \frac{V_{jb}}{\Phi_0}} \tag{6.11}$$

where V_{jb} is the junction to substrate voltage, we can define an average capacitance between 0 and 3.3 V as

$$C_{j\text{-}av} = \frac{Q_j(3.3\text{ V}) - Q_j(0\text{ V})}{3.3\text{ V} - 0\text{ V}} \tag{6.12}$$

Using equation (6.12) and $\Phi_0 = 1.0$ V gives

$$C_{j\text{-}av} = 0.63C_{j0} \tag{6.13}$$

for a 3.3-V power supply. For junctions being charged from 0 to 2.1 V, this will underestimate the capacitance; for junctions being discharged from 3.3 to 1.0 V, this will overestimate the junction capacitance. Repeating the calculation for a 5-V power-supply voltage, which recently was the standard, but is now becoming less popular, gives

$$C_{j\text{-}av} = 0.56C_{j\text{-}0} \tag{6.14}$$

[2]Recently, modern simulation programs have been substantially improved in this area; the relevant manuals should be carefully read with respect to how junction capacitances are modeled.

INTERCONNECT CAPACITANCES

The third major source of load capacitance is due to the interconnecting lines between transistors and logic gates. Normally, whenever possible, metal is used for interconnecting MOS circuits; although for short distances, polysilicon or a diffusion region might sometimes be used. The parasitic capacitance of a metal interconnect has two components. The first is the parallel plate capacitance between the interconnect and the silicon substrate. The second is a fringing capacitance that goes from the edge of the interconnect to the substrate and to adjacent interconnects. Both of these components are shown in Fig. 6.4.

The parallel plate capacitance is dominant and can be simply calculated using the formula

$$C_{plate} = \frac{\varepsilon_{ox}A}{H} \tag{6.15}$$

where A is the area of the interconnect (that is, its width multiplied by its length) and H is the thickness of the oxide between the interconnect and the conductive substrate. This ignores the capacitance due to the top plate of the conductor. A *fudge factor* can be included to approximately account for this to give the formula for the plate capacitance as (Sakurai and Tamaru, 1983).

$$C_{plate} = 1.15\frac{\varepsilon_{ox}A}{H} \tag{6.16}$$

The fringing capacitance becomes more important as dimensions shrink. It may be as large as 50% of the plate capacitance for smaller dimensions. This capacitance is relatively independent of geometry, and is often determined by empirical measurements of a particular process. Alternatively, the formulas of Sakurai and Tamaru (1983) may be used where the fringing capacitance from both edges *per unit length* is given by

$$2C_{edge} \cong 2.8\varepsilon_{ox}\left(\frac{T}{H}\right)^{0.222} \tag{6.17}$$

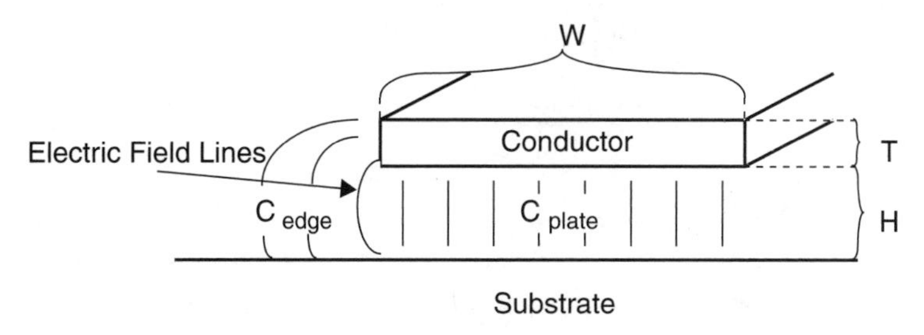

Figure 6.4 The bottom plate and fringing capacitance of interconnect.

where T is the thickness of the interconnect and H is the height of the conductor above the conductive substrate. This capacitance represents a lower bound on the capacitance to the substrate as dimensions get small. The total capacitance of a single line *per unit length* is now approximated by

$$C_L = \varepsilon_{ox}\left[\frac{W}{1.15H} + 2.8\left(\frac{T}{H}\right)^{0.222}\right] \tag{6.18}$$

Example 6.1

What is the capacitance to the substrate, per unit length, of metal that is 3 μm wide, 0.5 μm thick, and is 0.5 μm from the substrate.

Solution: Using H = 0.5 μm, W = 3 μm, and $\varepsilon_{ox} = 3.45 \times 10^{-2}$ fF/μm in (6.16) gives C_{plate} = 0.238 fF/μm. Also, using T = 0.5 μm in (6.17) gives $2C_{edge}$ = 0.097 fF/μm. Thus, the total capacitance per micrometer is

$$C_{total} = C_{plate} + 2C_{edge} = 0.335\text{-fF}/\mu\text{m} \tag{6.19}$$

Note that the total fringing capacitance $2C_{edge}$ is approximately 40% of C_{plate}. Also note that a 1000-μm interconnect would introduce a 335-fF load.

There is another component of interconnect capacitance that can be important; this is the capacitance between adjacent interconnects. This situation for two interconnects is shown in Fig. 6.5. The capacitive loading per unit length is now greater due to the coupling between the conductors. The additional capacitive loading per unit length due to the coupling capacitance can be found using the empirically derived formula (Sakurai and Tamaru, 1983)

$$C_{cpl} = \varepsilon_{ox}\left[0.03\left(\frac{W}{H}\right) + 0.83\left(\frac{T}{H}\right) - 0.07\left(\frac{T}{H}\right)^{0.222}\right]\left(\frac{S}{H}\right)^{-1.38} \tag{6.20}$$

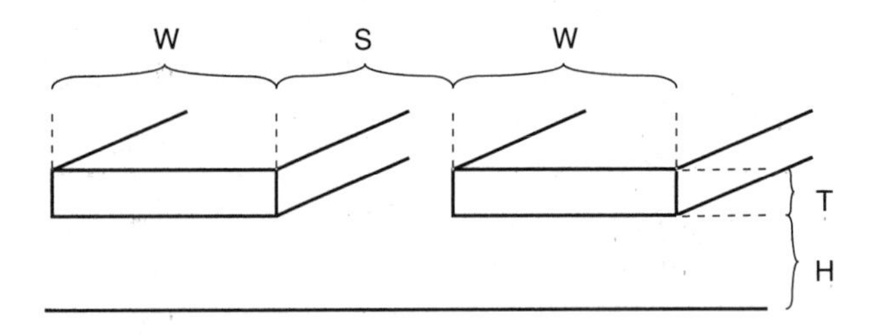

Figure 6.5 Parasitic capacitance between two conductors.

and the total capacitance per unit length due to the plate, fringing, and coupling capacitance is now given by

$$C_{total} = C_L + C_{cpl} \tag{6.21}$$

where C_L is the capacitance of a single line given by (6.19).

Example 6.2

For the same dimensions as in Example 6.1 and a 0.6 μm distance between two conductors, find the coupling capacitance per unit length.

Solution: Using (6.20), we have

$$C_{cpl} = \varepsilon_{ox}(0.18 + 0.83 - 0.07)0.778 = \varepsilon_{ox}0.731 = 0.0252 \text{ fF}/\mu\text{m}$$

When there is a bus with more than two parallel lines, the coupling capacitance from line to line is still given by (6.21), but now the total capacitive loading of lines surrounded on both sides by other lines is given by

$$C_{total} = C_L + 2C_{cpl} \tag{6.22}$$

DELAY OF RC LADDER STRUCTURE

Many complicated digital circuits can be modeled using RC ladder structures as shown in Fig. 6.6. It is possible to find a moderately simple formula for the step response of this circuit. To do this, the approximate transfer function of the circuit of Fig. 6.6 is first found. This is easy to do, if one works from the output to the input. For example, we have

$$\begin{aligned} V_2 &= V_1 + I_1R_1 \\ &= V_1 + sR_1C_1V_1 \\ &= V_1(1 + sR_1C_1) \end{aligned} \tag{6.23}$$

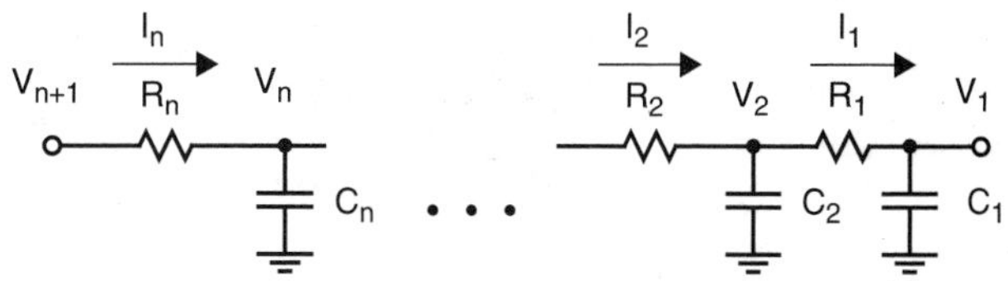

Figure 6.6 An RC network that approximates many MOS networks.

Using an iterative argument, it is possible to show (see Problem 6.6) that

$$V_{n+1} = V_1(1 + s\tau) + \text{higher} - \text{order terms} \tag{6.24}$$

where

$$\tau = \sum_{i=1}^{n}\left(C_i \sum_{j=i}^{n} R_j\right) \tag{6.25}$$

Thus, the RC circuit of Fig. 6.6 has a transfer function that can be approximated by a first-order transfer function with a time constant given by (6.25). This approximate time constant is *the sum of the time constants due to each capacitor, where the time constant due to each capacitor is simply the capacitor multiplied by the total resistance between that capacitor and the input.* This time constant can then be used in the formula for the step response of a first-order circuit. Since it is also known that RC circuits have only real poles, this approximation is usually fairly good.

Example 6.3

Using equation (6.25), what is the $t_{+70\%}$ rise time of the circuit of Fig. 6.7. Compare your result to that obtained using SPICE.

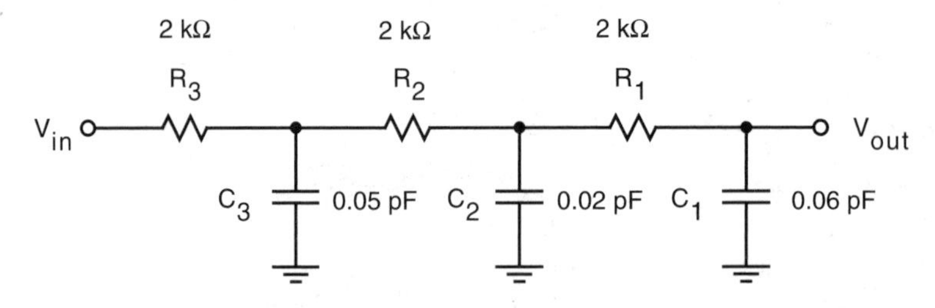

Figure 6.7 The circuit used in Example 6.3.

Solution: The time constant due to C_3 is simply $R_3\,C_3$, that due to C_2 is $(R_2 + R_3)\,C_2$, and that due to C_1 is $(R_1 + R_2 + R_3)\,C_1$. Thus,

$$\tau = R_3C_3 + (R_2 + R_3)C_2 + (R_1 + R_2 + R_3)C_1 = 0.54 \text{ ns} \tag{6.26}$$

The $t_{+70\%}$ time is simply 1.2 τ or 0.65 ns. This compares favorably with the $t_{+70\%}$ time obtained using SPICE, which is 0.636 ns.

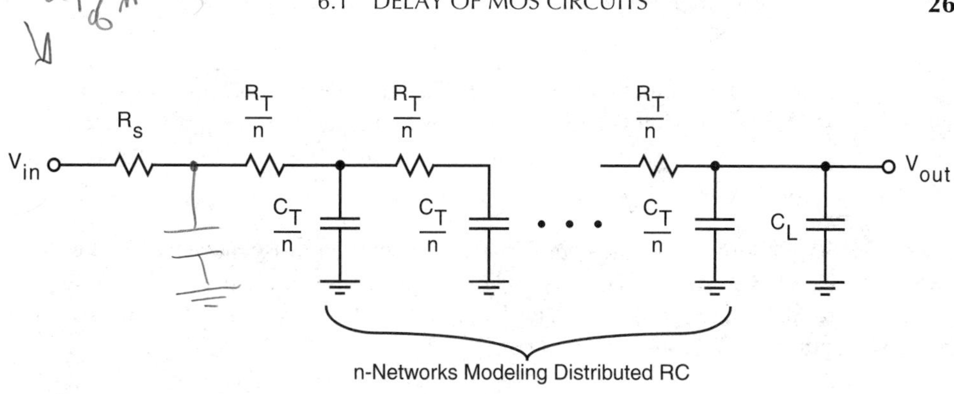

Figure 6.8 The model for driving a transmission line with a source resistance R_S and a load capacitor C_L.

Another circuit often encountered in MOS ICs is a distributed RC delay line being driven by a gate with output impedance R_S and being loaded at the end by C_L. This situation arises when a logic gate is driving a very long interconnect (the interconnect is modeled by the RC delay line). This situation can be approximated by the circuit shown in Fig. 6.8, where the distributed RC line has been approximated by n separate RC networks each having resistance R_T/n and capacitance R_T/n. C_T and R_T are the total capacitance and resistance of the RC delay line, respectively. The approximate time constant of this network is

$$\tau = \left(R_S + \frac{R_T}{n}\right)\frac{C_T}{n} + \left(R_S + \frac{2R_T}{n}\right)\frac{C_T}{n} + \cdots + \left(R_S + \frac{nR_T}{n}\right)\frac{C_T}{n} + (R_S + R_T)C_L$$

$$= \left[R_S C_T + \frac{R_T C_T}{n^2}\frac{n(n+1)}{2} + (R_S + R_T)C_L\right] \tag{6.27}$$

As n gets large and the circuit of Fig. 6.8 more accurately models reality, the time constant of the RC line [i.e., the second term in (6.27)] becomes equal to $(R_T C_T)/2$. Thus, the time constant of the circuit in Fig. 6.8 becomes approximately equal to

$$\tau = R_S C_T + \frac{R_T C_T}{2} + (R_S + R_T)C_L \tag{6.28}$$

This formula is useful in the analysis of the delays in long buses or interconnects in memories; a digital IC designer might consider memorizing it.

One of the ramifications of equation (6.25) is that as the number of stages goes up, the delay increases quadratically. This is why one should seldom have more than five transmission gates in series (one of the structures that is usually modeled by an RC ladder structure), without inserting a saturating logic circuit between them. The recommended practice for logic circuits having many series transmission gates is to interpose an inverter after every three gates.

DELAY OF CMOS LOGIC GATES

Now that we have a formula for the approximate time constant of RC networks, and we know how to estimate the parasitic capacitances and equivalent resistances, we need only approximate the critical paths through CMOS gates by equivalent RC networks. This is usually possible and is best illustrated through use of an example.

Example 6.4

Consider the CMOS *nand* gate shown in Fig. 6.9. It will be assumed that the gate is driving two inverters, where each inverter consists of an n-channel and a p-channel transistor that have dimensions of 3 μm/0.6 μm. Furthermore, it will be assumed that initially A is a "0", B is a "1", and V_{out} = "1".

Solution: The gate will be analyzed for its "1" to "0" delay when A goes high. The first step is to estimate the parasitic capacitances due to the junctions. The capacitance C_{p1} is due to the junction at the drain of Q_1, which is also the source of Q_2. Almost certainly this will be a shared junction without a contact. Its area is probably 3 by 0.6 μm or 1.8 μm^2. Its periphery is also probably 1.2 μm. These dimensions assume the gates of Q_1 and Q_2 are separated by 0.6 μm. Assuming that for n-channel transistors, we have $C_{j\text{-}0} = 0.54\ \text{fF}/\mu\text{m}^2$ and $C_{j\text{-}sw\text{-}0} = 0.15\ \text{fF}/\mu\text{m}$ gives

$$C_{p\text{-}1\text{-}0} = C_{j\text{-}0} \times 1.8\ \mu\text{m}^2 + C_{j\text{-}sw\text{-}0} \times 1.2\ \mu\text{m} = 1.15\ \text{fF}$$

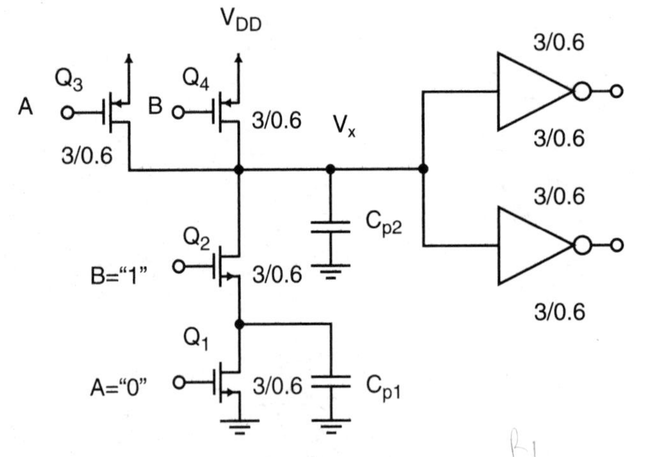

Figure 6.9 The *nand* gate being analyzed for its transient response.

After adjusting for its average value, we get $C_{p-1} = 0.63(1.15) = 0.73$ fF. The parasitic capacitance C_{p-2} consists of the load capacitance of the two inverters added to the junction capacitances of the drain of Q_2 and the sources of Q_3 and Q_4. The drain of Q_2 is an unshared junction with a contact so its area is probably 3 x 1.5 μm and its periphery is probably 6 μm. Together these give a parasitic capacitance at 0 V of

$$C_{j\text{-}0} \times 4.5\ \mu\text{m}^2 + C_{j\text{-}sw\text{-}0} \times 6\ \mu\text{m} = 3.33\ \text{fF}$$

The sources of Q_3 and Q_4 most likely share a junction with a contact. Using typical design rules, this would normally mean the gates are 1.8 μm apart; thus, the area is 3 x 1.8 μm and the periphery is 3.6 μm. Using the SPICE parameters from Chapter 1 for p-channel transistors, we now have $C_{j\text{-}0} = 0.93\ \text{fF}/\mu\text{m}^2$. Thus, the junction capacitance of the p-channel junction at 0 V bias is

$$C_{j\text{-}0} \times 5.4\ \mu\text{m}^2 + C_{j\text{-}sw\text{-}0} \times 3.6\ \mu\text{m} = 5.56\ \text{fF}$$

The total junction capacitance at the output, after scaling for large-signal operation, is now 0.63(3.33 + 5.56) = 5.6 fF. Added to this is the load capacitance of the inverters. Since this is a load of four transistor gates, with each gate being 3 x 0.6 μm, this represents a load of $4C_{ox}$ x 3 μm x 0.6 μm. Using $C_{ox} = 3.4\ \text{fF}/\mu\text{m}^2$, gives a load capacitance due to the inverters of 24.5 fF. Adding the junction capacitance to the load capacitance gives

$$C_{p2} = 5.6 + 24.5 = 30.1\ \text{fF} \tag{6.29}$$

We can now use the concept of equivalent resistors to estimate the delay. When A goes high, C_{p-1} will be discharged through the equivalent resistance of Q_1. Also, C_{p-2} will be discharged through the equivalent resistance of Q_1 in series with the equivalent resistance of Q_2. Thus, the time constant for the discharge of the output node is

$$\tau = R_{eq-1}C_{p-1} + (R_{eq-1} + R_{eq-2})C_{p-2} \tag{6.30}$$

Using equation (4.28), we also have

$$R_{eq-1} = R_{eq-2} = \frac{2.5}{\mu_n C_{ox}(3/0.6)(3.3 - 0.8)} = 1.60\ \text{k}\Omega \tag{6.31}$$

Thus the time constant is

$$\tau = 1.60(0.76) + 3.20(30.1) = 97.5\ \text{ps} \tag{6.32}$$

Multiplying this by 1.2 to get the Δ70% fall time gives

$$t_{-70\%} = 117\ \text{ps} \tag{6.33}$$

The fall time obtained using SPICE is 152 ps. This accuracy is not great, but is about as good as can be expected for the comparatively simple analysis method used.

When there are many gates in cascade, the total delay is simply the sum of the individual delays of each gate. This rule is fairly accurate as long as the gain of the individual gates in the transition region is greater than four or five.

USING SPICE FOR R_{eq} AND C_j CALCULATIONS

In previous sections, we have seen how to use approximate RC equivalent circuits to estimate the delay of MOS circuits. In addition, approximate formulas have been given for calculating the resistances and capacitances needed in the RC models. In this section, it will be shown how SPICE can be used to develop more accurate formulas for the resistances and capacitances. This need be done only once for a given process and requires only a few analyses of fairly simple circuits. This section will be based on the transistor model parameters given in Chapter 1, but could be done with any models.

The first step is to find formulas for equivalent resistances. This is done by analyzing an inverter driving a very large linear capacitor. For our example, we will assume an inverter composed of an n-channel and a p-channel transistor, with both having a W/L of 3 μm/0.6 μm. The load will be taken as a 1 nF capacitor, an extremely large size for integrated circuits. After SPICE analysis, we find the inverter has $t_{-70\%}$ and $t_{+70\%}$ fall and rise times of 2.02 and 4.80 μs, respectively. We can then use

$$R_{eq\text{-}n} = \frac{t_{fall}}{1.2C_L} \tag{6.34}$$

and

$$R_{eq\text{-}p} = \frac{t_{rise}}{1.2C_L} \tag{6.35}$$

to get $R_{eq\text{-}n}$ = 1.68 kΩ and $R_{eq\text{-}p}$ = 4.00 kΩ for 3-μm-wide n- and p-channel transistors, respectively. Equations (6.34) and (6.35) can be normalized to formulas for 1-μm-wide transistors after which the value for any width transistor is simply found by taking

$$R_{eq\text{-}n} = \frac{5.04\ \text{k}\Omega}{W} \tag{6.36}$$

and

$$R_{eq\text{-}p} = \frac{12.0\ \text{k}\Omega}{W} \tag{6.37}$$

where W is in micrometers.

The next step is to find formulas for the n-channel junction capacitances. We first set $C_L = 0$, take the n-channel drain area very large, and redo the analysis. For example, using $A_j = 10^4\ \mu m^2$ we get $t_{-70\%} = 5.6$ ns and $t_{+70\%} = 16.9$ ns, respectively. Using

$$C_j = \frac{t_{-70\%}}{1.2R_{eq\text{-}n}} \tag{6.38}$$

and

$$C_j = \frac{t_{+70\%}}{1.2R_{eq\text{-}p}} \tag{6.39}$$

gives $C_j = 2.78$ and 3.52 pF, respectively. Dividing by the 10^4-μm^2 junction area and averaging gives

$$C_{j\text{-}n} = 0.32\ \text{fF}/\mu\text{m}^2 \tag{6.40}$$

The same procedure can be followed to get

$$C_{j\text{-}p} = 0.59\ \text{fF}/\mu\text{m}^2 \tag{6.41}$$

Continuing in a similar way, simulations are run where the periphery of the junctions are taken very large to get simple formulas for the sidewall capacitances. Using PD = 1 mm for the n-channel transistor resulted in fall and rise times of 0.258 and 0.656 ns from SPICE analyses. Using formulas similar to (6.38) and (6.39) gives

$$C_{j\text{-}sw\text{-}n} = 0.13\ \text{fF}/\mu\text{m} \tag{6.42}$$

In an identical manner, the sidewall capacitance for the p-channel transistor was found to be

$$C_{j\text{-}sw\text{-}p} = 0.13\ \text{fF}/\mu\text{m} \tag{6.43}$$

The final step is to obtain formulas for the capacitive load due to transistor gates. For this case, an inverter should be loaded by a single very large inverter. Using a load of an inverter with 100-μm-wide p- and n-channel transistors gave fall and rise times of 1.18 and 2.82 ns. Using formulas similar to (6.38) and (6.39), and making the assumption that the gate capacitance of n-channel and p-channel transistors are equal, gives

$$C_g = 0.29\ \text{fF}/\mu\text{m} \tag{6.44}$$

expressed as capacitance per unit gate width.

Comparing the formulas just derived using SPICE to those used for hand analysis in the previous section, we see the greatest errors in the previous section resulted from the use of (6.31), which was based on (4.28) for equivalent resistances approximating transistors (i.e., R_{eq}). This formula was originally derived for older processes that had 5-V power-supply voltages. This formula is less accurate for more modern short-channel processes that suffer to a greater degree from second-order effects such as mobility degradation, velocity saturation, and series junction resistances. Based on the SPICE analyses just described, a more accurate formula for the process used in this text would be

$$R_{eq} = 4R_{TR} = \frac{4}{\mu_n C_{ox}(W/L)(V_{DD} - V_{tn})} \tag{6.45}$$

The actual fudge factor for other processes would probably differ, but the methodology just presented should allow accurate fudge factors for any process to be quickly found with just a few simulations.

6.2 Input/Output Circuits

So far, we have seen how to realize logic gates using MOS technology and how to arrive at reasonable approximations for the delay through these gates. To realize actual integrated circuits, it is still necessary to learn how to get the signals on and off chip. An important problem in getting signals on chip is how to protect the gates of transistors against *pinholes* caused by shorts due to electrostatic charge buildup. The major problem in getting signals *off-chip* is how to drive the large capacitive off chip loads at high speeds. Sometimes, it is necessary to have bidirectional input/output (I/O) pins that can be either inputs or outputs. This section will deal with all of these subjects.

Input-Protection Circuits

The input impedance of an MOS gate is on the order of 10^9 Ω and greater. This makes it very easy to electrostatically charge a gate node to over 1000 V. When this happens, the gate oxide *breaks down* and typically shorts the gate to the source or drain. This sort of breakdown is catastrophic. To guard against it, input protection circuits are typically placed between input pads and MOS gates. There are many varieties of these protection circuits but the basic ideas are all similar. They typically consist of some circuit that nondestructively breaks down before the gates short out. One example is shown in Fig. 6.10. Diodes D_1 and D_2 are used to clamp the input signal to within 0.7 V of V_{DD} or ground. The resistor R is used to further minimize any

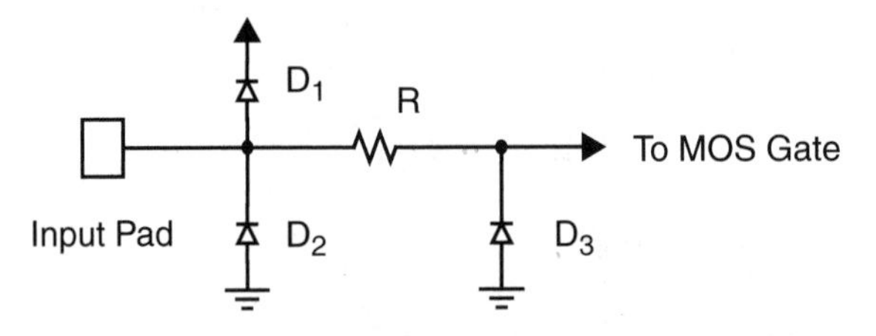

Figure 6.10 A commonly used CMOS input protection circuit.

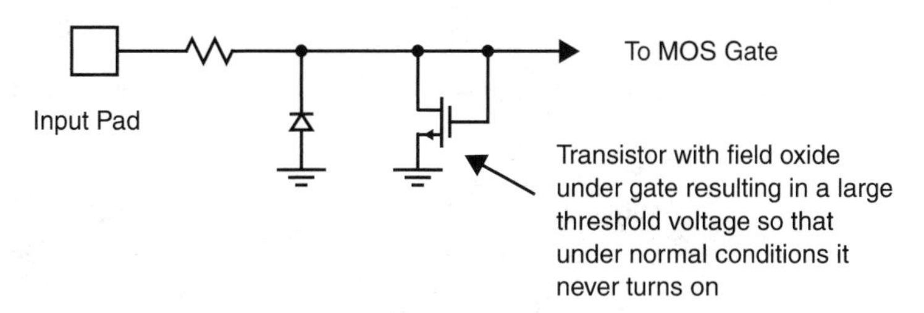

Figure 6.11 An alternative input protection circuit.

input current in the event D_3 breaks down or conducts. The resistor is usually realized by a long n^+ diffusion region that might surround the contact pad. This merges the resistor and the diodes D_2 and D_3 into a single distributed structure.

An alternative input protection circuit is shown in Fig. 6.11. In this case, a diode-connected MOS transistor that has a very large threshold voltage is used. Thus, it is normally off unless very large voltages are encountered (i.e., around 20 V or larger depending on the technology). The transistor is realized by using a metal gate placed over a region of field oxide under which no field implant exists. The field oxide acts as the gate oxide and because it is quite thick the threshold voltage of the transistor is quite high; under normal operation it would never turn on; however, with electrostatic discharge it clamps the gate voltage to be no more than approximately 20 V.

When simulating CMOS logic circuits, it is important to simulate the input protection circuitry, as well as the on-chip circuitry, since it is often the major limitation on the maximum clock rate. The design of input protection circuits that do not limit the speed of digital ICs is a research area that is beyond the scope of this text.

Output Circuits and Driving Large Capacitors

The largest capacitive loads encountered in MOS circuits typically occur at the chip I/O. This capacitive load ranges from about 1 to around 30 pF or sometimes larger. To be able to drive this large a load, a very big output buffer or inverter is required. Ideally, an inverter with transistor widths of 500 to 1000 μm would be used. Unfortunately, this inverter would then greatly load down any preceding gates. To maximize the speed, an additional inverter or two, also having larger transistors, should be placed between standard logic and the output buffer. Thus, a standard output circuit normally consists of a chain of inverters with ever-increasing dimensions.

The obvious question is how many inverters should one include and how much should the transistor widths increase as one goes from inverter to inverter. The optimum solution to this problem has been found (Mead and Conway, 1980). Optimally, the size of the inverters should increase by a factor of e (i.e., 2.72) from inverter to inverter. Also, the optimum number of inverters is such that the input capacitance of the last inverter is about $1/e$ that of the load capacitance. Although this is the optimum, if one takes the ratio somewhat larger, area is saved and the increase in delay above the optimum case is small. Typically, an increase in inverter size, from inverter to inverter in the chain, by a ratio of 4 to 5 is used. For example, a typical

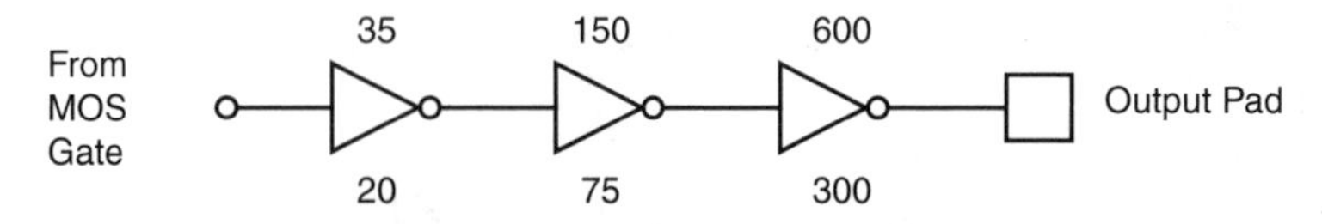

Figure 6.12 A typical output driver.

output driver as shown in Fig. 6.12 might be used. The numbers by the inverters indicate the widths of the p-channel and n-channel transistors (with the p-channel transistors being larger). This output buffer can be used for clock speeds up to 100 MHz and more. Output buffers at speeds higher than this are beyond the scope of this text.[3]

Note that no protection is required on the output pad as it is connected only to transistor junctions and not to transistor gates. However, the currents from the power supplies to output drivers can be considerable and therefore one must ensure that the power-supply interconnect is wide enough so that metal electromigration does not occur. This is a long-term phenomena that can occur whereby the current actually causes the metal to flow, eventually causing a break in the interconnect (Wolf, 1990). A wider power-supply interconnect also helps to minimize glitches on the power-supply lines when the output buffers change state. This is a major problem for high-speed digital circuits. *It is recommended that separate I/O pads be used for the power-supply connections to the output drivers as opposed to power-supply connections to other logic gates and circuitry.*

THREE-STATE OUTPUTS

Very often, integrated circuits require outputs that can be configured either as inputs or outputs. Normally, when configured as inputs, these I/O pins should have a very high impedance. This leads to the name *three-state outputs*; these I/O pins can have a "0" output, a "1" output, or a high-impedance output.

A very-simplistic way of making a three-state output circuit is shown in Fig. 6.13, where a transmission gate has been used. When the output is enabled, the transmission gate would be *on* and the output would be equal to the signal Data. When Enable is low, the transmission gate would be *off* and the output would be high impedance. Although, this approach is very simple, it is *not recommended* for a simple reason: when the output in enabled, the load capacitance has to be driven through a minimum of two series transistors, one from the transmission gate and one from the inverter. This makes the output circuitry slow and/or requires very wide output transistors, which then would represent large loads on the circuitry connected to the Enable and $\overline{\text{Data}}$ inputs.

A better approach is shown in Fig. 6.14. In this case, some logic has been added before single transistors (Q_1 and Q_2) placed between the output pad and the supply voltages. If Enable

[3]They may consist of fully differential limited-voltage-swing circuits where the output pins are being driven by transistor drain junctions and terminating resistors are on the chips being driven.

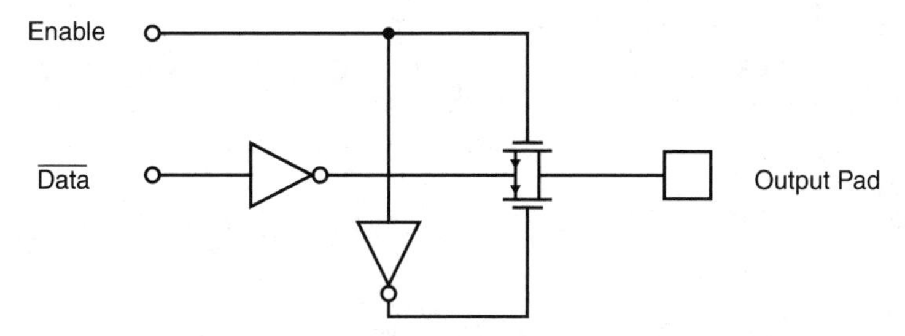

Figure 6.13 A possible (but poor) choice for a three-state output circuit.

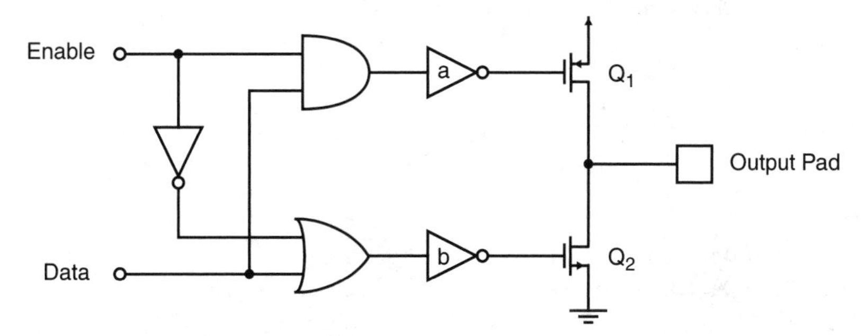

Figure 6.14 An improved three-state output circuit.

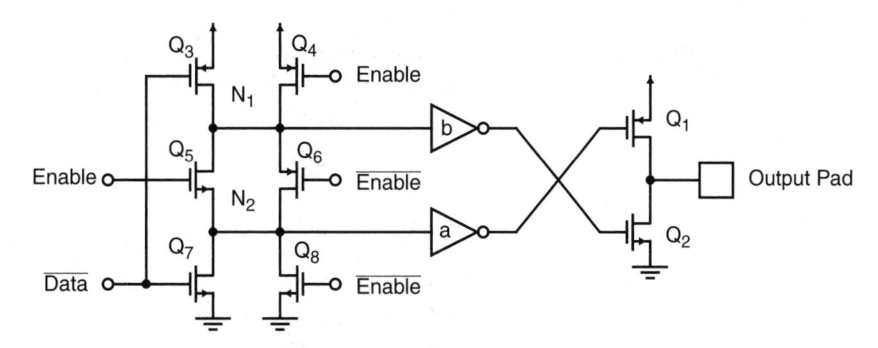

Figure 6.15 An efficient realization of the circuit of Fig. 6.14.

is a "0", the *and* gate's output will be a "0" and the *or* gate's output will be a "1". This guarantees that both Q_1 and Q_2 are off and the output will be high impedance. If Enable is a "1", then either Q_1 or Q_2 will be on, pulling the output to V_{DD} or ground, respectively, through a single transistor. In addition, inverters a and b have been added to switch the large output transistors more quickly, as was outlined in the last section.

A particularly efficient realization of the circuit of Fig. 6.14 is shown in Fig. 6.15. When Enable is a "1", the transmission gate composed of Q_5 and Q_6 is on. This connects the nodes

N_1 and N_2 together. Also Q_4 and Q_8 are off. In this case, the gate composed of Q_3 to Q_8 operates as an inverter controlled by Q_3 and Q_7 with the output equal to Data. When Enable is a "0", the transmission gate composed of Q_5 and Q_6 is off. Also, Q_4 and Q_8 are on. This clamps node N_1 to V_{DD} and N_2 to ground, thus ensuring that both Q_1 and Q_2 are off and the output is high impedance.

Bonding Pads and Wires

A very significant limitation on the speed of modern digital circuits is due to parasitics at the IC I/O interfaces. These parasitics include parasitic capacitances due to the on-chip bonding pads, off-chip package, and printed circuit board interconnects. In addition, bonding wires have finite inductances, which also limit the speed.

The parasitic capacitance of a bonding pad is easily found using the approximate formula (Sakurai and Tamaru, 1992)

$$C_{pad} = \varepsilon_{ox}\left[1.15\frac{A_{pad}}{H} + 1.4\left(\frac{T}{H}\right)^{0.222}P_{pad}\right] \quad (6.46)$$

where A_{pad} is the bonding pad area, P_{pad} is the bonding pad periphery, H is the height of the bonding pad above the conductive silicon substrate, and T is the thickness of the metallization of the bonding pad. Using this formula with typical dimensions, one finds that the parasitic capacitance of bonding pads are on the order of 0.2 to 0.4 pF for a 100 × 100-μm pad. Normally, at any IC interface, there will be parasitic capacitance due to the pads of two chips; this means a practical minimum value for the load capacitance at any chip I/O is on the order of 1 pF, a much larger capacitance than the loads normally encountered inside an IC. *This makes driving signals on and off chips typically slower than driving signals on chip.*

Example 6.5

Find the parasitic capacitance of a 100 × 100 μm pad that is realized with a 0.5-μm-thick metallization layer that is 2 μm above the substrate.

Solution: Using (6.46), we have

$$C_{pad} = \varepsilon_{ox}[5.75\times10^{-3} + 0.41\times10^{-3}] = 0.21\ \text{pF}$$

Also, the package itself will often add a few tenths of a pico-Farad. This will vary from pin to pin and package to package; the manufacturer's specifications should be consulted to estimate this component. Finally, the parasitic capacitance of printed circuit board interconnect can be significant and often dominant. Again, this will vary depending on the type of material and the dimensions used, but a typical number might be 1.5 to 2 pF/in. of 0.03-in. wide interconnect 0.06 in. above a ground plane for a fiberglass epoxy printed circuit board (Motorola, 1983).[4]

Another significant parasitic at an IC interface is the inductance of the bonding wires. This inductance is approximately given by

$$L_{wire} = 1.1\,\text{nH} \times (\text{length in millimeters}) \tag{6.47}$$

and is fairly independent of the bonding wire diameter (to a first-order approximation). The length of bonding wire varies depending on the package and position of the pin on the package, but a normal lower limit at the time of publication might be 3–5 nH for bonding wire inductances unless newer high-speed packages are used.[5] This, along with the parasitic capacitances at the interface, imposes an upper limit on the frequency of signals that can be transmitted on and off an integrated circuit. Together they realize a second-order LC bandpass filter that can have quite small damping factors causing ringing. The resonant frequency of such a circuit is given by

$$\omega_{resc} = \frac{1}{\sqrt{L_{wire} C_{total}}} \tag{6.48}$$

where C_{total} is the sum of the parasitic capacitances at both sides of the bonding wire. An upper limit on the frequency of digital signals that may be transmitted through such an interface would be $\omega_{resc}/2$ and often would be substantially lower than this.

Example 6.6

Assume the bonding pad, package, and interconnect capacitances are 2 pF and the bonding pad inductance is 5 nH. What is an upper limit on the frequency of signals that can be transmitted on and off chip.

Solution: Using (6.48), we have

$$\omega_{resc} = \frac{1}{\sqrt{L_{wire} C_{total}}} = 1 \times 10^{10}\ \text{rad/s}$$

---

[4]The signal propagation delay of interconnect on a printed circuit board is on the order of 150 ps/in.

[5]At the present, flip-chips with ball-grid I/O connections are quickly gaining in popularity. With this technology, multiple ICs are often mounted on a single ceramic substrate, which may then be connected to a printed circuit board. This packaging technology supports much higher speed connections than bonding wires.

Therefore the upper limit on digital signals at this interface would be less than

$$f_{max} = \frac{1}{2}\frac{\omega_{resc}}{2\pi} = 0.8 \text{ Ghz}$$

Another serious limitation caused by bonding wires is that they, along with the total on-chip capacitance between V_{DD} and ground, realize a resonant circuit that can be excited by the supply current of I/O drivers, particularly those that switch at a periodic rate. *Adding on-chip bypass capacitance does not alleviate this problem, it merely changes the resonant frequency.* A partial solution is the inclusion of on-chip bypass capacitances with each capacitor having a finite series resistance to decrease the damping factor at the power-supply node. Further discussion regarding this subject is beyond the scope of this text.

6.3 Bibliography

M. Annaratone, *Digital CMOS Circuit Design*, Kluwer, 1986.

M. Elmasry, ed., *Digital MOS Integrated Circuits, II,* IEEE Press, 1991.

L. Glasser and D. Dopperpuhl, *The Design and Analysis of VLSI Circuits*, Addison-Wesley, 1985.

D. Hodges and J. Jackson, *Analysis and Design of Digital Integrated Circuits*, McGraw-Hill, 1988.

M. Mano, *Digital Design,* Prentice Hall, 1984.

C. Mead and L. Conway, *Introduction to VLSI Systems*, Addison-Wesley, 1980.

Motorola Inc., Mecl System Design Handbook, 1983.

J. Rabaey, *Digital Integrated Circuits, A Design Perspective*, Prentice Hall, 1996.

T. Sakurai and K. Tamaru, "Simple Formulas for Two- and Three-Dimensional Capacitances, *IEEE Transactions on Electron Devices*, ED-30 (2) 183–185, February 1983.

H. Veendrick, "Short-Circuit Dissipation of Static CMOS Circuitry and Its Impact on the Design of Buffer Circuits," *IEEE Journal of Solid-State Circuits*, SC-19 (4) 468–473, 1984.

N. Weste and K. Eshragian, *Principles of CMOS VLSI Design: A Systems Perspective*, Addison-Wesley, 1983.

S. Wolf, *Silicon Processing for the VLSI Era, Vol. 2—Process Integration*, Lattice Press, 1990.

6.4 Problems

For the problems in this chapter, assume the following transistor parameters:

- npn bipolar transistors:

  $\beta = 100$

  $V_A = 80$ V

  $\tau_b = 13$ ps

  $\tau_s = 4$ ns

  $r_b = 330\ \Omega$

- n-channel MOS transistors:

  $\mu_n C_{ox} = 190\ \mu A/V^2$

  $V_{tn} = 0.7\ V$

  $\gamma = 0.6\ V^{1/2}$

  $r_{ds}\ (\Omega) = 5000L\ (\mu m)/I_D\ (mA)$ in active region

  $C_j = 5 \times 10^{-4}\ pF/(\mu m)^2$

  $C_{j\text{-}sw} = 2.0 \times 10^{-4}\ pF/\mu m$

  $C_{ox} = 3.4 \times 10^{-3}\ pF/(\mu m)^2$

  $C_{gs(overlap)} = C_{gd(overlap)} = 2.0 \times 10^{-4}\ pF/\mu m$

- p-channel MOS transistors:

  $\mu_p C_{ox} = 50\ \mu A/V^2$

  $V_{tp} = -0.8\ V$

  $\gamma = 0.7\ V^{1/2}$

  $r_{ds}\ (\Omega) = 6000L\ (\mu m)/I_D\ (mA)$ in active region

  $C_j = 6 \times 10^{-4}\ pF/(\mu m)^2$

  $C_{j\text{-}sw} = 2.5 \times 10^{-4}\ pF/\mu m$

  $C_{ox} = 3.4 \times 10^{-3}\ pF/(\mu m)^2$

  $C_{gs(overlap)} = C_{gd(overlap)} = 2.0 \times 10^{-4}\ pF/\mu m$

6.1 Estimate the junction areas and peripheries of each node of a two-input traditional CMOS *nand* gate. Assume all transistors have W/Ls equal to $4\ \mu m/0.5\ \mu m$.

6.2 Estimate the junction capacitances of each of a two-input CMOS *nand* gate based on your answer to Problem 6.1.

6.3 Estimate the worst-case rise and fall times of a two-input CMOS *nand* gate based on your answers to the previous two problems. Assume the load capacitance if 50fF.

6.4 Estimate the junction areas and peripheries of each node of a two-input traditional CMOS *nor* gate. Assume all n-channel transistors have W/Ls equal to $2\ \mu m/0.5\ \mu m$ and all p-channel transistors have W/Ls equal to $8\ \mu m/0.5\ \mu m$.

6.5 Estimate the junction capacitances of each of a two-input CMOS *nor* gate based on your answer to Problem 6.4.

6.6 Estimate the worst-case rise and fall times of a two-input CMOS *nor* gate based on your answers to the previous two problems. Assume the load capacitance if 50fF.

6.7 Estimate the junction area at every node of the following circuit.

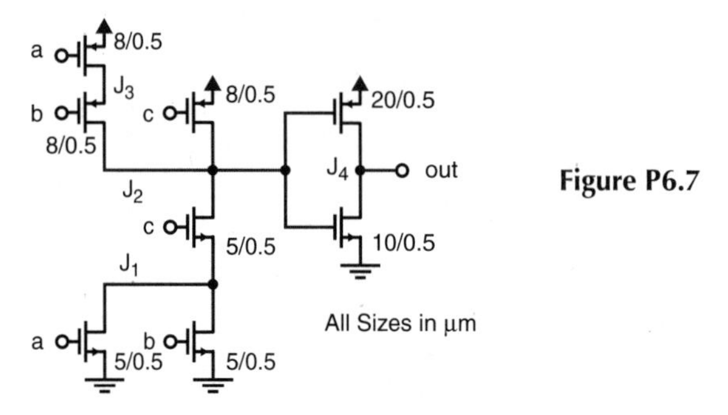

Figure P6.7

6.8 Estimate the junction periphery at every node of the circuit shown in Fig. P6.7.

6.9 Estimate the junction capacitance at every node of the circuit shown in Fig. P6.7.

6.10 Estimate all of the junction areas and peripheries at the labelled nodes of the circuit in Fig. P6.10. Give your answer in tabular form. Next, estimate the parasitic capacitances at the labelled nodes. Assume all the p-channel transistors have sizes of 5 μm/0.5 μm and all n-channels have sizes of 2 μm/0.5 μm. State specifically all assumptions made.

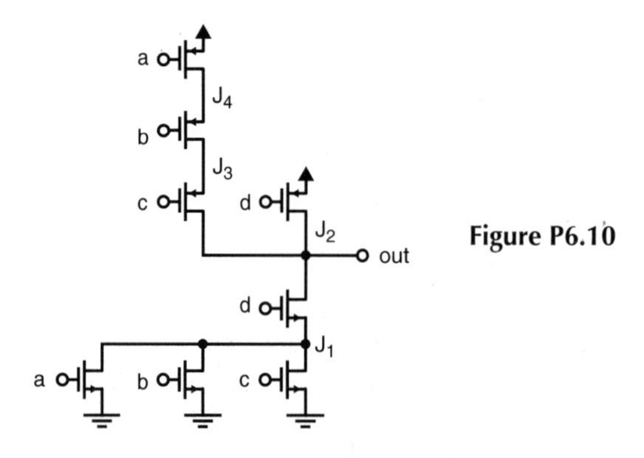

Figure P6.10

6.11 Repeat Example 6.1 but assume the interconnect is 1 μm above the silicon surface.

6.12 Repeat Example 6.2 but assume the interconnect is 2 μm above the silicon surface.

6.13 Assume there are two parallel interconnect lines and simultaneous to one going high, the other goes low. Give a formula for an equivalent capacitance to ground, as seen by one line, that models the apparent parasitic capacitance between the two lines.

6.14 Repeat the previous problem but for finding the apparent parasitic capacitance of an interconnect going high when two adjacent lines on each side of the first line are simultaneously going low.

6.15 For the RC network of Fig 6.6, prove that the coefficient of the first-order term of the denominator of the transfer function is given by equation (6.25).

6.16 Repeat the analysis of Example 6.3, but assume all resistors are 100 Ω and $C_1 = 20\text{fF}$, $C_2 = 40\text{fF}$, and $C_3 = 30\text{fF}$.

6.17 Verify the analysis done for Problem 6.16 using SPICE.

6.18 Estimate the worst-case time from when the input changes to the 70% rise and fall times at the input to the inverter of the circuit of Problem 6.7. Assume $C_{J\text{-}1} = 7\text{fF}$, $C_{J\text{-}2} = 60\text{fF}$ (this includes the input capacitance of the inverter), and $C_{J\text{-}3} = 20\text{fF}$.

6.19 A CMOS inverter is driving two identical inverters through a 40 μm metal line that is 2 μm wide and 0.5 μm above the surface of the IC. The thickness of the metal is also 0.5 μm. All the n-channel transistors are 5 μm/0.5 μm and the p-channel transistors are 8 μm/0.5 μm. Estimate the total load capacitance of the first inverter. You may ignore junction capacitance, but do not ignore the fringing capacitance of the interconnect. What is the 70% rise and fall times of the first inverter?

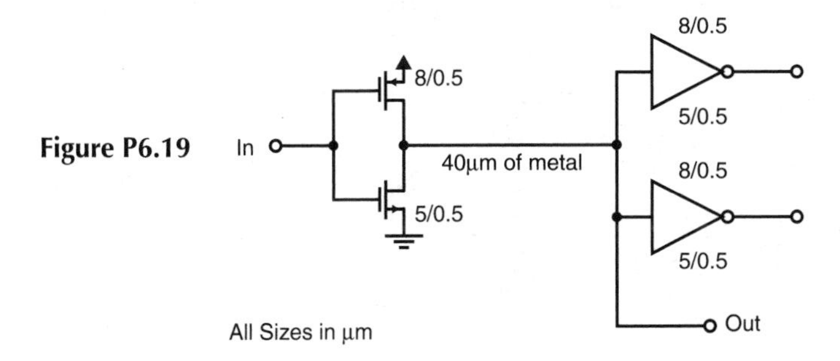

Figure P6.19

6.20 A polysilicon interconnect, 1.6 μm wide, having 25 Ω/□, and separated from the surface of the polysilicon by 0.5 μm (the thickness of the polysilicon is 0.4 μm), has a transistor gate (5 μm/0.5 μm) connected every 30 μm. Estimate the delay down the length of the line if the total length of the line is 3000 μm.

6.21 An n-channel only transmission gate has a CMOS inverter load. The channel transmission gate is 5 μm / 0.5 μm. The p-channel in the inverter is 5 μm/0.5 μm and the n-channel transistor in the inverter is 10 μm/0.5 μm. Next, for two transmission gates in series, estimate the delay times.

6.22 Verify your answer to the previous problem using SPICE.

6.23 Estimate the parasitic capacitance of a 75 μm x 75 μm bonding pad that is 1 μm above the surface of the silicon.

6.24 The CMOS output buffer of Fig. 6.12 has n-channel and p-channel widths of 20 μm/35 μm, 75 μm/150 μm, and 300 μm/600 μm with all transistors having 0.5 μm lengths. The total load capacitance is 10 pF. Assuming the buffer is being driven by a CMOS *nand* gate having 5 μm wide transistors, estimate the delay for each stage and the overall delay from the input of the *nand* gate to the final output for a positive-going input. Give your answer in tabular form.

CHAPTER 7

Latches, Flip-flops, and Synchronous System Design

We have now seen a number of different ways to realize logic gates using *CMOS* technology. All of the logic gates described thus far have been *acyclic* logic gates. That is, they have no memory or clock signals; output changes in response to input changes occur almost immediately. In this chapter, certain types of digital memory circuits will be covered. These circuits are often called *cyclic* logic circuits. These include latches, flip-flops, registers, and memory storage cells. A latch consists of a single memory location with some access circuitry. A flip-flop in an IC typically consists of two cascaded latches and perhaps some additional input logic. A register usually consists of a larger number of special-purpose latches. An IC memory consists of one or more arrays of specialized latches, called memory storage cells, with additional circuitry to access individual cells. System-level issues when designing synchronous[1] digital circuits will be discussed. In addition, two examples of synchronous logic-circuit design will be presented.

[1]A synchronous logic circuit is one that has all flip-flops clocked by a single global clock. This is a recommended design style as it is less prone to errors compared to alternative design styles.

7.1 CMOS Clocked Latches

STATIC CMOS DIGITAL LATCHES

A digital latch is the simplest digital storage element. The most common way to store a digital signal is to use two cross-coupled inverters, having positive feedback, as is shown in Fig. 7.1a and b. Two cross-coupled inverters have two stable states: inverter *a* has a "1" output and inverter *b* has a "0" output, or, alternatively, inverter *a* has a "0" output and inverter *b* has a "1" output. One of these states will be somewhat arbitrarily defined to represent storing a "1", whereas the other state is defined to represent storing a "0".

In addition to having two cross-coupled inverters for digital storage, a latch requires some means of accessing it in order to set its value. One method, based on transmission gates, which have already been described in Chapter 5, is shown in Fig. 7.2.[2] This latch is accessed when Clk is high. At this time, the positive feedback of the cross-coupled inverters is *broken* since transmission-gate X_2 is *off*. Also, X_1 is connected to the input of inverter *a*. The output of inverter *a* will be equal to $\overline{D}$ and the output of inverter *b* will be equal to D. After Clk goes low, the contents of the latch will be dependent on the value of D just before Clk went low.

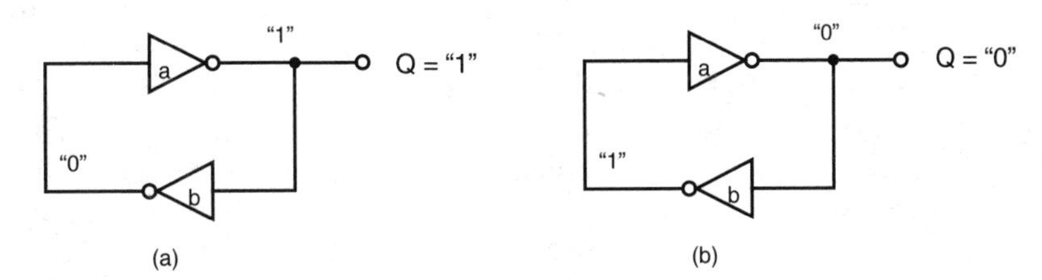

Figure 7.1 (a) Using two cross-coupled inverters to store a "1" and (b) to store a "0".

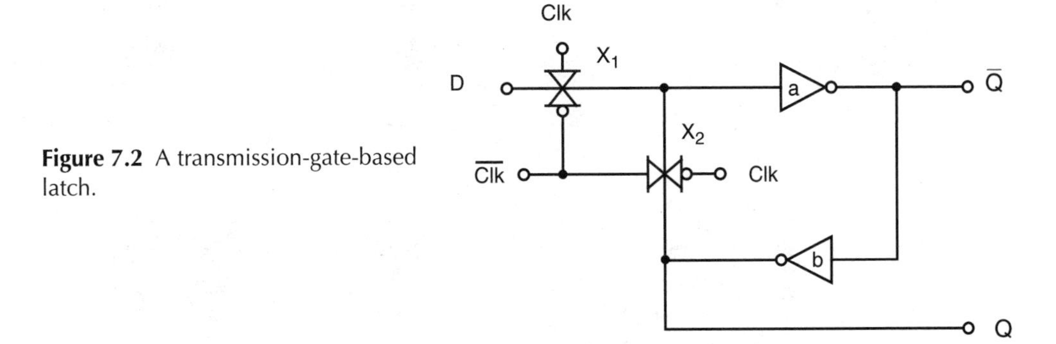

Figure 7.2 A transmission-gate-based latch.

[2]The transmission gates are now being represented by symbols with the circle representing the connection to the gate of the p-channel transistor.

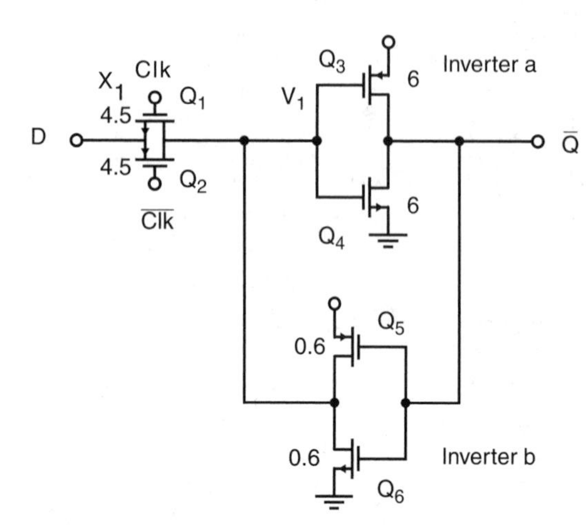

Figure 7.3 A simplified latch with hysteresis.

A modern variation of the latch of Fig. 7.2 is shown in Fig. 7.3. In this variation, transmission gate X_2 has been eliminated and the positive feedback is not *broken* during the time the latch is being accessed. Rather, the transistors have been sized so that when Clk is high, the input signal through X_1 dominates over the positive feedback through inverter *b*. Reasonable widths are shown for all transistors. In order for the signal through X_1 to dominate when it is on, the widths of the transistors of inverter *b* have been taken very small, at their minimum width (assuming a 0.6 μm minimum width). Thus, assuming the output impedance of the gate connected to D is not too large, the input signal can easily overcome the output of inverter *b* to change the state of the latch. By not breaking the positive feedback loop, hysteresis has been added to the latch. This slows the response slightly, but improves the noise immunity, particularly for slow-moving input signals. For this reason, the modified latch is often used when the input signal is coming from *off chip*. This latch is also often used in modern pipelined circuits as it requires less layout area than the latch of Fig. 7.2. One of the limitations of this latch is that only inverter *a* can drive other circuits without excessively loading down the latch. Thus, it is an *inverting latch*; its output is the inverse of its input and a *noninverting* output is not available.

Example 7.1

What are the input voltages required for a "0" to "1" and a "1" to "0" transition of the output voltage of the latch of Fig. 7.3? Verify your answers using SPICE.

Solution: This is one of the more difficult examples encountered so far, but it is tractable if some simplifying approximations are made. It illustrates a methodology that can often be used to calculate threshold voltages of other examples of MOS circuits having positive feedback. The following assumptions will be used:

1. It is assumed the latch changes state when V_1 is approximately equal to the threshold voltage of inverter *a*.

2. For the "1" to "0" output transition, just before the transition, when V_1 is slightly less than the threshold voltage of inverter *a*, we will assume the output of inverter *a* is at V_{DD}, and Q_5 is *off.* Since the gain of inverter *a* is large, this should not cause too much error.

3. Since the widths of Q_1 and Q_2 are so much larger than those of Q_5 or Q_6, we will assume the drain-source voltages of Q_1 and Q_2 are small; Q_1 and Q_2 are *hard in the triode region*, and these transistors can be approximated by resistances.

The solution now follows. The equation for the threshold voltage of inverter *a* was derived in Chapter 4 to be

$$V_{TH\text{-}a} = \frac{V_{tn} + (V_{DD} + V_{td})\sqrt{(\mu_p(W/L)_3/\mu_n(W/L)_4)}}{1 + \sqrt{(\mu_p(W/L)_3/\mu_n(W/L)_4)}} \tag{7.1}$$

which for $(W/L)_3$, = $(W/L)_4$, V_{tn} = 0.8 V, V_{tp} = −0.9 V, and $\mu p/\mu n$ = 130/545, gives $V_{TH\text{-}a}$ = 1.32 V. Next, for V_1 = 1.32 V, and assuming $V_{GS\text{-}6} \cong 3.3$ V, we have

$$I_{D\text{-}6} \approx \mu_n C_{ox}\left(\frac{W}{L}\right)_6\left[(V_{GS\text{-}6} - V_{tn})\ V_1 - \frac{V_1^2}{2}\right] = 457\ \mu A \tag{7.2}$$

Now, approximating Q_1 by a resistor of size

$$r_{ds\text{-}1} \cong \frac{1}{\mu_n C_{ox}(W/L)_1(V_{GS\text{-}1} - V_{tn})} \tag{7.3}$$

where $V_{GS\text{-}1} = V_{DD} - V_1$, gives $r_{ds\text{-}1}$ = 601 Ω. In a similar manner, we get $r_{ds\text{-}2}$ = 1.04 kΩ and $(r_{ds\text{-}1} \| r_{ds\text{-}2})$ = 381 Ω. Thus, the voltage across the transmission gate x_1 is $V_{DS\text{-}1} \cong I_{D\text{-}6}(r_{ds\text{-}1} \| r_{ds\text{-}2})$. This implies,

$$V_{TH\text{-}2} = V_{TH\text{-}a} + V_{DS\text{-}1} = 1.49\ V \tag{7.4}$$

For the output "0" to "1" transition, we assume Q_6 is off and Q_5 is conducting with $V_{GS\text{-}5} \cong 3.3$ V. This gives $I_{D\text{-}5} = 102\ \mu A$ and

$$V_{TH\text{-}1} = V_{TH\text{-}a} - I_{D5}(r_{ds\text{-}1} \| r_{ds\text{-}2}) = 1.28\ V \tag{7.5}$$

Thus, the hysteresis is $V_{TH\text{-}2} - V_{TH\text{-}1} = 0.21$ V.

The transfer curves of the SPICE simulation of the latch are shown in Fig. 7.4, for both the "1" to "0" transitions and the "0" to "1" transitions. The threshold voltages were 1.39 and 1.34 V, respectively, with a hysteresis of 0.05 V, which is smaller than that predicted. Most of the error is due to having a smaller effective width for Q_5 and Q_6 than was used in the analysis due to three-dimensional effects that have a noticeable effect on decreasing the current for such narrow transistors. Other

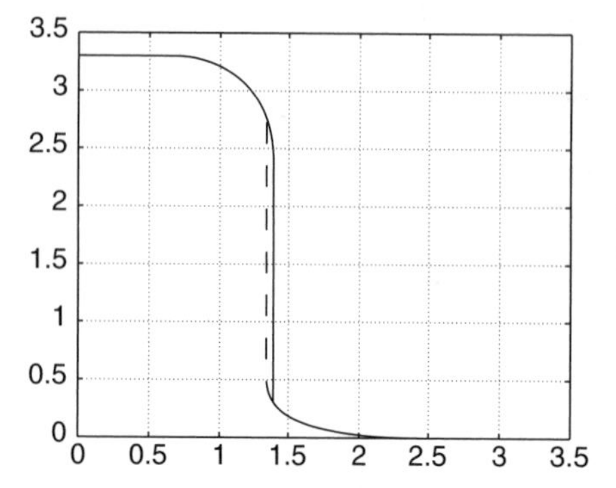

Figure 7.4 The transfer curves obtained using SPICE for the latch of Fig. 7.3.

sources of inaccuracy include assuming $V_{GS\text{-}6}$ is 3.3 V when it is actually a little less than V_{DD} as V_1 gets close to $V_{TH\text{-}a}$, as well as short-channel effects that were ignored.

A variation on the latch shown in Fig. 7.3 is shown in Fig. 7.5. This latch uses a single n-channel input transistor, Q_1, for a pass transistor to access the latch. This does not pass a full "1" logic value due to the voltage drop through the n-channel pass transistor Q_1 when transmitting high logic values. By including Q_2, the logic level at the input to the inverter is regenerated to be equal to V_{DD} when a "1" is being input. Q_2 would normally be taken with a relatively small W/L; it must be taken small enough so that when a "0" is being transmitted through Q_1, the node V_1 is pulled below the threshold voltage of the inverter composed of Q_4 and Q_5. When the latch is not being accessed, Q_3 is turned *on*; this supplies positive feedback when a "0" is being stored at the input to the inverter.

Another alternative latch that is often used in CMOS technology is based on the cross-coupled *nor* circuit shown in Fig. 7.6. This circuit is the basis for latches in many different technologies.

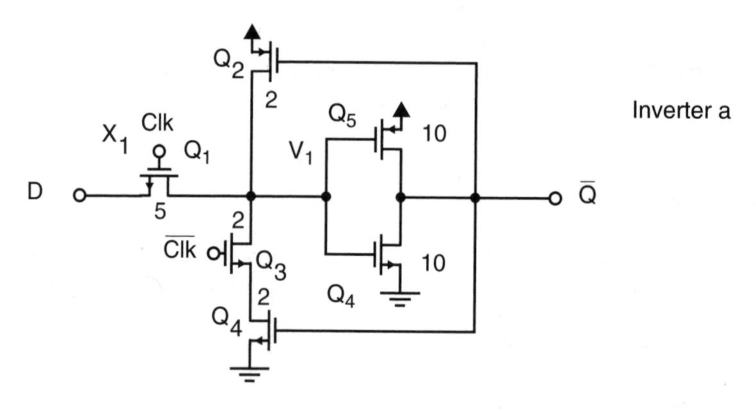

Figure 7.5 An alternative clocked latch.

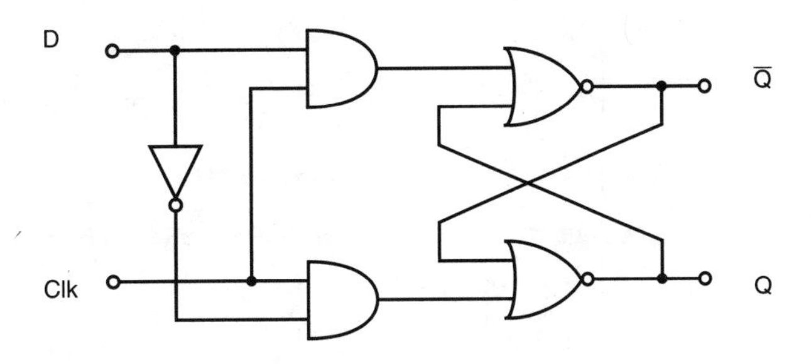

Figure 7.6 A latch based on cross-coupled *nor* gates.

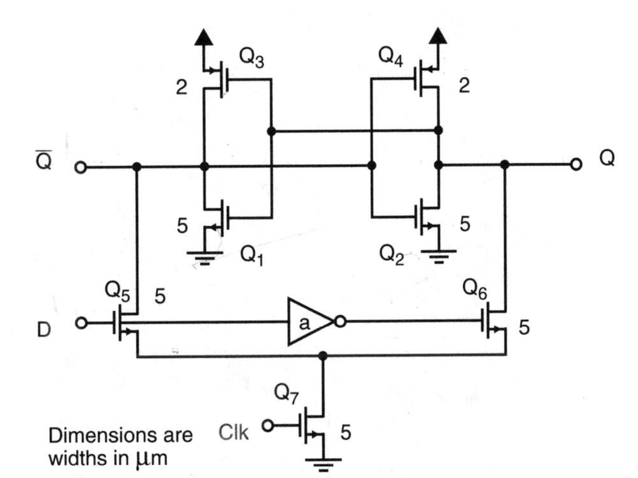

Figure 7.7 A CMOS realization of the latch of Fig. 7.6.

A CMOS realization of this latch is shown in Fig. 7.7. It is easy to see that again the latch has a core of cross-coupled inverters and extra circuitry for accessing the core.

For the latch of Fig. 7.7 to work correctly, the transistors must be properly sized. To see this, consider the latch with initially Q = "0", $\overline{Q}$ = "1", D = "1", and Clk changing from a "0" to a "1". When this happens, a conductive path between $\overline{Q}$ and ground through Q_5 and Q_7 forms. If the widths of Q_5 and Q_7 are large enough compared to that of p-channel transistor Q_3, then $\overline{Q}$ will be pulled low. For the latch shown, the effective width of Q_5 in series with Q_7 is a 2.5-μm-wide transistor. Since the mobility of n-channel transistors is 2.5 to 3 times that of p-channel transistors, these widths are adequate. Only when $\overline{Q}$ has been *pulled low* will Q_4 turn *on* and Q_2 turn *off*. This, in turn, will now cause Q to change to a high voltage. Thus, $\overline{Q}$ must first change *low* and then one inverter delay later Q will change to a high voltage. The opposite is true for a "1" to "0" transition of Q.

It is possible to eliminate this one inverter delay, at the expense of increased power dissipation, if *pseudo-NMOS* loads are used, and differential inputs and clock signals are available. A latch that illustrates this is shown in Fig. 7.8. When this latch is not being accessed, $\overline{\text{Clk}}$ is high, and Q_5, Q_6, and Q_8 are all guaranteed to be *off*. Also, Q_7 is *on* and its drain voltage is guaranteed to be low. The core is now composed of two *pseudo-NMOS* inverters in a positive-feedback loop.

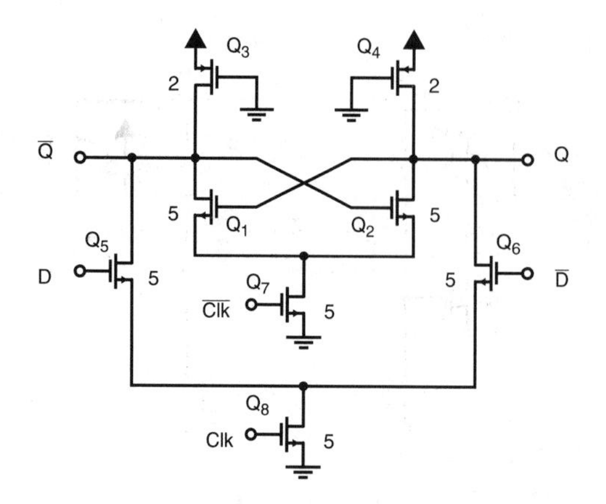

Figure 7.8 A pseudo-NMOS biphase latch.

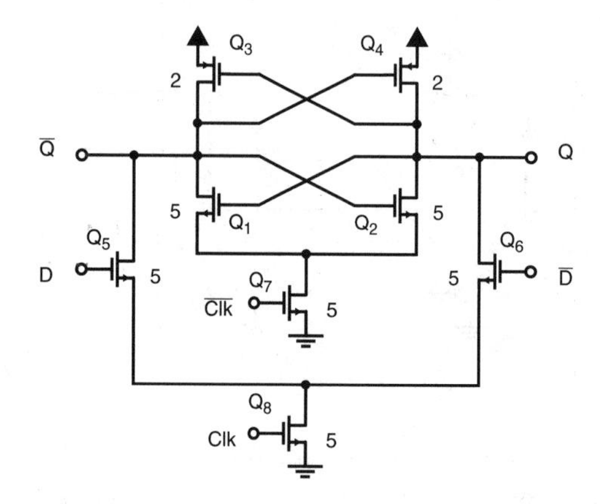

Figure 7.9 A compromise biphase latch.

When Clk next goes high, Q_7 goes *off* and the positive-feedback loop is broken. Also, at this time, Q_8 turns *on*, and either Q_5 or Q_6 will be *on* depending on the input signal. Assuming that D is a "1" and $\overline{Q}$ is initially a "1", the latch now operates as follows: when Clk goes high a conductive path is formed between $\overline{Q}$ and ground, which begins to pull $\overline{Q}$ low. At the same time, since Q_7 goes *off*, the initially low impedance path between Q_2 and Q_7 goes to high impedance. Since $\overline{D}$ is a "0", Q_6 also is high impedance. This allows p-channel transistor, Q_4, to immediately start charging output Q to a high voltage at the same time as output $\overline{Q}$ is being discharged. It is not necessary for $\overline{Q}$ to first go low before Q can be charged.

This type of latch is often called a *biphase* latch because it requires two clock phases, Clk and $\overline{Clk}$. It is the standard way of building the highest frequency divide-by-two circuits, almost irrespective of the technology used. This circuit is a critical component of frequency dividers used in high-speed satellite communication systems.

A compromise between the previous two latches is shown in Fig. 7.9. This latch has the cross-coupled p-channel loads similar to Fig. 7.7; this results in no d.c. power dissipation. Similar to the latch of Fig. 7.8, the cross-coupled n-channel transistors are disabled during the track phase when the latch is being written into; this significantly decreases the rise-time of a

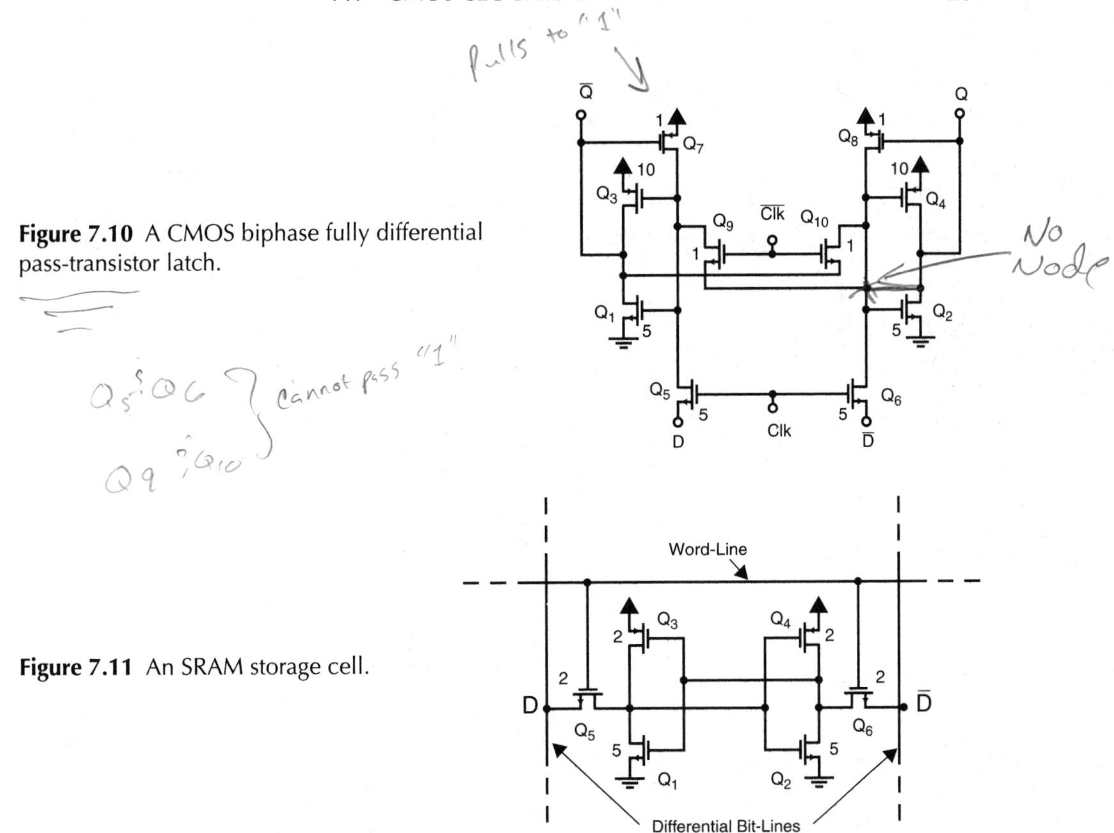

Figure 7.10 A CMOS biphase fully differential pass-transistor latch.

Figure 7.11 An SRAM storage cell.

positive-going output during transistions. The trade-off is a two phase clock is required; this is usually a justified compromise. This latch is a good choice to include in a library of general-pupose digital cells.

One final example of a CMOS latch will be described; this latch[3] combines elements of the biphase latch just described and the alternative latch of Fig. 7.5. It is shown in Fig. 7.10. It has the *biphase* feature that both sides change states at the same time; however, unlike the previous latch, it does not dissipate d.c. power. It has differential buffered outputs, unlike the latch of Fig. 7.5, and thus is particularly useful in applications in which differential logic is being used.

Static Random-Access Memory Cell

A popular storage element used in integrated static random-access memories (*SRAMs*) is similar to the clocked latch of Fig. 7.7. These memories consist of one or more arrays of SRAM storage cells and associated circuitry whereby it is possible to write into or read from any selected cell of a given array. A popular storage cell is shown in Fig. 7.11. These storage cells

---

[3]This latch was invented by the author and further significantly improved by then Postdoctoral Research Associate Ning Ge.

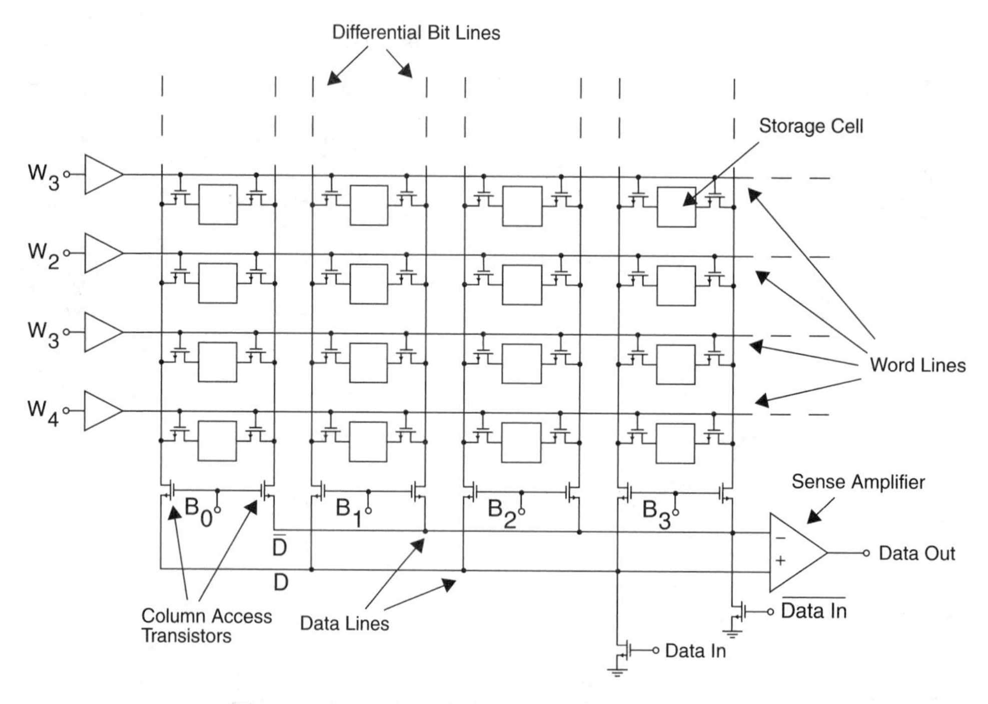

Figure 7.12 A simplified portion of an SRAM array.

would be included in a large array as shown in a simplified manner in Fig.7.12. Normally, only a single SRAM cell would be accessed from a particular array at a given time. If it was desired to read or write a number of bits (i.e., a word) at the same time, then a cell from a different array would be accessed for each bit of the word. When accessing a particular cell of an array, then only one of the *word lines,* W_i, and only one of the *column lines*, B_i, will be high at a given time. The word line that is high will turn on the access transistors of all the memory cells in the row connected to that word line. In Fig. 7.11, the access transistors are Q_5 and Q_6. When the access transistors are turned *on*, each memory cell in the row is connected to its individual pair of differential *bit lines*. In addition, only one pair of differential bit lines will be connected to the *data lines* at the bottom of the array. The memory cell at the intersection of the row with the high word line and the column with the high bit line is the selected cell. What happens next depends on whether the selected memory cell is being written into or read from.

If the cell is being written into, then one of the data lines will be pulled low and the other data line will be left high, depending on the value of the data that is being written into the cell. If the cell is being read from, then the signals DataIn and $\overline{\text{DataIn}}$ will be both low and the *sense amplifier* will respond to the value stored in the cell.

For the SRAM cell to operate correctly, the ratio of the transistor widths must be chosen reasonably. To be specific, the W/Ls of the access transistors, Q_5 and Q_6, must be taken smaller than the W/Ls of the n-channel cross-coupled transistors, Q_1 and Q_2 in Fig. 7.11, by a ratio of 2.5 to 3. In addition, the W/Ls of the p-channel transistors in the cross-coupled inverters, Q_3

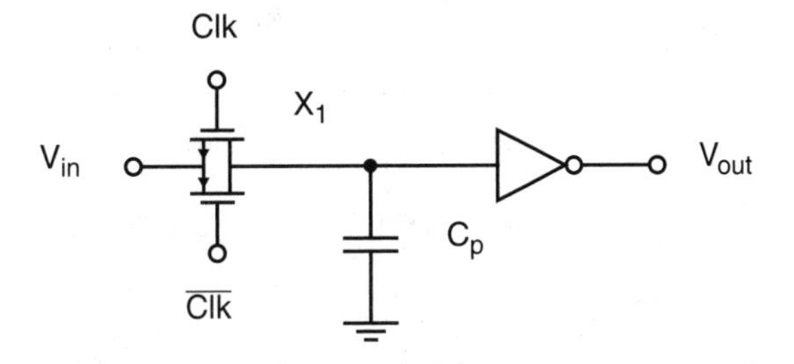

Figure 7.13 A transmission-gate-based dynamic latch.

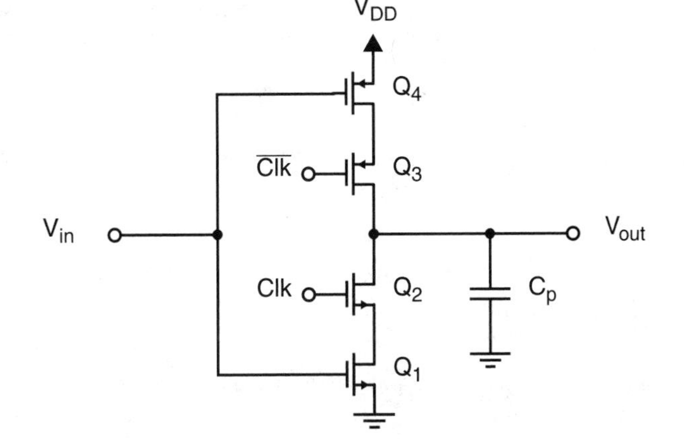

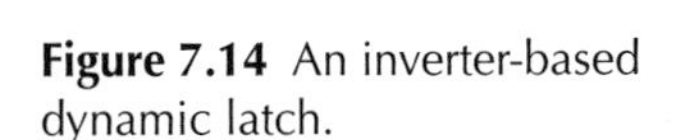

Figure 7.14 An inverter-based dynamic latch.

and Q_4, must also be taken smaller than the W/Ls of the n-channel transistors in the cross-coupled inverters by approximately the same ratio.

This has been only a very brief introduction to SRAM storage cells and memories, which are covered in much greater detail in Chapter 11.

Dynamic CMOS Latches

It is possible to realize latches using capacitors rather than positive feedback for the digital storage. An example is shown in Fig. 7.13. The capacitor C_p is used to store the input value just before Clk goes low and transmission gate X_1 turns off. The capacitor C_p normally is not explicitly included, but is the parasitic input capacitance of the inverter and the junction capacitance of the transmission gate.

The dynamic latch must be refreshed periodically, otherwise its value can become corrupted due to the leakage currents of the transmission-gate junctions. This imposes a minimum frequency on the clock signal Clk. For this reason, dynamic latches are used only when it is known a priori that they will always be clocked. They also make debugging circuits much more difficult, as it is not possible to stop the clock signals while internal IC node voltages are being probed. As a result, most designers do not use dynamic latches unless the area savings, or slightly higher speed that sometimes results, is critical.

Another example of a dynamic latch that has been gaining in popularity is shown in Fig. 7.14. This latch is basically a three-state inverter and is often called a *clocked inverter.* If Clk is low

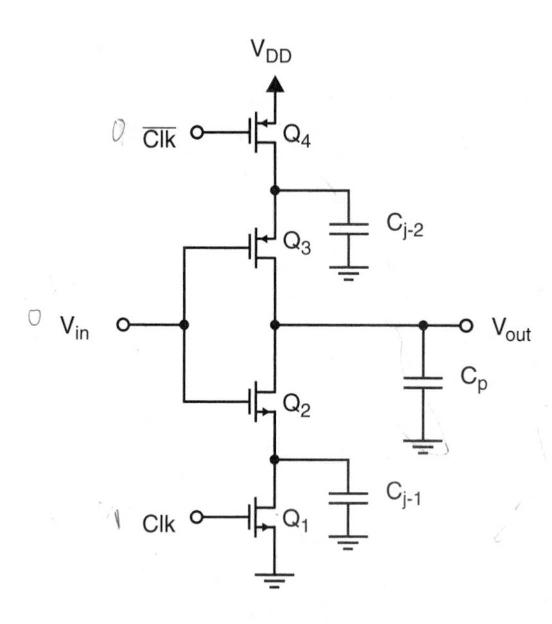

Figure 7.15 A poor realization of an inverter-based latch.

(and therefore $\overline{Clk}$ is high), the output is high impedance; in this case, the output value will be the inverse of the input value just before Clk went low. This output value will be stored on C_p, a parasitic capacitance that is due primarily to the input capacitance of the next stage and partially to the junction capacitance at the drains of Q_2 and Q_3.

One of the reasons for the popularity of the inverter-based latch of Fig. 7.14 is that it requires less layout area than the transmission-gate-based latch of Fig. 7.13. This is true because Q_1 and Q_2 can share a common junction area that does not contain a contact. The same is true for Q_3 and Q_4.

An inexperienced designer is warned against realizing the dynamic clocked latch as is shown in Fig. 7.15, as has often been done in the past. This realization has a charge-sharing problem after Clk has gone low, if V_{in} then changes. To see this problem, consider the case in which V_{in} is low just before Clk goes low. This will cause C_p to be charged to V_{DD}, a "1". Also, consider the parasitic junction capacitances, C_{j-1} and C_{j-2}, shown in Fig. 7.15. Before Clk goes low, C_{j-1} will be discharged to 0 V, whereas C_{j-2} will be charged to 3.3 V. Next, consider what happens if V_{in} goes high after Clk has gone low. This will turn Q_3 off, and turn Q_2 on, which connects C_{j-1} to C_p. The charge on C_{j-1} and C_p will be shared; since C_{j-1} was originally discharged, the charge sharing will decrease the voltage on C_p. This type of detrimental charge sharing does not occur in the preferential latch of Fig. 7.14.

The detrimental charge sharing just seen is a simple example of a problem that a designer must always be wary of when designing dynamic circuits. There will be additional examples of this in Chapter 9 on advanced CMOS design techniques.

7.2 Flip-flops

Flip-flops are the traditional storage elements used to realize synchronous logic circuits. They save the state of the machine from one clock cycle to the next. They are also used to break up

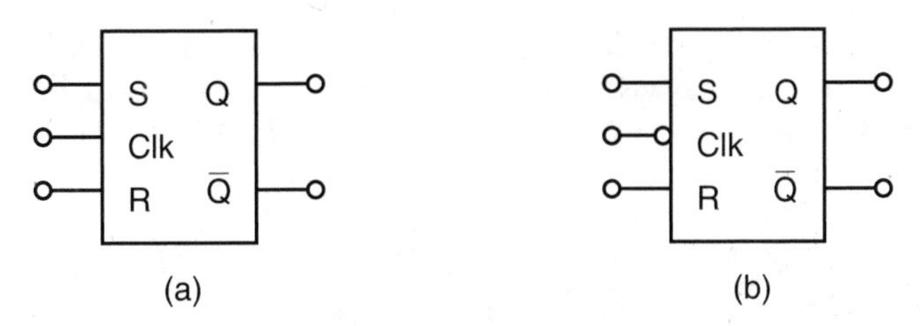

Figure 7.16 Commonly used symbols for an SR flip-flop that (a) changes at a positive-going clock edge and (b) changes at a negative-going clock edge.

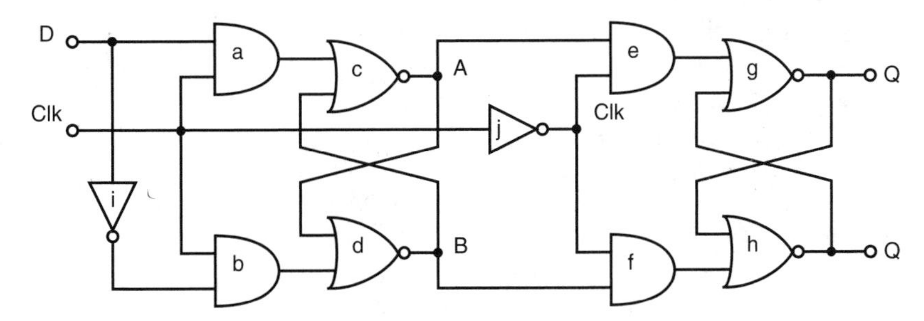

Figure 7.17 The logic diagram of a master–slave D flip-flop.

any feedback loops around a cyclic logic circuit to prevent the logic from having *race conditions* or oscillations. Flip-flops normally have signal inputs and a clock signal. Most flip-flops also have a differential output that is almost always denoted Q and $\overline{Q}$. The output values of a flip-flop change only just after a clock edge. The change is dependent on the input signals just before the clock changes. Some flip-flops also have asynchronous set and reset inputs. If either of these inputs becomes high, it immediately controls the internal storage and the output of the flip-flop, independent of the synchronous inputs and the clock input.

Common techniques for realizing flip-flops are a *master–slave* approach whereby two clocked latches are cascaded, or an *edge-sensitive* approach; the latter is not popular for integrated-circuit design because it requires clock waveforms having fast rise and fall times.

Common types of flip-flops are SR flip-flops, D flip-flops, JK flip-flops, and toggle or T flip-flops. *D flip-flops are the preferred type for integrated circuit applications.* Figure 7.16a shows a commonly used symbol for an SR flip-flop that has output changes just after a positive-going clock edge, whereas the small circle on the clock input of the SR flip-flop shown in Fig. 7.16b indicates that its outputs change just after a negative-going clock edge. SR flip-flops are not very popular in ICs because they have an indeterminate state when both inputs are high.

D Flip-flop

Figure 7.17 shows a logic diagram of a possible realization of a D flip-flop, which is the flip-flop perhaps most often used in integrated circuits. The implementation shown is based on cross-coupled *nor* latches similar to those shown previously in Fig. 7.6. The first latch is clocked by Clk, whereas the second latch is accessed on the opposite phase by $\overline{\text{Clk}}$. Thus, it is not possible for a signal to propagate from the input through both latches at one time; either the first (i.e., the *master latch*) or the second (i.e., the *slave latch*) will be deactivated and in *hold mode*. This technique of using a cascade of two latches, clocked on the opposite phases, is called a master–slave approach. This is the most commonly used approach in ICs to realize flip-flops.

The D flip-flop shown is described by the characteristic Table 7.1.

Table 7.1 implies that after the flip-flop is clocked, its output will be equal to whatever value D had just before the clock changed. To see that the flip-flop of Fig. 7.17 realizes this function, consider initially that D and Clk are both "1"s. At this time, the master latch is being accessed, but the slave is isolated. The value of A will be $\overline{\text{D}}$ and B will be equal to D. If the input D changes, both A and B will reflect that change, but the output of the flip-flop will not change since the second or slave latch is isolated by the $\overline{\text{Clk}}$ = "0" inputs to the *and* gates *e* and *f*. When Clk goes low, the master latch becomes isolated from the D input by *and* gates *a* and *b* both having "0" outputs. The contents of the master latch will reflect the D input just before the clock went low. One inverter delay after Clk goes to a "0", $\overline{\text{Clk}}$ will go to a "1", which now connects the slave latch to the master. At this time, the Q output of the slave latch will change, if need be, to be equal to $\overline{\text{A}}$, and $\overline{\text{Q}}$ will be equal to $\overline{\text{B}}$. Thus, after the clock has gone low, Q_{n+1} will be equal to the value of D just before Clk went low. If D changes after Clk has gone low, its change is ignored as the master is now isolated. D flip-flops are one of the most commonly used flip-flops inside ICs.

SR Flip-flop

Figure 7.18 shows a master–slave realization of a very similar flip-flop, the SR flip-flop. This flip-flop is described by the characteristic Table 7.2. This type of flip-flop can either be set, reset, or left in its previous state. If S is a "1" while Clk is high, then Q_{n+1} will be a "1" after Clk goes low. If R is a "1" while Clk is high, then Q_{n+1} will be a "0". If neither S nor R go

Table 7.1 The Characteristic Table for a D Flip-flop

D	Q_{n+1}
0	0
1	1

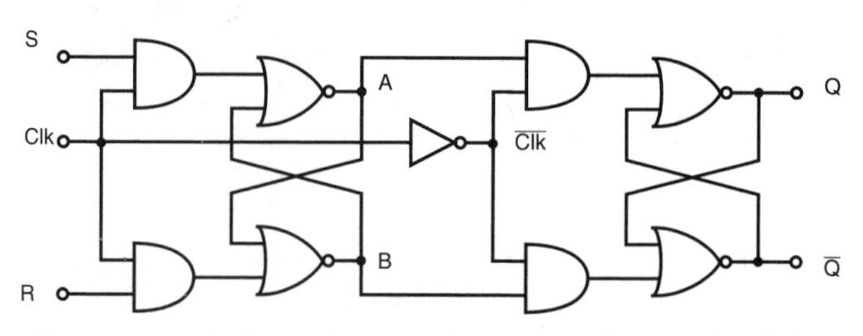

Figure 7.18 The logic diagram of a master–slave SR flip-flop.

high while Clk is a "1", then Q_{n+1} will keep the same value, Q_n, that it had just before Clk goes low. The situation in which both S and R are high at the time Clk goes low is considered to be an illegal stage; if this ever happens, then the state of the output after Clk goes is considered to be *indeterminate*; it could end up in either state.

The SR flip-flop of Fig. 7.18 has a major limitation; it can be affected by the inputs not only at the instance just before Clk goes low but any other time that Clk is high. To see this, consider the case in which the flip-flop is initially in a "0" state; A is a "1", B is a "0", Q is a "0", and $\overline{Q}$ is a "1". Now, assume that after Clk goes high both S and R are supposed to remain low, but for some reason there is a noise glitch on S, and it temporarily goes high well before Clk goes low. This will switch the state of the master latch, which will also cause the slave to change later when Clk goes low, even though, by that time, both S and R are low. To get around this problem, and to make the flip-flop's output dependent on the values of S and R just when it goes low, edge-sensitive flip-flops were developed.

EDGE-TRIGGERED SR FLIP-FLOP

An edge-sensitive SR flip-flop is shown in Fig. 7.19. When Clk is high, the outputs of *and* gates *a* and *b* are affected by inputs S and R. This has no effect on the cross-coupled latch composed of *nor* gates *g* and *h*. The high clock signal also guarantees that the outputs of *nor* gates *e* and *f*, and therefore the inputs to the cross-coupled latch are both "0". When Clk goes low, then *nor* gates *e* and *f* pass the results to the latch. A short time later, the delay time through inverters *a* and *b* to be exact, *and* gates *a* and *b* are disabled by the output of inverter *b*

Table 7.2 The Characteristic Table of an SR Flip-flop

S	R	Q_{n+1}
0	0	Q_n
0	1	0
1	0	1
1	1	Indeterminate

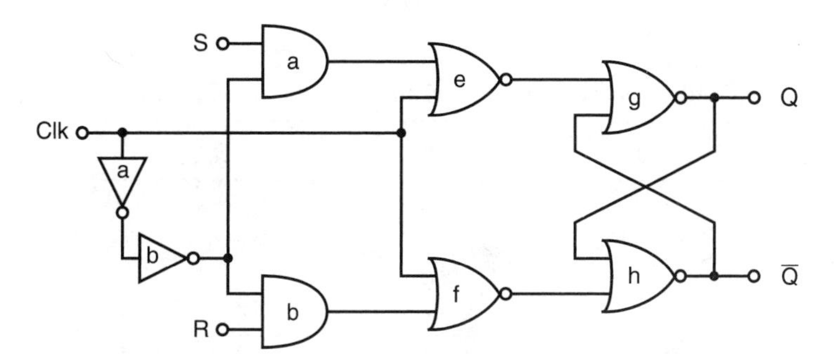

Figure 7.19 A negative edge-sensitive SR flip-flop.

going low. Thus, the only time the S and R inputs can pass through to the latch is during the negative edge of Clk, hence the flip-flop is called an edge-sensitive flip-flop. These types of flip-flops were very popular in the days of SSI digital integrated circuits; however, they do have some shortcomings. For them to work correctly, it is necessary that the clock edges be sharp and that the threshold voltage of inverter *a* be larger than the threshold voltage of *nor* gates *e* and *f*. Also, the delay time through inverters *a* and *b* is critical; it must be long enough to allow the inputs to propagate through to the latch. It must also be short enough so that the inputs are disabled before any feedback signals from the outputs of the latch through external logic can cause the input signals, S and/or R, to change. For all of these reasons, *the use of edge-sensitive flip-flops is not recommended in ICs.*

Thus, master–slave SR flip-flops can have problems with noise, whereas edge-sensitive SR flip-flops have problems with clock and timing requirements. Together, these indicate that *SR flip-flops are not recommended for general use inside an IC.*

JK Flip-flop

A flip-flop that is commonly used in ICs, although not as often as D flip-flops, is the JK flip-flop. This type of flip-flop is often used for synchronous machines or counters. Table 7.3 is a characteristic table for a JK flip-flop.

This is equivalent to the following:

1. If J is a "1" and K is a "0", the flip-flop will be set.
2. If K is a "1" and J is a "0", the flip-flop will be reset.
3. If both J and K are "0"s, the state will not change.
4. If both J and K are "1"s, the flip-flop will change state or toggle.

Thus, the JK flip-flop is similar to an SR flip-flop, except the "11" input state is now defined. It is this additional state that makes the flip-flop so useful for counters.

It is both useful and interesting to write the truth table for Q_{n+1} as a function of J, K, and Q_n, the state of the slave latch just before the master latch goes into latch mode. The truth table and Karnaugh map are shown in Figs. 7.20 and 7.21. The truth table is based on the characteristic table in which all possible values of Q_n have been added. The logic function that needs to

Table 7.3 The Characteristic Table for a JK Flip-flop

J	K	Q_{n+1}
0	0	Q_n
0	1	0
1	0	1
1	1	$\bar{Q}_n$

J	K	Q_n	Q_{n+1}
0	0	0	0
0	0	1	1
0	1	0	0
0	1	1	0
1	0	0	1
1	0	1	1
1	1	0	1

Figure 7.20 The truth table for a JK flip-flop.

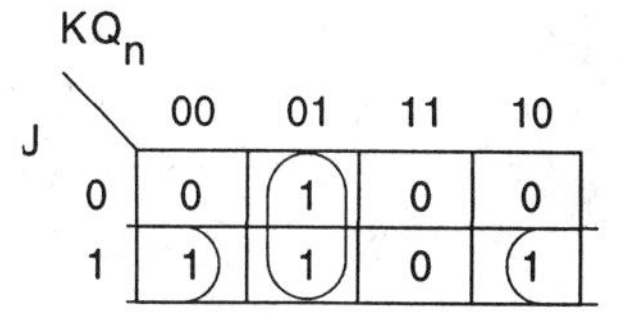

Figure 7.21 The Karnaugh map reduction of the truth table of the JK flip-flop.

be implemented to realize a JK flip-flop is found by *or*ing all the input values that give a "1" for Q_{n+1}. Working our way from the top of the table to the bottom, this gives

$$Q_{n+1} = Q_n\bar{J}\bar{K} + J\bar{K}\overline{Q_n} + J\bar{K}Q_n + JK\overline{Q_n} \tag{7.6}$$

Each of these terms is called a *prime implicant* for the logic function for Q_{n+1}. However, it is possible to find a much simpler function that is equivalent to (7.6) using Karnaugh map reduction (Roth, 1985). For the JK flip-flop, this results in

$$Q_{n+1} = J\overline{Q_n} + \bar{K}Q_n \tag{7.7}$$

The above logic simplification procedure has been gone through in moderate detail because it is important for digital integrated-circuit designers to be familiar with the major techniques used in designing logic machines, such as Karnaugh map reduction, state-machine design, and asynchronous-logic design, and this is an example of where these techniques are useful.

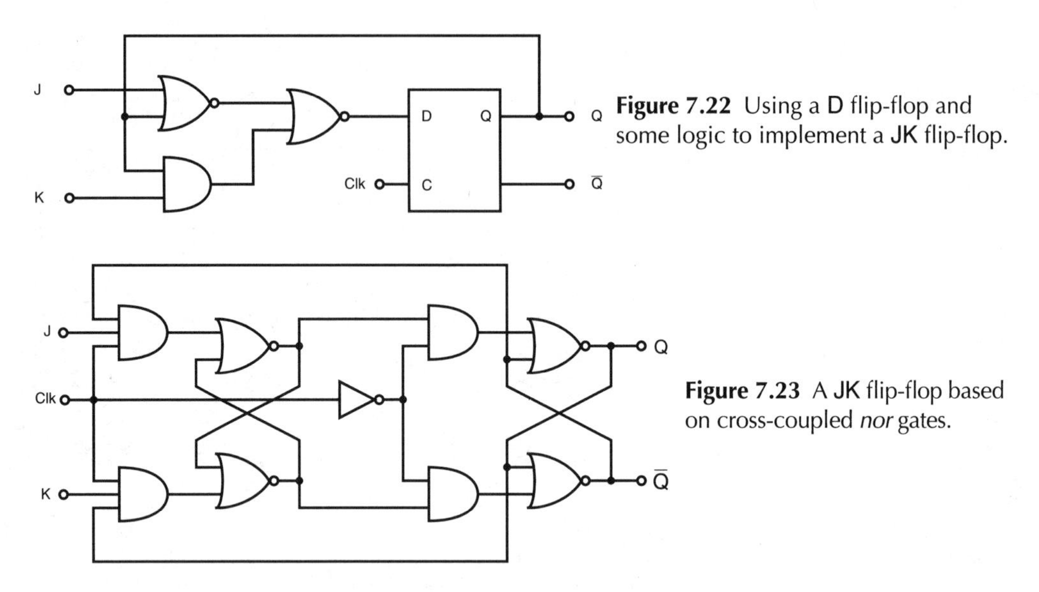

Figure 7.22 Using a D flip-flop and some logic to implement a JK flip-flop.

Figure 7.23 A JK flip-flop based on cross-coupled *nor* gates.

For simple logic functions of three or less inputs, it is sometimes possible to reduce the functions by inspection and the application of logic manipulation rules. For example, rewriting (7.6) in a different order, and applying the associative[4] rule, gives

$$\begin{aligned} Q_{n+1} &= J\overline{Q}_n\overline{K} + J\overline{Q}_n K + \overline{K}Q_n\overline{J} + \overline{K}Q_n J \\ &= J\overline{Q}_n(\overline{K} + K) + \overline{K}Q_n(\overline{J} + J) \\ &= J\overline{Q}_n(1) + \overline{K}Q_n(1) \\ &= J\overline{Q_n} + \overline{K}Q_n \end{aligned} \tag{7.8}$$

the same as (7.7). Although this straightforward method is adequate for reducing (7.6), more formalized methods are normally necessary for reducing functions of four or more inputs.

Equation (7.7) is the basis for most integrated-circuit realizations of JK flip-flops. It allows one to easily convert a D flip-flop to a JK flip-flop. All that is necessary is to add a 2-to-1 multiplexor to the input of a D flip-flop; this multiplexor inputs $\overline{K}$ if Q_n is a "1", otherwise it inputs J. A possible logic circuit that implements a JK flip-flop in this manner is shown in Fig. 7.22. Although this particular circuit has been used commercially, it is often possible to incorporate the multiplexor into the master stage of the D flip-flop in a more efficient manner. For example, Fig. 7.23 shows an alternative circuit where the multiplexor has been incorporated into the master stage of the cross-coupled *nor*-gate-based D flip-flop that was shown earlier in Fig. 7.17, with little increase in complexity. This particular configuration is the basis for many CMOS JK flip-flops, *but it does suffer from a similar limitation as*

---

[4]Indeed, most algorithms for simplifying logic functions are mainly applying the associative rule in a systematic, intelligent way.

Table 7.4 The Characteristic Table for a Toggle or T Flip-flop

T	Q_{n+1}
0	Q_{n+1}
1	$\overline{Q}_{n+1}$

the SR flip-flop of Fig. 7.18, that is, glitches on the inputs that occur well before the negative edge of Clk can cause errors. For this reason, prudent IC designers often restrict themselves to using D flip-flops only.

T FLIP-FLOP

Another popular flip-flop, the T or toggle flip-flop, has a single input and is described by the characteristic Table 7.4. This flip-flop, which is often used for counters, can be realized by connecting the J and K inputs of a JK flip-flop together.

All of the popular flip-flops can easily be modified to incorporate asynchronous set and/or reset inputs. For master–slave inputs, this normally means adding some extra circuitry to both the master and the slave latch so that the asynchronous inputs will dominate independent of the clock and other inputs. These inputs are often used to initialize the state of digital ICs at the time the power is first applied. Normally, a set or reset input is required, but seldom both. In the next sections we will see examples of how set and/or reset inputs can be added to CMOS flip-flops.

7.3 CMOS Flip-flops

As was mentioned, a master–slave D flip-flop is easily realized by cascading two latches that are clocked on opposite phases. Figure 7.24 shows one example of this, where the latch based on cross-coupled *nor* gates from Fig. 7.6 is used. The flip-flop has also had asynchronous set and reset inputs added. In cases in which only one of these asynchronous inputs is required, the unnecessary transistors used for the other asynchronous input can be left off. During operation, when the state of the master is changing state, it is necessary to *pull* one side *low* before the other side can go *high*. For this to be possible, it is necessary to size the transistors correctly, similar to what is necessary for the clocked latches. Also, shown in Fig. 7.24 are two additional inverters used to buffer the output. These are sometimes needed because of the large capacitive loads that flip-flops often have to drive. Figure 7.25 shows a slightly modified flip-flop, where the D flip-flop of Fig. 7.24 has been converted into a JK flip-flop based on the approach of Fig. 7.22. To simplify the circuit somewhat, no asynchronous set or reset has been shown, but when required, these can be easily added as in Fig. 7.24. Note that only a couple of transistors need to be added to realize the JK function. Also note that the load transistors of the master latch had to be taken somewhat smaller to still allow the n-channel drive networks, which now consist of three series transistors, to dominate when changing the state of the latch. This flip-flop is easily converted into a toggle flip-flop simply by connecting J and K together and relabeling them T.

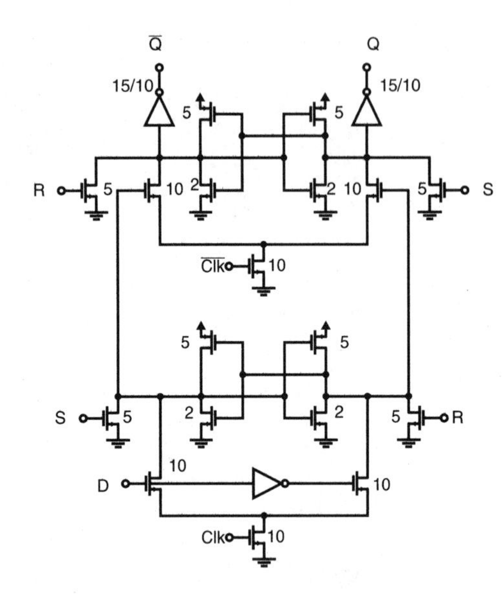

Figure 7.24 A CMOS D flip-flop with asynchronous set/reset and output buffers.

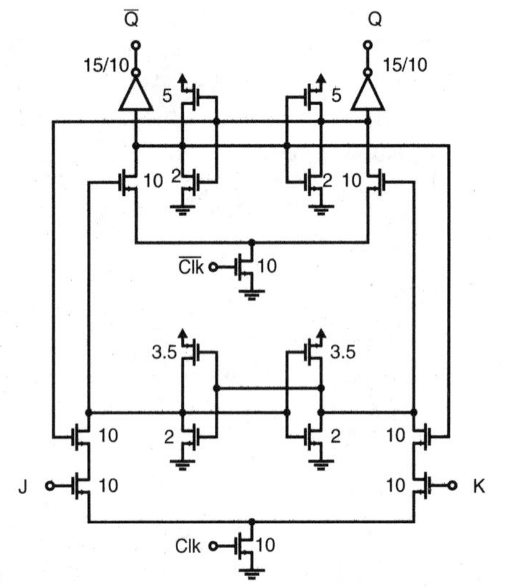

Figure 7.25 A CMOS JK flip-flop.

Both of the flip-flops just presented can be easily converted to high speed *biphase* flip-flops by changing the loads to pseudo-NMOS loads and adding the additional clock input transistors to break up the positive feedback when the latches are being changed in a similar manner as has been done in Fig. 7.8. A compromise flip-flop based on the latch of Fig. 7.9 where only the cross-coupled n-channel transistors are disabled during write phases is shown in Fig. 7.26. This would be a reasonable choice for inclusion in a standard library.

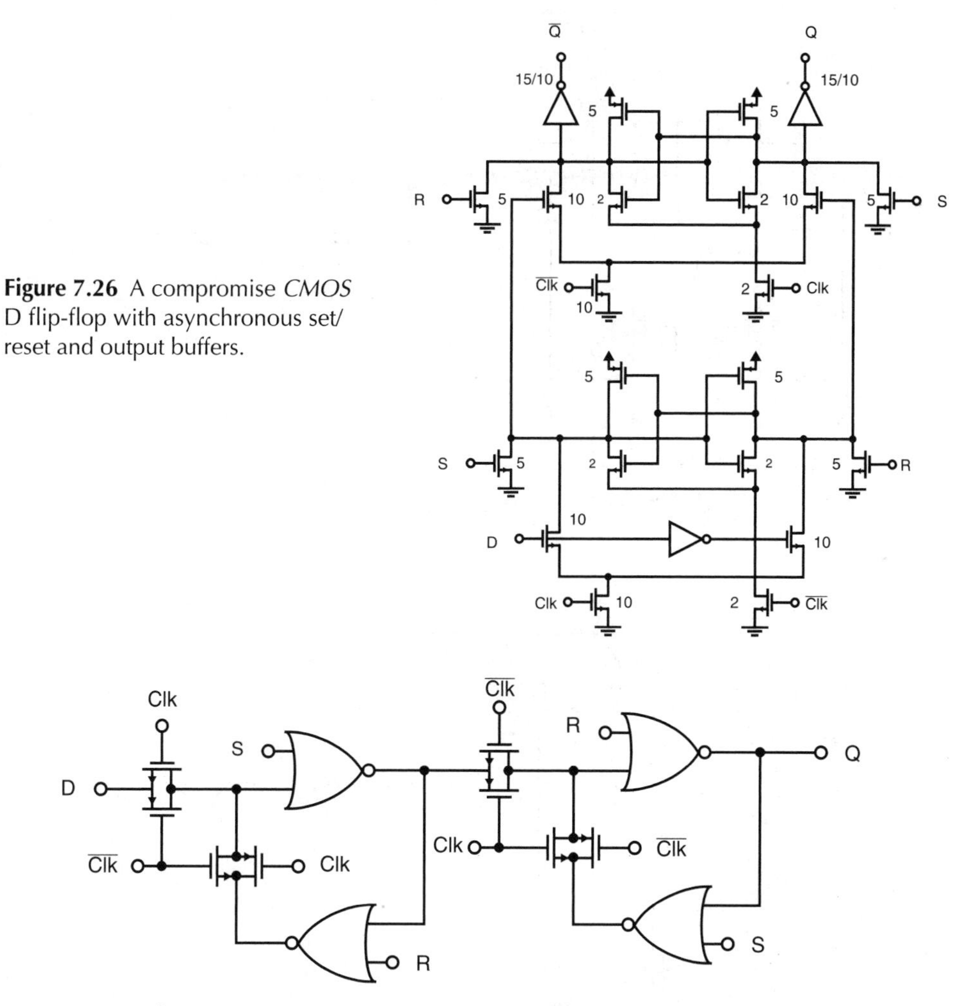

Figure 7.26 A compromise *CMOS* D flip-flop with asynchronous set/reset and output buffers.

Figure 7.27 A transmission-gate-based master–slave D flip-flop.

Transmission-gate-based flip-flops are also popular choices for integrated flip-flops, particularly for D flip-flops. These, similar to cross-coupled *nor*-gate flip-flops, are again easily realized by cascading two latches having opposite clock phases. Figure 7.27 shows an example of this in which asynchronous set and reset have been added to the latch of Fig. 7.2 by replacing inverters with *nor* gates.

An example of a transmission-gate-based JK flip-flop is shown in Fig. 7.28. This flip-flop is based on placing a multiplexor in front of a D flip-flop to realize equation (7.7). This implementation has not been as popular as that of Fig. 7.25, although the reason is unknown.

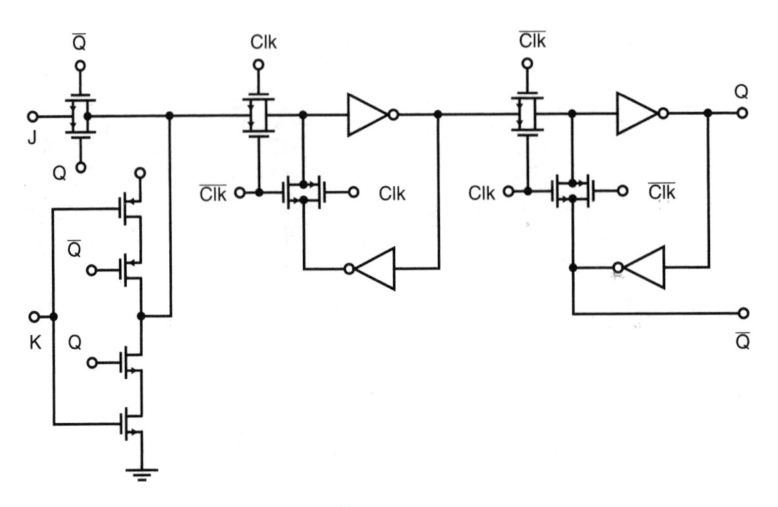

Figure 7.28 A transmission-gate-based JK flip-flop.

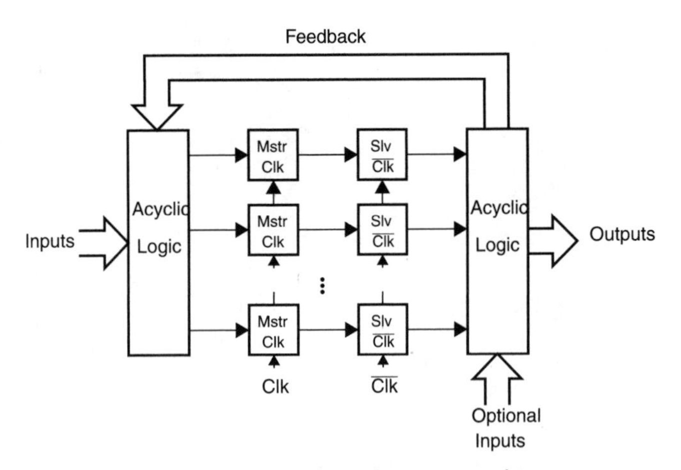

Figure 7.29 A typical synchronous machine.

7.4 Synchronous System Design Techniques

The design of synchronous digital circuits is well covered in a number of other sources such as Roth (1985) and Mano and Kime (1997) and is not the intended subject matter of this text; however, a few details specific to their successful realization in ICs will be covered.

An example of a typical synchronous logic circuit is shown in Fig. 7.29. It consists of acyclic logic and flip-flops. As was mentioned, it is recommended that all flip-flops in an IC be master–slave flip-flops, normally D type, although sometimes JK flip-flops are used for counters.

As is shown in Fig. 7.29, every feedback loop is broken up by both a master and a slave latch, which are clocked by complementary clocks. *In general, every feedback path must*

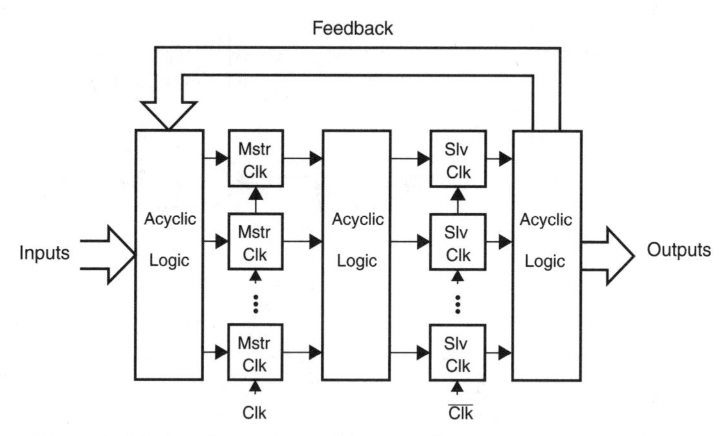

Figure 7.30 An alternative architecture for synchronous machines.

contain an even number of clocked latches, with alternate latches clocked on opposite phases. Not shown in Fig. 7.29 are simple latches that are often included at the inputs and outputs of an IC. This is particularly true for high-speed ICs, where it greatly eases the timing problems at the interface between ICs. Also not shown in Fig. 7.29 is the additional circuitry that a prudent designer typically adds for testing the internal logic. This will be described in Chapter 13.

Sometimes in an IC the master and slave latches of a D flip-flop are separated and acyclic logic is placed between them as is shown in Fig. 7.30. This technique is very useful in minimizing the number of logic delays in the acyclic logic between any two latches, and thus maximizes the speed of the circuit. It is somewhat similar to a ubiquitous technique for designing logic circuits where *latency* (or total delay time through a circuit) is not important, but where the *throughput rate* is critical; this technique is called *pipelining*.

Pipelined Systems

In this technique, a large acyclic logic circuit, having many delays, is broken into a number of smaller acyclic circuits that have smaller delays. Clocked latches are also inserted between the acyclic circuits. This procedure is shown symbolically in Fig. 7.31a and b.

Figure 7.31a consists of three cascaded acyclic circuits. The time delay of the circuit is the total delay through the networks A, B, and C. The inputs D_0, D_1, . . . , D_n must be stable for longer than the delay through all three networks. If the output of acyclic network C is to be clocked into a register, then the shortest allowable period for the clock used for the register is the *total* delay through all three acyclic networks.

An alternative approach is shown in Fig. 7.31b. In this pipelined approach, registers consisting of clocked latches (not master–slave flip-flops) are interposed between the acyclic circuits. Every alternate register of latches is clocked by complementary clocks. That is, the first set of latches is clocked by Clk, the second set by $\overline{\text{Clk}}$, the third set by Clk, and so on. The minimum clock period is now the maximum delay of any of the smaller subnetworks and a latch. This

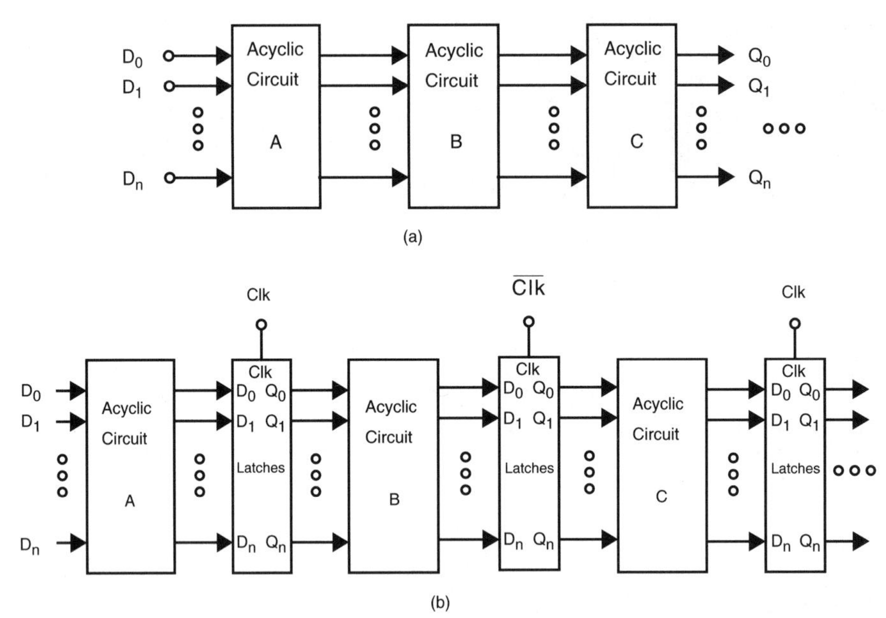

Figure 7.31 (a) A large acyclic network consisting of three cascaded circuits and (b) the same network pipelined by clocked latches interposed between subnetworks.

allows a faster clock to be used, particularly if the delay through the smaller subnetworks is restricted to be only two or three gate delays.

Note that the total delay time through a pipelined network, from when the inputs to the network are stable to when the final outputs are valid, is not decreased by pipelining. Rather, the total delay (or latency) is increased because the clock period must be greater than the largest delay of any stage. The total delay through the whole network, or the latency, is then the sum of this maximum delay for each stage. What is improved by pipelining is the maximum clock frequency, and therefore the throughput rate, because all of the stages are working at the same time. After the outputs of acyclic network A are latched, then network A can immediately be reused with a new set of inputs, before the outputs of the total circuit are valid. This technique of pipelining is almost always used now for high-speed signal-processing circuits and microprocessors.

One of the most commonly used latches for pipelining static circuits is the transmission-gate-based latch shown earlier in Fig. 7.27. For dynamic circuits, the inverter-based latch of Fig. 7.14 is a popular choice. As will be seen in the next chapter, this latch is less sensitive to overlapping clock waveforms than transmission-gate-based dynamic latches.

System Clock Issues

An important principle of design in an IC is that all internal clocks be directly derived from a *single master clock* and not from acyclic logic. Otherwise, the chance of incurring errors due to glitches is quite high. In applications in which this rule is not followed, if possible, a flip-flop should be used between any acyclic logic used for deriving a new clock, and any flip-flops to which it is applied. The interjecting flip-flop should be clocked by the master clock. In addition, circuits that do not use a single master clock should be carefully simulated using a timing simulator, as subtle errors are possible.

The problem comes primarily from intermediate states that can cause an unexpected glitch. The interposition of a flip-flop between the acyclic logic and the new clock signal helps to ignore these intermediate states. This principle will be illustrated through the use of an example.

Example 7.2

Consider the case in which it is desired to clock a shift register every time a four-bit counter enters the "1000" state.

Solution: A naive solution to this problem is shown in Fig. 7.32, where it is assumed all flip-flops change state on positive going clock edges. The idea is that whenever the counter enters the state "$Q_3Q_2Q_1Q_0$" = "1000", the output of the *and* gate will go positive, which, assuming the shift register clocks on a positive-going edge, will advance the shift register. However, consider the case in which the counter changes from "$Q_3Q_2Q_1Q_0$" = "1001" to "1010". It is possible that Q_0 will change to "0" before Q_1 changes to "1" causing the counter to temporarily enter the intermediate state "1000"; this incorrectly advances the counter.

A much better solution is shown in Fig. 7.33 where the state "0111" is detected. When the counter enters this state, the output of the *and* gate goes to a "1". The next time Clk goes high, the output of the D flip-flop goes high for one period only. This then advances the shift register. The circuit is insensitive to any intermediate states since these occur after the positive clock edge, whereas the D flip-flop is sensitive only to its input just before the clock goes high.

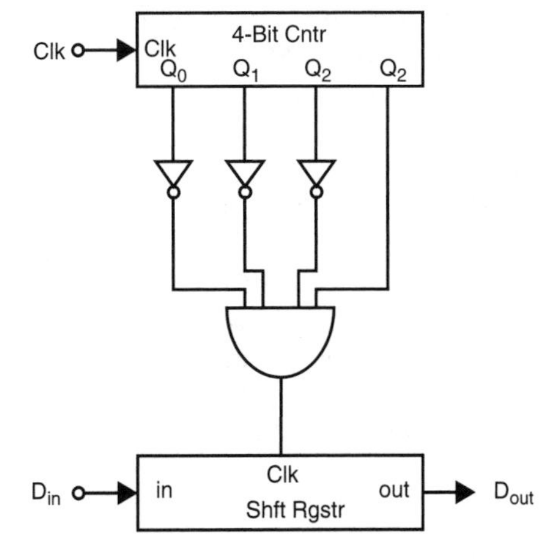

Figure 7.32 A dangerous example of generating an on-chip clock.

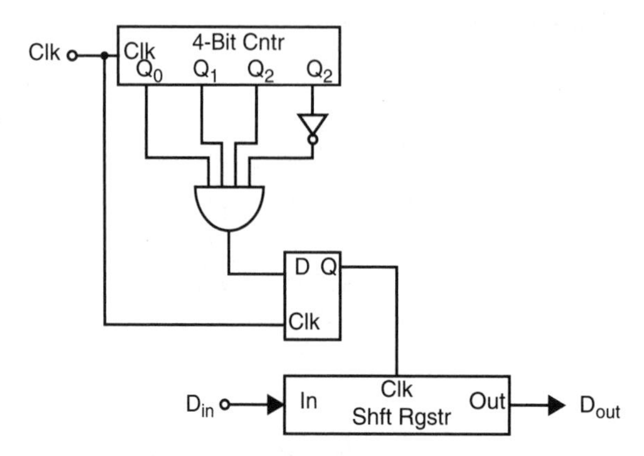

Figure 7.33 An improved method for generating an on-chip clock.

Very similar errors can occur when using asynchronous set and reset inputs, particularly in situations such as realizing programmable counters. Again, the safer solution is to always take the signal used to set or reset the counter (or flip-flops) from a separate flip-flop that is clocked by the master clock.

Even when clocks that are derived on chip are taken from the outputs of flip-flops and no extraneous glitches are present, there still is a possibility of subtle errors occurring. An example of this is illustrated in the following example.

Example 7.3

Consider the circuit of Fig. 7.34a. This circuit is a serial-to-parallel converter. The serial "Data_In" signal is right shifted through the four D flip-flops at the top of the figure, once each period of the signal Clk. It is intended that one clock period after the asynchronous external signal Strb goes low, the contents of the serial shift register will be loaded into the four-bit parallel shift register immediately below it. The signal Int_Clk will be high only when both inputs to the *nor* gate are "0". For this to be true, Strb must have been high (for x_2 to be a "0") and gone low (for x_1 to be a "0"). The timing diagram of the various signals is shown in Fig. 7.34b, assuming all master–slave D flip-flops change at a negative-going edge.

Note that the internally generated clock Int_Clk goes negative one *nor* gate delay after the negative edge of the master clock, Clk. Since the outputs of the serial shift register change at the negative edges of Clk, it is highly possible that the outputs of the serial shift register will be changing while they are being sampled by the parallel shift register. This illustrates the basic problem that internally generated clocks, in this case Int_Clk, can be delayed with respect to the master clock.

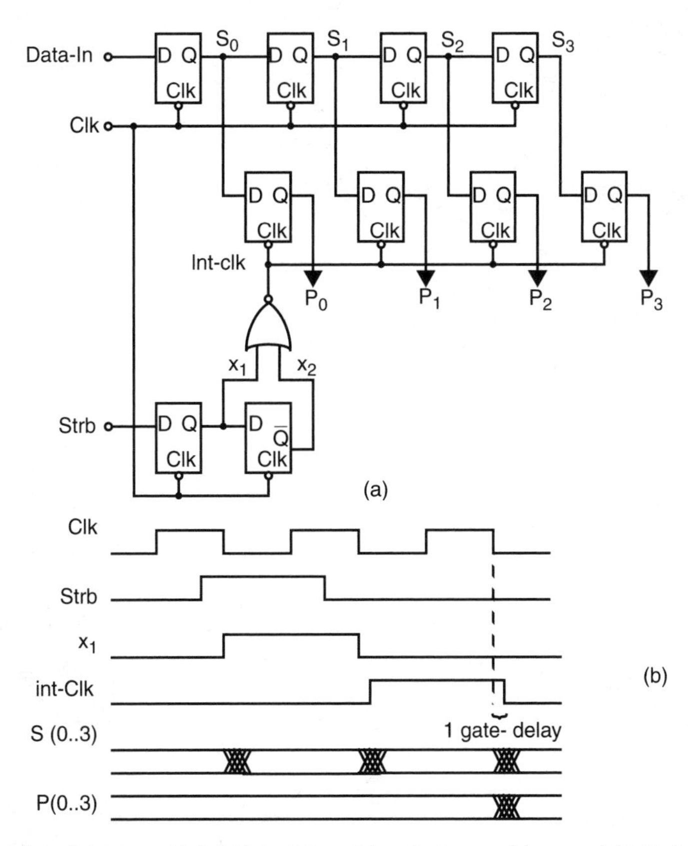

Figure 7.34 (a) A serial-to-parallel shift register with a timing problem and (b) its timing diagram.

A possible change to the circuit of Fig. 7.34a, to eliminate the problem, is to change the D flip-flops used to generate Int_Clk to flip-flops that change for positive-going edges. This causes the parallel shift register to load its contents just after a positive-going edge of Clk, a time when the outputs of the serial shift register are not changing.

Again, *internally generating clocks can sometimes result in subtle errors and at a minimum should be very carefully considered and simulated.*

Master Clock Distribution

The distribution of clocks in an IC is one of the most critical design issues, particularly for high-speed ICs. Normally, a single master clock will be distributed, which will be used to drive local buffers. The local buffers then drive the flip-flops. This minimizes the loading on the master clock. An example of a local buffer that is commonly used is shown in Fig. 7.35. This cross-coupled *nor*

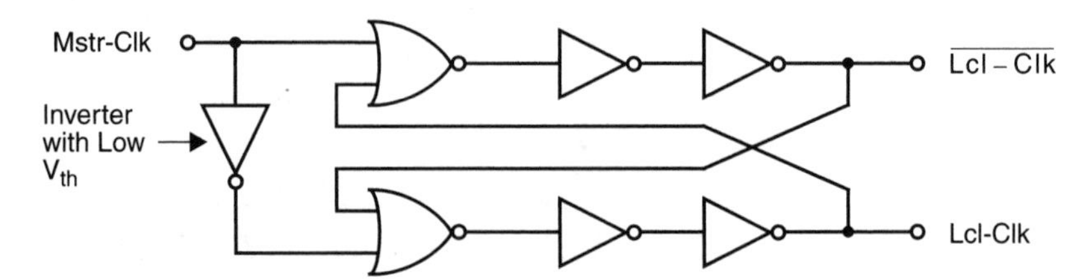

Figure 7.35 A clock-buffer circuit suitable for regenerating clocks and providing nonoverlapping outputs.

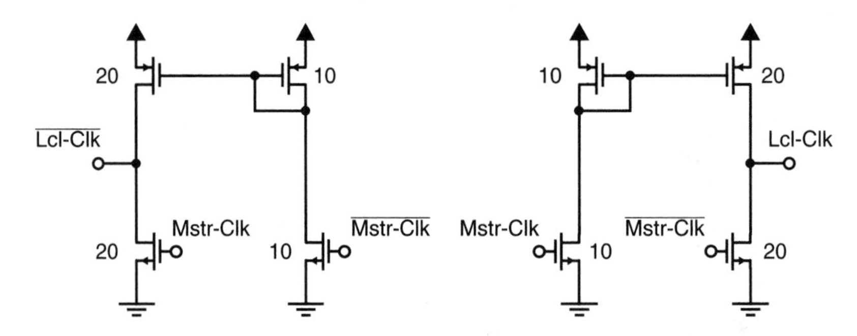

Figure 7.36 A differential clock buffer.

circuit generates two-phase nonoverlapping local clocks from a single-phase master clock. The nonoverlap time is the delay through a *nor* gate and the two output inverters. Normally, the output inverters will be sized progressively larger in order to drive a larger load. When only a small number of flip-flops are being driven locally, then the output inverters may be eliminated.

One of the advantages of the clock-regeneration circuit of Fig. 7.35 is that as long as the gate-threshold voltage of the first inverter is less than that of the *nor* gates, the positive feedback around the *nor* gates gives the clock-generator hysteresis. This gives very fast output waveforms with sharp edges even when the master clock does not have sharp edges. This is often needed for the proper operation of dynamic logic, which can be dependent on having clocks with sharp edges.

An alternative technique is to use a differential master clock. Despite the larger area required, this allows the generation of local clocks having very nearly a 50% duty cycle even when the master clock is heavily loaded. Another advantage is that *differential master clocks inject less noise into the substrate*; this is an important consideration for mixed-mode ICs that contain analog circuits in addition to digital circuits. An example of a CMOS clock-regeneration circuit for this case is shown in Fig. 7.36. This circuit consists of two *current comparators*. The outputs are *high* or *low* depending on the differential input signals rather than the common-mode voltages of the differential master clock. This again gives much sharper output edges for slowly moving input voltages, but now the output waveforms are symmetric rather than nonoverlap-

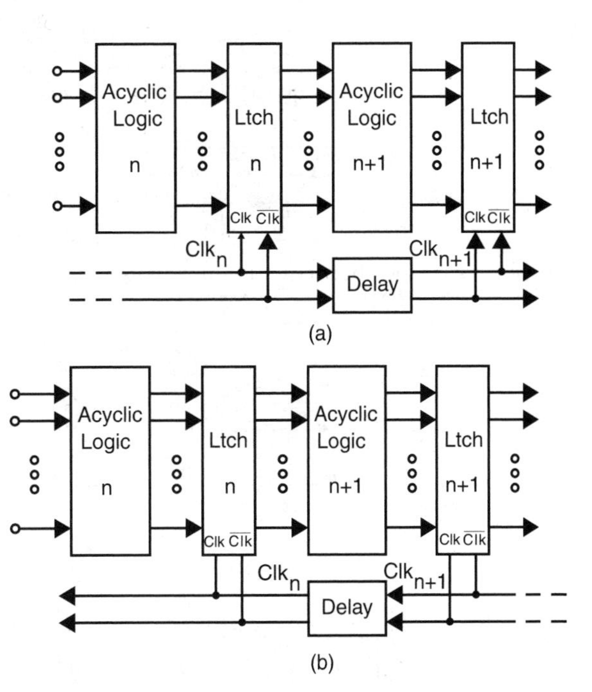

Figure 7.37 (a) An error-prone order for distributing the clock and (b) the correct order for distributing the clock.

ping. The circuit of Fig. 7.36 is a simple example of a *sense amplifier.* More advanced examples of sense amplifiers will be described in Chapter 11, where it will be seen how they can be used to quickly read the contents of a semiconductor memory. Besides minimizing the load on the master clock and giving sharper local clocks, they allow for smaller voltage swings of the master clock, which significantly increases the speed. Indeed, modern high-speed CMOS circuits will often operate with master clock voltage swings of less than 1 V.

Another consideration when designing the clock distribution system of an integrated circuit is the order in which it is distributed around the IC. The golden principle here is that *clock signals should be transmitted in the opposite direction of the data-signal propagation;* flip-flops and latches further along the pipeline should receive the clock first. To see why this is required, consider the pipelined systems shown in Fig. 7.37a and b. In both cases, it is assumed the clock is being fed to two stages of the pipeline with a slight delay between them. In Fig. 7.37a it is first fed to the earlier stage, whereas in Fig. 7.37b it is first fed to the later stage.

In the case of Fig. 7.37a, the outputs of latch n will start to change when Clk_n goes high. Slightly after this, the inputs of the acyclic logic network n + 1 will start to move. If the delay through the acyclic network is less than the delay of the clock, latch n + 1 will still be in track mode at this time due to the delay of the clock. In this case its contents will be affected by the changes of acyclic network n, happening when Clk_n goes high. This is incorrect operation as the contents of latch n + 1 should be dependent only on the output of acyclic network n before Clk_n goes high.

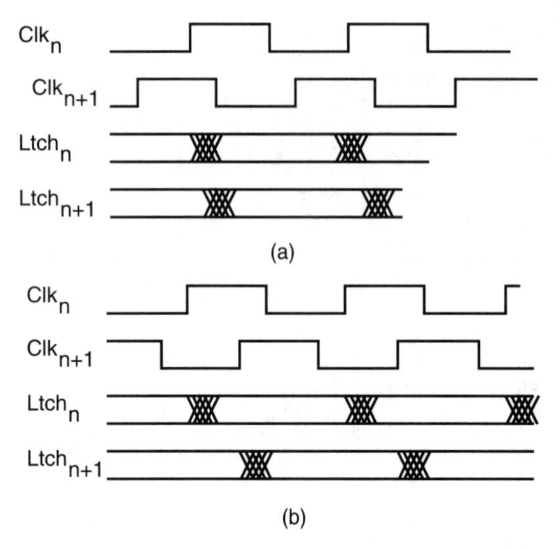

Figure 7.38 The timing diagrams for (a) the circuit of Fig. 7.37a and (b) the circuit of Fig. 7.37b.

In the correct distribution system, shown in Fig. 7.37b, Clk_{n+1} goes low before Clk_n goes high. Thus, latch $n + 1$ goes into latch mode before the output of latch n even begins to move. This circuit works correctly even when the acyclic networks have no delay. The timing diagrams for the two cases are shown in Fig. 7.38, and it is seen that the circuit of Fig. 7.37b never has the outputs of the two latches moving at the same time.

In summary, it has been seen that the distribution of clocks in a digital IC is nontrivial and must be done carefully if the IC is to function correctly. Besides the problems we have described, there are other more subtle problems that can cause errors that are beyond the scope of this text. Successful designers carefully design and review their clock distribution system using both hand and computer simulations. The importance of this task cannot be overemphasized.

7.5 Synchronous System Examples

In this section, two examples of synchronous machines are described. The first example describes the design of a Gray-code counter. This is a counter in which only a single bit changes at each transition from one state to the next. The second example describes the design of a very simple register-based controller. This example is intended as an introduction to design methodologies that are the basis for controllers and microcomputers. Although both examples are simple, allowing them to be described without taking an unreasonable amount of time, they are indicative of methodologies that can be used for much more elaborate designs. These examples would normally be covered near the end of a first-level course in digital IC design, but can also be safely skipped for instructors who would rather spend the time covering more advanced subjects presented later in the text.

GRAY-CODE COUNTER

Example 7.4

To illustrate synchronous machine design, a simple example of the design of a four-bit Gray-code counter will be described.

Solution: A Gray-code counter has one and only one output bit change every active clock edge, as it changes from one state to the next. Because of this property, Gray-code counters are often used as the basis for state machines. The states that a four-bit Gray-code counter goes through sequentially are given in Table 7.5.

The first step in the design is the state transition table. This table lists the present state, the next state, and the required values for the J and K inputs in order to achieve the necessary state change. The resulting state transition table for our example is shown in Table 7.6 where d stands for do not care. The next step is to use Karnaugh maps to find simplified functions for each of the inputs. The Karnaugh maps are shown in Fig. 7.39.

Table 7.5 The Sequential Outputs of a Four-Bit Gray-Code Counter

State	Outputs
0	0000
1	0001
2	0011
3	0010
4	0110
5	0111
6	0101
7	0100
8	1100
9	1101
10	1111
11	1110
12	1010
13	1011
14	1001
15	1000

Table 7.6 The State Transition Table for a Four-Bit Gray-Code Counter

Present state	Next state	J_3	K_3	J_2	K_2	J_1	K_1	J_0	K_0
0000	0001	0	d	0	d	0	d	1	d
0001	0011	0	d	0	d	1	d	d	0
0011	0010	0	d	0	d	d	0	d	1
0010	0110	0	d	1	d	d	0	0	d
0110	0111	0	d	d	0	d	0	1	d
0111	0101	0	d	d	0	d	1	d	0
0101	0100	0	d	d	0	0	d	d	1
0100	1100	1	d	d	0	0	d	0	d
1100	1101	d	0	d	0	0	d	1	d
1101	1111	d	0	d	0	1	d	d	0
1111	1110	d	0	d	0	d	0	d	1
1110	1010	d	0	d	1	d	0	0	d
1010	1011	d	0	0	d	d	0	1	d
1011	1001	d	0	0	d	d	1	d	0
1001	1000	d	0	0	d	0	d	d	1
1000	0000	d	1	0	d	0	d	0	d

There are many possible ways for implementing the logic before the flip-flops. For flip-flop-0, one possibility is to use a three-input differential-output exclusive-or circuit as shown in Fig. 7.40. This single gate can be used for producing the inputs to both J_0 and K_0. For flip-flop-1, we need $J_1 = (Q_3Q_2 + \overline{Q_3}\overline{Q_2})\overline{Q_0}$ and $K_1 = (\overline{Q_3}Q_2 + Q_3\overline{Q_2})Q_0$. Now assume Q_0 was a "1" for the moment. Then a single two-input differential-output exclusive-nor gate could be used to produce the inputs for both J_1 and K_1. All that is needed is to modify the gate so both J_1 and K_1 are low if Q_0 is a "0". A single gate that achieves this is shown in Fig. 7.41.

In a similar manner, single gates can be found that implement the logic to produce the inputs for J_2 and K_2, as well as for J_3 and K_3. These are shown in Fig. 7.42a and b, respectively. Note how by using a little ingenuity and imagination, the number of transistors can be considerably decreased. Although these ad hoc methods can be used to decrease area or increase speed, they are more prone to errors than more systematic methods and thus should be used with caution. The JK flip-flops required to finish the example can be realized in a similar manner to the circuits of Fig. 7.25 or Fig. 7.28. It is left as an exercise for the reader to determine reasonable transistor sizes for the gates just presented.

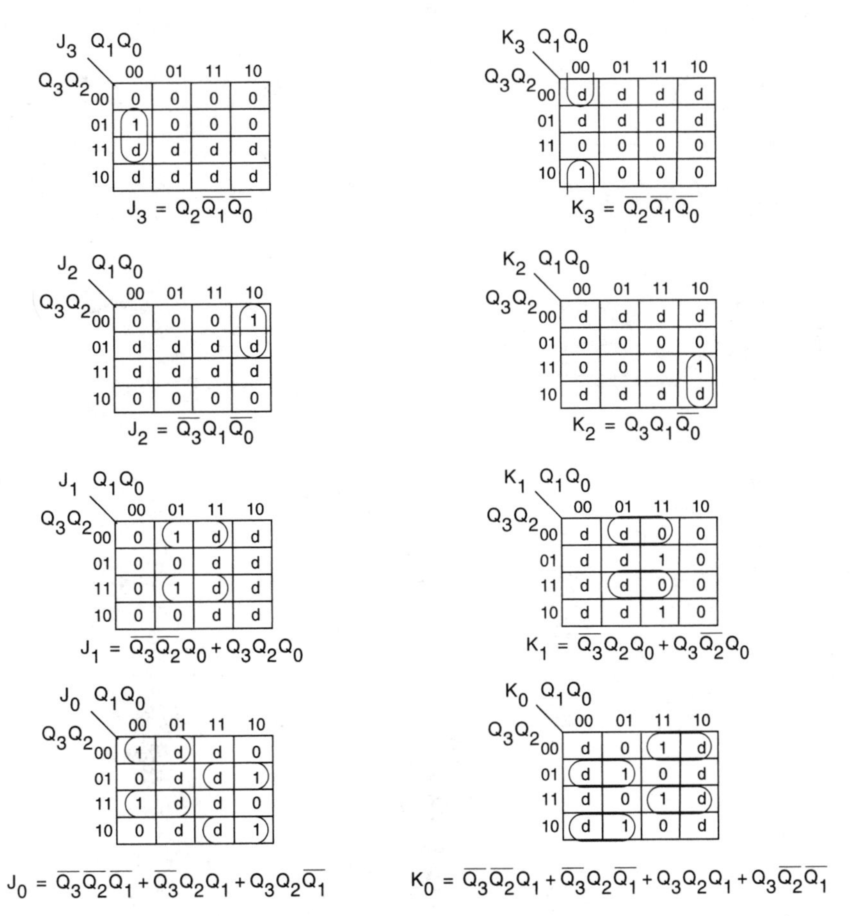

Figure 7.39 The Karnaugh map reductions for the logic for each **JK** flip-flop of a four-bit Gray-code counter.

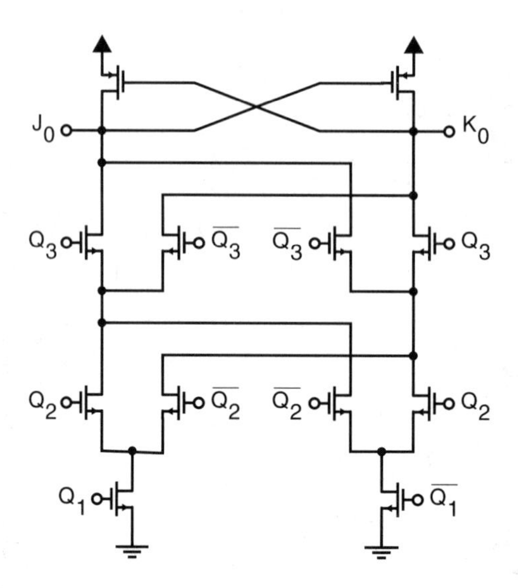

Figure 7.40 A single gate that realizes the logic for both J_0 and K_0.

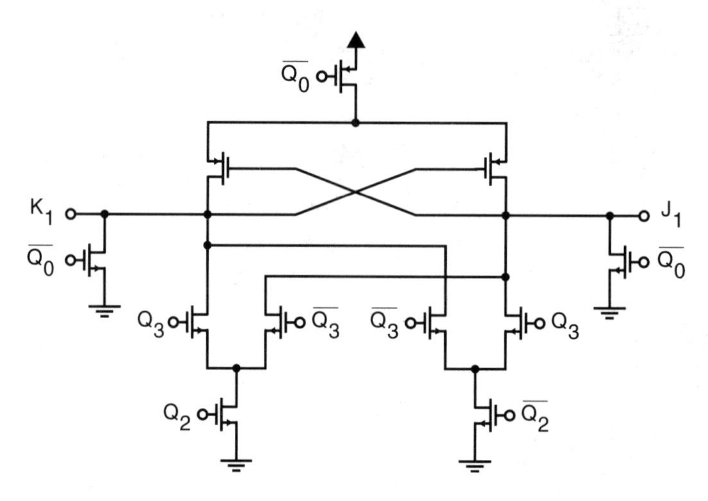

Figure 7.41 A single gate that realizes the logic for both J_1 and K_1.

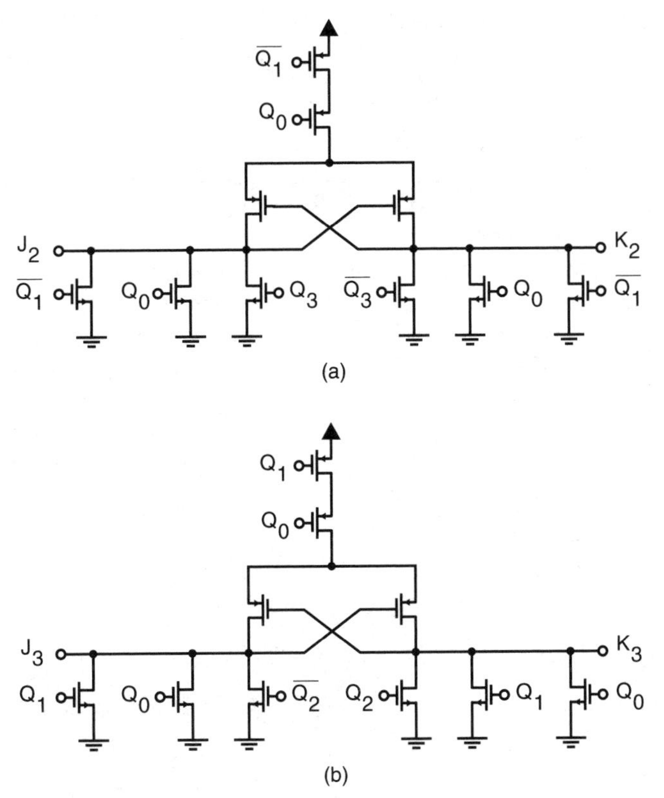

Figure 7.42 Logic gates for producing (a) J_2 and K_2 and (b) J_3 and K_3.

REGISTER-BASED CONTROLLERS

In the previous section, it was shown how state machine techniques could be used to realize a four-bit Gray-code counter. These techniques are adequate for relatively simple logic circuits. For more complicated logic functions, state machine design techniques become unwieldy; a preferred approach is to design a machine that is more general purpose and perhaps programmable. This was the rationale behind what eventually evolved into digital computers. In this section, an introduction to some of the design techniques involved in realizing very simple computers and controllers based on manipulating digital words stored in registers is given.

To illustrate some of the methodology behind controller-based designs, a simplified example will be presented.

Example 7.5

A simple controller based on the architecture shown in Fig. 7.43 will be described. This controller has a single input bus that can be operated on in a number of different ways in order to produce the output bus, which is stored in a register. Most functions operate not just on the input bus, but also on the value that is currently stored in the output register.

The logic function is performed by a multifunction logic unit. This logic unit is an acyclic circuit that has many different output buses. Each output bus represents a different transformation of the inputs and the register outputs. A multiplexor chooses a particular transformation output based on a control input. In this example, there is a register-based controller that has a single input bus (A). This might come from a value stored in a random-access memory. Also, the contents of the output register may be stored in a memory location (after they are valid) as well.

The outputs of the logic unit are defined as follows:

1. Store the input bus (A) in the output register.
2. Store the current value of the register (R) back into the output register.
3. Store the output-register bus back into the output register after shifting it left by one bit, (R_{-1}).
4. Store the output-register bus back into the output register after shifting it right by one bit, (R_{+1}).
5. Store the input A after inverting all bits.

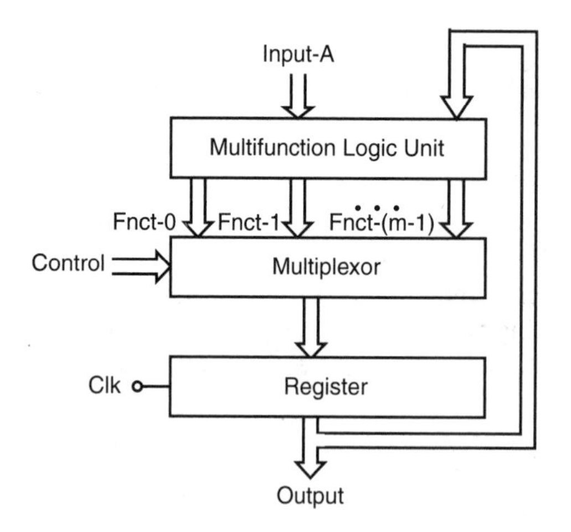

Figure 7.43 A register-based controller.

6. Perform the logic *or* function bit by bit between the input A and the current contents of the output register and store the results back into the output register.
7. Perform the logic *and* function bit by bit between the input A and the current contents of the output register and store the results back into the output register.
8. Perform the logic *ex-or* function bit by bit between the input A and the current contents of the output register and store the results back into the output register.

These operations are adequate for realizing most desired logic functions, although the resulting implementation may not be the most efficient; this example is designed to illustrate the methodology simply rather than be the most efficient realization. A *real* machine would probably include additional operations such as alternative shift operations, addition, subtraction, and perhaps multiplication.

The particular order of functions that the logic unit implements is determined by what is sequentially applied to the Control inputs. All possible functions implemented by the logic unit are implemented in parallel, with there being one output bus per function. The multiplexor then determines which function is being output during any particular clock cycle. Depending on the function control inputs to the multiplexor, a particular multifunction logic unit output will be chosen and its value will be selected by the multiplexor and stored in the output register. The function inputs to the multiplexor would normally come from successive locations in a read-only instruction memory, the external world, or a combination of the two.

Thus, the register-based controller is a general-purpose machine in which one design can be used for many different applications. All that would need to be changed for different applications are the contents of the read-only instruction memory locations that are sequentially sent to the multiplexor.

In this simple example, we assume a four-bit register-based controller is to perform the specified functions depending on the three inputs to the multiplexor.

The multiplexor will consist of four 8-to-1 multiplexors, one for each bit. The logic unit will also consist of four subunits in which each *bit slice* will have four inputs and eight outputs. The inputs for the ith bit slice will be A_i, R_i, R_{i-1}, and R_{i+1}. A possible design for the ith bit slice of the logic unit, multiplexor, and register is shown in Fig. 7.44. This design realizes the logic unit as a number of parallel fully differential circuits. The multiplexor is realized as a 3-input to 1-of-8-output decoder. Depending on the function desired, one and only one of the differential circuits will be enabled. The differential logic circuit that is enabled will control the differential output bus,

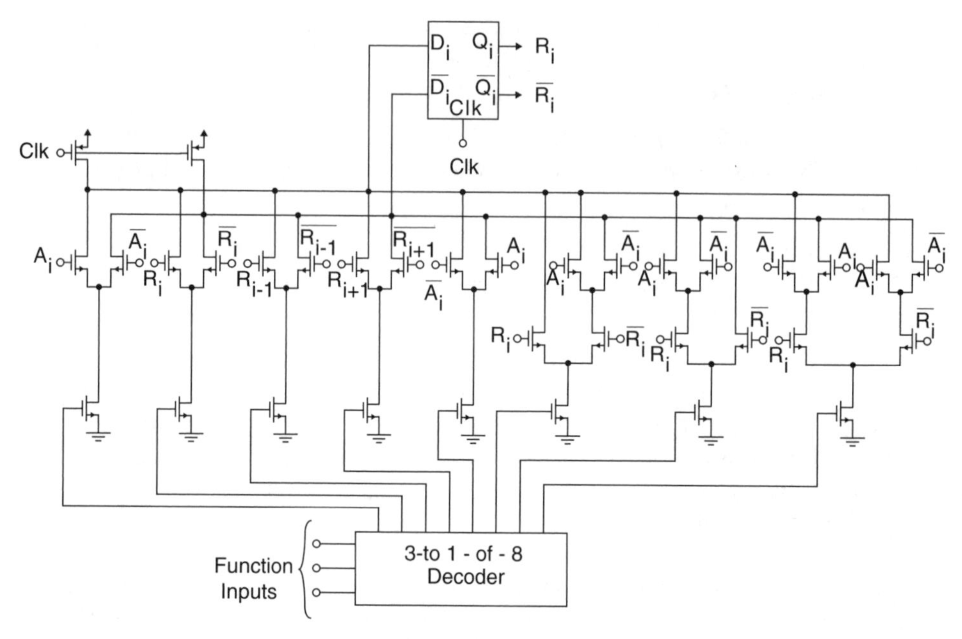

Figure 7.44 A bit slice of the logic unit and multiplexor suitable for use in Example 7.5.

which will be connected to the D_i and $\overline{D}_i$ inputs of the ith D flip-flop. A possible choice for the D flip-flop might be the circuit shown previously in Fig. 7.24. By having differential inputs, the inverter at the input of the D flip-flop can be eliminated.

The loads of the logic unit are dynamic. During the time when Clk = "0", the master stage of the D flip-flops will be in latch mode and thus isolated from the logic-unit outputs. At this time, the outputs of the logic unit are precharged high, dynamically. If the 3-to 1-of-8 decoder is realized using Domino CMOS logic (which will be described in a later chapter), then all of the differential logic circuits of the logic units will have "0"s coming to the transistors that select them, and thus there will be no d.c. paths to ground at this time. By having a dynamic precharge of the logic-unit output, very large p-channel precharge transistors can be used, resulting in a very fast precharge despite the large number of transistors connected to the output.

This example is almost too simple for a real application, but has been chosen to introduce modern digital IC methodologies. It is illustrative of a couple of modern design methodologies; these include the manipulation of data based on fixed-length words taken from registers being transformed by an arithmetic-logic unit and the use of dynamic logic.

7.6 Bibliography

M. Annaratone, *Digital CMOS Circuit Design*, Kluwer, 1986.
M. Elmasry, ed., *Digital MOS Integrated Circuits, II,* IEEE Press, 1991.
L. Glasser and D. Dopperpuhl, *The Design and Analysis of VLSI Circuits*, Addison-Wesley, 1985.
D. Hodges and J. Jackson, *Analysis and Design of Digital Integrated Circuits*, McGraw-Hill, 1988.
M. Mano, *Digital Design,* Prentice Hall, 1984.
M. Mano and C. Kime, *Logic and Computer Design Fundamentals*, Prentice Hall, 1997.
C. Mead and L. Conway, *Introduction to VLSI Systems,* Addison-Wesley, 1980.
J. Rabaey, *Digital Integrated Circuits, A Design Perspective*, Prentice Hall, 1996.
C. Roth, *Fundamentals of Logic Design*, West, 1985.
T. Sakurai and K. Tamaru, "Simple Formulas for Two- and Three-Dimensional Capacitances, *IEEE Transactions on Electron Devices*, ED-30 (2), 183–185, February 1983.
J. Uyemura, *Circuit Design for CMOS VLSI*, Kluwer, 1992.
H. Veendrick, "Short-Circuit Dissipation of Static CMOS Circuitry and Its Impact on the Design of Buffer Circuits," *IEEE Journal of Solid-State Circuits*, SC-19 (4), 468–473, 1984.
N. Weste and K. Eshragian, *Principles of CMOS VLSI Design: A Systems Perspective*, Addison-Wesley, 1983.

7.7 Problems

For the problems in this chapter, assume the following transistor parameters:

- npn bipolar transistors:
  $\beta = 100$
  $V_A = 80\ \text{V}$
  $\tau_b = 13\ \text{ps}$
  $\tau_s = 4\ \text{ns}$
  $r_b = 330\ \Omega$

- n-channel MOS transistors:
  $\mu_n C_{ox} = 190\ \mu\text{A/V}^2$
  $V_{tn} = 0.7\ \text{V}$
  $\gamma = 0.6\ \text{V}^{1/2}$
  $r_{ds}\ (\Omega) = 5000 L\ (\mu\text{m})/I_D\ (\text{mA})$ in active region
  $C_j = 5 \times 10^{-4}\ \text{pF}/(\mu\text{m})^2$
  $C_{j\text{-}sw} = 2.0 \times 10^{-4}\ \text{pF}/\mu\text{m}$
  $C_{ox} = 3.4 \times 10^{-3}\ \text{pF}/(\mu\text{m})^2$
  $C_{gs(overlap)} = C_{gd(overlap)} = 2.0 \times 10^{-4}\ \text{pF}/\mu\text{m}$

- p-channel MOS transistors:
  $\mu_p C_{ox} = 50\ \mu A/V^2$
  $V_{tp} = -0.8\ V$
  $\gamma = 0.7\ V^{1/2}$
  $r_{ds}\ (\Omega) = 6000L\ (\mu m)/I_D\ (mA)$ in active region
  $C_j = 6 \times 10^{-4}\ pF/(\mu m)^2$
  $C_{j\text{-}sw} = 2.5 \times 10^{-4}\ pF/\mu m$
  $C_{ox} = 3.4 \times 10^{-3}\ pF/(\mu m)^2$
  $C_{gs(overlap)} = C_{gd(overlap)} = 2.0 \times 10^{-4}\ pF/\mu m$

7.1 Modify the latch of Fig. 7.2 so that it has an additional *Reset* input that for normal operation is held low, but if at any given time goes high, will cause the latch to be placed into the "0" state irrespective of other inputs.

7.2 Repeat Example 7.1 but assume the lengths of all transistors are 0.5 μm and the transistor widths are $W_1 = W_2 = 10\ \mu m$, $W_3 = 10\ \mu m$, $W_4 = 15\ \mu m$, $W_6 = 2\ \mu m$, and $W_3 = 2\ \mu m$.

7.3 Modify the latch of Fig. 7.5 so that it has an additional *Reset* input that for normal operation is held low, but if at any given time goes high, will cause the latch to be placed into the "1" state irrespective of other inputs.

7.4 For the *CMOS* latch of Fig. 7.7, assuming all transistor lengths are 0.5 μm and the transistor widths are as shown, find analytically the gate threshold voltages for "0" to "1" and "1" to "0" transitions of D assuming Clk is high.

7.5 For the pseudo-NMOS biphase latch of Fig. 7.8, assuming all transistor lengths are 0.5 μm and the transistor lengths are as shown, find analytically the gate-threshold voltages for "0" to "1" and "1" to "0" transitions of D assuming Clk is high.

7.6 Is it possible to modify the pseudo-NMOS biphase latch of Fig. 7.8 so that it has no d.c. power dissipation without losing the feature that both outputs change immediately after the D and $\overline{D}$ inputs change? Give the reasons for your answer.

7.7 Estimate the d.c. power dissipation of the pseudo-NMOS biphase latch of Fig. 7.8.

7.8 For the *clocked latch* of Fig. 7.15, assume $C_p = 50$ fF and assume $C_{j\text{-}1} = C_{j\text{-}2} = 20$ fF. Further assume that when Clk was high, V_{in} was a "0" and that later, after Clk went low, V_{in} went high. Estimate the change in the output voltage V_{out}.

7.9 Design a master–slave D flip-flop similar to that shown in Fig. 7.17, but instead of using *and-nor* gates use *or-nand* gates.

7.10 Modify the flip-flop of Fig. 7.19 so that it changes state at a positive-going edge of the clock.

7.11 Modify a D flip-flop to be a T flip-flop by feeding Q and $\overline{Q}$ to a two-input multiplexor at the input of the D flip-flop that is controlled by the T input. Show a transistor-level realization of the multiplexor based on traditional CMOS logic.

7.12 Show realizations of the multiplexor of Problem 7.11 based on using *CMOS* transmission gates. Show realizations of the multiplexor of Problem 7.11 based on using fully differential CMOS logic.

7.13 Modify the JK flip-flop of Fig. 7.25 so that it includes asynchronous *set* and *reset* inputs. Give reasonable dimensions for all transistors.

7.14 Modify the JK flip-flop of Fig. 7.28 so that it includes asynchronous *set* and *reset* inputs. Give reasonable dimensions for all transistors.

7.15 Modify the JK flip-flop of Fig. 7.28 to have an inverting clocked latch similar to that shown in Fig. 7.5 that is active when Q is high.

7.16 Give a transistor-level implementation of the clock buffer of Fig. 7.35 with reasonable transistor sizes. All transistor lengths should be 0.5 μm and the widths of the n-channel transistors of the last inverters should be 100 μm.

7.17 Give a logic-gate realization of a three-bit Gray-code counter, similar to that described in Example 7.4, but that counts backward.

7.18 Design a *CMOS* four-bit programmable counter that has a clock input and two control inputs. The output should be high for one clock period each k clock period where k is determined by the two control inputs according to the following table. T flip-flops should be used.

Control inputs

C_1	C_0	k
0	0	8
0	1	10
1	1	11
1	0	16

7.19 Give transistor-level realizations of the blocks found in Problem 7.17.

7.20 Modify the register-based controller of Fig. 7.43 so that it includes an instruction that will take the contents of the register, increment it, and then store it back in the register.

7.21 Modify the register-based controller of Fig. 7.43 so that it includes an instruction that will take the contents of the register, decrement it, and then store it back in the register.

CHAPTER 8

Bipolar and BiCMOS Logic Gates

Historically, bipolar integrated circuits used to be much more popular than MOS integrated circuits, particularly for small-scale logic circuits. There were two major reasons for this: first, bipolar-junction transistors (BJT) originally could be manufactured more reliably than MOS transistors, and second, they were faster. As the reliability of MOS transistors improved, and as integrated circuits became more complex, which made the lower power and smaller size of MOS logic more important, the popularity of BJT logic decreased; however, BJT technology is still popular for the highest frequency logic circuits. Also, due to the recent emergence and growing popularity of BiCMOS ICs, where both MOS and bipolar gates can be realized in the same IC, bipolar logic design is growing in importance.

The main reason for the high speed of BJTs, and therefore BJT logic circuits, is the high transconductance of BJTs. Remember from Chapter 3 that for a BJT we have the transconductance, g_m, given by

$$g_m = \frac{I_C}{kT/q} = \frac{I_C}{0.026\ V} \tag{8.1}$$

At room temperature a bipolar transistor has a transconductance of approximately 4 mA/V for a 0.1-mA bias current. This is more than an order of magnitude larger than the typical transconductances of MOS transistors having comparable sizes. This large transconductance results in a large unity current gain frequency (i.e., f_t) for the transistor. In addition, it allows parasitic capacitances to be charged and discharged quickly. Finally, it also allows BJT logic gates to be operated with small voltage changes, which is another major reason why modern bipolar logic circuits are so fast. An additional reason for the high speed of BJT digital integrated circuits is that the intrinsic device capacitances tend to be larger than those in MOS circuits, which in turn minimizes the effects of interconnect capacitances.[1]

Previously, most BJT logic gates consist of npn transistors, resistors, and possibly diodes. Since integrated pnp transistors tend to be large and slow, they are seldom used, although this could change in the future. The large size of the resistors is one of the reasons why BJT logic requires larger IC areas than MOS logic; the other reason is the area required for collector isolation from the substrate.

The first type of bipolar logic that was developed was called *resistor–transistor logic* (RTL). This evolved into *diode–transistor logic* (DTL), and then into *transistor–transistor logic* (TTL). Until recently, TTL was the most popular logic family, particularly for small-scale integrated circuits. However, TTL logic is never or seldom used now. For this reason, the descriptions of RTL, DTL, TTL, and other seldom used bipolar logic families, such as integrated-injection logic (I^2L), are not covered in this book.

This chapter will describe the currently most popular BJT logic families for high-speed design. These include emitter-coupled logic (ECL) and current-mode logic (CML). In this section, fully differential logic design will be emphasized. It is particularly important for CML circuits to provide noise immunity despite the small signal swings used. Also covered in this chapter will be slightly advanced subjects such as CML output buffers and properly terminating ECL circuits.

After discussing CML, BiCMOS logic gates will be described. These are gates that employ both CMOS and bipolar transistors in an effort to obtain the speed of bipolar logic while using CMOS transistor techniques to minimize power.

8.1 Emitter-Coupled Logic Gates

Emitter-Coupled Differential Pairs

The most important building block in modern bipolar logic families is the emitter-coupled differential pair shown in Fig. 8.1. The differential pair operates much like a current switch, as we

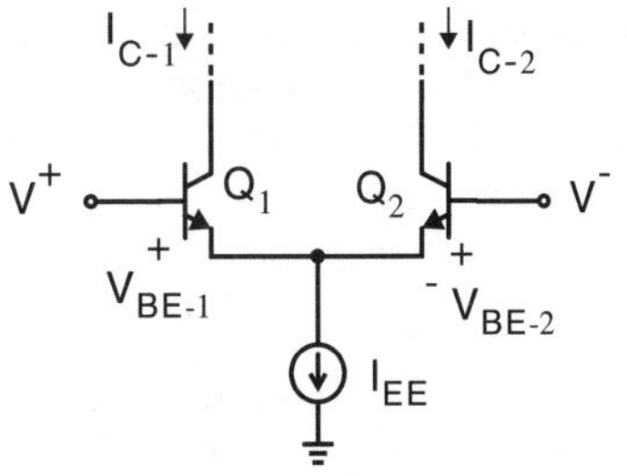

Figure 8.1 A bipolar differential pair.

[1]The larger intrinsic capacitances do slow the circuits compared to MOS circuits but what really matters is the transistor f_ts, which are approximately given by g_m/C_{be} and the large g_m, more than compensates for the large C_{be}.

shall see. When V^+ is much greater than V^-, then all of the bias current I_{EE} is diverted through Q_1; in the opposite case, all of the bias current I_{EE} is diverted through Q_2.

This circuit's large-signal behavior can be analyzed by first recalling the exponential relationship for a bipolar transistor,

$$I_C = I_S e^{(V_{BE}/V_T)} \tag{8.2}$$

Writing an equation for the sum of voltages around the loop between V^+ and V^-, we have

$$V^+ - V_{BE\text{-}1} + V_{BE\text{-}2} - V^- = 0 \tag{8.3}$$

Taking the ratio of $I_{C\text{-}1}/I_{C\text{-}2}$, assuming $I_{S\text{-}1} = I_{S\text{-}2}$, and using (8.3), we can write

$$\frac{I_{C\text{-}1}}{I_{C\text{-}2}} = e^{[(V^+ - V^-)/V_T]} = e^{(V_{id}/V_T)} \tag{8.4}$$

where V_{id} is defined to be the difference between V^+ and V^-. In addition, we can also write

$$\alpha I_{EE} = I_{C\text{-}1} + I_{C\text{-}2} \tag{8.5}$$

where the constant α is defined as the ratio $I_{C\text{-}i}/I_{E\text{-}i} = \beta/(\beta + 1)$, which is slightly less than one since the base currents are not zero. Combining (8.4) and (8.5), we have

$$I_{C\text{-}1} = \frac{\alpha I_{EE}}{1 + e^{(-V_{id}/V_T)}} \tag{8.6}$$

$$I_{C\text{-}2} = \frac{\alpha I_{EE}}{1 + e^{(V_{id}/V_T)}} \tag{8.7}$$

A plot of these two currents with respect to V_{id} is shown in Fig. 8.2 where we note that the currents are split equally at $V_{id} = 0$ and saturate near either I_{EE} or 0 for differential input voltages

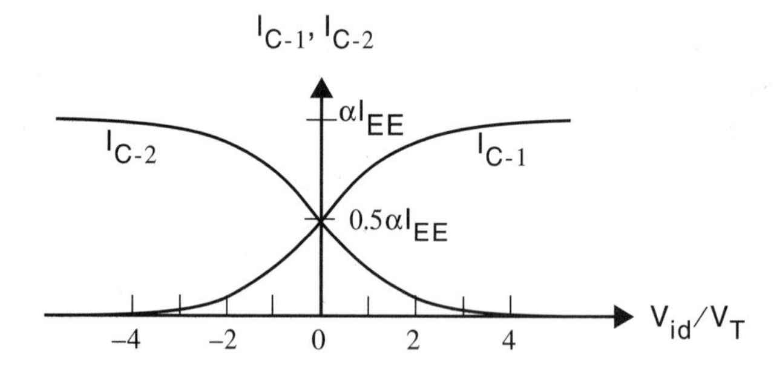

Figure 8.2 Collector currents for a bipolar differential pair.

approaching $4V_T$ (i.e., around 100 mV). *This means that for a differential-input voltage as small as 100 mV the bias current essentially goes through one transistor only. Thus, for differential input voltages greater than 100 mV, the bias current,* I_{EE}, *has switched to either the collector of* Q_1 *or the collector of* Q_2 *depending on the sign of the differential input voltage.*

EMITTER-COUPLED LOGIC

Emitter-coupled logic (ECL) is the fastest commonly available discrete small-scale integration (SSI) logic family. There are other faster bipolar families [such as current-mode-logic (CML)], but these typically are not available as discrete gates. There are two principles behind ECL: never let transistors saturate and keep voltage changes less than 0.8 V, or so.

Consider where a third bipolar transistor has been added to a differential pair as shown in Fig. 8.3. Assume further that the base of Q_2 is connected to a d.c. bias voltage, V_{ref}, and that the bases of Q_{1A} and Q_{1B} are connected to single-ended logic inputs. If the input voltages V_A and V_B are more than 0.1 V below V_{ref}, they are considered to represent "0"s; if they are more than 0.1 V above V_{ref}, they are considered to represent "1"s. Consider first the case in which both inputs are "0"s. Both of the input transistors Q_{1A} and Q_{1B} will be *off* and all of I_{EE} will be diverted through Q_2. In this case, I_1 will be zero and I_2 will be equal to I_{EE} (ignoring finite base currents). However, if either V_{1A} or V_{1B} is more than 0.1 V above V_{ref}, then the bias current will be diverted through either Q_{1A} or Q_{1B} (or both if both inputs are "1"s), and I_1 will be equal to I_{EE}; also, I_2 will be zero. Thus, the current will be diverted to the left if either V_a or V_b represents a "1", otherwise, it is diverted to the right.

This configuration is used to realize an ECL logic gate as shown in Fig. 8.4. The gate consists of an *emitter-coupled* section followed by *emitter-follower* buffers. The gates normally are connected between ground and a minus supply (V_{EE}) to get better power-supply noise rejection. This results because the signal paths have much smaller impedances to the positive supply than they do to the negative supply. Thus, by having the logic signals referenced to a *quiet* ground connection, which is the most positive voltage, the circuits are less susceptible to noise. The *pull-down* resistors of the emitter followers (R_3 and R_4) are normally *off-chip termination resistors*. They will often be either 50-Ω resistors connected to –2 V or 2-kΩ resistors connected to –5.2 V.

In earlier ECL gates, the current source I_B would be realized by a resistor, although in more modern families it would be a transistor connected to realize a constant current source.

The basic operation of the ECL gate is quite simple. Assuming V_A and V_B both are less than V_{ref} by around 0.4 V, then both Q_{1A} and Q_{1B} will be *off*. This will divert I_B through Q_2 to R_2.

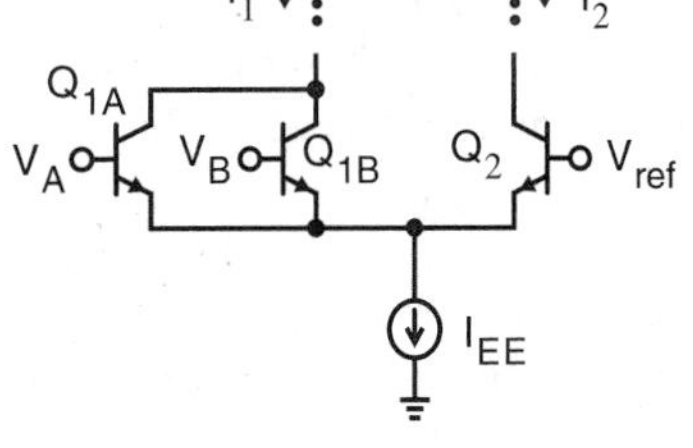

Figure 8.3 Adding an additional bipolar transistor to a differential pair to realize logic functions.

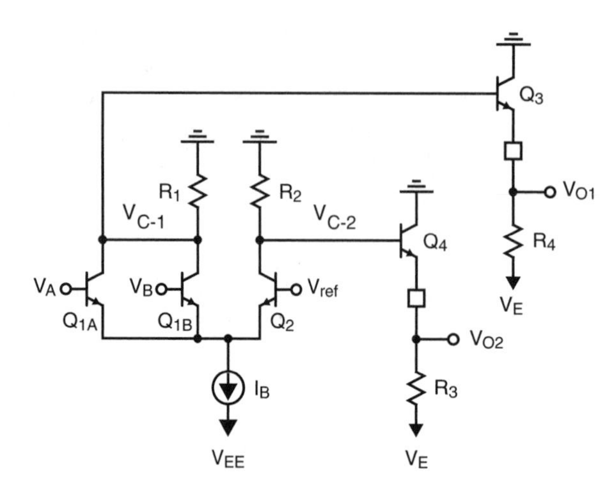

Figure 8.4 An emitter-coupled logic (ECL) gate.

Since both Q_{1A} and Q_{1B} are *off*, there will be no current through R_1 and the collector voltage of Q_{1A}, $V_{C\text{-}1}$, will be high at 0 V. The collector voltage, $V_{C\text{-}2}$, will be low at

$$V_{C\text{-}2} \cong -I_B R_2 \tag{8.8}$$

if we ignore base currents. Normally, $I_B R_2$ will be taken equal to approximately 0.8 V. The output voltages V_{o1} and V_{o2} will each be a base-emitter voltage drop below the appropriate collector voltages. Assuming the base-emitter voltages are about 0.75 V for typical ECL current levels, then V_{o1} will be at –0.75 V, which corresponds to a "0", and V_{o2} will be at –1.55 V, which corresponds to a "1". The reference voltage, V_{ref}, will be chosen about half way between these extremes or around –1.15 V.

Next, if either V_A and/or V_B goes to a voltage corresponding to a "1", which again is approximately 0.75 V, then the current will be diverted not through Q_2 but through Q_{1A} and/or Q_{1B}. Q_2 will *shut off* and $V_{C\text{-}2}$ will go to 0 V, implying V_{o2} goes to –0.75 V. Since we now have current going through R_1, $V_{C\text{-}1}$ will be at –0.8 V (assuming $R_1 = R_2$) and V_{o1} will now be at –1.55 V. Thus, the logic function corresponding to V_{o2} is the *or* function. That is $O_2 = A + B$ and $O_1 = \overline{O_2} = \overline{A + B}$.

Note that if one of the input transistors is conducting, its collector-base voltage is approximately –0.05 V, whereas if both input transistors are off, their collector-base voltages will be about 1.55 V. Thus, the transistors are both in the *active region*, but never *saturated*, irrespective of whether they are conducting or not. This, in conjunction with the relatively small voltage changes (as compared to typical CMOS values), makes ECL logic fast.

Because the gates have *emitter-follower* output buffers, the gates can be very fast for output "0" to "1" transitions. If the output pull-down resistors are 50 Ω connected to –2 V, they are also quick for "1" to "0" transitions, however if the pull-down resistors are 2 kΩ connected to –5.2 V, then the "1" to "0" transition can be considerably slower.

Example 8.1

Figure 8.5 shows one of the early ECL gates where I_B was realized by a resistor. This gate was a member of what was called the ECL 10K series. Assuming $V_{BE\text{-}on} = 0.75$ V, and that 2 kΩ external pull-down resistors connected to –5.2 V are used, calculate the output voltages for both inputs having voltages equal to –1.8 V, and separately for one input voltage equal to –0.9 V and the other input voltage equal to –1.8 V. Also find the gate-threshold voltage and the gain at $V_A = V_{TH}$ when V_B is a "0". Why were R_1 and R_2 not taken equal? What is the power dissipation of the gate?

Solution: With both V_A and V_B at –1.8 V, we have Q_{1A} and Q_{1B} both *off* and Q_2 is conducting. Thus,

$$V_{E\text{-}2} = V_{B\text{-}2} - 0.75\text{ V} = -2.05\text{ V} \tag{8.9}$$

Note that

$$V_{BE\text{-}1A} = V_{BE\text{-}1B} = V_{B\text{-}1} - V_{E\text{-}2} = -1.8\text{ V} - (-2.05\text{ V}) = 0.25\text{ V} \tag{8.10}$$

and both transistors are *off*. Continuing, we have

$$I_{R\text{-}3} = V_{R\text{-}3}/R_3 \tag{8.11}$$

$$\Rightarrow I_{R\text{-}3} = \frac{V_{E\text{-}2} - (-5.2\text{ V})}{R_3} = \frac{3.15\text{ V}}{779\ \Omega} = 4.04\text{ mA} \tag{8.12}$$

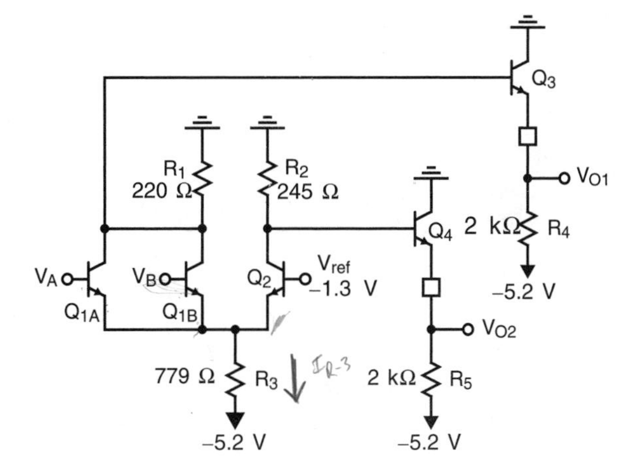

Figure 8.5 An ECL 10K series gate.

Since $I_{C\text{-}2} \cong I_{E\text{-}2} = I_{R_2}$, we have

$$V_{C\text{-}2} \cong -R_2 I_{R\text{-}3} = -0.99 \text{ V} \tag{8.13}$$

This implies $V_{o2} = V_{C\text{-}2} - 0.75 \text{ V} = -1.74 \text{ V}$. Since Q_{1A} and Q_{1B} are *off*, $V_{C\text{-}1A}$ will be approximately 0 V ignoring base currents. Thus, V_{o1} will be at $V_{C\text{-}1A} - 0.75 \text{ V} = -0.75 \text{ V}$.

As V_A becomes positive, eventually Q_{1A} will start to conduct and thus divert some of the current of R_3 away from Q_2. This begins to happen when V_A is approximately 0.1 V below V_{ref} (which is equal to –1.3 V). As V_A increases, Q_{1A} conducts more and more until when V_A is approximately 0.1 V above V_{ref}, it is conducting almost all of the current coming from R_3.

When V_A exactly equals V_{ref}, $I_{R\text{-}3}$ will be evenly divided between Q_{1A} and Q_2. This is the approximate gate-threshold voltage for one input only changing.

For this condition, a small-signal analysis of the stage can be done to find the gain. Using $g_{m\text{-}1A} = g_{m\text{-}2} = g_m$, we have the small-signal currents (that is the changes in the collector currents about their operating points) given by

$$\begin{aligned} i_{C\text{-}1A} &\cong g_m \frac{V_A - V_{ref}}{2} \\ i_{C\text{-}2} &\cong -g_m \frac{V_A - V_{ref}}{2} \end{aligned} \tag{8.14}$$

and

$$\begin{aligned} V_{C\text{-}1A} - V_{C\text{-}2} &= i_{C\text{-}1A} R_1 + (-i_{C\text{-}2}) R_2 \\ &= (V_A - V_{ref}) \frac{g_m}{2} (R_1 + R_2) \end{aligned} \tag{8.15}$$

Using

$$g_m = \frac{I_{R\text{-}3}/2}{V_T} = 77.7 \text{ mA/V} \tag{8.16}$$

gives

$$\frac{V_{C\text{-}1A} - V_{C\text{-}2}}{V_A - V_{ref}} = -18 \tag{8.17}$$

The total differential gain is split approximately evenly between the collector voltages of Q_{1A} and Q_2 since an increase in current flow through R_1 is matched by an equal decrease in current flow through R_2. Thus, the gain to V_{o1} will be about half of the differential voltage gain, or around –9. When V_A reaches –0.9 V, we have Q_2

and Q_{1B} completely cut off. $V_{E\text{-}1A}$ is now $V_A - 0.75$ V = −1.65 V. Thus, I_{R_3} is now given by

$$I_{R\text{-}3} = \frac{V_{E\text{-}1A} - (-5.2)}{R_3} = 4.56 \text{ mA} \tag{8.18}$$

Notice it is now larger than it was for the case of V_A being a "0". Continuing and ignoring base currents, we have

$$V_{C\text{-}1A} \cong -I_{R\text{-}3} R_1 = -1.00 \text{ V} \tag{8.19}$$

and

$$V_{o1} = V_{C\text{-}1A} - 0.75 = -1.75 \text{ V} \tag{8.20}$$

Also, since Q_2 is off, V_{o2} = 0 V − 0.75 V = −0.75 V. The transfer curve that corresponds to this analysis has been sketched in Fig. 8.6. Note that for V_A larger than −1.3 V, there still is some negative gain, although it is small.

The reason for taking R_1 smaller than R_2 is to compensate for the larger current through R_3 when V_A is high as opposed to when it is low. The power dissipation for V_A low is easily found by summing all the currents and multiplying by the power-supply voltage. For V_A low, we have

$$I_{R\text{-}4} = \frac{-0.75 \text{ V} - (-5.2 \text{ V})}{2 \text{ k}\Omega} = 2.23 \text{ mA} \tag{8.21}$$

and

$$I_{R\text{-}5} = \frac{-1.75 \text{ V} - (-5.2 \text{ V})}{2 \text{ k}\Omega} = 1.73 \text{ mA} \tag{8.22}$$

Since $I_{R\text{-}3}$ was previously found to be 4.04 mA, the total current is 8.0 mA and the power dissipation is 41.6 mW. This is very large for a single logic gate, which is one of the major limitations of ECL.

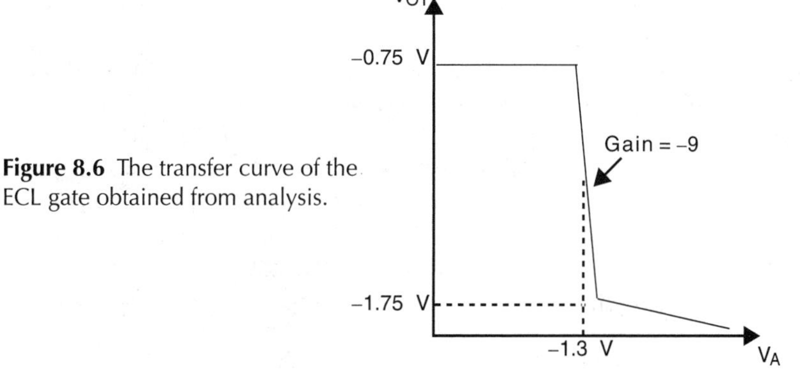

Figure 8.6 The transfer curve of the ECL gate obtained from analysis.

Although ECL has proven to be very popular for high-speed circuit design, it does have some shortcomings. The major limitation is its high-power dissipation. The 42-mW power dissipation found in Example 8.1 would constrain an ECL IC to have not more than approximately 500 gates without heat sinks being used assuming typical IC packages. *This high-power dissipation is a very serious limitation.*

Another major limitation is that its waveform tends to be very noisy, which is due to a number of factors. One factor is the emitter-follower buffers on the outputs of the gates. Emitter followers tend to ring when driving capacitive loads. This is because at high frequencies the output impedance of their emitters approximates a lossy inductor (Grey and Meyer, 1993). This inductance forms a resonant circuit with the load capacitance causing overshoot, particularly for positive-going signals. Some of the more recent ECL families have tried to minimize this by placing a small on-chip resistor in series with the emitters. This does help but also slows the circuits somewhat.

Another source of noise is the large current spikes on the power supplies and ground during the transients. Since the voltage swings are already small, these can easily cause errors. To minimize these spikes ECL circuits are normally connected between 0 and –5.2 V as opposed to 5 and 0 V. This helps, since the gain from the most positive power supply to internal signal nodes of ECL gates is greater than from the most negative power supplies. Thus, by taking the more positive power supply as ground, it is hoped that it will have wider interconnect lines and smaller impedances between the gates than would otherwise be the case. It is also very important to use many good bypass capacitors to help minimize power-supply noise.

Terminating Emitter-Coupled Logic

An important consideration when using ECL is true for any logic that operates at high frequencies, meaning at clock rates of 100 MHz and more. At these frequencies, where the rise and fall times become less than 1 ns, the time it takes a signal to propagate along a printed circuit board (PCB) connection becomes important. *The speed of an electrical signal along a PCB connection is on the order of 190 ps per inch.* Thus a 3-in. PCB connection, if not properly terminated, will have reflections returning from the far end in around 1 ns. Unless the rise and fall times are significantly greater than 1 ns, these reflections will cause considerable *noise spikes*.

A possible method to terminate an ECL gate and eliminate these reflections would be to include a resistor in series with the output where the size of the resistor would be a little smaller than the *characteristic impedance* of the PCB interconnect trace. (This might be on the order of 70 Ω and is specified by the PCB manufacturer.) An example of this *series termination* is shown in Fig. 8.7. The reason for the terminating resistor being smaller than the *characteristic impedance* of the board is to account for the nonzero output impedance of the emitter-follower buffer of the ECL gate. This sort of termination is used when all of the gates being driven are *located close together at the far end of the PCB trace.* If the gates being driven are distributed along the PCB interconnect line, then often the load resistor of the emitter follower is placed at the end of the PCB line as shown in Fig. 8.8. This topology is called *parallel termination.* This section is very brief as microwave considerations, such as terminating transmission lines, are beyond the scope of this text; irrespective, its importance should not be minimized.

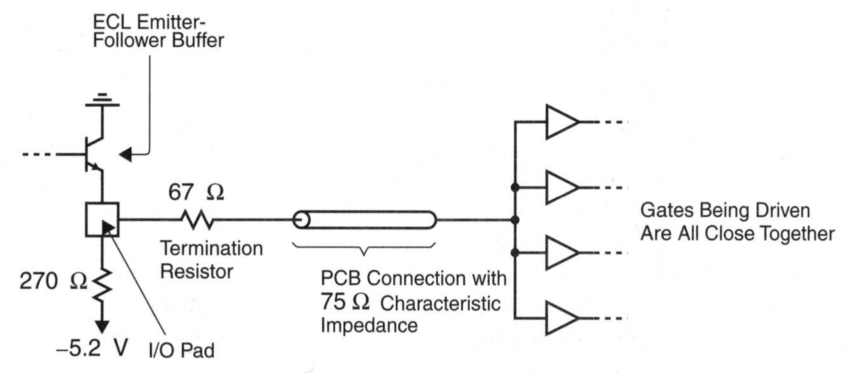

Figure 8.7 Terminating an ECL output to minimize microwave reflections.

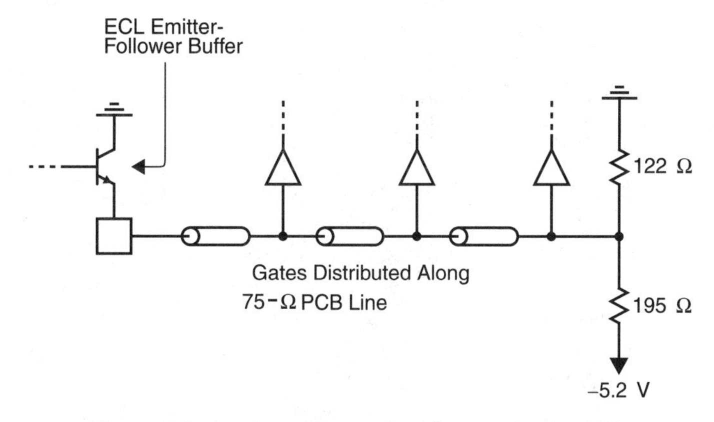

Figure 8.8 An alternative method for terminating ECL.

Temperature Sensitivity

A third possible problem with ECL is the temperature sensitivity of the base-emitter junctions. As the temperature increases, junction voltages decrease by –2 mV/°C. At very high or very low temperatures, this could cause the voltage levels to be substantially different than those expected. Given the small differences between the "0" and "1" voltage levels, this could be a potential problem. This problem has been largely solved by designing the circuits that produce the voltage reference, V_{ref}, so as to adjust V_{ref} with temperature so that all final output voltage levels are relatively temperature insensitive. Some of the details on how this is done can be found in Taub and Schilling (1977). It will not be covered here as it also is beyond the scope of this book.

It was in response to these problems, and also in response to the desire to integrate more and more circuitry running at ever higher speeds, that ECL evolved into what is now the most important logic family for high-speed bipolar logic design, that is, *current-mode logic* or CML.

8.2 Current-Mode Logic

Current-mode logic (CML) is a second generation of ECL. As such, it has many features in common with ECL, such as never allowing transistors to saturate and making large use of emitter-coupled differential pairs. However, it has a number of features that differentiate it from traditional ECL. Primary among these is the use of more complicated gates that realize greater functionality for a given power dissipation. This is achieved by realizing logic at common-mode voltage levels different than traditional ECL voltages. In addition, since most of the gates are driving smaller on-chip loads only, it is possible to bias the gates at lower current levels without appreciably degrading the speed. This is particularly true for modern, small, self-aligned and oxide-isolated, bipolar transistors. Typical bias currents range from 0.1 to 1 mA as opposed to 1 to 10 mA for ECL. This does not cause an excessive increase in gate delay as the transistors and on-chip capacitive loads are normally much smaller than those used for traditional discrete ECL.

Another important difference is much less use of power-hungry emitter followers at the gate outputs, that tend to ring. When they are used, they are biased at substantially lower current levels than was traditionally typical for ECL. This is particularly true for gates that drive only on-chip loads, but is starting to become prevalent even for driving off chip at the highest frequencies. When level shifts are required, they often are achieved by using low-power emitter followers at the inputs of the gates rather than the outputs.

The third important difference is a much more extensive use of fully differential logic design. In fully differential logic, an input consists of its true and complement values. Also, every gate produces an output and its complement. The functioning of a gate is dependent on the differential value of an input voltage and is insensitive to the common-mode or absolute voltage of an input. Signals as they are propagated through logic circuits might undergo a number of different level shifts, and thus, at different places in the circuits have different common-mode voltages; irrespective of the common-mode voltages, the logic values are determined by differential voltages only. Normally, the common-mode level of signal voltages will change by a base-emitter voltage drop as common-mode voltages change from one level to the next. For example, if one assumes a base-emitter voltage drop is equal to 0.7 V, the signal voltages at the top level might be 0 and –0.4 V, the signal voltages of the second level down might be –0.7 and –1.1 V, the signal voltages of the third level down might be –1.4 and –1.8 V, and so on. Some designers prefer not to use the top level for speed considerations as this level is not buffered by emitter followers at the output. This will be discussed in greater detail later in this section.

Fully differential logic has a number of important advantages: primary among these is it is relatively immune to noise, particularly power supply and ground noise, in addition to its high speed. This is because the noise may effect the common-mode voltage of an input signal but has little effect on the differential voltage of an input (or output). Due to the differential nature of the circuits, the currents through the power-supply connections have relatively small transients when the gates switch states. Largely due to this greater noise tolerance, fully differential logic can run at smaller voltage changes than ECL. Absolute voltage changes of only 0.4 V (or 0.8 V differential) are typical, with changes sometimes as small as 0.2 V (0.4 V differential) being used for the fastest circuits. Another important advantage of differential logic is that an inversion can be realized by simply interchanging the two lines of an input without incurring any delay. This is particularly important in arithmetic circuits such as adders or multipliers where it substantially decreases the carry-generation delay by eliminating an inverter delay. The delay of the CML carry-generation circuit is only slightly greater than the delay of a simple CML inverter.

Fully differential logic does have some shortcomings: it requires two interconnect lines for every signal connection, which takes more area; also, it is not good for realizing gates having many inputs—four inputs are about the maximum. Irrespective of these shortcomings, differential logic is becoming increasingly popular for very high-frequency logic design for almost all logic families and technologies including MOS and GaAs.

Current-Mode Logic Gates

An example of a CML *exclusive-or* gate is shown in Fig. 8.9. It has two differential inputs: the pair of input signals A and $\overline{A}$ together represents one logic input signal and the pair of input signals B and $\overline{B}$ represents the second logic input. The input pair A, $\overline{A}$ is connected to the differential pairs Q_5, Q_6 and Q_7, Q_8. The input pair B, $\overline{B}$ goes to the differential pair Q_3, Q_4. Since this differential pair is *stacked under* the differential pairs that A, $\overline{A}$ are connected to, the common-mode voltage of the input to Q_3, Q_4 must be lower than that of Q_5, Q_6 or Q_7, Q_8. This common-mode voltage difference is achieved by emitter followers Q_1 and Q_2. These realize a *voltage drop* of a base-emitter junction or approximately 0.7 V. The voltage levels of the top-level inputs might be between 0 and –0.4 V. Thus, the voltage levels to Q_3, Q_4, at the second level down, might be –0.7 or –1.1 V. Note that these voltage levels are all different than typical ECL levels; if CML and ECL levels are to be interfaced, then interface circuits are required.

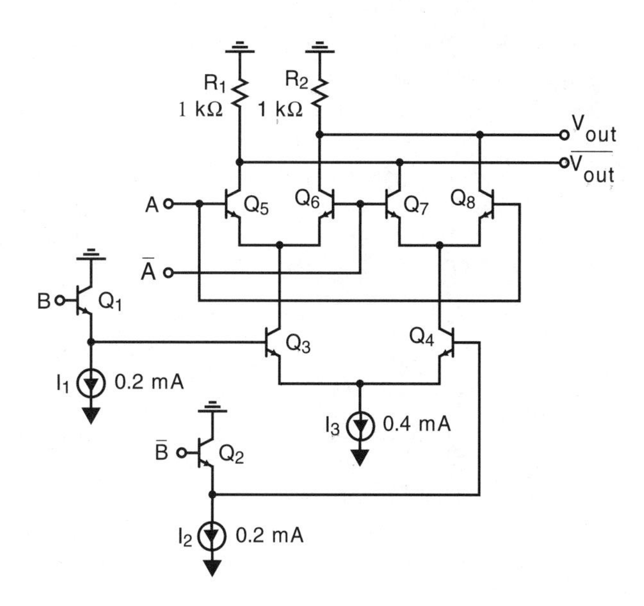

Figure 8.9 A CML two-input *exclusive-or* gate.

The operation of the gate is quite simple. If A and $\overline{B}$ are both "1"s, then V_A will be 0 V and $\overline{V_A}$ will be –0.4 V. After the level shift of emitter followers Q_1 and Q_2, we have $V_{B\text{-}3} = -0.7$ V and $V_{B\text{-}4} = -1.1$ V. This causes Q_3 to be *on* and Q_4 to be *off*, which diverts the bias current, I_3, to differential pair Q_5, Q_6, and guarantees that differential pair transistors Q_7, Q_8 are both *off*. Thus, we can ignore Q_4, Q_7, and Q_8 as they are all guaranteed to be *off*. Now, since A is a "1", $V_{B\text{-}5}$ is 0.4 V above $V_{B\text{-}6}$, and I_3, which is going through Q_3, is also diverted through Q_5 to R_1, since Q_6 is *off*. Thus,

$$\overline{V_{out}} = V_{C\text{-}5} = -I_3R_1 = -0.4 \text{ V} \tag{8.23}$$

and

$$V_{out} = V_{C\text{-}6} = 0 \text{ V} \tag{8.24}$$

which implies the output is a "1".

If A now changes, to a "0", then Q_5 turns *off*, Q_6 turns *on*, and $I_{C\text{-}3}$, which is approximately equal to I_3, is directed to R_2 instead of R_1. This causes V_{out} to go to a "0". Alternatively, if instead of A going to a "0", we leave A at a "1" but change B to a "0", we see $V_{B\text{-}3}$ is –1.1 V and $V_{B\text{-}4}$ is –0.7 V. This causes Q_3 to turn *off* and Q_4 to turn *on*. Thus, the bias current I_3 is now diverted to the differential pair Q_7, Q_8 rather than Q_5, Q_6, which are now guaranteed to be *off*. Note that the inputs A and $\overline{A}$ are interchanged for differential pair Q_7, Q_8, as opposed to differential pair Q_5, Q_6. So now when A is a "1", I_3 is diverted to Q_8 and V_{out} = "0", whereas if A = "0", V_{out} = "1". Based on the description of the operation of the gate just presented, the truth table shown in Table 8.1 results. On examination of the truth table of this gate, we see V_{out} is a "1" if and only if A or B is a "1", but not when both A and B are "1"s. Thus, $V_{out} = A \oplus B$ and $\overline{V_{out}} = \overline{A \oplus B}$.

In general, we see that *the drive networks of CML gates consist of differential pairs operating as switches only, except the level-shift networks at the gate inputs and outputs; when the input signals change, the drive-network bias current is diverted from one side of a differential pair to the other side.* Any transistors stacked on top *of* a collector having no current are guaranteed to be *off*. Also, all voltage changes are small at 0.4 V, only. Thus, the gate is based on switching currents rather than voltages, hence the name current-mode logic. *Since voltage changes on the nodes are minimized, the charging and discharging of capacitances are minimized and the speed of the gate is optimized!* This is one of the major reasons for the high speed of CML gates.

Table 8.1 The Truth Table of the CML Gate of Fig. 8.9

A	B	V_{out}
0	0	0
0	1	1
1	0	1
1	1	0

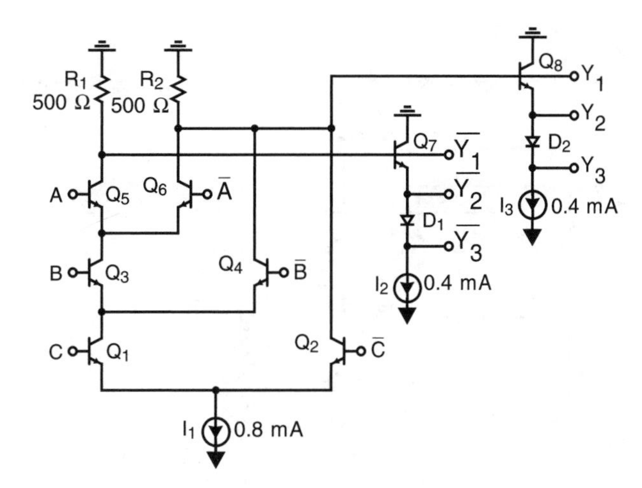

Figure 8.10 A general-purpose CML gate that can realize the *and, nand, or,* and *nor* functions.

This general configuration of an *exclusive-or* circuit has become quite popular recently. Indeed, similar, fully differential CMOS *exclusive-or* gates, based on the bipolar CML gate, have also been gaining in popularity.

Another example of a CML gate is shown in Fig. 8.10. This is a general-purpose three-input gate, where it is assumed each input is at a different common-mode level. It has been assumed that any required level shifts of the input signal have been provided for by previous circuits. Also, this gate has three outputs: Y_1 and $\overline{Y_1}$ at the top voltage level (–0.4 or 0 V), Y_2 and $\overline{Y_2}$ at the second voltage level (–1.1 or –0.7 V), and Y_3 and $\overline{Y_3}$ at the third voltage level (–1.8 or –1.4 V). Note that the outputs at the second and third levels are buffered by the emitter followers (Q_7 and Q_8), whereas the outputs at the top level are unbuffered. This is one reason why designers may decide to never use the top level. Also, if outputs at the fourth level are required, additional diodes in series with D_1 and D_2 could be included. These would be inserted between the existing diodes and the current sources labeled I_2 and I_3.

The gate shown in Fig. 8.10 is the most commonly used fully differential CML gate. It is a general-purpose gate that can be used to realize the *and, nand, or,* or *nor* functions. To see this, first consider when A, B, and C are all "1"s. In this case, I_1 is diverted to R_1, $\overline{Y_1}$ is low, and Y_1 is high. If any one or more of the inputs changes to a "0", then the drive-network bias current, I_1, will be *diverted* to R_2 and the differential output will be "0". Thus,

$$Y_1 = A \cdot B \cdot C \tag{8.25}$$

and the gate realizes the *and* function. To realize the *nand* function, the outputs are simply interchanged. Thus, the collector voltage of Q_5 becomes Y_1 and the collector voltage of Q_6 becomes $\overline{Y_1}$, and $Y_1 = \overline{ABC}$. Next, consider when the inputs are interchanged rather than the outputs. Thus, A goes to Q_6, $\overline{A}$ goes to Q_5, B goes to Q_4, $\overline{B}$ goes to Q_3, etc. In this case, we have

$$Y_1 = \overline{A} \cdot \overline{B} \cdot \overline{C} = \overline{A + B + C} \tag{8.26}$$

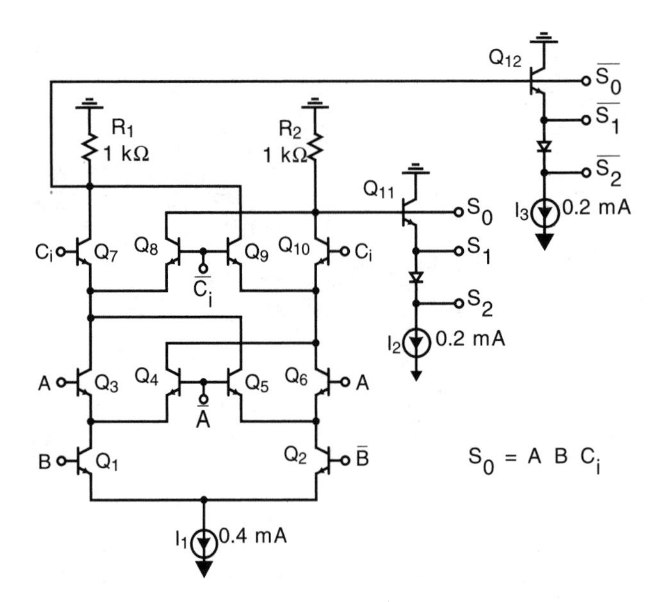

Figure 8.11 A CML gate for realizing the sum-generate function of a full adder.

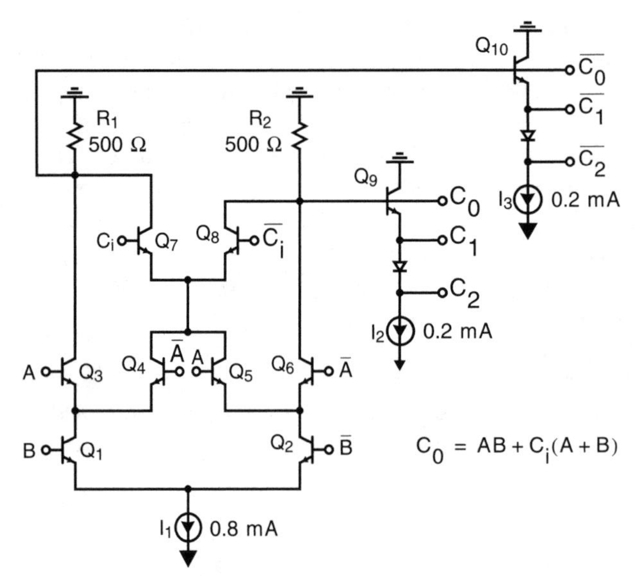

Figure 8.12 A CML gate for realizing the carry-generate function of a full adder.

by *de Morgan's theorem.* Thus, the gate realizes the *nor* function. To realize the *or* function, the outputs are again simply interchanged. Another interesting feature of this gate is how collectors at a low level, such as the collector of Q_2, can be connected directly to resistor loads; *level-shift circuitry is needed only for level shifting down in voltage, not for level shifting up in voltage.*

Two additional examples of CML gates are shown in Figs. 8.11 and 8.12. These logic gates, together, can be used to realize a *full adder.* The gate of Fig. 8.11 is a three-input *exclusive-or* gate based on principles similar to the two-input *exclusive-or* gate of Fig. 8.9, but with the level-shift emitter followers placed at the gate outputs rather than at the gate inputs. This gate realizes the *sum-generate function*, that is,

$$S_o = A \oplus B \oplus C_i \tag{8.27}$$

The gate in Fig. 8.12 realizes the *carry-generate function*,[2] that is,

$$C_o = AB + C_i(A + B) \tag{8.28}$$

Note, that except for the level-shifting networks on the outputs and the load resistors, both gates are composed of differential pairs only that are used to divert the bias current in one direction or another. Also, note again that sometimes a collector is connected to a level higher than the one immediately above it. An example of this is in Fig. 8.12 where the collectors of Q_3 and Q_6, transistors at the second level, completely bypass the first level and are connected directly to the load resistors.

The *full adder* just presented is one of the fastest full-adder realizations available for a number of reasons. Two of these reasons we have already mentioned; the transistors never saturate and the voltage swings are small. A third reason is that the *critical-path delay* through the full adder is only the delay of a *single* gate. This is possible only when fully differential logic is used, whereby inversions can be obtained simply by interchanging the differential signal wires. Furthermore, the delay for inputs at the top level, in this case *carry-in* inputs, is almost the same as that of a simple *fully differential inverter.* This makes the *carry-generation time* very small, which in turn allows for very fast *ripple-carry adders.* To further optimize the delay of the *carry-generate circuit,* it has been biased at a higher current level than the *sum-generate circuit* in the examples shown.

One item worth mentioning is that the loading of outputs taken at the top level should be minimized, as this output is unbuffered. For example, this can be achieved by using the full adders just presented to realize a large ripple-carry adder. In this example, the ripple-carry time is critical. Thus, the load on the carry-generate circuits should be minimized. The output of the carry-generate circuit is normally fed to inputs of the carry-generate and sum-generate circuits of the next stage. If it is fed to the bottom-level inputs of the non-time-critical sum-generate stage, then this stage will be buffered from the carry-generate outputs and only a single differential pair (Q_1, Q_2) of the sum-generate circuit need be driven. When the carry-generate circuit is fed to the next carry-generate circuit, it must be fed into the top level, as this delay is critical and the delay through a CML gate is shortest from the top level inputs. In general *the delay increases for inputs at lower levels; critical-path inputs should always be connected to the top level of a CML gate.*

[2]This circuit was independently invented by the author in 1984 while consulting for Hughes Aircraft Research Laboratories, as well as by many others.

One final rather subtle point should be mentioned regarding CML gates: an additional reason for not using the top level for connecting signals between gates is that it helps minimize the collector-base capacitances of top-level transistors. This is the case because the collector-base bias voltages of the highest level differential-pair transistors are increased if the top level is not used. These larger bias voltages make the depletion collector-base capacitances smaller because of the nonlinear voltage-–capacitance relationship of these depletion capacitances. This helps to minimize the parasitic collector-base junction capacitances at the high-impedance resistor-load nodes. This is an important consideration that may be better appreciated after reviewing the section on depletion capacitances in Chapter 3.

Example 8.2

Give a fully differential CML realization of a single gate that realizes the function Y = AB + CD. It is not necessary to show the level-shifting circuits.

Solution: A CML realization of the required function is shown in Fig. 8.13. This gate is based on combining the drive networks of two *and* gates.

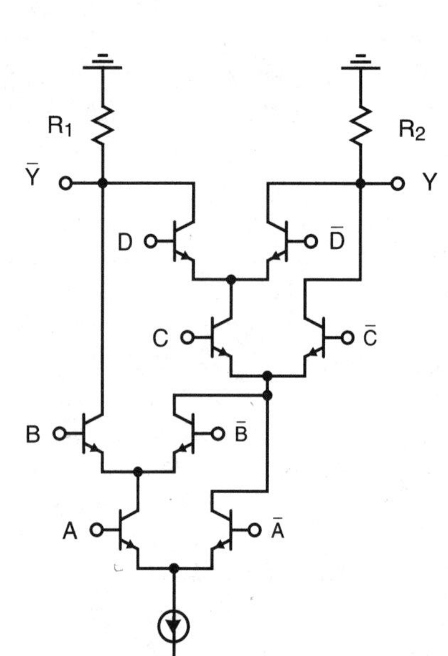

Figure 8.13 A fully differential realization of the function Y = AB + CD.

Example 8.3

Using SPICE, find the propagation delay of the carry-generate circuit shown in Fig. 8.12. Repeat for an input at each of the three levels. This should be done assuming three circuits are connected as a ring oscillator, with all of the interconnections being at the same level. Also, assume the lowest level of each circuit has 0.3-pF load capacitors to represent the loading of full-adder circuits.

Solution: The simulation of logic circuits connected as a ring oscillator is one of the best ways to characterize the delay of acyclic logic circuits as it accurately represents the input rise and fall times and the typical loading encountered in actual use. Figure 8.14 shows a possible interconnection at the top level that can be used for the SPICE simulation. Notice that the input voltages at the second level correspond to "0" inputs, whereas the inputs at the third level are all "1"s except for the input to the first carry-generate circuit, which switches from a "0" to a "1". This is used to initialize the simulation. With the B inputs of the first circuit being at the "0" level, the outputs of the first circuit are guaranteed to be "0". This initializes the outputs of the second and third circuits to "0"s as well. When the B inputs of the first circuit go to the "1" voltage levels, then oscillation begins. The loop inversion required to obtain negative feedback, which is needed for the circuits to oscillate, is achieved

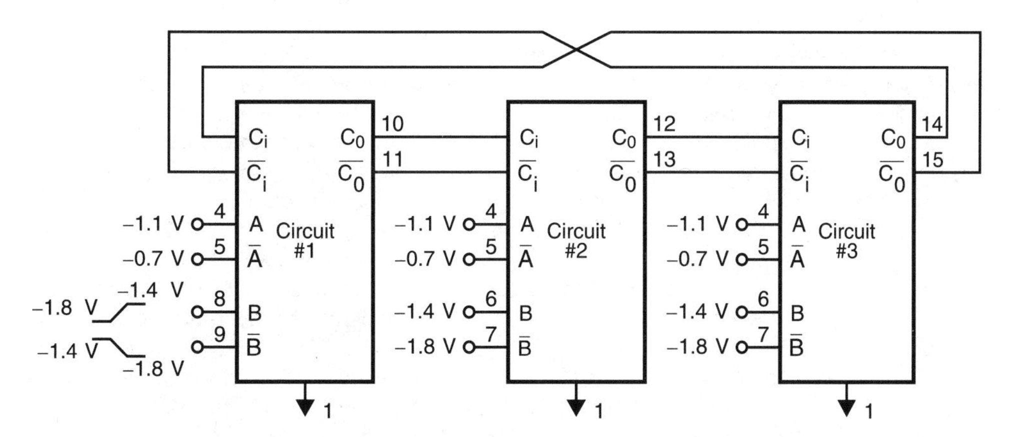

Figure 8.14 Connecting the carry-generate circuits of Fig. 8.12 as a ring oscillator for a SPICE simulation to determine the gate delays.

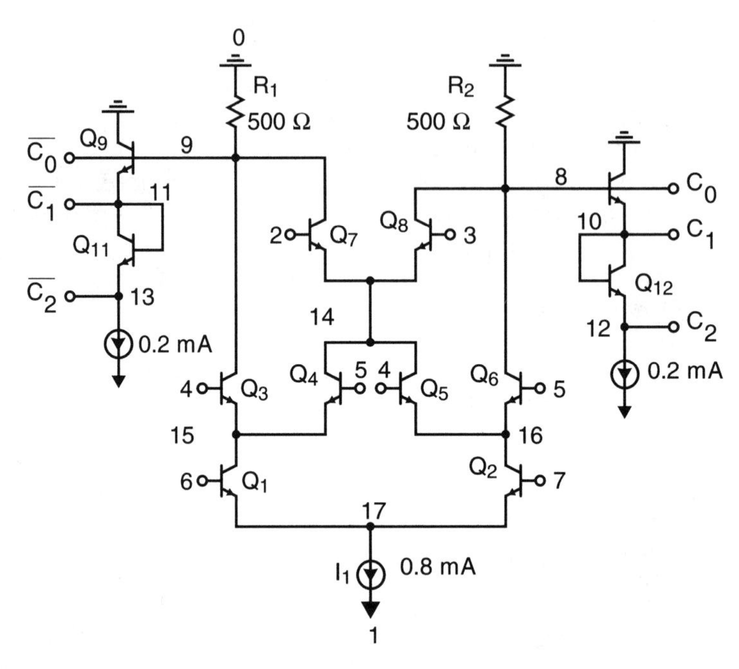

Figure 8.15 The node labels used in a SPICE simulation of the carry-generate circuit.

by interchanging the outputs of the third circuit before feeding them back to the inputs of the first circuit.

This simulation is moderately complicated compared to previous simulations. It is made simpler by using the subcircuit facility available in SPICE. The carry-generate circuit is described in a subcircuit called c_gen. The node labels used for describing this subcircuit in SPICE are shown in Fig. 8.15. Note that the diodes have been implemented as transistors, the way they are actually realized in ICs. A listing of the input file used for the simulation is shown in Fig. 8.36. A plot of the $\overline{C}_0$ output of the first circuit is shown in Fig. 8.16a. The frequency of the oscillation is 778 MHz. Since each period corresponds to six gate delays, the average delay of the circuit is seen to be 214 ps. The outputs of the simulation at the second and third levels are shown in Fig. 8.16b and c, respectively. The frequencies of oscillation are seen to be 400 and 308 MHz, which correspond to gate delays of 416 and 542 ps. It is seen that by interconnecting the carry generation of a ripple-carry adder at the top level, the carry generate is almost twice as fast as would be the case for interconnecting the circuits at the second level, despite the fact that there is no buffering at this level. This important consideration has been overlooked in many commercial ICs. Indeed, there have even been examples of commercial carry-generate circuits interconnected at the third level; this would make the carry generate almost 2.5 times slower than necessary.

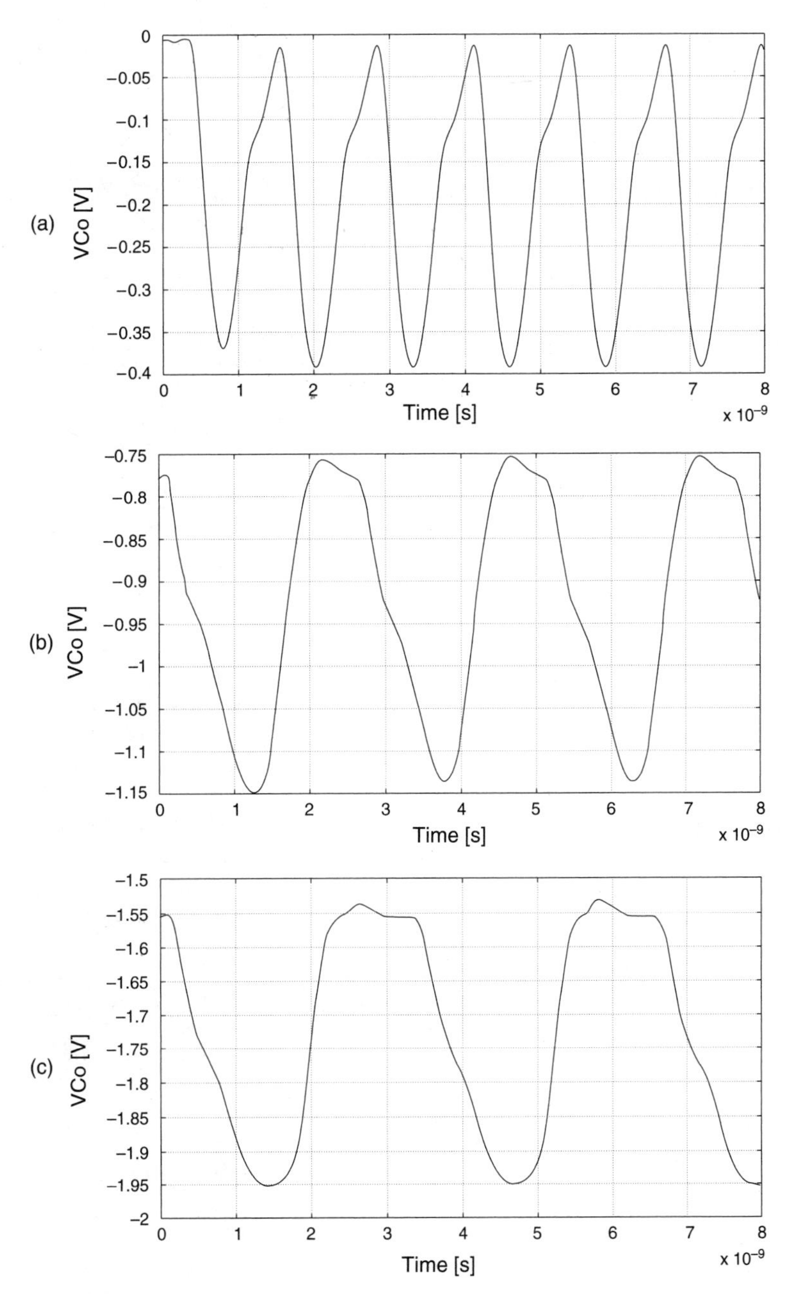

Figure 8.16 The transient response of the carry-generate circuits connected as a ring oscillator at the (a) top level, (b) middle level, and (c) bottom level.

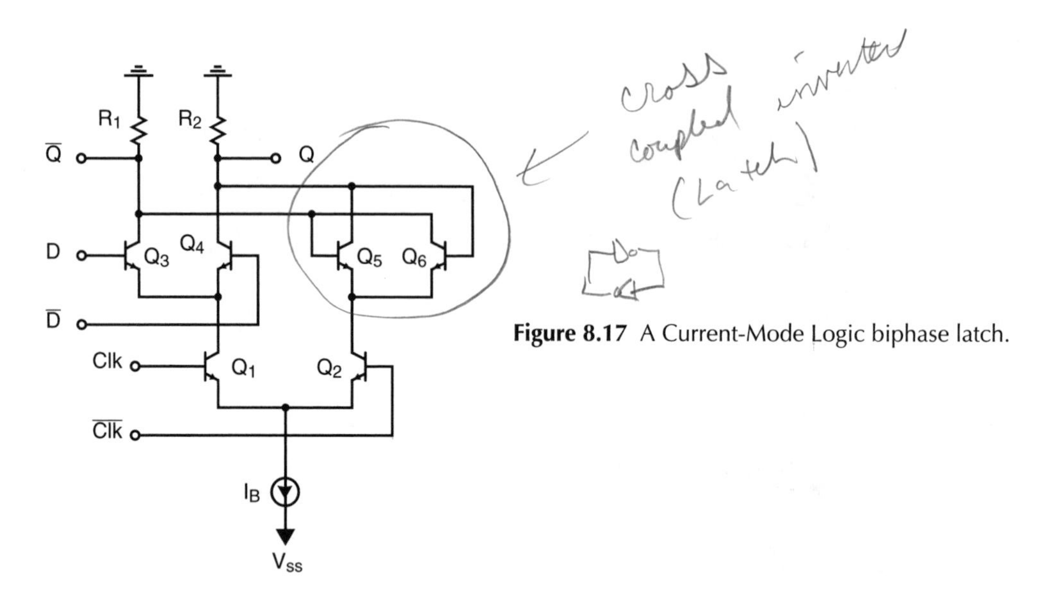

Figure 8.17 A Current-Mode Logic biphase latch.

CURRENT-MODE LOGIC LATCHES

A CML digital latch is shown in Fig. 8.17. This latch is a *biphase latch*, that is, the positive feedback used to store the digital value is only enabled during one phase of the clock, in this case $\overline{\text{Clk}}$. During the phase when Clk is *high*, the latch is in track mode and operates as a simple open-loop differential amplifier.

The operation of the CML latch is straightforward. Consider first when Clk is high. In this case, the bias current is diverted through Q_1 to the differential pair consisting of Q_3 and Q_4; the differential pair consisting of Q_5 and Q_6 is guaranteed to be *cut off* and can be ignored. In this case, Q_3 and Q_4 operate as the input transistors of a differential amplifier; they cause the amplified difference between D and $\overline{\text{D}}$ to appear between Q and $\overline{\text{Q}}$. Thus, Q will be the same logic value as D. Next, when Clk goes low, the differential pair consisting of Q_3 and Q_4 are *shut off* and I_B is *diverted* through Q_5 and Q_6. These are cross-coupled in a positive-feedback loop and thus when Clk is low, the circuit has two stable states, either Q *high*, with Q_6 *on* and Q_5 *off*, or Q *low*, with Q_5 *on* and Q_6 *off*. This positive feedback will retain whatever state the outputs were placed in just before Clk went low.

Example 8.4

Using SPICE estimate the minimum time required for Clk to remain high when the CML latch is changing states. Assume $I_B = 0.2$ mA, $R = 2$ kΩ, and the differential-pair output is loaded by another differential pair biased at 0.2 mA.

The node labeling used for the input to SPICE is shown in Fig. 8.18. Also shown is the additional differential pair used for loading the latch. A listing of the input file used for the simulation is shown in Fig. 8.37. It should be noted that the nodeset command is used to initialize the latch when it is originally in its latch mode with Clk low. SPICE was then run many times with the duty cycle of Clk being decreased each time until the latch stopped changing state. Shown in Fig. 8.19 is a plot of the output nodes 6 and 7 and the emitter nodes 8 and 9 for the smallest duty cycle

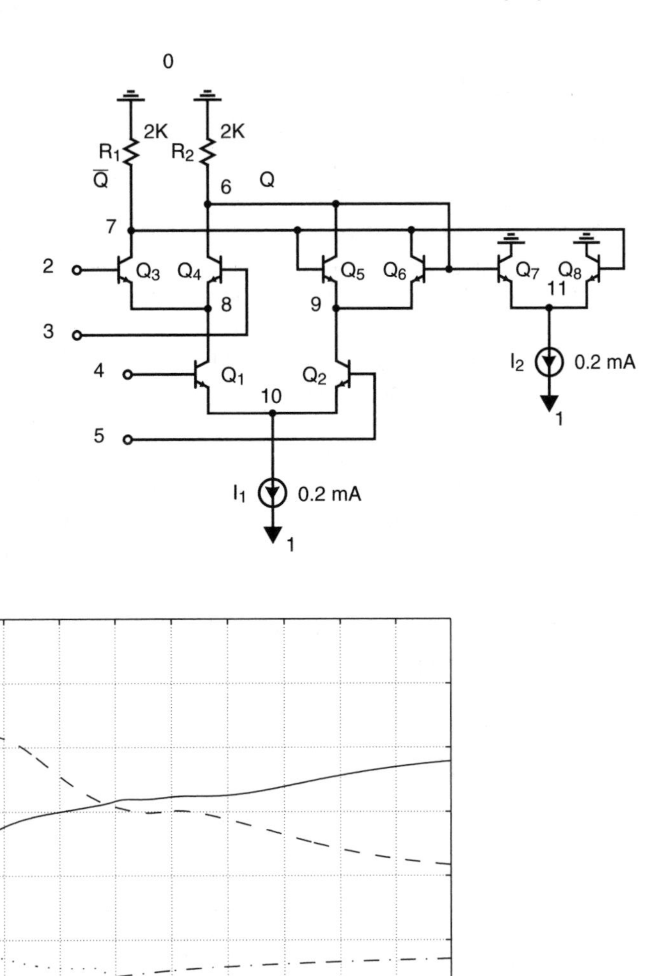

Figure 8.18 The node labeling used for the SPICE simulation of the CML latch.

Figure 8.19 The simulated latch output voltages and emitter voltages for Clk having a 0.7 ns duty cycle.

found in which the latch correctly changed states. It can be seen that it takes about 0.24 ns for Q_3 and Q_4 to turn *on* and another 0.52 ns for node 6 to get larger than node 7. The total minimum 50% duty cycle required for Clk is 0.76 ns, which corresponds to a 1.5-ns clock or a maximum clock frequency of over 660 MHz (for transistor models corresponding to a 10-GHz transistor unity-gain frequency technology).

Current-Mode Logic Flip-Flops

CML flip-flops, like CMOS flip-flops, are typically realized in ICs as *master–slave flip-flops* in which each stage is clocked on the opposite phase. A popular CML D flip-flop is realized this way by cascading two of the latches of Fig. 8.17 as is shown in Fig. 8.20. Often output emitter-follower buffers might also be included when large *fan-outs* are anticipated.

A similar JK flip-flop can be realized by adding an additional differential pair to the master latch and moving the clock inputs to the third level down. In addition, the outputs of the slave stage, after being shifted down to the second level, are fed back to determine which differential input (i.e., J or K) controls the master stage. This is shown in Fig. 8.21. If the flip-flop output is $\overline{Q}$, then the first differential pair with J and $\overline{J}$ controls the master stage; otherwise the second differential pair with K and $\overline{K}$ controls the master latch. Note that this second differential pair has its collectors interchanged, as compared to the first differential pair, in order to realize an inversion. Therefore, when the flip-flop is in the "0" state, the next state is the same as the input J, whereas if the flipflop is in the "1" state, the next state is the same as the inverse of $\overline{K}$, which is the required functionality. Also, note that the flip-flop outputs are now buffered outputs *one level down.*

Differential Current-Mode Logic to Single-Ended Current-Mode Logic

As mentioned previously, one of the major limitations of differential CML is that the *fan-in* or number of inputs a gate can have is limited to a maximum of about four or five. (Gates with five

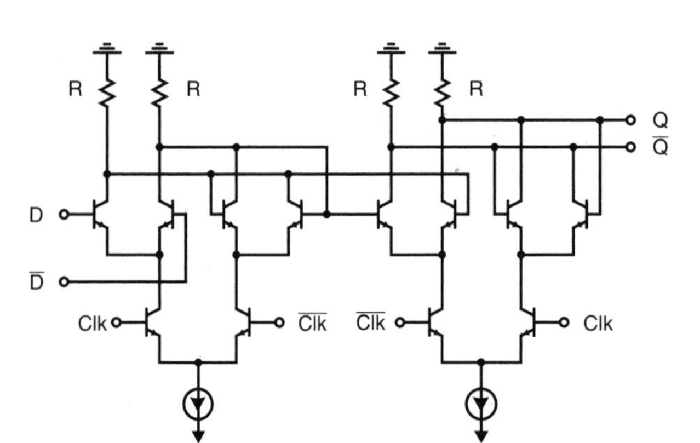

Figure 8.20 A CML D flip-flop.

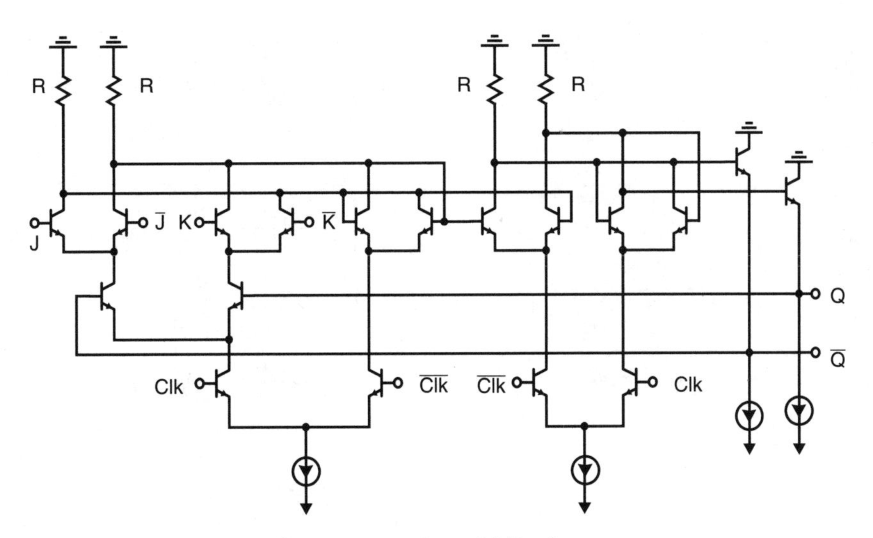

Figure 8.21 A CML JK flip-flop.

inputs would be very slow for input changes at the bottom level.) However, there are some times when gates with more inputs than this are required. Examples of this might be an *address decoder.* When a gate with many inputs is required, a *single-ended input* gate is usually used. To derive signals to drive this gate, the differential output of a CML gate is usually converted to a single-ended signal, with a larger voltage swing, for example, 0.6 V.

The differential-to-single-ended conversion is relatively easy to realize by a CML gate with no inputs at the top level, just using one of the outputs. The larger voltage swing is usually realized by taking the load resistors 50% larger than would normally be the case. This is the reason why inputs at the top level are not permitted; with a voltage swing this large, transistors at the top level could possibly enter the saturation region.

After the differential-to-single-ended conversion, it is possible to use an ECL like gate or, alternatively, a single-ended self-biased gate with hysteresis. An example of this is shown in Fig. 8.22. Assume that all inputs are at the second level down and that the signal swings have been increased to 0.6 V by increasing the gain of the previous stage. This corresponds to a "0" being represented by –1.3 V and a "1" being represented by –0.7 V, which are symmetric about –1V. To see how this gate works, first consider that all the inputs are at low voltages, less than –1.1 V, so that transistors Q_1 to Q_8 are all *off.* Thus, $V_{C\text{-}8}$ will be approximately 0 V and I_1 will be diverted through Q_9 to R_3. Thus, V_{out} will be given by

$$V_{out} = -I_1 R_3 = -0.6 \text{ V} \tag{8.29}$$

Also, we have

$$\begin{aligned} V_{B\text{-}9} &= V_{C\text{-}8} - V_{BE\text{-}10} - I_2 R_2 \\ &= 0 \text{ V} - 0.7 \text{ V} - 0.4(0.5) = -0.9 \text{ V} \end{aligned} \tag{8.30}$$

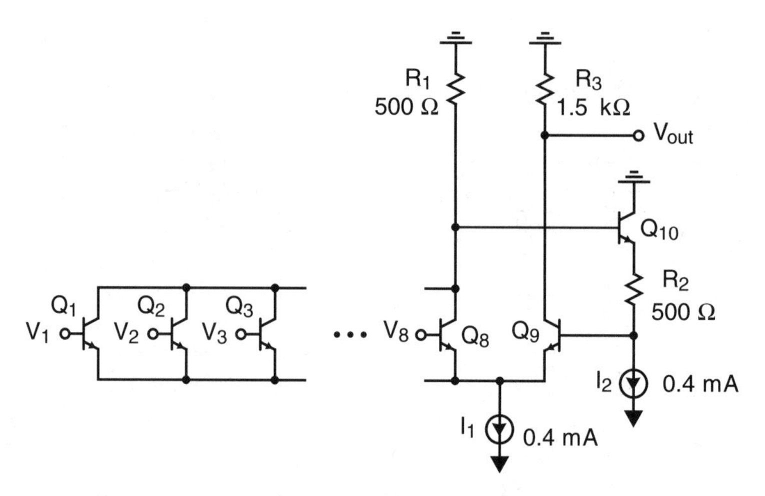

Figure 8.22 An eight-input self-biased gate with hysteresis.

Next, assume one of the inputs, V_1 to V_8, goes to a voltage slightly larger than –0.9 V. This will cause some of I_1 to be diverted through R_1. This will bring $V_{C\text{-}8}$ to a lower voltage, which, in turn, causes $V_{B\text{-}9}$ to go to a lower voltage. This positive feedback decreases the base-emitter voltage of Q_9, which diverts more of I_1 toward R_1. Once enough current is going to R_1 so that the loop gain is larger than 1, the positive feedback causes the gate to quickly change state forcing all of I_1 to go to R_1 and Q_9 to turn *off*. When this happens, we have

$$V_{C\text{-}8} \cong -I_1 R_1 = -0.2\ \text{V} \tag{8.31}$$

and

$$\begin{aligned} V_{B\text{-}9} &= V_{C\text{-}8} - V_{BE\text{-}10} - I_2 R_2 \\ &= -0.2\ \text{V} - 0.7\ \text{V} - 0.4(0.5) = -1.1\ \text{V} \end{aligned} \tag{8.32}$$

Thus, the self-generated reference voltage ($V_{B\text{-}9}$) has now moved from –0.9 V down to –1.1 V. Also, at this time, V_{out} will be at 0 V (i.e., a "1"). To switch V_{out} back to –0.6 V, we need all of the inputs to change to voltages less than –1.1 V. Thus, the gate has a 0.2 V hysteresis; greater than –0.9 V on one input is needed to change to a "1" output; less than –1.1 V on all inputs is needed to change to a "0" output. These switching levels are compatible with CML logic having a 0.6 V swing at the second level. At this level, a "1" is –0.7 V and a "0" is –1.3 V, which are symmetric about –1.0 V, the middle of the hysteresis curve. The transfer function for this circuit is shown in Fig. 8.23. By having hysteresis, we have increased our typical noise margins to 0.4 V, similar to ECL, while keeping our voltage swings at 0.6 V. The hysteresis also makes the gate less sensitive to slowly moving noisy inputs.

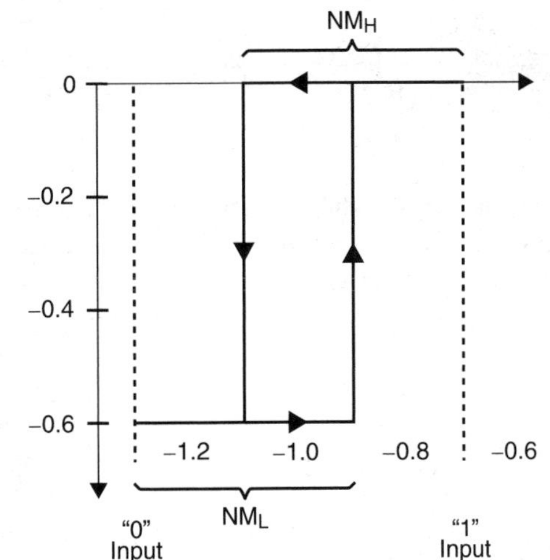

Figure 8.23 The transfer curve of the single-ended CML gate shown in Fig. 8.22.

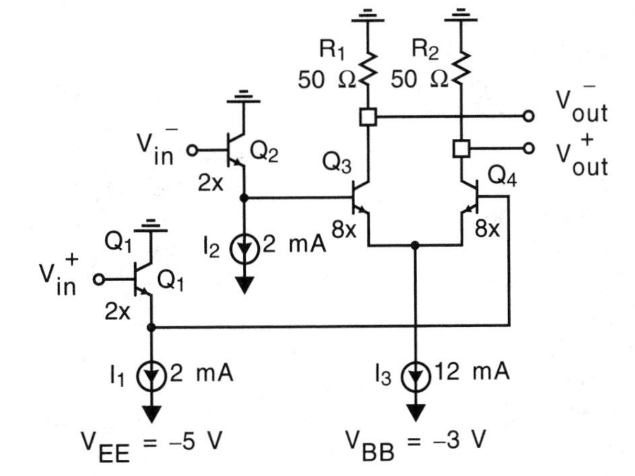

Figure 8.24 A CML output buffer.

CURRENT-MODE LOGIC BUFFERS

Driving off-chip loads at data rates of 100 MHz and above, without introducing serious ringing, is difficult to do, particularly if emitter-follower outputs are used. This problem becomes greatly exacerbated at clock frequencies of 1 GHz and above. At these very high frequencies, emitter-follower outputs are seldom used unless ECL compatibility is required. Increasingly, an open-collector differential pair with off-chip loads is being used as an output buffer. An example of such a buffer is shown in Fig. 8.24. It consists of two emitter followers followed by a differential pair. The emitter followers, Q_1 and Q_2, are biased at larger currents than the typical levels of an on-chip CML gate, in this example at 2 mA. To accommodate the higher currents, they are also realized using larger transistors, perhaps twice as large. The differential pair consisting of Q_3 and Q_4 is biased at even higher levels, in this case at 12 mA. In order to support these current levels, the transistors of the differential pair are taken even larger and would typically consist of a number of transistors connected in parallel. Typical numbers might be four or eight transistors. The collectors of the differential-pair transistors are fed directly to the output pads. The load resistors of the differential pair are off-chip resistors (R_1 and R_2) that would normally be matched to the characteristic impedance of the printed circuit board. In addition, the current source of the differential pair, I_3, would normally be connected to a separate power supply, having its own pin, to minimize the injection of noise from the switching of the output buffer into other logic on the IC. It might also be a smaller voltage to minimize on-chip power dissipation, as shown.

8.3 BiCMOS

We have seen that bipolar logic design is preferable for realizing very fast digital circuits, particularly when large capacitive loads are to be driven. We have also seen that CMOS logic has much less power dissipation, and therefore an IC with a larger number of gates can be included. Recently, processing advances have allowed both bipolar gates and CMOS gates to be included in the same IC. These processes are often called BiCMOS processes (Alvarez, 1989). They allow the majority of the logic gates to be CMOS gates, which minimizes the power dissipation. Any high-speed buffers, precharge circuits, or critical gates can selectively use bipolar transistors to maximize speed.

The use of BiCMOS technology has been increasing, particularly for high-speed circuits. It is presently being used extensively to realize static random-access memories, microprocessors, wireless communication circuits, gate arrays, hard-disk interfaces, and other ICs.

A BiCMOS process (Haveman et al., 1987; Horken et al., 1989) can be developed by extending either an advanced CMOS process or a bipolar process, since, as time goes on, the processes are becoming more and more similar. Most often, CMOS processes have been extended to be able to realize BJT transistors. This might involve an extra 4 or 5 masks for a total of around 15 to 20 masks.

Normally, a BiCMOS process will be an *epitaxial* process. Before the epitaxial layer is grown, the p^- substrate is normally doped n^+ under the regions in which npn transistors and p-channel transistors will reside. It might also be doped p^+ under the regions in which the n-channel transistors will reside. An epitaxial layer will then be grown that might be n or almost intrinsic (or sometimes even p). By having an epitaxial process, latch-up is minimized (which is described in Chapter 3), the series collector resistance of npn transistors is minimized, and the ICs are less sensitive to radiation.

Some other modifications to a modern CMOS process necessary to realize bipolar transistors include the formation of the p-base region, the formation of a deep n^+ region used for connecting the buried collector n^+ region to the surface of the IC, and the formation of the shallow, heavily doped n^+ emitter region. A typical cross section of a BiCMOS process is shown in Fig. 8.25. The actual processing steps used in realizing a modern BiCMOS IC are beyond the scope of this book; the interested reader is referred to Haveman et al. (1987) and Alvarez (1989).

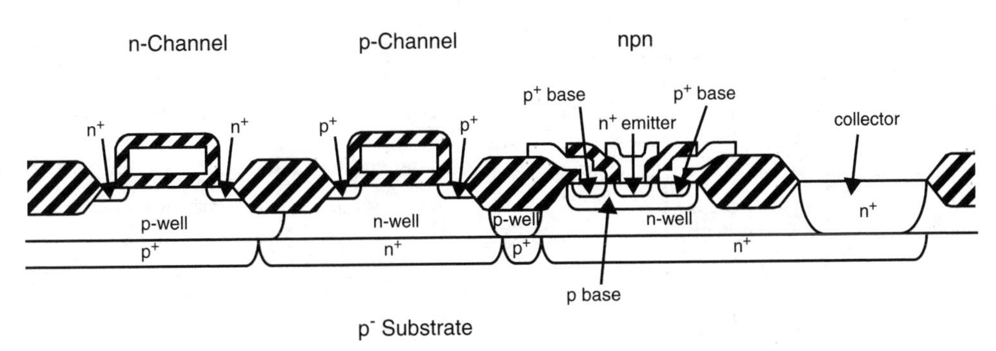

Figure 8.25 Cross-sectional view of a twin-well BiCMOS IC.

A modern BiCMOS process will always have n-channel and p-channel MOS transistors and vertical npn BJTs. Possibly, the process might include resistors, vertical pnp transistors, or even *double-polysilicon* capacitors, although these latter extended processes are not too common due to their expense.

BiCMOS Logic Gates

BiCMOS logic gates are a cross between CMOS logic gates and traditional TTL gates. They normally consist of two stages, with the first stage being used to realize the logic functionality and the second stage being a largely bipolar output buffer. Because of the bipolar output buffers, BiCMOS gates are very fast when driving large capacitive loads. This is often useful in applications such as gate arrays in which the fan-out can be large and there might be large parasitic load capacitances due to long interconnects generated by *auto routers*. Also, in applications in which long buses need to be quickly precharged, the bipolar output buffers can result in much faster operation than would be the case with CMOS logic only. Thus, BiCMOS is becoming popular in applications such as fast static memories and microprocessors. When the fan-out and capacitive loads are small, then BiCMOS gates can be slower than CMOS gates due to their extra complexity. Also, the area of BiCMOS logic gates can be significantly larger than equivalent CMOS gates; *thus, they should be used only when large capacitive loads justify the additional expense.*

An example of a BiCMOS inverter is shown in Fig. 8.26. The included inverter would be a standard CMOS inverter. To understand the operation of the complete BiCMOS inverter, consider first when V_{in} is a "0". In this case, Q_1 will be *off* and can be ignored. Also, the output of the inverter will be a "1", which is equal to V_{DD}, nominally 5 V.[3] This will turn Q_2 on *hard*, which pulls the base of Q_4 to ground turning it *off*. The BJT Q_3 acts as an emitter follower having a 5 V input. Its output will be one V_{BE} drop below this, or around V_{DD} – 0.5 V, assuming

[3]At present, the most popular power-supply voltage for CMOS circuits is 3.3 V, but BiCMOS gates typically still use 5 V power-supply voltages.

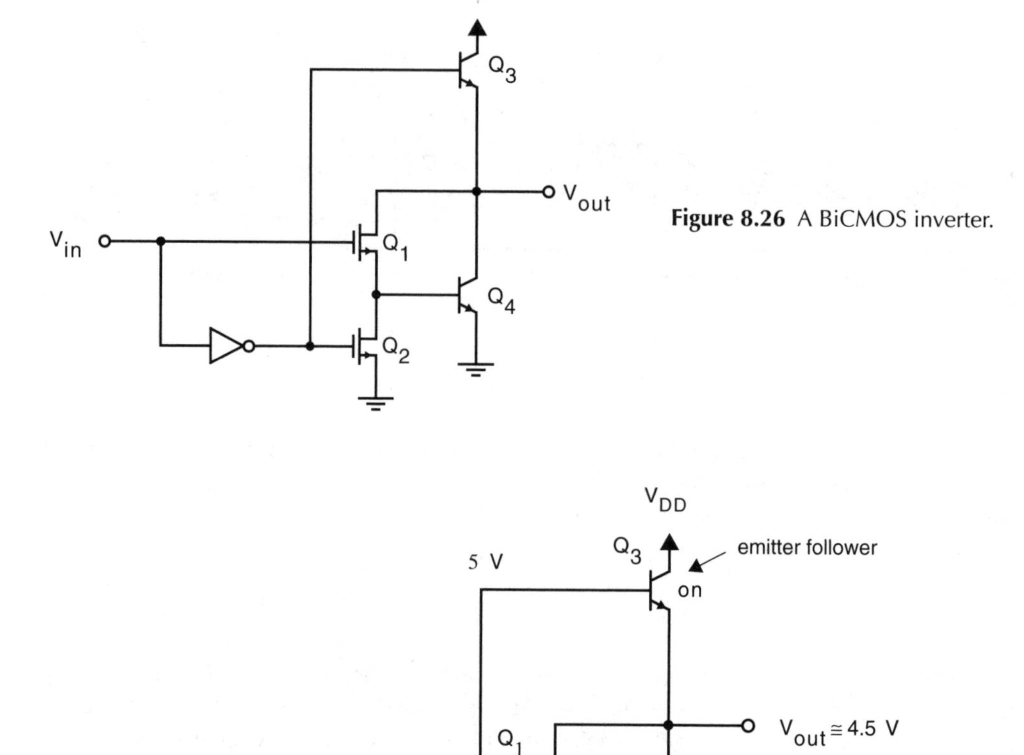

Figure 8.26 A BiCMOS inverter.

Figure 8.27 The voltages of the BiCMOS inverter for V_{in} = "0".

there is little d.c. current required by the load. The various voltages of the nodes of the BiCMOS inverter are shown in Fig. 8.27 for the case when V_{in} is a "0".

Consider next the case when V_{in} goes *high* to 5 V. The output of the inverter will go *low* to approximately 0 V. This will quickly turn Q_2 and Q_3 *off*, and they can be ignored. Also, the gate of Q_1 will be at 5 V, which turns Q_1 *on hard*. This quickly turns BJT Q_4 *on*, as well, causing the output to discharge very quickly. As the output gets *low*, transistor Q_1 enters the *triode region* and is approximately equivalent to a resistor. An approximately equivalent circuit for V_{out} *low* is shown in Fig. 8.28 where all of the *off* transistors have been removed and Q_1 has been replaced by a resistor. Notice that turning Q_1 *on* has effectively connected the base of Q_4 to its collector in a *diode* connection. This quickly discharges the output node as if a *diode* with incremental resistance $1/g_{m4}$ had been placed between the output and ground. As the output

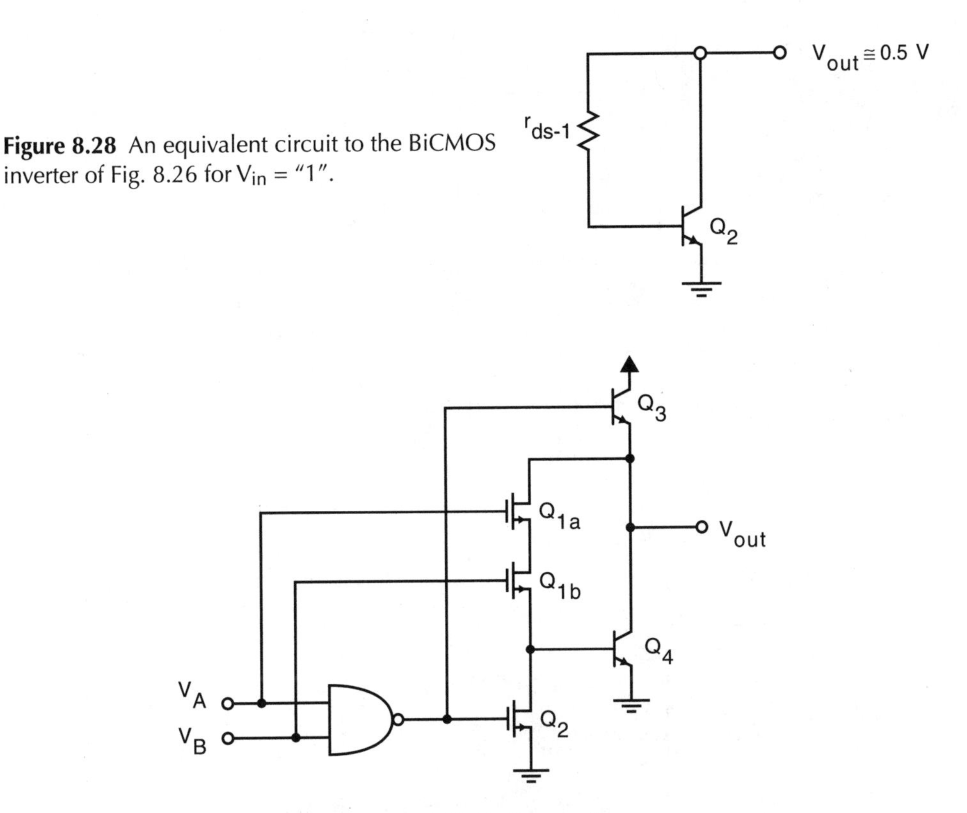

Figure 8.28 An equivalent circuit to the BiCMOS inverter of Fig. 8.26 for V_{in} = "1".

Figure 8.29 A BiCMOS *nand* gate.

gets to around 0.5 V, Q_4 will turn *off* assuming there is no d.c. load current. Thus, Q_4 never enters the *saturation region.*

Next, consider the transition when V_{in} goes back to a "0", Q_1 quickly turns *off* and Q_2 turns *on*. This connects the base of BJT Q_4 to ground, which in turn discharges its *minority base charge* and, in turn, quickly turns Q_4 *off.* Also, during the transition, the cascade of the inverter and the BJT emitter follower, Q_3, quickly charges the output to a "1", even in the presence of a large capacitive load.

It should be noted that the typical output voltage for a "1" is around 4.3 to 4.5 V, whereas a typical output voltage for a "0" is around 0.5 to 0.7 V, depending on whether there is any d.c. output current. Thus, the output voltage swing is less than is the case for traditional CMOS logic circuits. *Also, if the BiCMOS gates feed directly into traditional CMOS gates that have transistors with threshold voltages close to zero, then the CMOS gates may have finite d.c. power dissipation because of the limited voltage swing of the BiCMOS output and subthreshold currents of MOS transistors in traditional CMOS gates.*

A BiCMOS *nand* gate can be realized by replacing the inverter of Fig. 8.26 with a traditional CMOS *nand* gate and replacing Q_1 with two series transistors. The resulting gate is shown in Fig. 8.29. Its operation is similar to the inverter in Fig. 8.26. If both inputs are *high*, the *nand*

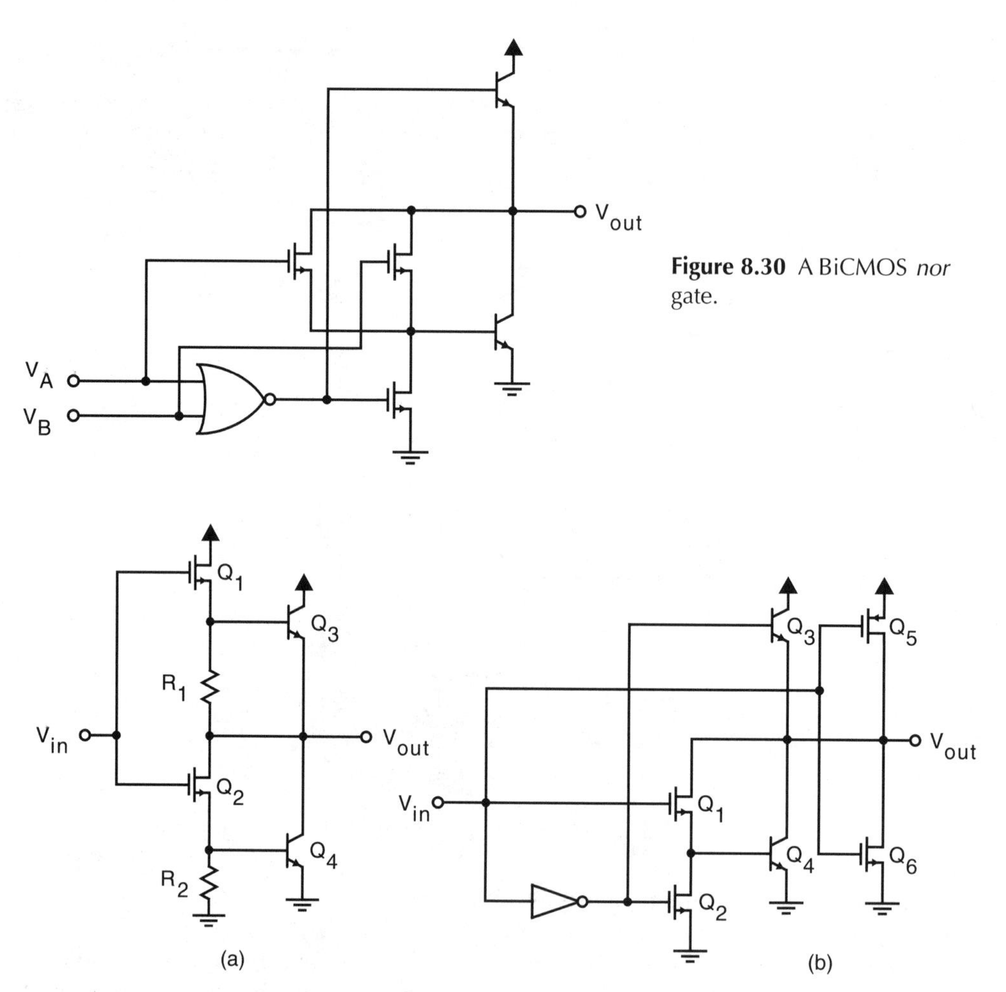

Figure 8.30 A BiCMOS *nor* gate.

Figure 8.31 Full-swing BiCMOS inverters implemented using (a) passive resistors and (b) a parallel output inverter.

gate is *low* turning Q_2 and Q_3 *off*, Q_{1a} and Q_{1b} *on*, and thereby connecting Q_4 as a diode to pull the output *low*. If either input is *low*, then the *nand* gate output will be *high* turning Q_2 and Q_3 *on*, and, therefore, Q_4 *off*. The base-collector connection of Q_4 is broken so it no longer acts as a diode. Q_3 now operates as an emitter follower and pulls the output voltage *high*. A very similar BiCMOS *nor* gate can be realized as shown in Fig. 8.30. It should be noted that the preceding two gates require quite a large area as compared to CMOS gates and should therefore be used only sparingly. Similar to the BiCMOS inverter, these gates have a limited output voltage swing of approximately between 0.5 and 4.5 V.

There have been many modifications proposed for increasing the output voltage swing of BiCMOS logic gates. Figure 8.31a and b show two alternatives that have been proposed. In

Fig. 8.31a, during an output "0" to "1" transition, Q_1 pulls the base of Q_3 *high,* which in turn pulls the output *high.* After the output reaches about 4.3 V, the output charging current from Q_3 quickly decreases, but assuming no d.c. load current, the output will still be slowly *pulled high* by the series connection of Q_1 and R_1. During an output "1" to "0" transition, Q_2 comes *on,* as in the inverter of Fig. 8.26, connecting Q_4 in a diode connection and quickly discharging the output to around 0.7 V. Then, Q_4 turns *off* while the output is still slowly pulled low through the series combination of R_2 and Q_2. Also, during this transition, the base of Q_3 is *pulled low* through R_1.

In the circuit of Fig. 8.31b, a CMOS inverter composed of Q_5 and Q_6 has been added in parallel with output transistors Q_3 and Q_4. During an output transition, most of the output transition current is supplied by either npn transistors Q_3 or Q_5 until the output gets within 0.7 V of ground or V_{DD} and the appropriate BJT transistor begins to turn *off.* The output will then continue to be slowly charged or discharged closer to either 0 or 5 V by either Q_6 or Q_5 depending on the transition involved. In another variation on these full-swing BiCMOS inverters, the resistors R_1 and R_2 of Fig. 8.31a were replaced by transistors biased in the triode region.

For all of these full-swing BiCMOS gates, the additional circuitry takes up more area, yet adds little to the transition speeds. The output voltages will eventually get close to either 0 or 5 V, but the final 0.7 V transition is very slow once the npn output drivers have stopped conducting as the sourcing or sinking capability of the additional circuitry is very limited. The only real advantage is that the d.c. power dissipation caused by subthreshold currents is reduced in succeeding CMOS gates, but this power savings is normally small and probably does not warrant the extra complexity in the author's opinion.

Alternative BiCMOS Approaches and Circuits

The BiCMOS gates presented in the previous section are much faster than CMOS gates when driving large capacitive loads. When a gate has a small load or *fan-out,* the speed gain is very minimal and the additional complexity is seldom warranted. Thus, a general design philosophy for BiCMOS ICs is to use just CMOS circuits for the majority of the gates. BiCMOS gates would be used only for gates that must drive large loads or be very fast. When very fast gates are required, then current-mode logic connected between 0 and 5 V followed by a CML to CMOS converter is recommended. A simple CML to CMOS converter is shown in Fig. 8.32. The differential input signal to the converter should be one or two levels below the top level to ensure input transistors Q_1 and Q_2 never saturate. Q_3 is used to guarantee that common-source transistors Q_4 and Q_5 are *on* all of the time; however, one of these transistors will be conducting a larger current because of the differential voltage developed across resistors R_1 and R_2. The simple current mirror consisting of transistors Q_6 and Q_7 forms an active load that serves the function of a *differential-to-single-ended converter* and amplifier so that the output voltage levels are large enough to drive standard CMOS logic.

The npn BJT transistors can also be used for any circuits that require fast precharging. This is particularly useful for precharging buses in memory circuits, microprocessors, or programmable-logic arrays. In addition, npn differential pairs are very useful as the first stage of sense amplifiers in memory circuits. Finally, npn transistors can be very effective when configured as *common-base amplifiers* to sense small voltage changes on nodes that have large parasitic

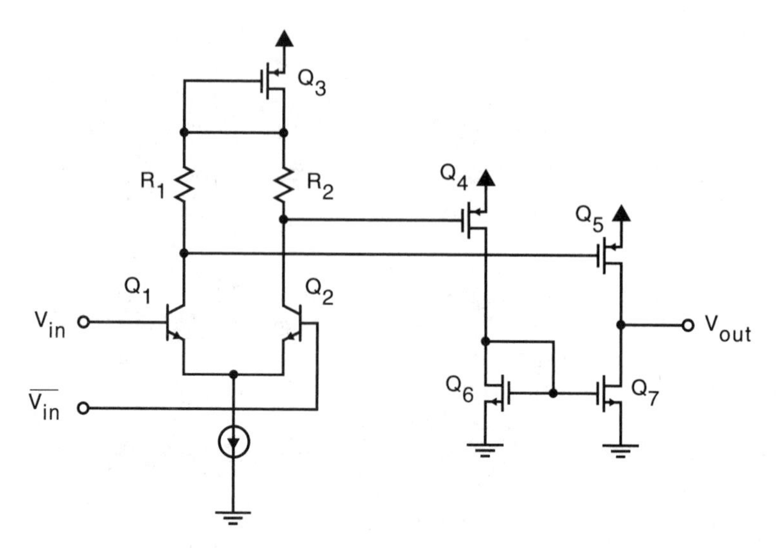

Figure 8.32 A CML to CMOS converter circuit.

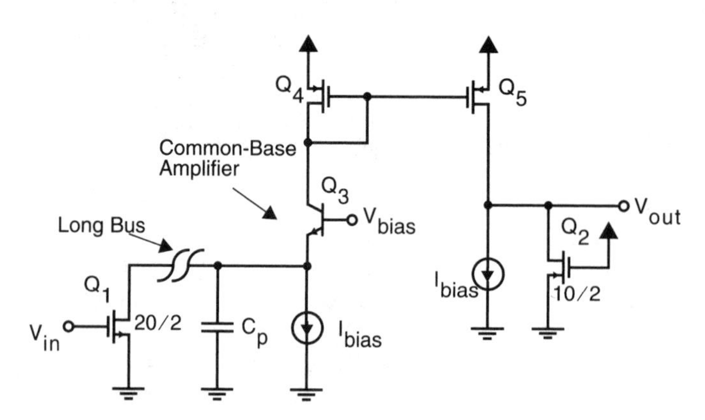

Figure 8.33 Using a BiCMOS common-base amplifier to minimize the effect of the parasitic capacitance of a bus.

capacitances. When configured as common-base stages, they limit the voltage swings across the parasitic capacitances, thereby limiting the effects of the capacitances. An example of this application is shown in Fig. 8.33. Q_1 is connected through a long bus having a large parasitic capacitance to the common-base amplifier composed of Q_3 and I_{bias}. The collector current of Q_3 is mirrored by the simple current mirror composed of Q_4 and Q_5 and compared to a reference current composed of a current source equal to I_{bias} plus the drain current of Q_2, which is

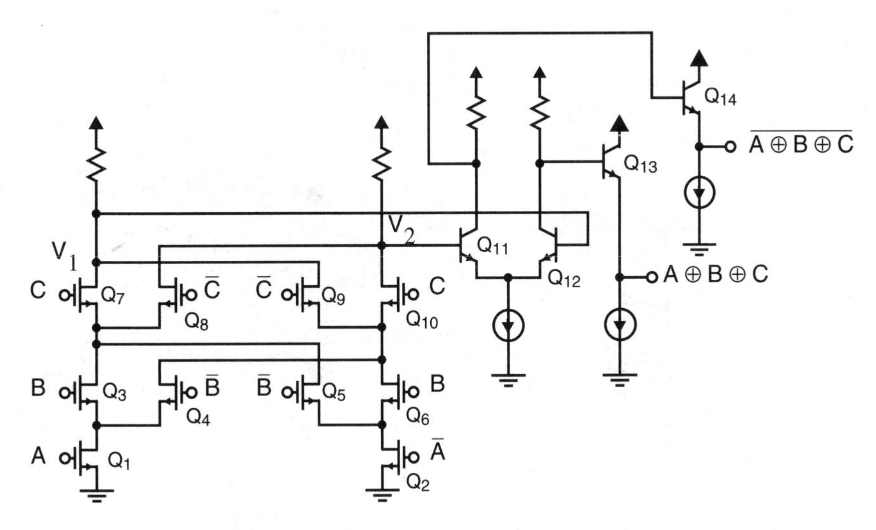

Figure 8.34 A DCVSL three-input *exclusive-or* circuit from Liang et al. (1991).

half the size of Q_1. If V_{in} is a "1", the collector current of Q_3 will be equal to I_{bias} plus the drain current of Q_1. This will be larger than the reference current and V_{out} will be high. Otherwise, if V_{in} = "0", $I_{C\text{-}3} = I_{bias}$ only, which is less than $I_{ref} = I_{bias} + I_{D\text{-}2}$ and V_{out} will be pulled low. When V_{in} changes, the voltage change on the long bus might only be 0.1 V or so and the effect of the large parasitic capacitance of the bus is thereby minimized.

Recently, a number of new types of BiCMOS gates have been proposed. For example, a BiCMOS differential cascade-voltage-switch logic gate (DCVSL) (Liang et al., 1991) is shown in Fig. 8.34. The idea behind this gate is that the transistor sizes in the n-channel differential switch network can be taken relatively small. This results in the differential voltage between V_1 and V_2 being quite small, but this small differential voltage is still easily sensed by the BJT differential pair consisting of Q_{11} and Q_{12} and amplified to CML logic levels. Q_{13} and Q_{14} act as emitter-follower outputs and are optional. The small voltage changes help maximize speed. The circuit also operates as a CMOS to CML converter.

Another BiCMOS alternative is to modify the first stage of the differential gate of Fig. 8.34 to be a differential split-level (DSL) gate by adding the common-base transistors Q_1 and Q_2 shown in Fig. 8.35. These transistors limit the voltage swings in the n-channel differential drive network resulting in smaller delays. This circuit also functions as a CMOS to CML converter.

The alternatives that have been described are only some of the possibilities available with BiCMOS processing. Many other alternatives are being proposed on a regular basis as designers become more familiar with this relatively new technology and learn how to take advantage of its capabilities. It is predicted that in the future, BiCMOS could become a dominant technology for both analog and digital high-speed ICs.

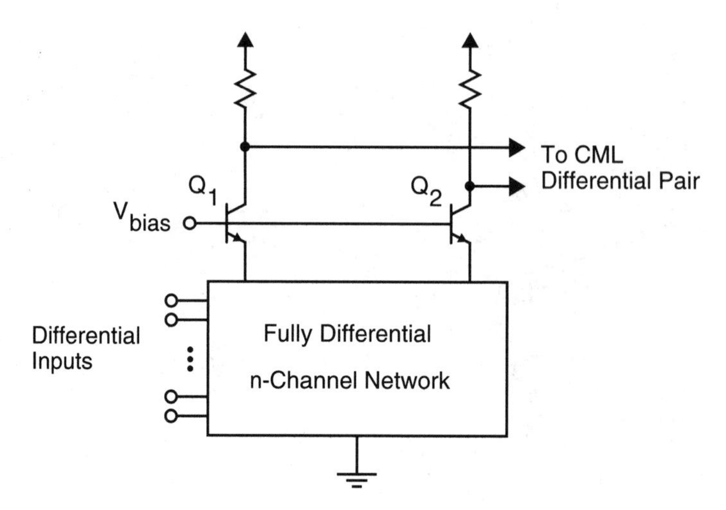

Figure 8.35 A BiCMOS realization of a DSL gate.

8.4 SPICE Simulations

The SPICE input files are shown in Figs. 8.36 and 8.37.

```
* A Carry Generate Circuit connected as a ring oscillator with the carry-
* propagate connected at the top level.
*
.MODEL N1 NPN (BF=80 BR=1 IS=2.1e-17 RB=550 RE=30 RC=55 TF=30p TR=2.4n
+ CJE=28f CJC=23f CJS=49f VJE=0.9 VAF=75)
*
.SUBCKT c_gen 2 3 4 5 6 7 10 11 1
Q1 15 6 17 N1
Q2 16 7 17 N1
Q3 9 4 15 N1
Q4 14 5 15 N1
Q5 14 4 16 N1
Q6 8 5 16 N1
Q7 9 2 14 N1
Q8 8 3 14 N1
Q9 0 9 11 N1
Q10 0 8 10 N1
Q11 11 11 13 N1
Q12 10 10 12 N1
```

Figure 8.36 The input file used to simulate the oscillator composed of CML carry-generate circuits connected at the top level in Example 8.3.

```
*
R1 0 9 500
R2 0 8 500
*
I1 17 1 0.8m
I2 13 1 0.2m
I3 12 1 0.2m
*
Cl1 13 0 0.3p
Cl2 12 0 0.3p
*
.ENDS c_gen
*
X1 15 14 4 5 8 9 10 11 1 c_gen
X2 10 11 4 5 6 7 12 13 1 c_gen
X3 12 13 4 5 6 7 14 15 1 c_gen
*
V2P 4 0 dc -1.1
V2N 5 0 dc -0.7
V3P 6 0 dc -1.4
V3N 7 0 dc -1.8
*
VSS 1 0 -5.0V
V4P 8 0 dc. -1.8 PULSE(-1.8 -1.4 0n 0.1n 0.1n 8.9n 10n)
V4N 9 0 dc. -1.4 PULSE(-1.4 -1.8 0n 0.1n 0.1n 8.9n 10n)
*
.OPTION NOMOD POST INGOLD=2 NUMDGT=6 BRIEF ABSTOL=5e-10 CHGTOL=5e-1
*
.TRAN 0.005n 8n
.PRINT TRAN V(11)
*
.END
```

Figure 8.36 (continued).

```
A CML Latch
*
Q1 8 4 10 N1
Q2 9 5 10 N1
Q3 7 2 8 N1
Q4 6 3 8 N1
Q5 6 7 9 N1
Q6 7 6 9 N1
Q7 0 6 11 N1
Q8 0 7 11 N1
*
R1 0 7 2k
R2 0 6 2k
```

Figure 8.37 The SPICE input file used to simulate the CML latch of Fig. 8.18.

```
*
I1 10 1 0.2m
I2 11 1 0.2m
*
V1P 2 0 dc. 0
V1N 3 0 dc -0.4
V2P 4 0 dc. -1.1 PULSE(-1.1 -0.7 0n 0.05n 0.05n 0.70n 2n)
V2N 5 0 dc -0.7 PULSE(-0.7 -1.1 0n 0.05n 0.05n 0.70n 2n)
*
VSS 1 0 -3.3V
*
.NODESET V(6)=-0.4 V(7)=0.0 V(9)=-0.8
*
.MODEL N1 NPN ( BF=80 BR=1 IS=2.1e-17 RB=550 RE=30 RC=55 TF=30p
+ TR=2.4n CJE=28f CJC=23f CJS=49f VJE=0.9 VAF=75 )
*
.OPTION NOMOD POST INGOLD=2 NUMDGT=6 BRIEF
.TRAN 0.005n 2.0n
.PRINT TRAN V(6) V(7) V(8) V(9)
  .END
```

Figure 8.37 (continued).

8.5 Bibliography

A. Alvarez, ed., *BiCMOS Technology and Applications,* Kluwer, 1989.

P. Gray and R. Meyer, *Analog Integrated Circuits*, 3rd ed., John Wiley & Sons, 1993.

R. Haveman, R. Ekand, R. Haken, D. Scott, H. Tran, P. Fung, T. Ham, D. Farreau, and R. Virkus, "An 0.8 µmf 256 K BiCMOS SRAM technology," *Digest of Technical Papers, 1987, International Electron Devices Meeting*, 841–843, December 1987.

R.A. Horken, et al., "BiCMOS Process Technology," in *BiCMOS Technology and Applications*, A. Alvarez, ed., Kluwer, 1989.

S. Liang, D.H.K. Hoe, and C.A.T. Salama, "BiCMOS DCVSL Gate," *Electronics Letters,* 27 (4), 346–347, February 1991.

F. Pelayo, A. Prieto, A. Lloris, and J. Ortega, "CMOS Current-Mode Multilevel PLAs," *IEEE Transactions on Circuits and Systems*, 38 (4) 434–441, April 1991.

H. Taub and D. Schilling, *Digital Integrated Electronics*, McGraw-Hill, 1977.

N. Weste and K. Eshraghian, *Principles of CMOS VLSI Design,* Addison-Wesley, 1985.

8.6 Problems

For the problems in this chapter and all future chapters, assume the following transistor parameters:

- npn bipolar transistors:

  $\beta = 100$

  $V_A = 80$ V

  $\tau_b = 13$ ps

$\tau_s = 4$ ns
$r_b = 330\ \Omega$

- n-channel MOS transistors:

$\mu_n C_{ox} = 190\ \mu A/V^2$
$V_{tn} = 0.7$ V
$\gamma = 0.6\ V^{1/2}$
$r_{ds}\ (\Omega) = 5000L\ (\mu m)/I_D\ (mA)$ in active region
$C_j = 5 \times 10^{-4}\ pF/(\mu m)^2$
$C_{j\text{-}sw} = 2.0 \times 10^{-4}\ pF/\mu m$
$C_{ox} = 3.4 \times 10^{-3}\ pF/(\mu m)^2$
$C_{gs(overlap)} = C_{gd(overlap)} = 2.0 \times 10^{-4}\ pF/\mu m$

- p-channel MOS transistors:

$\mu_p C_{ox} = 50\ \mu A/V^2$
$V_{tp} = -0.8$ V
$\gamma = 0.7\ V^{1/2}$
$r_{ds}\ (\Omega) = 6000L\ (\mu m)/I_D\ (mA)$ in active region
$C_j = 6 \times 10^{-4}\ pF/(\mu m)^2$
$C_{j\text{-}sw} = 2.5 \times 10^{-4}\ pF/\mu m$
$C_{ox} = 3.4 \times 10^{-3}\ pF/(\mu m)^2$
$C_{gs(overlap)} = C_{gd(overlap)} = 2.0 \times 10^{-4}\ pF/\mu m$

8.1 For the bipolar differential pair, using (8.6) and (8.7), show that

$$I_{C\text{-}1} - I_{C\text{-}2} = \frac{e^{V_{id}/2V_T} - e^{V_{id}/2V_T}}{e^{V_{id}/2V_T} + e^{V_{id}/2V_T}} = \tanh\left(\frac{V_{id}}{2V_T}\right) \quad (8.33)$$

8.2 For a bipolar differential pair having a 40-mV differential input voltage, what is the ratio $I_{C\text{-}1}/I_{C\text{-}2}$?

8.3 For the ECL gate shown in Fig. 8.4, assume $I_B = 0.2$ mA, $R_1 = 4\ k\Omega$, $R_2 = R_3 = 8\ k\Omega$. Find the typical output high and low voltages (V_{OH} and V_{OL}), the gate threshold voltage (V_{TH}), and the d.c. power dissipation. Assume finite base currents are negligible and assume $V_{EE} = -5.2$ V.

8.4 Repeat Problem (8.3), but do not assume the finite base currents are negligible (assume $\beta = 100$).

8.5 What is the output impedance of the ECL gate shown in Fig. 8.4 for component values given in Problem (8.3)?

8.6 For the CML gate shown in Fig. 8.9, assume R_1 = 2 kΩ, $I_1 = I_2$ = 0.4 mA, and I_3 = 0.25 V. Find the typical output high and low voltages (V_{OH} and V_{OL}), and the total power dissipation. Assume the negative power-supply (V_{EE}) = –3.3 V.

8.7 Extend the general-purpose gate shown in Fig. 8.10 to a four-input gate.

8.8 Give a logic gate diagram that realizes the same logic function as the bipolar gate shown in Fig. P8.8. Assume $V_{B\text{-}1}$ = –0.3 V, $V_{B\text{-}2}$ = –1.1 V, and $V_{B\text{-}3}$ = –1.9 V, and that the input levels are either 0 or –0.6 V.

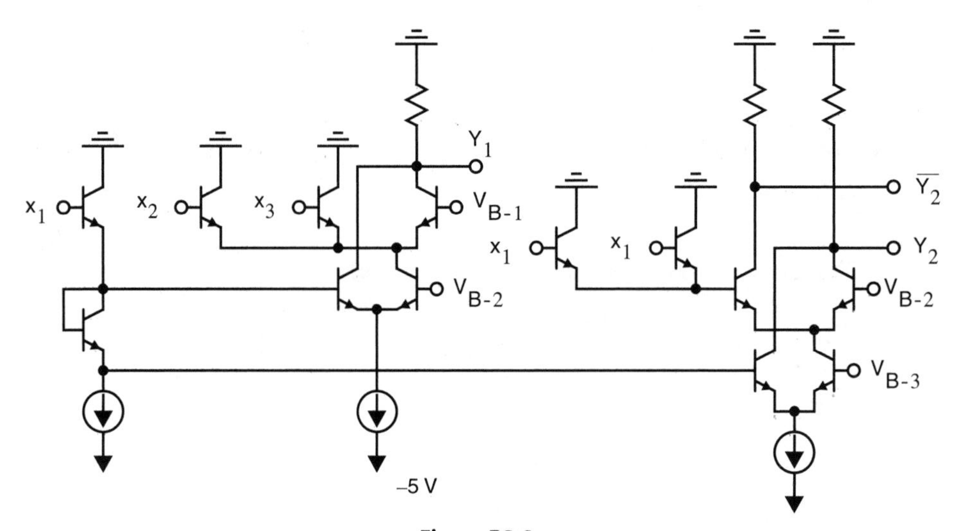

Figure P8.8

8.9 Design a CML circuit that uses only differential signals and realizes the function $Y = \overline{x_1(x_2x_3 + \overline{x_4})}$

8.10 Assuming the gate designed in Problem 8.7 has current sources requiring at least 0.5 V across them in order to operate adequately, what is the minimum power-supply voltage that is necessary for proper operation? Assume $V_{BE\text{-}on}$ = 0.8 V.

8.11 Using SPICE, simulate the gate designed in Problem 8.7 for the delay from the input voltage change to when the differential output voltage changes sign for each of the inputs. Use a 50 ps rise and fall time for the inputs.

8.12 Using CML, design a four-bit circuit that compares A and B inputs and outputs a "1" if A > B assuming both A and B are positive binary integers. Give reasonable values for all components.

8.13 Design a CML multiplexor where the output is equal to one of eight possible inputs dependent on three address inputs.

8.14 Design a CML static RAM memory cell.

8.15 Design a CML divide-by-six circuit where the output changes state once every three periods of a single input signal. Make sure the levels that connect the different blocks are correct.

8.16 What is the power dissipation and output logic levels of the CML output buffer shown in Fig. 8.24?

8.17 Modify the CML *single-ended to differential converter* of Fig. 8.22 so that it is compatible with input signals having 0.5 V signal swings and so that the hysteresis is 0.1 V.

CHAPTER 9

Advanced CMOS Logic Design

Up to this point, we have seen a number of techniques for realizing digital integrated logic circuits using both MOS and bipolar technologies. This chapter will describe many of the advanced logic design techniques that have become popular recently for state-of-the-art CMOS ICs. These techniques have been developed in response to the never-ending goal of increasing speed while minimizing silicon area and power dissipation. Most, but not all, of these techniques involve dynamic circuits of one type or another. Some of the techniques have already been introduced as examples, but, in many cases, they will be described again for completeness. As always, rather than trying to present a review of all of the design techniques that have been used, the principles involved will be emphasized with some of the more important techniques used as examples.

9.1 Pseudo-NMOS and Dynamic Precharging

As has been seen, traditional CMOS logic design normally requires as many p-channel transistors as n-channel transistors. This requires considerable area, particularly if the p-channel transistors are chosen to be wider than the n-channel transistors. In addition, due to the lower mobility of the p-channel transistors, the "0" to "1" transition of the logic gates may be slow if there are a number of series p-channel transistors between the output and the positive power supply. In response to these problems, modern CMOS design strives to minimize the number of p-channel transistors, particularly series p-channel transistors. Not only does this save on area, but it also minimizes the gate-load capacitance due to succeeding gates. Two techniques that are often used to achieve these goals involve using pseudo-NMOS loads or dynamic loads.

We have already seen many examples of how pseudo-NMOS loads are used, similar to the use of depletion transistor loads in an NMOS technology. One simply uses an n-channel drive network with a p-channel load transistor. The p-channel load transistor has its gate connected to ground, as is shown in Fig. 9.1.

When using pseudo-NMOS loads, it is necessary for the n-channel drive network to be able to overcome the p-channel load transistor. This limits the width of the p-channel load transistor. The load transistor width should be no greater than the smallest width of an n-channel transistor *equivalent to* the n-channel drive network when it is conducting similar to the size ratio shown in Fig. 9.1a. This relies on the mobility difference between the n-channel and p-channel transistors for obtaining a low enough voltage for a typical "0" output. This is an upper bound on the p-channel width, but it is not an optimum choice for two reasons. First, the typical output "0" voltage can often be quite large for this choice, sometimes almost 1 V. Second, the gate-threshold voltages are also high, often over 2 V for a 3.3-V power supply. A better choice is to take the p-channel load one-half the worst-case size (i.e., smallest size) of the equivalent n-channel transistor similar to the size ratio shown in Fig. 9.1b. This gives a gate-threshold voltage closer to $V_{DD}/2$ and also gives a smaller output "0" voltage, typically less than 0.3 V.

Figure 9.2 shows some d.c. transfer curves of pseudo-NMOS inverters, shown in Fig. 9.1, for the cases in which the p-channel load is equal to the equivalent n-channel drive transistor

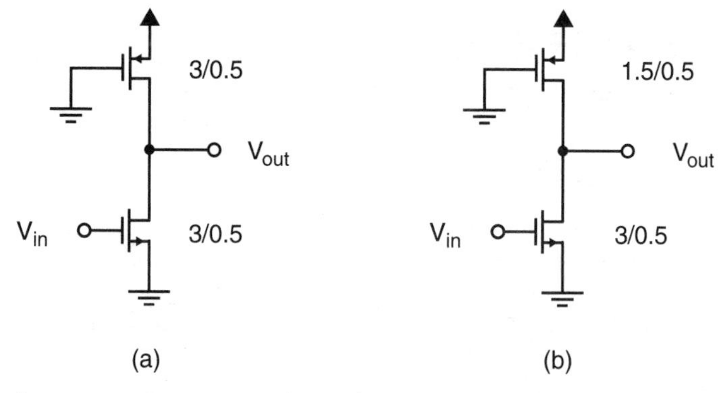

Figure 9.1 Some commonly used sizes for pseudo-NMOS inverters.

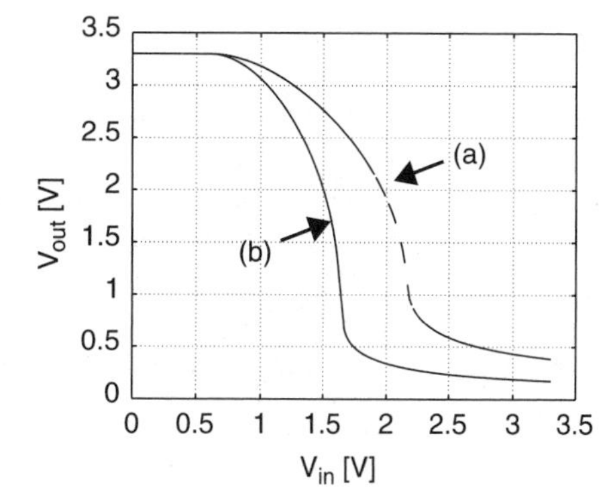

Figure 9.2 The transfer curves of the pseudo-NMOS inverters of (a) Fig. 9.1a with $W_p=W_n$ and (b) Fig. 9.1b with $W_p=W_n/2$.

(Fig. 9.2a) and for the case in which the load transistor is one-half the size of the equivalent transistor (Fig. 9.2b). The superior transfer curve for the latter case is evident.

One of the problems with pseudo-NMOS logic is that the gain at the gate-threshold voltage can sometimes be limited. This is because for V_{out} greater than $|V_{Tp}|$, or about 0.7 V, the p-channel load transistor will be in the triode region. This guarantees that the p-channel load is in the triode region when the output voltage is at the gate-threshold voltage. A variation of pseudo-NMOS logic that does not have this problem has the gate of the p-channel load connected to a voltage-reference circuit rather than ground, as is shown in Fig. 9.3. This generates a bias voltage that guarantees that the p-channel transistor is in the active region when the inverter output voltage is at the gate-threshold voltage. Note that this variation has the p-channel load transistor twice as wide as the equivalent n-channel driver transistor. Note also that one bias generation circuit can be used for a number of gates. The transfer curve for this inverter is shown in Fig. 9.4. The larger gain at the gate-threshold voltage is evident.

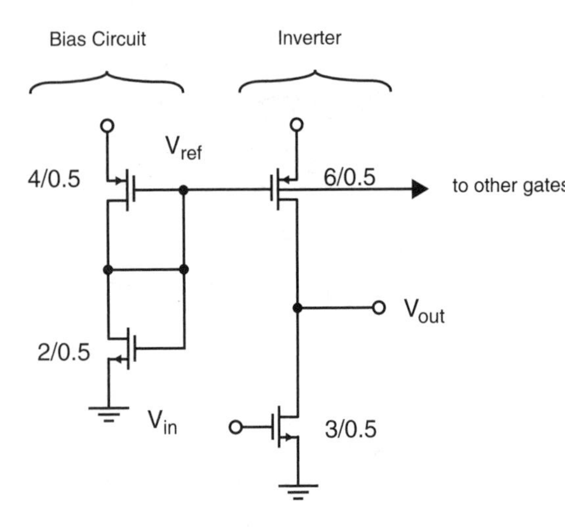

Figure 9.3 A variation of pseudo-NMOS that has larger gain when the output voltage is at the gate-threshold voltage.

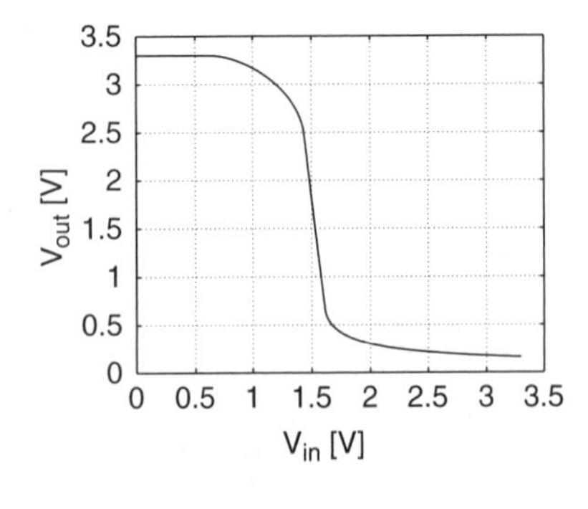

Figure 9.4 The transfer curve of the pseudo-NMOS variant shown in Fig.9.3.

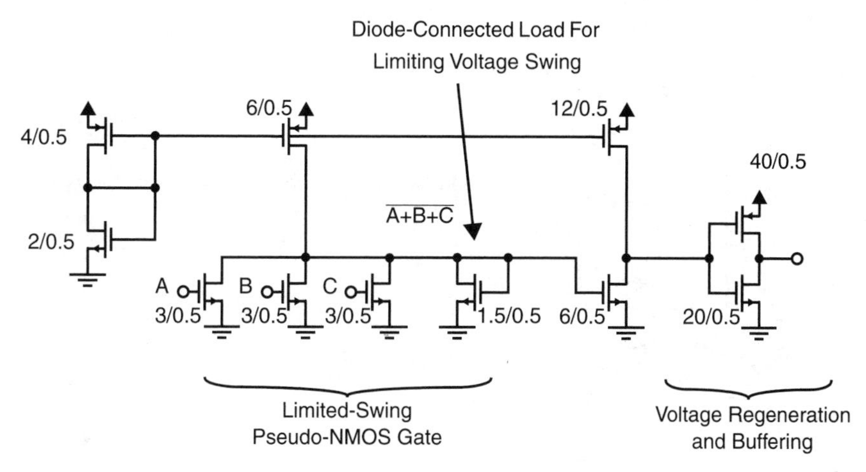

Figure 9.5 Pseudo-NMOS logic with voltage-limiting diode-connected loads similar to I^2L logic.

Another variation on pseudo-NMOS realizes logic gates very similar to I^2L logic[1] where *diode-connected* transistors are used to limit the voltage swings. By limiting the voltage swings, the speed is increased. An example of a 3-input *nor* gate is shown in Fig. 9.5. Note that the diode-connected transistor is one-half the width of the drive transistors. Also shown in Fig. 9.5 is a two-inverter buffer and voltage-regeneration circuit that can be used to interface *voltage-limited pseudo-NMOS* (VLPN) with regular CMOS logic. This type of logic is quite good for applications in which gates with many inputs are required such as decoders and programmable logic arrays (PLAs).

A major problem with pseudo-NMOS logic is that it dissipates d.c. power similar to standard NMOS logic. Thus, it is almost never used for a complete IC. Rather, it will be used selectively in an IC in which it can be of advantage. Examples of this are fully differential circuits that must operate very fast, such as latches and arithmetic circuits, and *nor* gates that have many inputs, such as sometimes found in decoders and PLAs.

Besides dissipating d.c. power, pseudo-NMOS logic has limited current available for charging the output during "0" to "1" transitions. An alternative technique, which does not have this problem, is to use *dynamic precharging*, as is shown in Fig. 9.6. In this logic gate, the pseudo-NMOS load is replaced by a p-channel transistor that operates as a *dynamic load.* The gate output voltage is first precharged high by turning the load transistor *on*. Next, during actual gate operation, the p-channel load transistor is turned *off* and the output voltage is either discharged by the n-channel driver network or left at a high voltage depending on whether the drive network is a low or high impedance. When it is high impedance, it is the parasitic capacitance present at the output node that actually keeps the output voltage high (for a short time) despite

[1]This used to be a popular bipolar logic family that is currently out of vogue. I^2L denotes *integrated injection logic.*

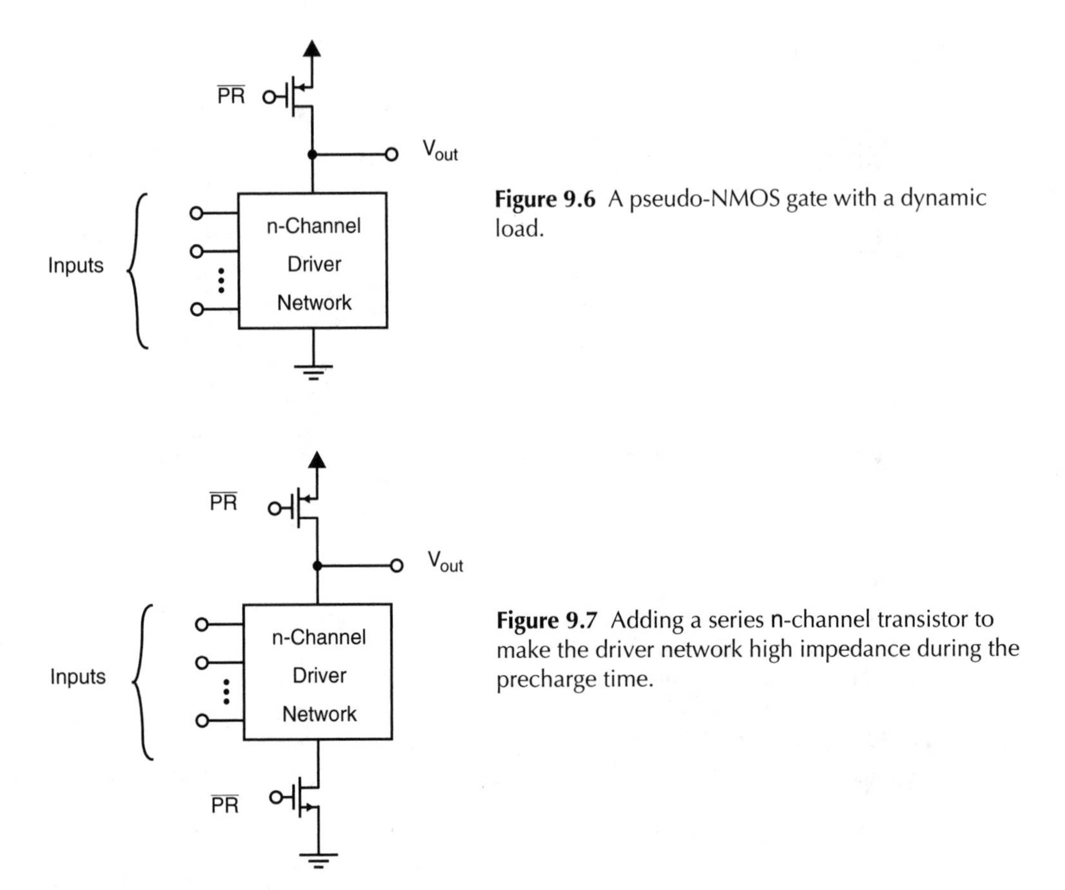

Figure 9.6 A pseudo-NMOS gate with a dynamic load.

Figure 9.7 Adding a series n-channel transistor to make the driver network high impedance during the precharge time.

the fact the node is a high-impedance node. For this gate, the width of the p-channel transistor can be taken much larger than for a standard pseudo-NMOS gate resulting in significantly faster "0" to "1" output transitions. This type of precharging is particularly popular when large capacitive loads have to be driven. Examples of this are long buses, or large PLAs, as will be described in the next chapter. This dynamic circuit still possibly dissipates power while the output is being precharged if any inputs have high values at this time. Normally, during the time the output is being precharged, the n-channel network should not be conducting. This would minimize power dissipation and make the precharge faster. The nonconductance of the n-channel network during the precharge time can sometimes be obtained by making sure all the inputs are "0" during the precharge time. This is usually not possible. When the inputs cannot be guaranteed low, then an n-channel transistor in series with the n-channel network can be added as is shown in Fig. 9.7. This is the basis for a logic family called *Domino-CMOS*, which will be described in the next section.

One of the limitations of dynamic loads is that an output "1" can be guaranteed valid only for a limited time. The reason for this is the leakage current of the junctions. If the leakage current of the n-channel junction is greater than that of the p-channel junction, then the output voltage might eventually discharge to a "0". The leakage current of a reverse-biased junction that is not close to breakdown is approximately given by (Uyemura, 1988)

$$I_{lk} \approx \frac{qA_j n_i}{2\tau_0} x_d \quad (9.1)$$

where A_j is the junction area, n_i is the intrinsic concentration of carriers in undoped silicon, τ_0 is the effective minority carrier lifetime, and x_d is the thickness of the depletion region. τ_0 is given by

$$\tau_0 \approx \frac{1}{2}(\tau_n + \tau_p) \quad (9.2)$$

where τ_n and τ_p are the electron and hole lifetimes. Also, x_d is given by

$$x_d \approx \sqrt{\frac{2\varepsilon_{si}}{qN_A}(\phi_0 + V_r)} \quad (9.3)$$

and n_i is given by

$$n_i \approx \sqrt{N_C N_V} e^{-E_g/kT} \quad (9.4)$$

where N_C and N_V are the density of states in the conduction and valence bands and E_g is the difference in energy between the two bands.

Since the intrinsic concentration n_i is a strong function of temperature (it approximately doubles for every 11°C temperature increase for silicon), the leakage current also is a strong function of temperature. Roughly speaking, it also doubles for every 11°C rise in temperature. Thus, at higher temperatures, it is much larger than at room temperature. This leakage current imposes a maximum time on how long a dynamically charged output can be left in a high-impedance state.

Another limitation of dynamic logic is that it makes the circuits very difficult (actually almost impossible) to probe when debugging them. Normally, when debugging a malfunctioning circuit, it is desirable to clock the system a number of times into a known state, stop it, and then probe internal nodes. This is not possible with dynamic loads only. They must always be clocked or their internal node values will be lost.

To eliminate the problems of charge leakage and the inability to stop the clock during debugging, modern designers often add pseudo-NMOS loads to dynamic loads as is shown in Fig. 9.8. The pseudo-NMOS load transistor will normally be taken minimum width and longer than minimum length, so as to minimize the power dissipation. Its function is to provide enough current to overcome the leakage current. The dynamic precharge transistor will be sized to have minimum length, and depending on how large the load capacitance is, could be

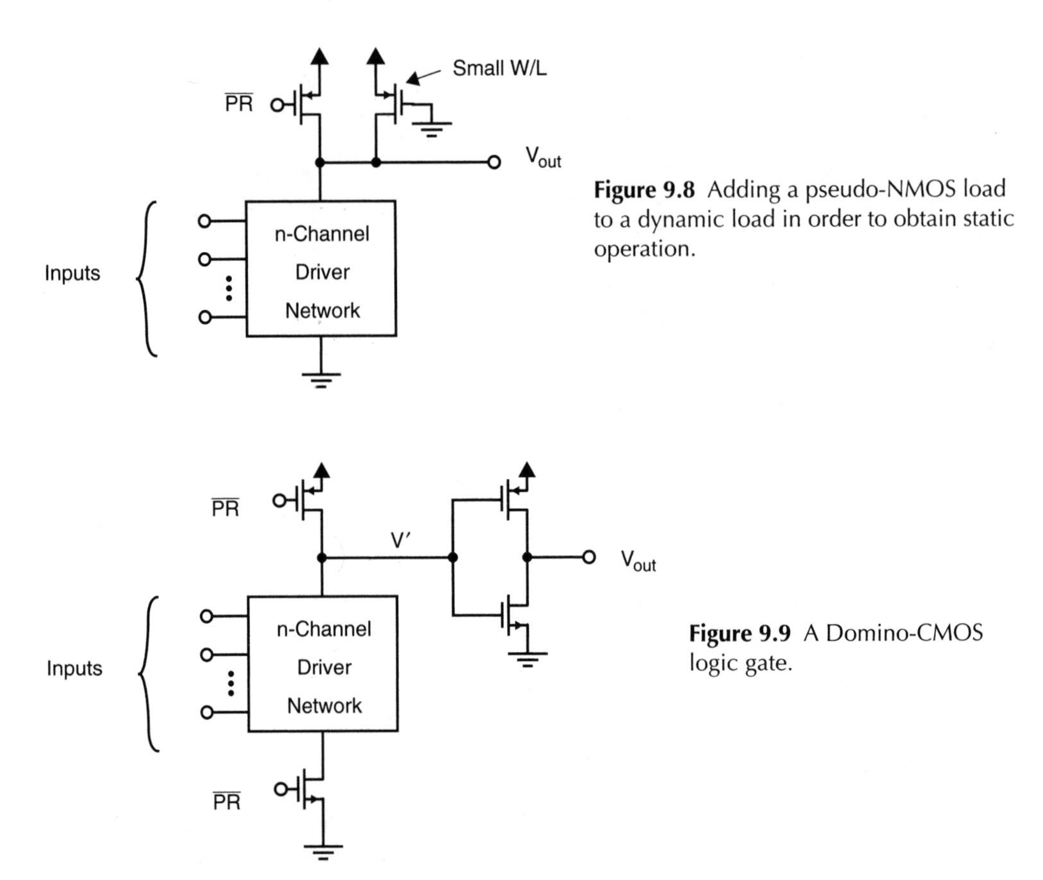

Figure 9.8 Adding a pseudo-NMOS load to a dynamic load in order to obtain static operation.

Figure 9.9 A Domino-CMOS logic gate.

quite wide. These pseudo-NMOS loads might be added only to selected gates that are intended to be probed to minimize power dissipation.

The combination of these two techniques along with the use of dynamic latches, pipelining, and differential circuit design has resulted in a number of different approaches to dynamic logic design that will be described in the next few sections.

9.2 Domino-CMOS Logic

One of the first popular approaches for dynamic CMOS logic design uses gates very similar to those of Fig. 9.7, but with inverters following them, as is shown in Fig. 9.9 (Krambeck et al., 1982). In this approach, a number of *Domino-CMOS logic* gates might be cascaded. All of them would be precharged together and then all of them would be placed in *evaluate mode* at the same time. During the *precharge phase*, the internal nodes (V') of all gates are charged to V_{DD}, which also causes all inverter outputs to go to 0 V. During the *evaluate phase*, the logic signals ripple through the cascade of Domino gates with some inverter outputs perhaps changing from 0 V to V_{DD}, whereas some inverter outputs will remain at ground.

Example 9.1

Design a logic circuit that compares two four-bit words, A and B. If the binary value of A is larger than B, then the single output should be a "1". Otherwise, the output should be a "0". The circuit should be composed of Domino-CMOS gates.

Solution: It is possible to do the comparison bit by bit starting at the least significant bit. The comparison circuit at the bit level will have three inputs A_i, B_i, and C_{in}, and one output C_{out}. The output of the ith bit slice, C_{out}, will be a "1" if considering only bits 0 to i $A_{0...i} > B_{0...i}$. The C_{out} of the final bit slice will be the overall output of the comparison logic function. The truth table for a single bit slice is shown in Table 9.1. The truth table of Table 9.1 can be minimized using a Karnaugh map as shown below.

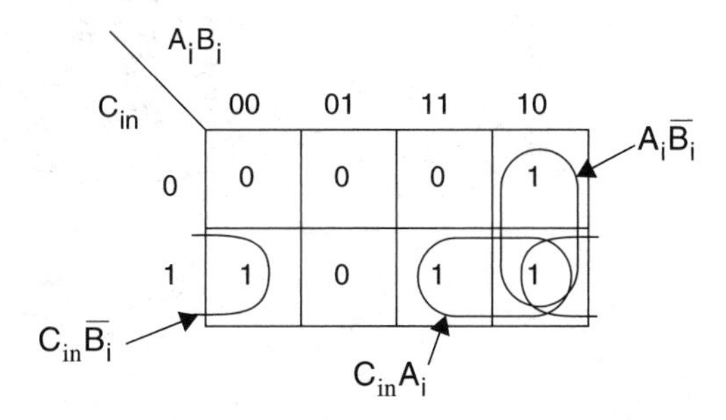

Table 9.1 The Truth Table for a Single-Bit Slice of the Ripple-Carry Comparison Circuit.

C_{in}	A_i	B_i	C_{out}	Comments (for bits 0 to i)
0	0	0	0	$A \le B$
0	0	1	0	$A < B$
0	1	0	1	$A > B$
0	1	1	0	$A \le B$
1	0	0	1	$A > B$
1	0	1	0	$A \le B$
1	1	0	1	$A > B$
1	1	1	1	$A > B$

We have

$$C_{out} = A_i\bar{B}_i + C_{in}\bar{B}_i + C_{in}A_i = A_i\bar{B}_i + C_{in}(A_i + \bar{B}_i) \quad (9.5)$$

This function is very similar to a carry-generate circuit except $\bar{B}_i$ is input rather than B_i. A Domino logic gate that implements this function is shown in Fig. 9.10. Cascading four of these gates and adding inverters to obtain $\bar{B}_i$ results in a four-bit comparator, as is shown in Fig. 9.11.

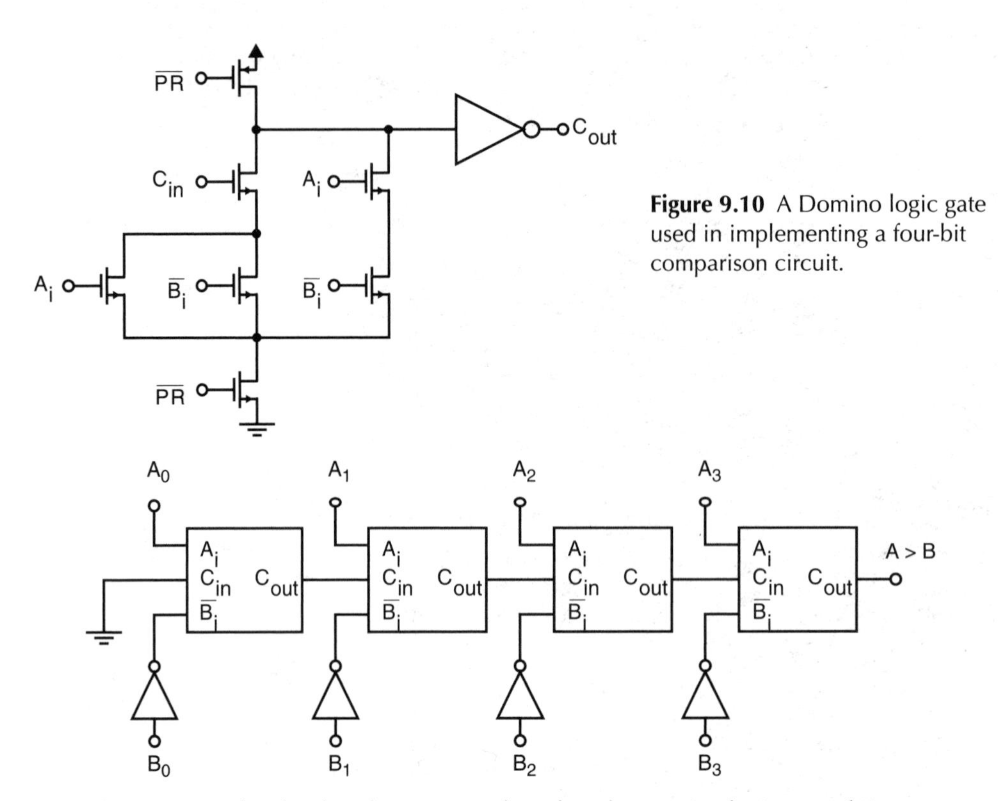

Figure 9.10 A Domino logic gate used in implementing a four-bit comparison circuit.

Figure 9.11 A four-bit digital comparator based on the Domino logic gate of Fig. 9.10.

It should be cautioned that the solution just presented will not work if the inputs come from the outputs of other Domino logic gates. This would imply two inverters would be between Domino gates, and, as we shall see, Domino logic works only with an odd number of inverters between the driver networks.

Domino Logic without Inverters

One of the major limitations of Domino logic is that only noninverting functions can be realized. The naive designer might think that an inverting function could be realized by simply eliminating the inverters at the output of the gates. However, if Domino gates are cascaded without the inverters, or with an even number of inversions between them, they quite-possibly will not work. To see this, consider the circuit of Fig. 9.12 in which two dynamic inverters are shown with no inverter separating them. Also assume that the input to the first gate, V_1, is a "1". First, both gates will be precharged to V_{DD}. Next, V_2 would be discharged to a "0" since the input of the first gate is a "1". Ideally, V_3 would be left at V_{DD} since its input is a "0". Unfortunately, it is highly possible that before V_2 becomes discharged, M_5 will be initially *on*, just after $\overline{PR}$ goes high, and V_3 will be *partially discharged.* A SPICE simulation of the circuit of Fig. 9.12 is shown in Fig. 9.13, where it is seen that the final voltage of V_3 is around 1.65 V for

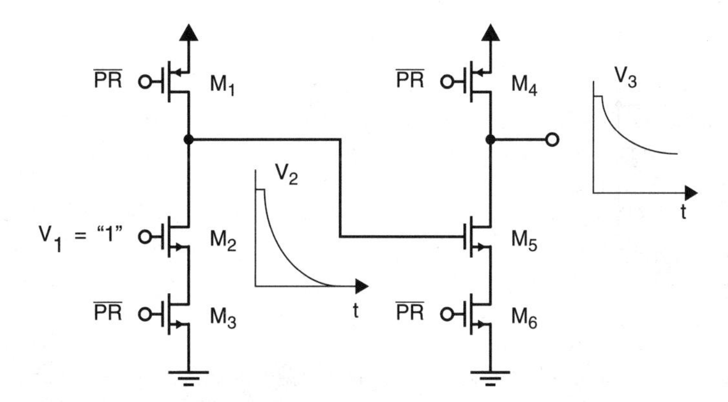

Figure 9.12 An incorrect method for connecting dynamic logic gates.

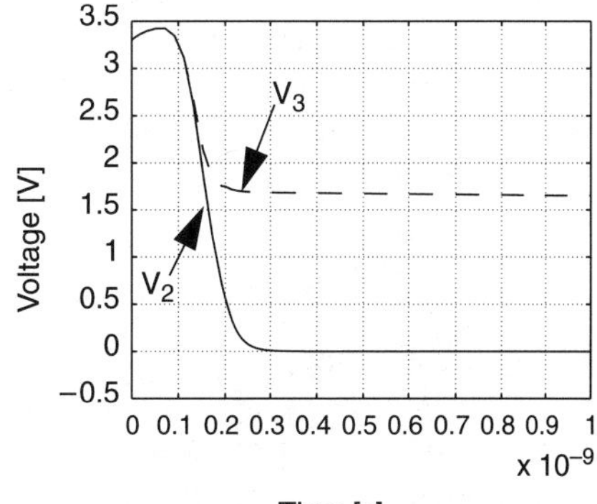

Figure 9.13 A plot of a SPICE simulation showing the undesirable partial discharge of V_3 in the circuit of Fig. 9.12.

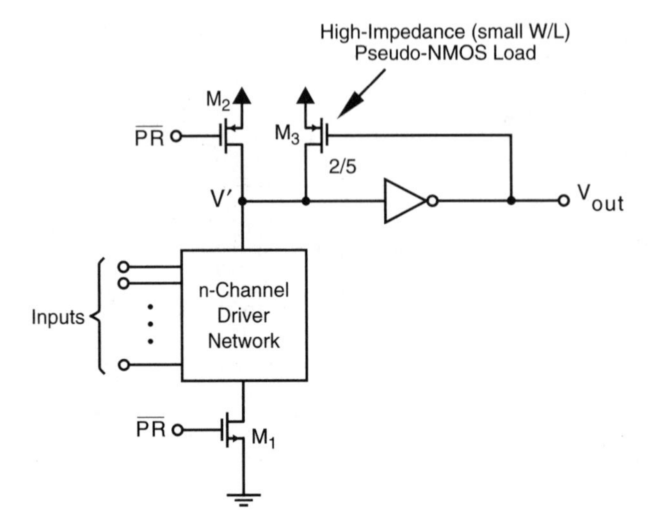

Figure 9.14 A static Domino-CMOS gate.

the parameters chosen (and a 3.3-V power-supply voltage). This final voltage is very dependent on the clock waveform, $\overline{PR}$, and on the relative capacitive loading of V_2 and V_3. In the simulation, the load on the second gate was assumed to be an identical gate. If the parasitic capacitance at V_2 had been larger, even a greater discharge would have occurred. Despite this limitation, Domino CMOS logic gates, and gates like them, have been effectively used in a number of applications.

Static Domino Logic

One modification of Domino logic that allows it to be used statically is to add pseudo-NMOS loads, similar to Fig. 9.8, but with the gate of the pseudo-NMOS load connected to the output inverter as is shown in Fig. 9.14. By connecting the pseudo-NMOS load, M_3, to the output inverter, the load is turned *off* and there is no d.c. power dissipation when the n-channel drive network is *on*. This also helps speed up the "0" to "1" transitions of V' somewhat. In general, this modification is well worth the slightly larger area required.

An additional advantage of including a pseudo-NMOS load is it adds hysteresis to the gate. Initially, the impedance of the n-channel drive network must be small enough to overcome the pseudo-NMOS load. After V' starts to move, the load is turned *off,* which increases the current available to discharge V'.

Charge Sharing of Domino Logic Gates

When using Domino logic gates, designers need to be wary of a subtle problem that can occur due to charge sharing. Charge sharing particularly causes a problem when the n-channel drive network is large with a number of transistors in series. To understand the problem, consider the Domino logic gate of Fig. 9.15. Consider first that during clock period n, A is a "0" and B, C,

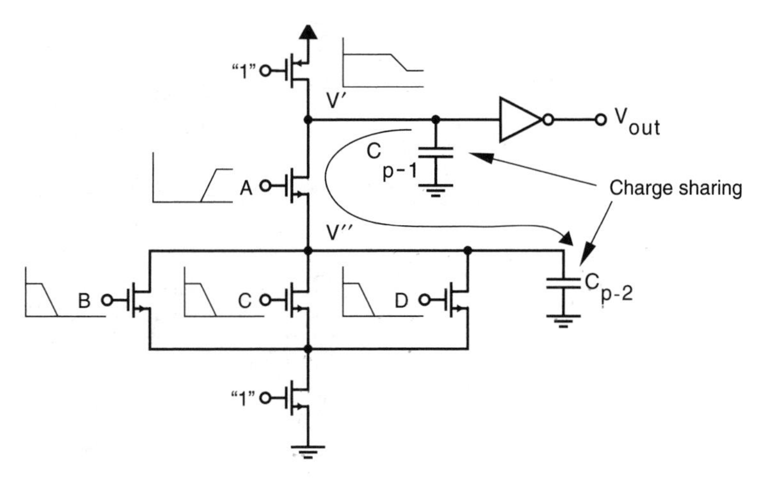

Figure 9.15 A Domino logic gate that can have charge sharing problems if $C_{p\text{-}2} > C_{p\text{-}1}$.

and D are all "1"s. With these inputs, after the *n*th evaluate phase, V' will remain at a high voltage whereas V'' will be discharged to 0 V. Next, during the n+1 precharge phase, all inputs will go to 0 V, assuming they come from other Domino logic gates. If during the n+1 evaluate phase, A then goes to a "1", while B, C, and D stay at "0"s, we would expect that V' would stay high and V_{out} would stay at "0". However, when A goes high it connects parasitic capacitances C_{p1} and C_{p2} together. Also, although $C_{p\text{-}1}$ is initially charged to V_{DD}, $C_{p\text{-}2}$ is initially charged to 0 V due to the inputs during the previous clock period n. This causes a charge sharing between $C_{p\text{-}1}$ and $C_{p\text{-}2}$ that causes V' to be discharged somewhat despite the fact that the impedance of the n-channel network is infinite. If the parasitic capacitance $C_{p\text{-}2}$ is larger than $C_{p\text{-}1}$, which can occur for gates having many inputs, V' may be discharged enough to cause an error. This type of error is difficult to predict at design time without very careful simulations using an accurate simulator. It is possible to modify Domino logic gates in order to eliminate this type of charge-sharing error by adding additional p-channel precharge transistors that charge nodes internal to the n-channel drive network that have large parasitic capacitances. For example, Fig. 9.16 shows how this can be done for the circuit of Fig. 9.15 where an additional transistor has been added to precharge V'' before each evaluate state irrespective of previous input values.

An additional method for minimizing the probability of a charge-sharing error occurring is to make the n-channel transistor in the output inverter the same width as the p-channel transistor. This will lower the inverter threshold voltage and increase the parasitic capacitance $C_{p\text{-}1}$, both of which minimize the chance of an error occurring.

Although this charge-sharing problem has been illustrated using a Domino logic gate, it is typical of a type of problem that can occur in many types of dynamic gates. For example, a similar problem was described previously for the dynamic latches of Fig. 7.13 in Chapter 7. The best method of eliminating errors due to it is to use careful forethought and extensive simulations.

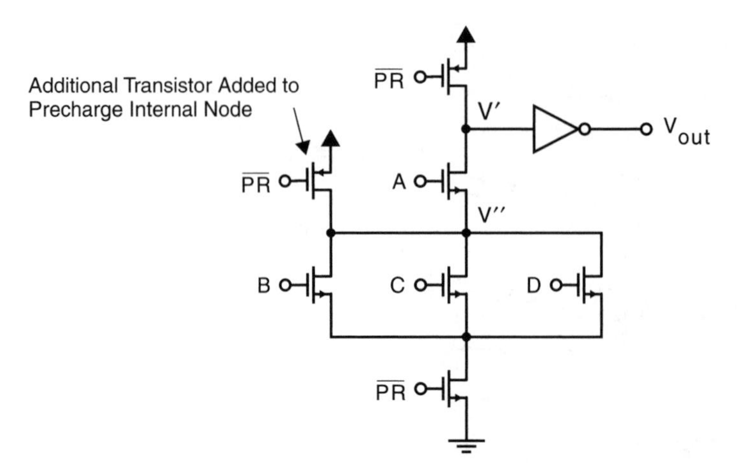

Figure 9.16 Precharging internal nodes to prevent charge-sharing problems in Domino logic gates.

MULTIPLE-OUTPUT DOMINO LOGIC CIRCUITS

A possible modification of Domino logic is to have multiple outputs from a single gate. Additional outputs can be generated by connecting additional output inverters to internal nodes (Hwang and Fisher, 1989). To prevent charge-sharing problems caused by the parasitic input capacitances of these additional inverters, the internal nodes to which the inverters are connected should be precharged in a manner similar to that shown in Fig. 9.16.

An example of a *multiple-output Domino logic gate* is shown in Fig. 9.17. This single gate simultaneously realizes $A \oplus B$, $A \oplus B \oplus C$, and $A \oplus B \oplus C \oplus D$. It is assumed that the inverted inputs do not come from Domino logic gates. Note that only some of the internal nodes need to be precharged because during PR = "1", the other nodes will still be pulled high irrespective of the input values.

Adding the extra inverters to the internal nodes does slow down the gates somewhat due to the increase in internal parasitic capacitances. Also, the need to precharge more nodes in addition to the larger parasitic capacitances of the internal nodes requires additional area for the p-channel transistors and places additional load on the clock driver. Still the overall area could be quite small due to the efficiency of using one gate for many outputs. Also, since the load on the preceding gates is only a single transistor, as opposed to many transistors typical for single-output conventional CMOS gates, the overall delay can be small. How practical this approach is in general has not been resolved. For some particular applications, it should be reasonable.

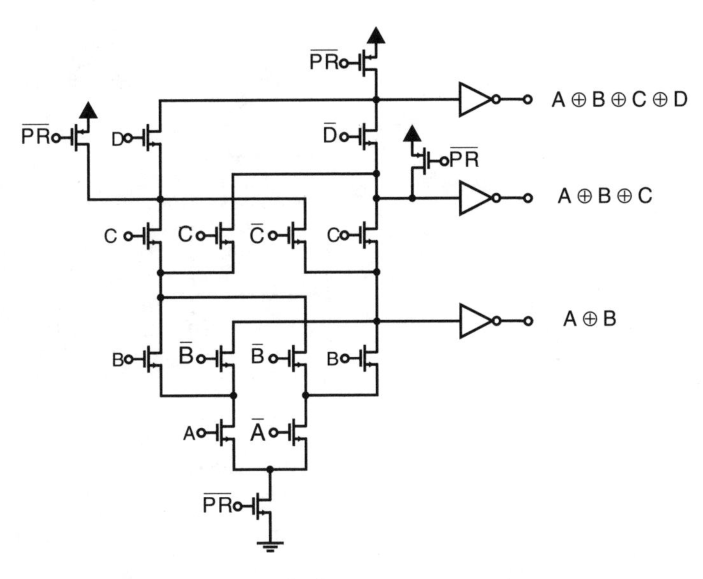

Figure 9.17 A multiple-output Domino logic gate.

9.3 No-Race-Logic

Perhaps, one of the major limitations of Domino logic is the restriction that an odd number of inverters must separate gates. There are two methods that can be used to eliminate this restriction. The first technique that can be used without any interceding inverters is to cascade Domino logic gates with n-channel driver networks with a complementary Domino logic gate that has p-channel driver networks, as is shown in Fig. 9.18. The Domino gate with the p-channel driver network has its output precharged low while the output of the Domino gate with the n-channel driver network is precharged high. During the evaluate phase, the outputs of the n-precharge stage will remain high or discharge low if the n-channel driver network becomes low impedance. Initially, the p-channel driver network is guaranteed to be high impedance (because all of its inputs are high). The output of the p-precharge stage will remain low or change to high if the p-channel driver network changes to a low impedance. Since all nodes either remain the same during the evaluate phase or make a single transition only, there is no problem with nodes being partially discharged at the beginning of the evaluate phase.

It is also possible to simplify the p-precharge stage of Fig. 9.18. Note that Q_4 is redundant since it is guaranteed that the p-channel driver network is high impedance during the precharge time, since all its inputs are “1”s. Eliminating Q_4 can significantly increase the speed of the p-precharge gate during its evaluate phase. *It is always possible to eliminate the transistor in series with the driver network from a dynamic gate when all the inputs to the driver network come from a dynamic gate that is of the opposite type (i.e., p-precharge gates following n-precharge gates or vice versa).*

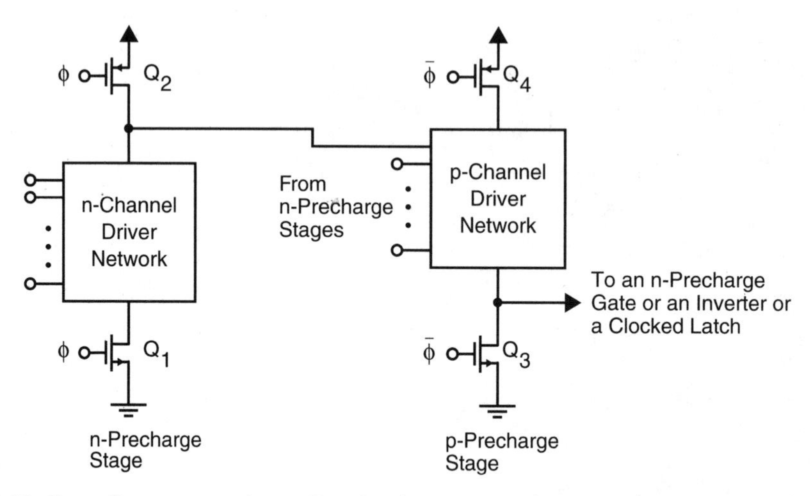

Figure 9.18 Cascading an n-precharge Domino logic gate with a p-precharge Domino logic gate.

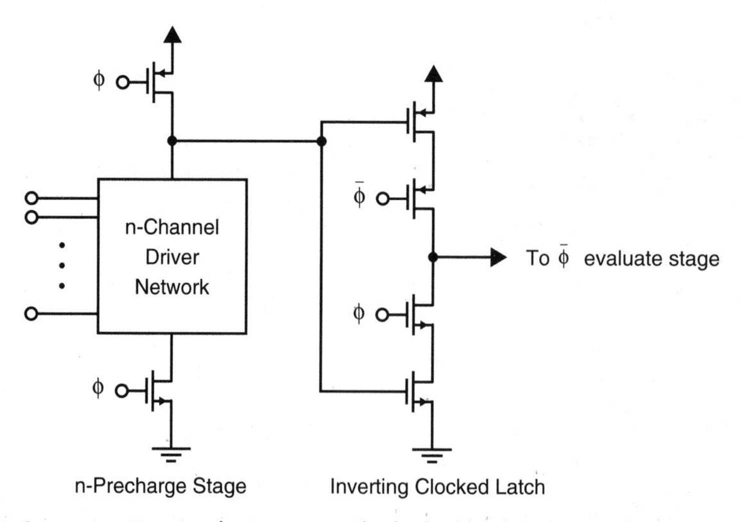

Figure 9.19 Separating Domino logic gates with clocked latches that are high impedance during the precharge phase.

Another means of eliminating the possibility of partially discharged nodes is to replace the inverters between Domino logic gates with an inverting clocked latch shown originally in Fig. 7.13 in Chapter 7 and shown again in Fig. 9.19. The clocked latch samples the output of the Domino logic gates during the evaluate phase and stores this value during the next precharge phase. The outputs of the clocked latches are connected to Domino logic gates that are precharged and evaluated on the opposite phase. That is, while the outputs of the clocked latches are being stored, the next stage of Domino logic gates is being evaluated. Thus, the only time

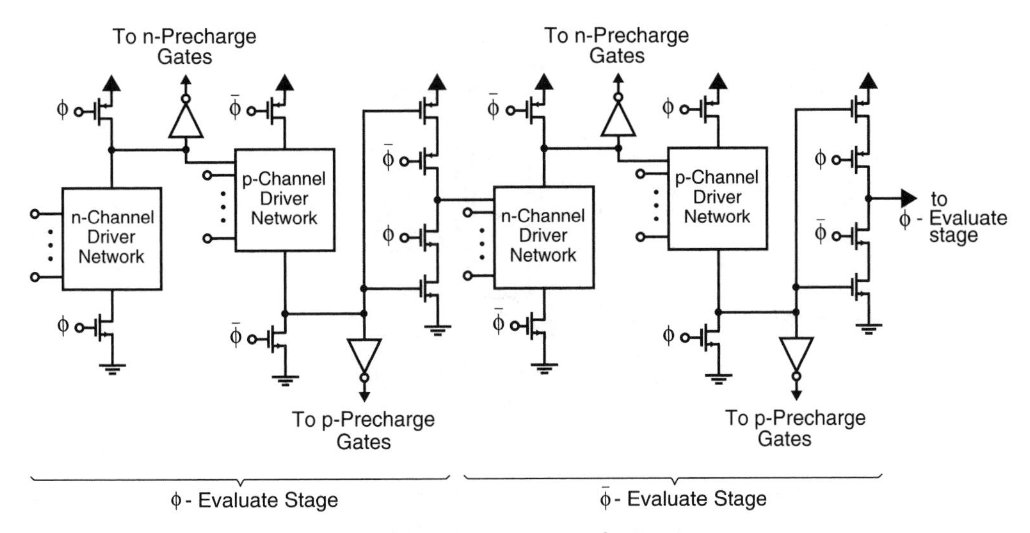

Figure 9.20 No-Race logic.

the outputs of the clocked latches will have transitions is during the precharge time of the next gates in the stage, a time that does not cause incorrect partial discharge of nodes. As long as the evaluate phase is interchanged at each stage of the pipeline, the values of all inputs to Domino gates at the beginning of evaluate phases are the same as at the end of the evaluate phases.

The combination of the above two techniques was first proposed in Goncalvez and De Man, (1983) and was called *No-Race* or *NORA logic*. This methodology is shown in Fig. 9.20. It is used to realize pipelined logic. Each stage of the pipeline is clocked on the opposite phase. The Domino gates of a particular stage can be n-precharge gates or p-precharge gates in any order, but gates of the same type must be separated by inverters. Although gates of both types are shown in each stage, it is permissible to have only one type of gate such as the n-precharge Domino gate, p-precharge Domino gate, or even a static gate.

When *static* gates are included, there are some constraints on the *number of inversions* if the NORA logic is to be guaranteed *race free* in the presence of clock skews. The constraints can be summarized as follows: if the input to a clocked latch comes from a static gate, then there must be an even number of static gates between the clocked latch and any dynamic circuits preceding them. Even without these constraints, most NORA circuits will work if clocks with fast transitions and no skew are used. Obviously, in order to maximize speed, the number of series p-channel transistors in the p-precharge gates must be minimized. It has been shown in Goncalvez and De Man (1983) that NORA logic is race free even with clock skew, as long as the clock edges are fast enough. *As with all clocked systems, the clock signals of the stages further down the pipeline should change at the same time or earlier than those of the stages that are earlier in the pipeline.*

An example of a NORA circuit is a serial (in time) full adder. It is taken from Yuan and Svensson, (1989) and shown in Fig. 9.21. This circuit calculates an additional bit each clock period. Initially, Start = "1", which guarantees that carry-in, C, will be a "0", at the time the

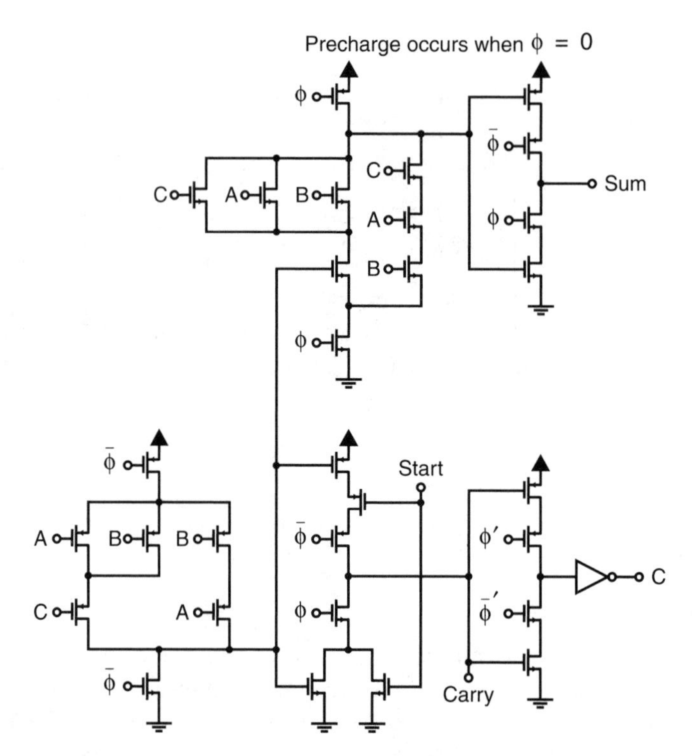

Figure 9.21 An example of a serial adder implemented using NORA logic.

addition starts. Note that the carry output C is fed back to the inputs of the first stage and that the loop has two clocked latches, the first one in hold mode when ϕ = "0" and the second one is in hold mode when $\overline{\phi}$ = "0".

Although NORA logic is race free with skewed clocks, it does place a minimum requirement on the slopes of the clock wave forms. A suitable clock generation circuit that can be used to regenerate clocks is shown in Fig. 9.22, where the clock signals with the prime superscripts should be used for the clocked latch that is active during $\overline{\phi}$ = "1". This clock generation circuit helps guarantee that two succeeding clocked latches in a pipeline are not both *on* at the same time; this guarantees race conditions do not occur.

9.4 Single-Phase Dynamic Logic

If one constrains a NORA stage to have only n-precharge gates, and not static gates, then a p-channel transistor can be eliminated from the clocked latch. To see this, consider the circuit of Fig. 9.23a when ϕ is an "0", the clocked latch should be in its high-impedance state where its output is being stored. At the same time, the n-precharge gate is being precharged. This guarantees that p-channel transistor Q_5 of the latch will be *off*. Since ϕ = "0", Q_4 is also *off* and

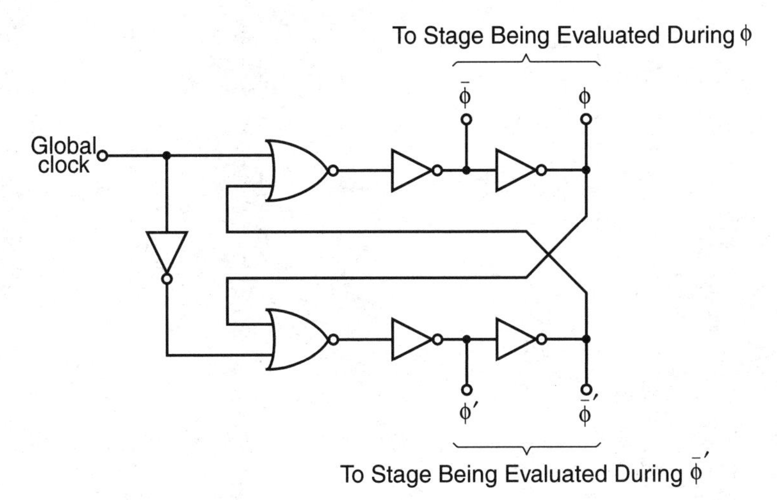

Figure 9.22 A circuit that can be used to locally generate clocks for NORA logic gates.

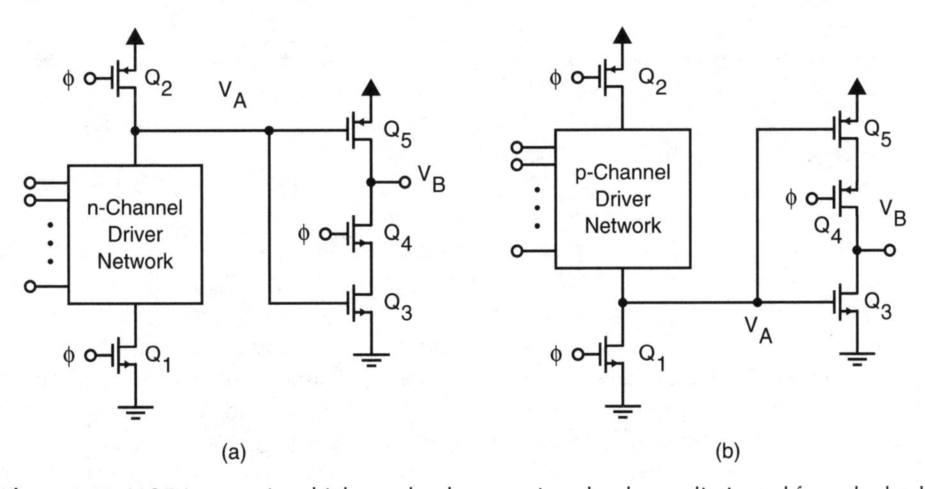

Figure 9.23 NORA stages in which a redundant transistor has been eliminated from the latch.

the latch is guaranteed to be in a high-impedance state, even though a p-channel transistor in the latch has been eliminated.

In a similar manner, if one constrains a NORA gate to have only p-precharge gates, then an n-channel transistor in the clocked latch can be eliminated, as is shown in Fig. 9.23b.

If one denotes the circuit of Fig. 9.23a as an n-block and the circuit of Fig. 9.23b as a p-block, then it is possible to realize a logic system of alternatively cascaded n-blocks and p-blocks that only requires a single clock and is insensitive to clock skew (Ji-Ren et al., 1987). For example, a dynamic master–slave D flip-flop that changes on the positive edge of a clock is

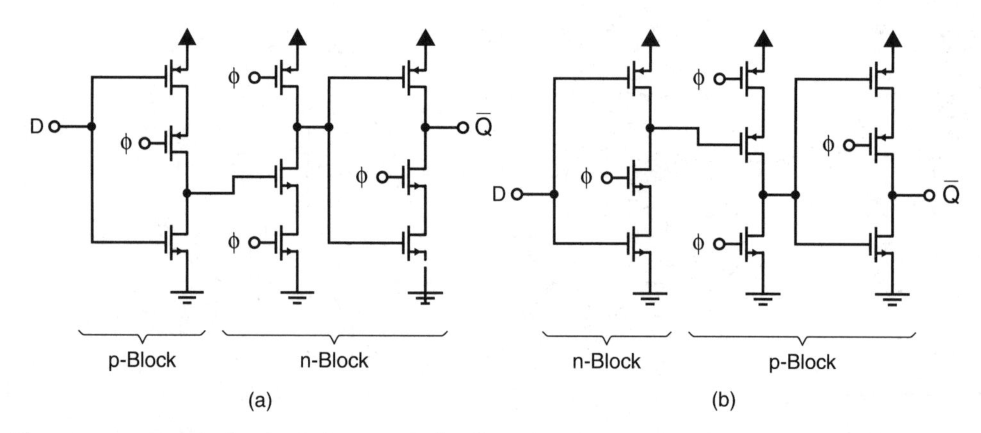

Figure 9.24 True-single-phase dynamic D flip-flops that change on (a), the positive clock edge and (b) the negative clock edge.

shown in Fig. 9.24a, whereas a dynamic D flip-flop that changes on the negative edge of the clock is shown in Fig. 9.24b (Yuan and Svensson, 1989).

Another example of a true single-phase clock (TSPC) circuit is the dynamic serial adder circuit taken from Yuan and Svensson (1989) and shown in Fig. 9.25. Note how the logic for the sum-generate circuit is realized using an n-precharge dynamic gate, whereas the logic for the carry-generate circuit is realized using a traditional static CMOS gate.

An alternative method for realizing TSPC pipelined circuits is to replace the latches of NORA circuits alternatively by the double latches of Fig. 9.26a and b (Yuan and Svensson, 1989). To see how the latch of Fig. 9.26a works, consider first the case V_{in} = "0", V' = "1", and V_{out} = "0" when ϕ goes low. If V_{in} changes after ϕ goes low, V', which is high, cannot change and, therefore, V_{out} does not change. Next consider the case V_{in} = "1", V' = "0", and V_{out} = "1" when ϕ goes low. If V_{in} now changes to an "0", V' will change to a "1" but now the second latch isolates the output and V_{out} will not change. A similar argument for the double latches of Fig. 9.26b shows that the first latch provides the isolation when V_{in} = "1" during the time ϕ = "1", whereas the second latch provides the isolation when V_{in} = "0" during the time ϕ = "1".

When using this approach to realize single-phase logic circuits, there is no constraint on having an even number of gates between latches in order to guarantee race-free operation, as there is in NORA logic.

At present, TSPC logic is perhaps the most popular approach for realizing dynamic logic circuits.

9.5 Differential CMOS

A design style that has been continuously gaining popularity is the use of fully differential logic gates (Heller et al., 1985; Chu and Pulfrey, 1987; Ng et al., 1996) similar to bipolar current-mode logic gates, particularly for high-frequency applications. There are two major

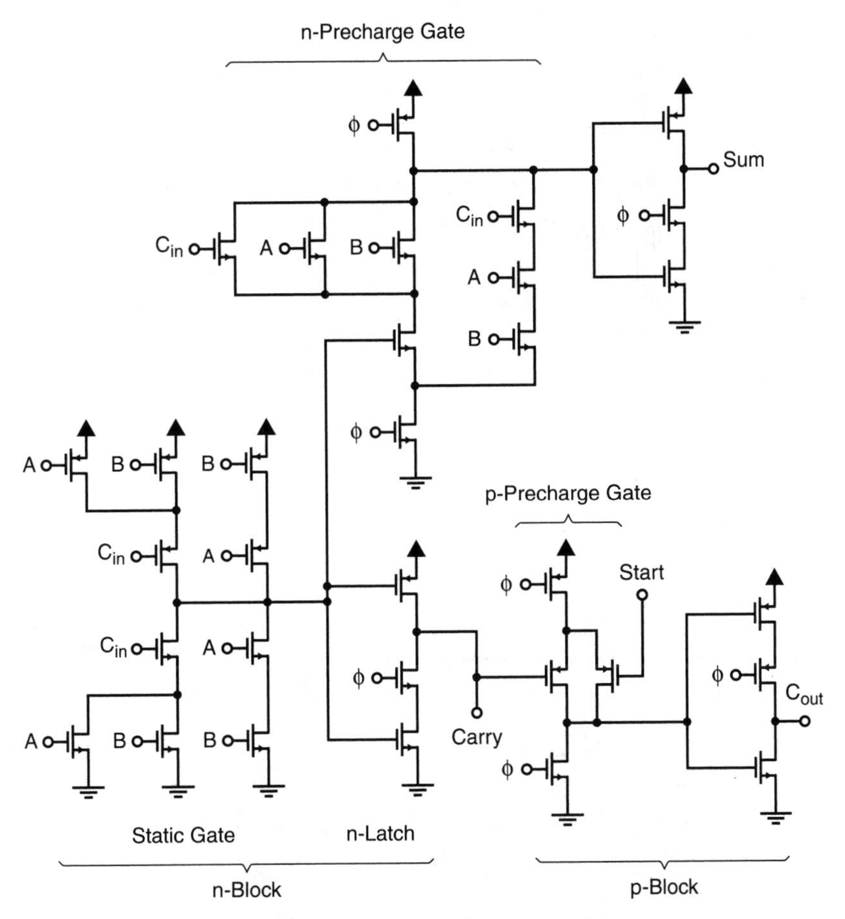

Figure 9.25 TSPC serial (in time) adder.

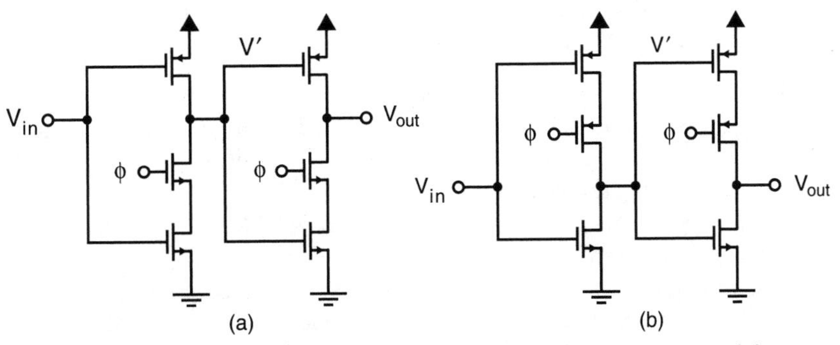

Figure 9.26 TSPC double latches that are transparent when (a) ϕ = "1" and (b) ϕ = "0".

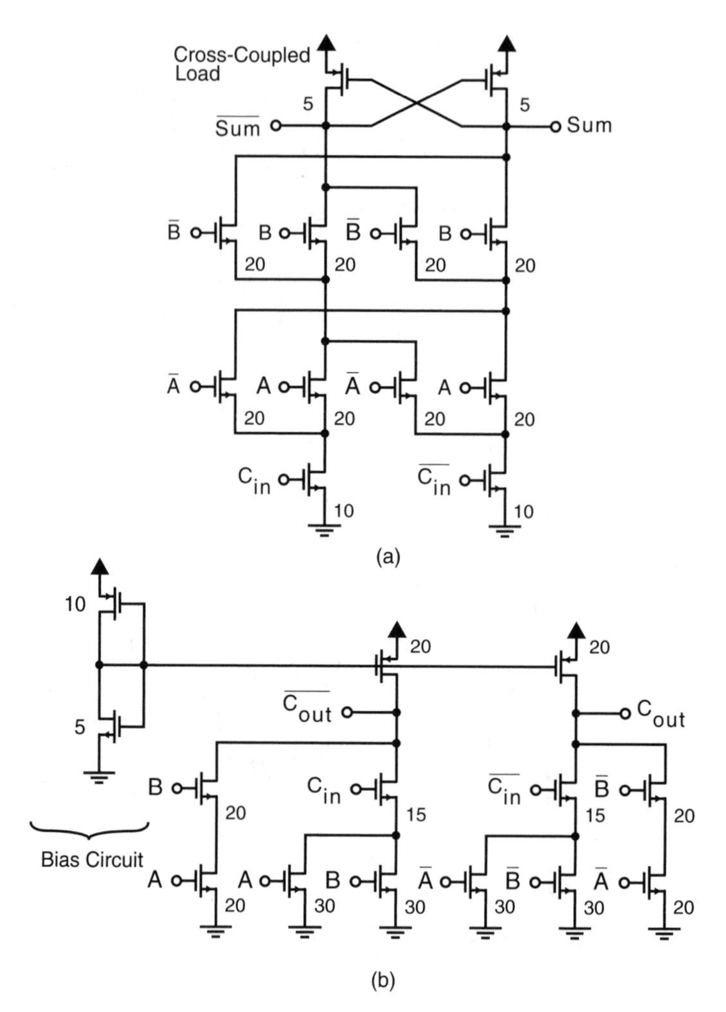

Figure 9.27 A fully differential full adder. (a) The sum-generate circuit and (b) the carry-generate circuit.

reasons for this. The primary reason is that inverted signals can be trivially obtained by simply interchanging the two connections of a differential signal. This often can eliminate a gate delay, which can considerably speed up some digital circuits, such as arithmetic circuits. The second reason is that the circuits are less sensitive to noise.

We have already seen some examples of fully differential CMOS logic gates in Chapter 5. The example of a full adder is repeated here and shown in Fig. 9.27a and b. This adder is a good illustration of many important considerations when trying to optimize the speed of logic gates. In particular, sizing considerations and also considerations in choosing which inputs to use for critical paths are illustrated. This adder has been optimized for a fast carry generate; this has been achieved in a number of ways. First, current source loads, similar to those shown in Fig. 9.3, have been used for the carry-generate circuit, whereas cross-coupled loads have been used for the sum-generate circuit. The cross-coupled loads have the advantage of no d.c. power dissipation, but when the differential outputs are changing, one side must first go low before the load of the opposite side turns *on* and *pulls* its output high. By having continuous loads for the carry-generate circuit, both outputs change simultaneously, almost immediately after the input changes that were responsible for the output changes. The disadvantage of the current source loads is of course the d.c. power dissipation. They would therefore be used only for a minimal number of gates in a critical path.

Another consideration has been which inputs have been used for C_{in} versus the inputs used for A and B. The choice has been made based on two considerations. First, the C_{in} inputs should be connected to transistors close to the outputs of the carry-generate circuits. This minimizes how many internal gate nodes need to change state during carry propagations. Second, the load on the carry-generate outputs has been minimized. It is for this reason that C_{in} has been connected to the bottom level inputs of the sum-generate circuit, which results in a longer delay through that circuit due to C_{in} changes, but means that each carry-generate output is loaded by only two transistors rather than by three transistors. For particular full adders where the delay from C_{in} to the Sum outputs is also critical, a different choice may be more optimum.

Finally, the device sizes have also been chosen to minimize the load on the carry-generate gate and optimize its speed. For this reason, wider transistor loads have been used for the carry-generate circuit as opposed to the sum-generate circuit. Also, the sizes of transistors to which C_{in} is connected have been minimized at the expense of the sizes of transistors in series with them. These must be taken wider in order to guarantee that the worst-case equivalent *pull-down* transistor is no smaller than the p-channel loads for both circuits.

Another advantage of using fully differential gates with n-channel driver networks is the number of p-channel transistors has been minimized without having to resort to dynamic techniques. By not using dynamic circuits, debugging and testing are simplified.

Altogether these considerations result in a carry-propagate delay that is only slightly larger than the delay of a single differential inverter! This is probably faster than could be achieved by using techniques such as carry lookahead (Mano and Kime, 1997).

This example has been used not only to illustrate the advantages of fully differential logic, but also to emphasize how important it is to choose the correct gate inputs and transistor sizes when minimizing the delay of a critical path.

DIFFERENTIAL SPLIT-LEVEL CMOS LOGIC

A variation of fully differential logic that has been gaining in popularity recently uses a load that is a compromise between a cross-coupled load with no d.c. power dissipation and a continuously-on load with d.c. power dissipation. It is called *differential split-level (DSL) logic* (Pfennings et al., 1985).

This compromise is shown in Fig. 9.28. This load is similar to the cross-coupled load of Fig. 9.27a with some common-gate or cascode transistors added. The cross-coupling of the p-channel load transistors is taken from the sources of the cascode transistors.

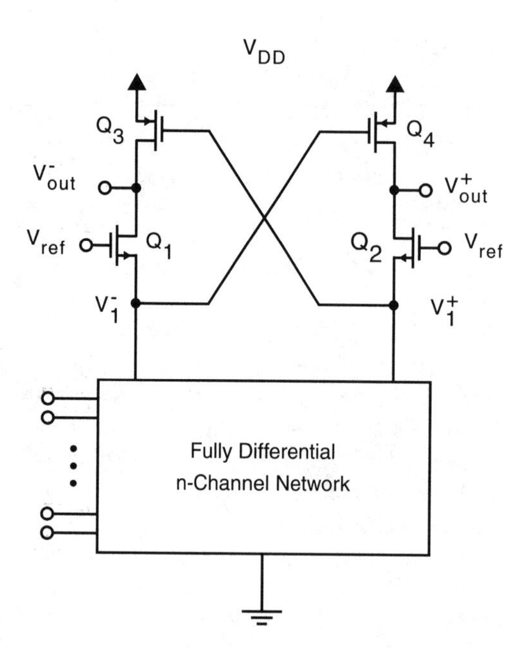

Figure 9.28 Fully differential split-level (DSL) logic.

To understand how the load functions, consider first that the n-channel drive network represents a low impedance to the node V_1^- and a high impedance to node V_1^+. Node V_1^- will be pulled low, close to 0 V. This turns p-channel transistor Q_4 *on hard,* which *pulls* V_{out}^+ to V_{DD}. However, due to the cascode transistor, Q_2, which is *off*, node V_1^+ will only be pulled up to $V_{ref} - V_{tn}$. Thus, cross-coupled load transistor Q_3 is still *on* but with a reduced effective gate voltage as compared to a pseudo-NMOS load. The current of Q_3 will go through transistor Q_1 and the n-channel network to ground. Cascode transistor Q_1 will be in its triode region and V_{out}^- will have a low voltage near ground.

Next, consider what happens when the n-channel driver network changes so it becomes a low-impedance path to node V_1^+ and high impedance to node V_1^-. Node V_1^+ immediately starts to change to a low voltage as eventually does V_{out}^+. However, at the same time, V_1^- starts to change to a high voltage because Q_3 was initially conducting although at a reduced level. As V_1^+ moves low, the current of Q_3 increases causing V_{out}^- and V_1^- to go to higher voltages, V_{out}^- to V_{DD} and V_1^- to $V_{ref} - V_{tn}$. Since V_1^- goes to a higher voltage, the current of Q_4 is reduced from its initial value, which allows V_{out}^+ and V_1^+ to be more quickly pulled down to around 0 V.

Thus, the loads have some of the features of both continuous loads and cross-coupled loads. Both outputs begin to change immediately, although quickly after the change begins there is a positive-feedback latching that helps to speed up the change. The loads do have d.c. power dissipation, but normally much less than pseudo-NMOS gates and usually less than the dynamic power dissipation of a traditional gate running at full speed. In addition, the gates are easily powered down when they are not needed by changing V_{ref} to 0 V.

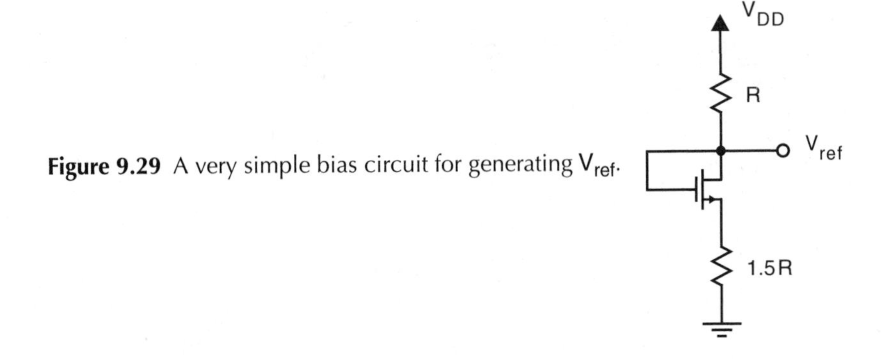

Figure 9.29 A very simple bias circuit for generating V_{ref}.

Another important advantage of DSL is that the nodes, V_1^-, V_1^+, and all the internal nodes of the n-channel driver network, have voltage changes between slightly greater than 0 V and $V_{ref} - V_{tn}$. This reduced voltage swing increases the speed of the logic gates, particularly for large n-channel driver networks or when there is a large parasitic capacitance at nodes V_1^- and V_1^+. An example of where this is particularly useful is when the DSL load is used for the bit lines of a memory, as we shall see in the next chapter. For a similar reason, *the DSL load is a good choice whenever there are differential bus lines, often between very high-speed ICs, or for digital transmission over twisted pairs.*

A final advantage of DSL is that the maximum drain-source voltage across the n-channel transistors is reduced by about one-half. This greatly minimizes the short-channel effects of these transistors and allows their channel lengths to be taken smaller than would normally be the case, resulting in an additional increase in speed.

Many different approaches are possible for generating the reference voltage V_{ref}. A very simple circuit is shown in Fig. 9.29. This circuit gives a V_{ref}, which results in $V_{ref} - V_{tn} \approx V_{DD}/2$. Another possibility, which is sometimes used, is to simply take $V_{ref} = V_{DD}$. This results in simplified circuitry but larger than optimum voltage swings and therefore slower gates. This latter choice is becoming increasingly popular for modern submicron technologies where a power-supply voltage of around 3 V is often used.

Differential Pass-Transistor Logic

A currently popular logic design methodology implements gates in a fashion similar to those shown in Fig. 9.27a, but replaces the n-channel drive network by a differential pass-transistor network that uses n-channel pass transistors only. It is called *differential pass-transistor logic* (Yano et al., 1990; Suzuki et al., 1993; Parameswar et al., 1996; Lai and Hwang, 1997). This logic methodology has a number of advantages compared to other approaches. First, pass-transistor networks for most required logic functions exist in which both sides of the cross-coupled loads are driven simultaneously; *it is not necessary to wait until one side goes low before the other side goes high.* This minimizes the time from when the inputs change to when the low-to-high transition occurs. This also removes the ratio requirements on the logic and has guaranteed functionality. In addition, the cross-coupled loads restore signal swings to full V_{DD} levels,

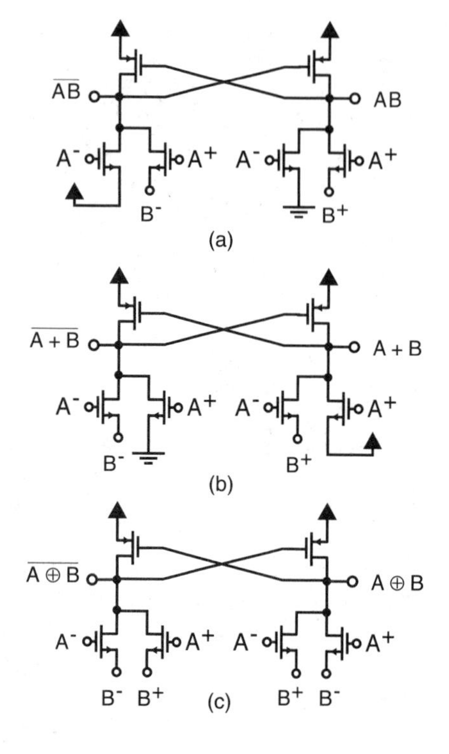

Figure 9.30 Some example logic gates based on cross-coupled loads and differential pass-transistor networks from Lai and Hwang (1997): (a) an *and/nand* gate, (b) an *or/nor* gate, and (c) an *xor/xnor* gate.

thereby eliminating the voltage drop that normally occurs when n-channel pass transistors are used. Some example gates taken from Lai and Hwang (1997) are shown in Fig. 9.30. As is the case for all pass-transistor gates, one cannot cascade more than about three gates having the signal flow through the pass transistors without seriously degrading the speed. By choosing the inputs carefully, however, one can usually avoid this problem by alternating gate inputs with source inputs as both types are available in most gates. If this restriction is not considered, performance suffers drastically.

A Karnaugh map-based procedure is presented in Lai and Hwang (1997) for realizing gates with additional functionality. For example, a gate that realizes the function $F = \overline{A}\overline{B}\overline{C} + C(A + B + D)$ is shown in Fig. 9.31.

9.6 Dynamic Differential Logic

Differential Domino Logic

Some recent design approaches, which have been gaining in popularity, combine differential logic design with dynamic techniques. For example, Fig. 9.32 shows how *differential Domino logic* gates can be realized, by simply using precharging techniques and adding inverters at the outputs (Heller et al., 1984; Grotjohn and Hoefflinger, 1986; Chu and Pulfrey, 1987). Also

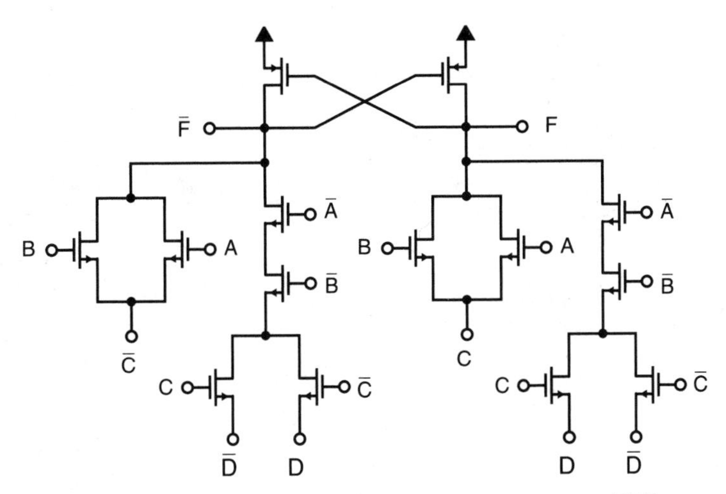

Figure 9.31 A differential pass-transistor realization of the function $F = \bar{A}\bar{B}\bar{C} + C(A + B + D)$ taken from Lai and Hwang (1997).

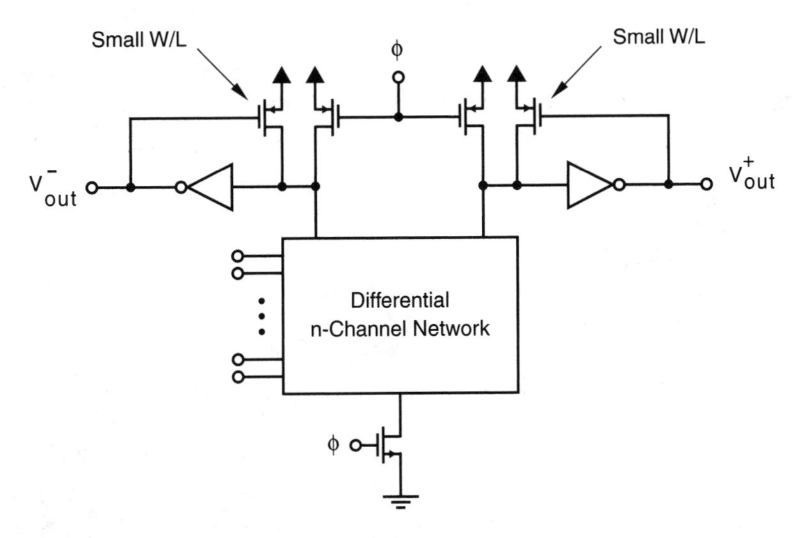

Figure 9.32 A differential Domino logic gate.

shown in Fig. 9.32 are some high-impedance *bleeder transistors* having small W/Ls that permit static operation.

This approach has many advantages. Its d.c. power dissipation is very small, whereas its speed is still quite good. Because of the buffers at the output, its output drive capability is also very good. Finally, and perhaps most importantly, one of the major limitations of Domino logic, the difficulty in realizing inverting functions, is eliminated because of the differential

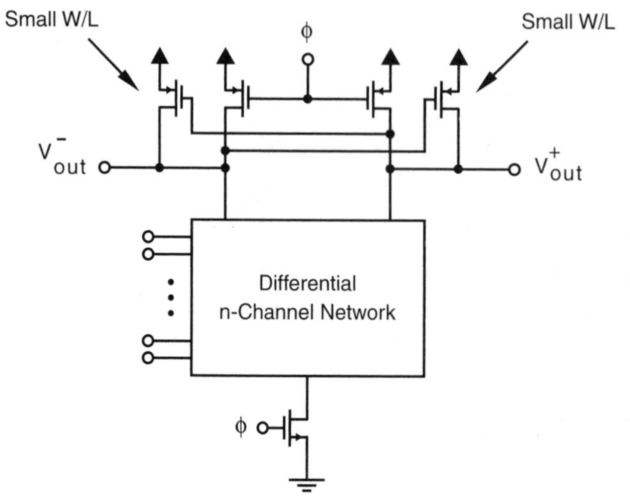

Figure 9.33 An alternative differential dynamic logic gate.

nature of the circuits. In differential Domino logic, this type of function is trivially implemented by realizing a noninverting logic function and simply interchanging the output lines.

When the fan-out is small, the inverters at the output can be eliminated and the inputs to the bleeder transistors can be taken from the opposite output as shown in Fig. 9.33. This alternative was reported as having very small delays and power-delay products in Ng et al. (1996).

Differential NORA Logic

Another example of combining dynamic techniques with differential circuits is to use clocked latches and precharge logic gates similar to NORA logic. It is often called *differential NORA logic* (Chu and Pulfrey, 1987). An example of this is shown in Fig. 9.34 where the output inverters have been replaced by clocked latches. A p-channel transistor has been eliminated in each clocked latch, since it is guaranteed that during the time the latches are in their high-impedance state, which is the same time the differential gates are being precharged, the nodes V_1^+ and V_1^- are at high voltages. Therefore, the clocked latches have a guaranteed high impedance to V_{DD} without including the additional clocked p-channel transistor.

When building a pipelined system using the differential NORA gates of Fig. 9.34, every second stage should be clocked using the complementary clock phase. *This design approach can result in very high clock rates and should be seriously considered, particularly for very fast, pipelined, arithmetic circuits when speed is critical.*

Regenerative Differential Logic

If logic circuits have large driver networks with many nodes and transistors in series, they can be very slow. The delay of pulling a node low through the driver network is quadratic with respect to the number of series transistors, as was shown in Chapter 4. One technique sometimes used to speed up circuits having large driver networks is to add positive-feedback

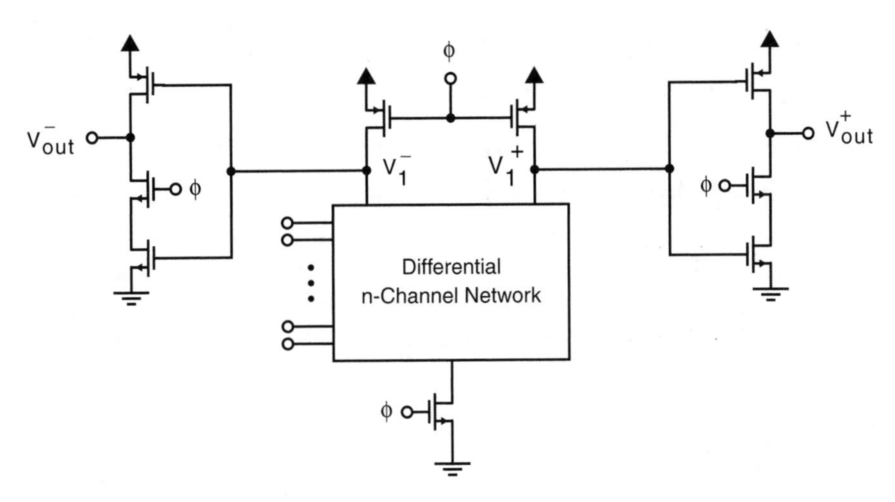

Figure 9.34 A differential NORA gate.

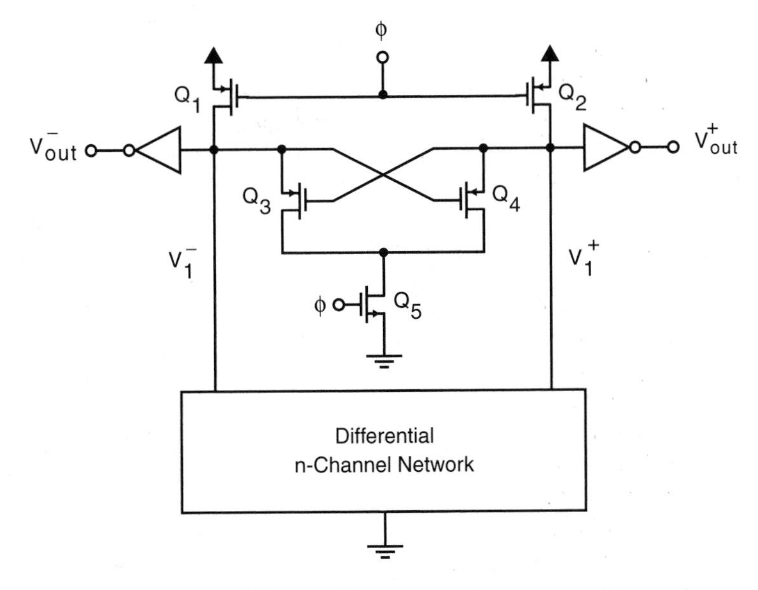

Figure 9.35 A differential logic gate using regenerative sensing.

sense amplifiers to sense and regenerate the output voltages after waiting only long enough for the gate output to develop a small differential voltage. This technique is very similar to using sense amplifiers to determine the contents of a random-access memory (RAM), as we shall see in the next chapter.

An example of using a positive-feedback regeneration circuit is shown in Fig. 9.35 (Grotjohn and Hoefflinger, 1986). During ϕ = "0", p-channel transistors Q_1 and Q_2 are turned *on*. These are normally wide transistors and pull both nodes V_1^+ and V_1^- to high voltages close

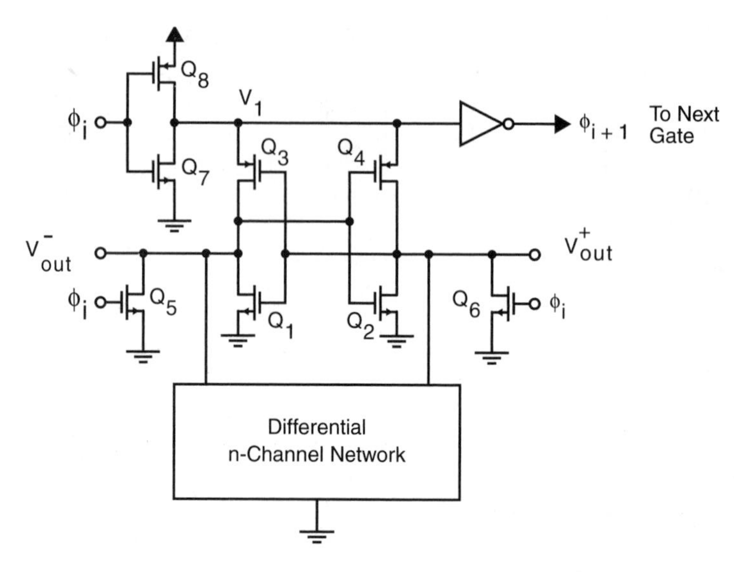

Figure 9.36 An alternative regenerative differential logic gate.

to V_{DD}. However, depending on which side of the differential n-channel tree network is a low impedance, either V_1^- or V_1^+ will be at a lower voltage. (The other node will be pulled all the way to V_{DD} by the p-channel load transistors.) Next, when ϕ goes to a "1", the dynamic p-channel load transistor is turned *off* and the dynamic sense amplifier *consisting* of Q_3, Q_4, and Q_5 is *powered-up.*[2] This cross-coupled circuit will quickly discharge whichever node, V_1^+ or V_1^-, was initially at a lower voltage. The other node, which was originally at V_{DD}, will be only slightly discharged. By using inverter output buffers that are designed to have low gate threshold voltages, it is easy to guarantee that the node that is left at a high voltage will be recognized as a "1", whereas the node that is discharged will be recognized as a zero.

This approach is especially useful when the n-channel network has many transistors in series. Examples of this include multi-input *exclusive-or* circuits that are required in applications such as parity-check and error-correction circuits or some built-in testing circuits. Another example is barrel-shifters.

An alternative approach for realizing a regenerative differential logic circuit is shown in Fig. 9.36 (Mead and Wawrzynek, 1985; Lu, 1988; Lu and Ercegovac, 1991). In this approach, all nodes are precharged low when ϕ_i = "1". When ϕ_i goes to "0", V_1 is pulled to a "1", which turns *on* the cross-coupled inverters consisting of, Q_1, Q_2, Q_3, and Q_4. This will quickly *pull* either V_{out}^+ or V_{out}^- high depending on which side of the differential n-channel network is low impedance.

[2]This dynamic sense amplifier is often used in dynamic RAMs and will be described in greater detail in the next chapter.

An interesting aspect of the gate of Fig. 9.36 is that assuming the devices are sized correctly, the signal $\phi_i +1$ will go low just after the outputs have reached their final values. If this signal is used as a clock signal for the next regenerative gate, then it will activate the next gate just when its inputs have become stable. Thus, we have a type of *self-timed logic*. It is expected that this type of logic will become more important in the future. Indeed, recently a complete microprocessor was realized using self timing logic (Jacobs and Brodersen, 1990).

9.7 Bibliography

K. Chu and D. Pulfrey, "A Comparison of CMOS Circuit Techniques: Differential Cascode Voltage Switch Logic Versus Conventional Logic," *IEEE Journal of Solid-State Circuits*, 22(4), 528–532, August 1987.

V. Friedman and S. Liu, "Dynamic Logic CMOS Circuits," *IEEE Journal of Solid-State Circuits*, 19, 263–266, April 1984.

N. Goncalvez and H. De Man, "NORA: A Racefree Dynamic CMOS Technique for Pipelined Logic Structures," *IEEE Journal of Solid-State Circuits*, 18(3), 261–266, June 1983.

T. Grotjohn and B. Hoefflinger, "Sample-Set Differential Logic (SSDL) for Complex High-Speed VLSI," *IEEE Journal of Solid-State Circuits*, 21(2), 367–369, April 1986.

L. G. Heller et al., "Cascoded Voltage Switch Logic: A Differential CMOS Logic Family," *ISSCC Digital Technical Papers,* 16–17, February 1984.

I. Hwang and A. Fisher, "Ultrafast Compact 32-bit CMOS Adders in Multiple-Output Logic," *IEEE Journal of Solid-State Circuits*, 24(2), 358–369, April 1989.

G. Jacobs and R. Brodersen, "A Fully-Asynchronous Digital Signal Processor," *IEEE Journal of Solid-State Circuits*, 26(5), 1526–1537, December 1990.

Y. Ji-Ren, I. Karlsson, and C. Svensson, "A True Single-Phase-Clock Dynamic CMOS Circuit Technique," *IEEE Journal of Solid-State Circuits*, 22(5), 899–901, October 1987.

R. Krambeck, C. Lee, and H. Law, "High-Speed Compact Circuits with CMOS," *IEEE Journal of Solid-State Circuits*, 17(3), 614–619, June 1982.

F. Lai and W. Hwang, "Design and Implementation of Differential Cascode Voltage Switch with Pass-Gate (DCVSPG) Logic for High-Performance Digital Systems," *IEEE Journal of Solid-State Circuits*, 32(4), 563–573, April 1997.

C. Lee and E. Szeto, "Zipper CMOS," *IEEE Circuits and Devices*, 10–16, May 1986.

S. Lu, "Implementation of Iterative Arrays with CMOS Differential Logic," *IEEE Journal of Solid-State Circuits*, 23, 1013–1017, August 1988.

S. Lu and M. Ercegovac, "Evaluation of Two-Summand Adders Implemented in ECDL CMOS Differential Logic," *IEEE Journal of Solid-State Circuits*, 26, 1152–1160, August 1991.

M. Mano and C. Kime, *Logic and Computer Design Fundamentals,* Prentice Hall, 1997.

C. Mead and J. Wawrzynek, "A New Discipline for CMOS Design: An Architecture for Sound Synthesis," *Proceedings of the 1985 Chapel Hill Conference: VLSI*, 87–104, 1985.

P. Ng et al., "Performance of CMOS Differential Circuits," *IEEE Journal of Solid-State Circuits*, 32(6), 841–846, June 1996.

A. Parameswar et al., "A Swing-Restored Pass-Transistor Logic-Based Multiply and Accumulate Circuit for Multimedia Applications," *IEEE Journal of Solid-State Circuits*, 31(6), 804–809, June 1996.

L. Pfennings et al., "Differential Split-Level CMOS Logic for Subnanosecond Speeds," *IEEE Journal of Solid-State Circuits*, 20(5), 1050–1055, October 1985.

J. Pretorius, A. Shubat, and C.A.T. Salama, "Latched Domino CMOS Logic," *IEEE Journal of Solid-State Circuits*, 21, 514–522, August 1986.

D. Rahakrishnan, S. Whitaker, and G. Maki, "Formal Design Procedures for Pass Transistor Switching Circuits," *IEEE Journal of Solid-State Circuits*, 20(2), 531–536, April 1985.

M. Suzuki et al., "A 1.5ns 32-b CMOS ALU in Double Pass-Transistor Logic," *IEEE Journal of Solid-State Circuits*, 28(11), 1145–1151, November 1993.

J. Uyemura, *Fundamentals of MOS Digital Circuits,* Addison-Wesley, 1988.

N. Weste and K. Eshraghian, *Principles of CMOS VLSI Design*, Addison-Wesley, October 1985.

K. Yano et al., "A 3.8ns CMOS 16×16-b Multiplier Using Complementary Pass-Transistor Logic," *IEEE Journal of Solid-State Circuits*, 25(2), 388–395, April 1990.

J. Yuan and C. Svensson, "High-Speed CMOS Circuit Technique," *IEEE Journal of Solid-State Circuits*, 24(1), 62–70, February 1989.

9.8 Problems

For the problems in this chapter, assume the following transistor parameters:

- npn bipolar transistors:
  $\beta = 100$
  $V_A = 80\ \text{V}$
  $\tau_b = 13\ \text{ps}$
  $\tau_s = 4\ \text{ns}$
  $r_b = 330\ \Omega$

- n-channel MOS transistors:
  $\mu_n C_{ox} = 190\ \mu\text{A/V}^2$
  $V_{tn} = 0.7\ \text{V}$
  $\gamma = 0.6\ \text{V}^{1/2}$
  $r_{ds}\ (\Omega) = 5000\text{L}\ (\mu\text{m})/I_D\ (\text{mA})$ in active region
  $C_j = 5 \times 10^{-4}\ \text{pF/}(\mu\text{m})^2$
  $C_{j\text{-sw}} = 2.0 \times 10^{-4}\ \text{pF/}\mu\text{m}$
  $C_{ox} = 3.4 \times 10^{-3}\ \text{pF/}(\mu\text{m})^2$
  $C_{gs(overlap)} = C_{gd(overlap)} = 2.0 \times 10^{-4}\ \text{pF/}\mu\text{m}$

- p-channel MOS transistors:
  $\mu_p C_{ox} = 50\ \mu\text{A/V}^2$
  $V_{tp} = -0.8\ \text{V}$
  $\gamma = 0.7\ \text{V}^{1/2}$

$r_{ds}\ (\Omega) = 6000L\ (\mu m)/I_D\ (mA)$ in active region
$C_j = 6 \times 10^{-4}\ pF/(\mu m)^2$
$C_{j\text{-}sw} = 2.5 \times 10^{-4}\ pF/\mu m$
$C_{ox} = 3.4 \times 10^{-3}\ pF/(\mu m)^2$
$C_{gs(overlap)} = C_{gd(overlap)} = 2.0 \times 10^{-4}\ pF/\mu m$

9.1 Analytically find the gate-threshold voltages of the pseudo-NMOS inverters shown in Fig. 9.1.

9.2 Analytically find the typical output low voltages of the pseudo-NMOS inverters shown in Fig. 9.1.

9.3 For the limited-swing pseudo-NMOS inverter shown in Fig. P9.3, find typical values for V_{OH} and V_{OL}. You may assume $V_{ref} = 1.65$ V.

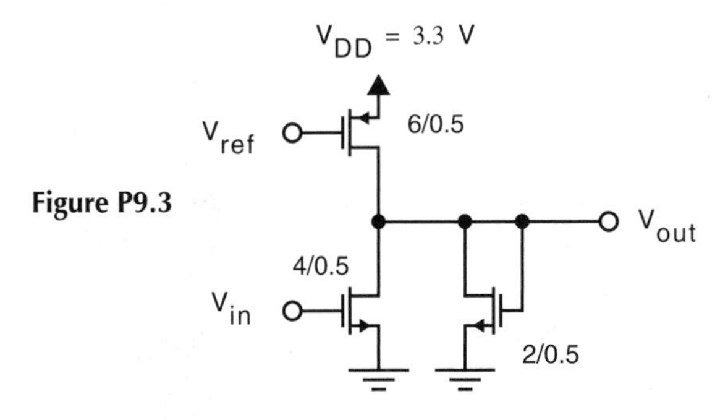

Figure P9.3

9.4 For the inverter shown in Fig. P9.3, find the gate-threshold voltage.

9.5 Design an 8-to-1 multiplexor using limited-swing pseudo-NMOS logic.

9.6 Design a four-bit input comparison circuit similar to Example 9.1, but using limited-swing pseudo-NMOS logic.

9.7 Design a CMOS full adder using Domino logic gates. You may assume inputs and their complements are available.

9.8 Design a CMOS parity-checker assuming Domino logic. Assume there are four inputs. If there is an even number of "1"s or "0"s, the output should be a "0", otherwise it should be a "1". Your solution should consist of a single building block that is replicated identically four times (i.e., once for each input bit). Assume inputs and their complements are available.

9.9 For the Domino *nand* gate shown in Fig. P9.9, and assuming the clock waveforms shown, estimate the voltage at node X. Assume the parasitic capacitance is 0.02 pF at every node except for node X, which has a parasitic capacitance of 0.05 pF.

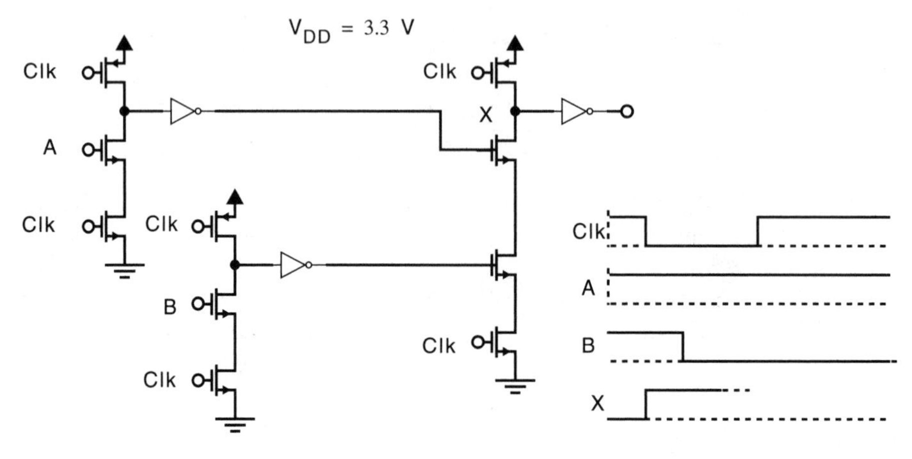

Figure P9.9

9.10 Design a multiple-output Domino gate that realizes a 3 to 1-of-8 decoder in which one and only one output will be high during the evaluate phase depending on the values of the three input signals.

9.11 Repeat Problem 9.10, but use NORA logic where the driver network is an n-channel network.

9.12 Using NORA logic, design a 6 to 1-of-64 output decoder. Give a hierarchical solution. Your solution should have at least two levels of gates.

9.13 For the D flip-flops of Fig. 9.24, give timing diagrams for every node for three periods when the output is connected back to the input and the flip-flops are initially in the "0" state with ϕ low.

9.14 For the latches of Fig. 9.26, give timing diagrams for every node for three periods when the latches are connected in cascade with the output of the second latch connected back to the input of the first latch. Assume the flip-flops are initially in the "0" state with ϕ high.

9.15 Using SPICE, simulate the carry-generate circuit of Fig. 9.27, but use the cross-coupled DSL load of Fig. 9.28. Assume $V_{ref} = V_{DD} = 3.3$ V. Try to come up with reasonable sizes for the load transistors. Discuss the rationale and trade-offs with respect to your choice of sizes.

9.16 Design a fully differential DSL inverter where V_{DD} = 3.3 V = V_{ref}. Give reasonable device widths. Estimate the voltages at all nodes for a "0" input.

9.17 Design a full adder using differential pass-transistor logic.

9.18 Modify the design of Problem 9.17 so that full-adder outputs are saved in a fully differential latch.

9.19 Design a four-bit input comparison circuit similar to Example 9.1, but using fully differential Domino logic.

9.20 Give a schematic for three stages of the carry-generate chain of pipelined full adders based on differential NORA gates.

9.21 Design a barrel shifter using self-timed logic whereby a 16-bit input can be shifted right an arbitrary number of places between 0 and 8. The bits on the left that are vacated should have "0"s shifted into them.

CHAPTER 10

Digital Integrated System Building Blocks

This chapter describes a number of digital integrated system building blocks. It is not intended to be all encompassing; rather, it is intended to give the reader exposure to a number of different circuits and methodologies. Some of the circuits described include *multiplexors, decoders, barrel shifters, counters, adders, subtractors, multipliers, and programmable logic arrays.*

10.1 Multiplexors and Decoders

A *multiplexor* is a circuit that makes the output equal to one of a number of possible data inputs as determined by address inputs. An example of a multiplexor was seen previously in Fig. 5.5 of Chapter 5, where n-channel *transmission gates* were used to realize an 8-to-1 multiplexor. Alternatively, this could be realized using CMOS transmission gates. Another alternative example of a multiplexor is shown in Fig. 10.1. This structure is somewhat similar to the transmission-gate multiplexor, except now the *tree* is only used to pull internal nodes *low* and a *pseudo-NMOS* load is used to *pull* internal nodes *high*. Alternatively, the pseudo-NMOS load could be replaced by a *dynamic load*. The multiplexor also has an output inverter that also serves the purpose as a buffer. If all the n-channel transistors in the inverted-tree network are taken as 10 μm wide, then a reasonable width for the pseudo-NMOS load might be 5 μm. This assumes all lengths are equal and taken as the minimum allowed.

A fourth alternative circuit for realizing an 8-to-1 multiplexor is based on using two levels of *nor* gates as shown in Fig. 10.2. Although this approach uses considerably more transistors, it can be laid out very densely using a *programmable logic array* (*PLA*)-like layout. The

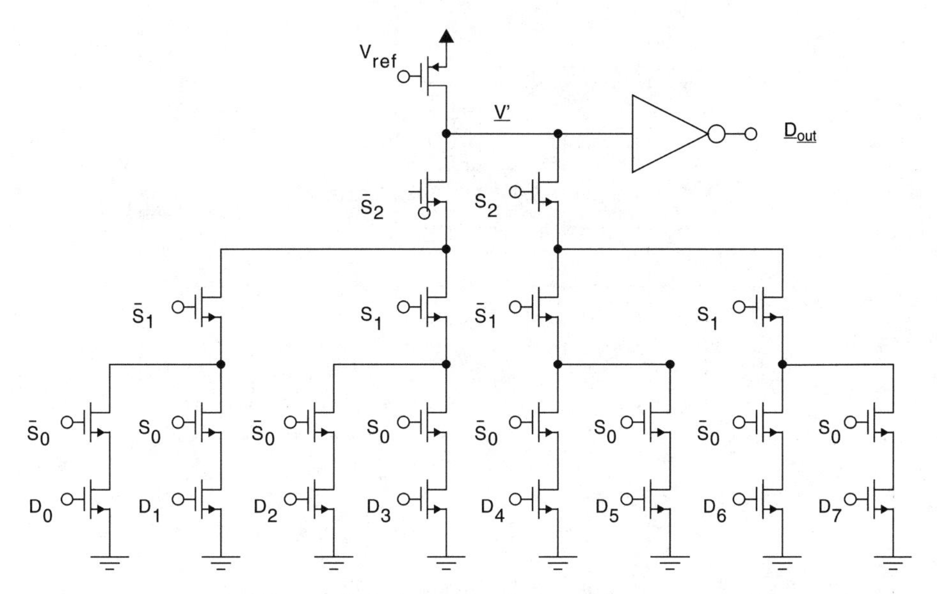

Figure 10.1 An *inverted-tree* multiplexor.

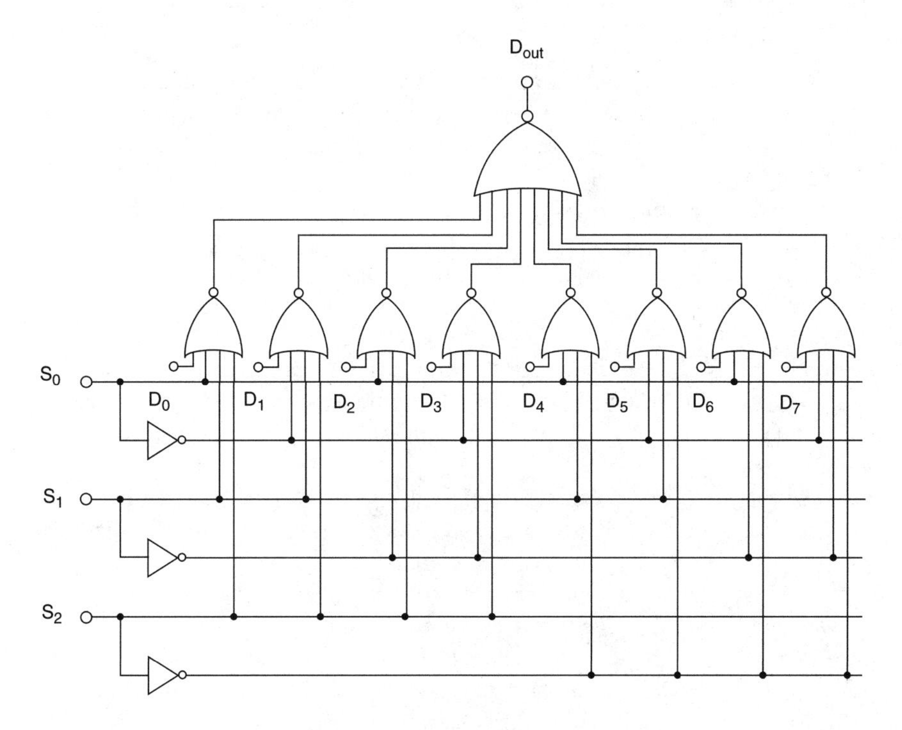

Figure 10.2 A *nor*-gate-based *multiplexor.*

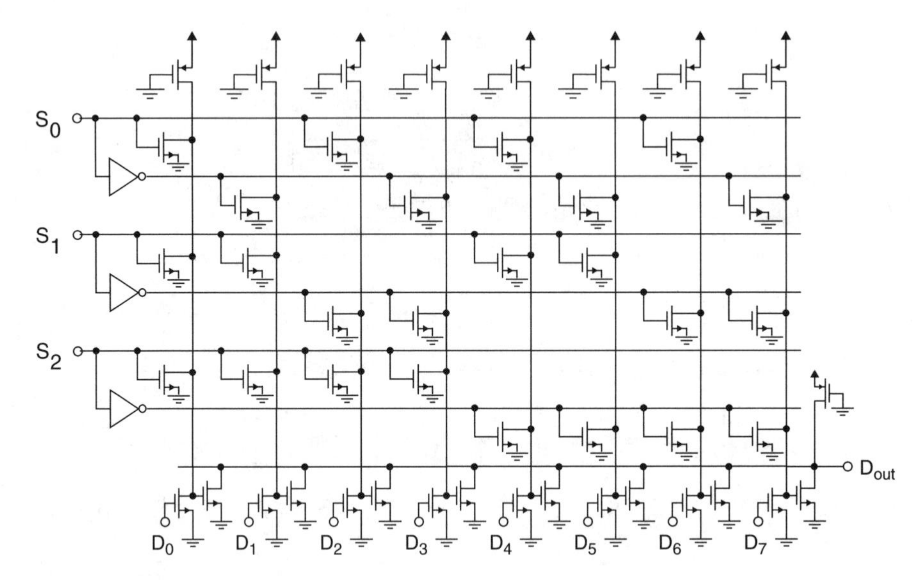

Figure 10.3 A PLA-like implementation of the circuit of Fig. 10.2.

transistor-level implementation of this circuit is shown in Fig. 10.3. It would be laid out in a very tight grid-like structure. Another advantage of this approach is that there are fewer n-channel transistors in series between any node and ground. Not only does this increase the speed of the circuit, but it also allows the n-channel transistors to be taken with smaller widths, which saves on area. As in the previous case, a modern design might have the pseudo-NMOS loads replaced with *dynamic* loads.

A fifth alternative, which combines a PLA-like structure and n-channel transmission gates, is shown in Fig. 10.4. Like the last example, the array of *nor* gates can be laid out in a very dense manner.

These represent just some of the alternatives for realizing multiplexors. Which one should be used in a given application will always depend on a trade-off between speed, layout area, and power. There seldom will be an absolute optimum and often one of a number of alternatives will suffice.

Bipolar current-mode multiplexors are almost always realized using a tree-like structure. For example, Fig. 10.5 shows a 4-to-1 multiplexor. Notice that only one of the differential pairs at the top level will be *powered-up* and its inputs will determine the output.

A circuit very similar to a multiplexor is a *decoder.* A decoder has n address inputs and 2^n outputs. Depending on the value of the address inputs, only one output will be high (or, in some variations, low). An example of a *3-to 1-of-8 decoder* based on a *tree structure* is shown in Fig. 10.6. In addition to not requiring many transistors, the output inverters can be taken wider and can therefore drive larger loads. Note that only one pseudo-NMOS load will

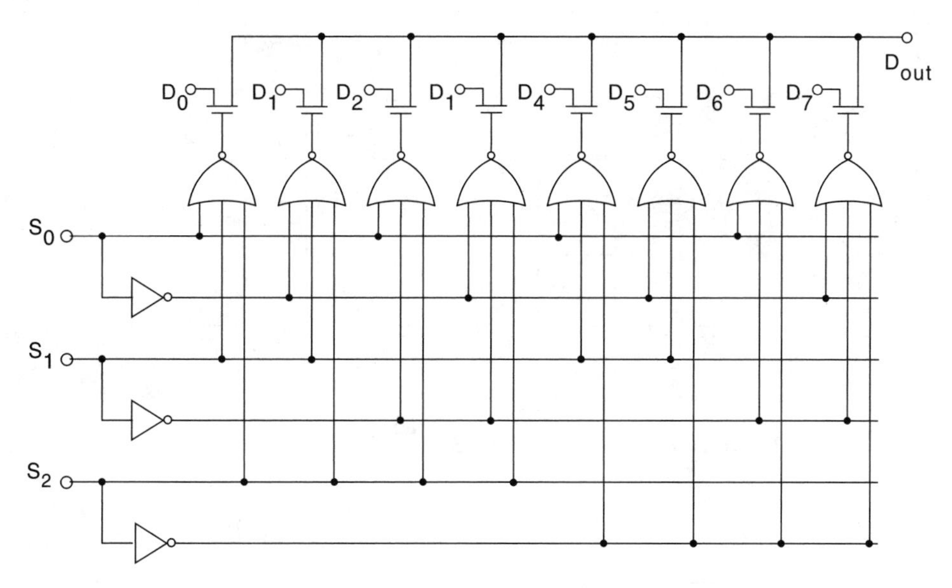

Figure 10.4 Another possibility for a multiplexor.

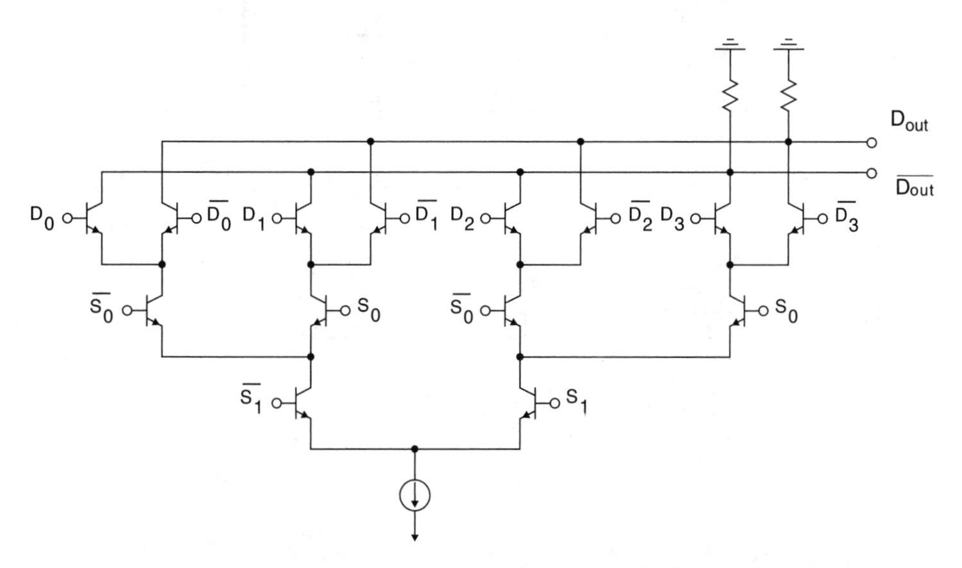

Figure 10.5 A current-mode 4-to-1 multiplexor.

be conducting at any given time, which helps minimize the power dissipation. Similar to realizing multiplexors, there are many different possibilities for realizing decoders. For example, if the transmission gates are removed from Fig. 10.4, a *nor*-gate-based decoder results. It should also be apparent that any decoder can easily be converted to a multiplexor by adding transmission gates to the outputs.

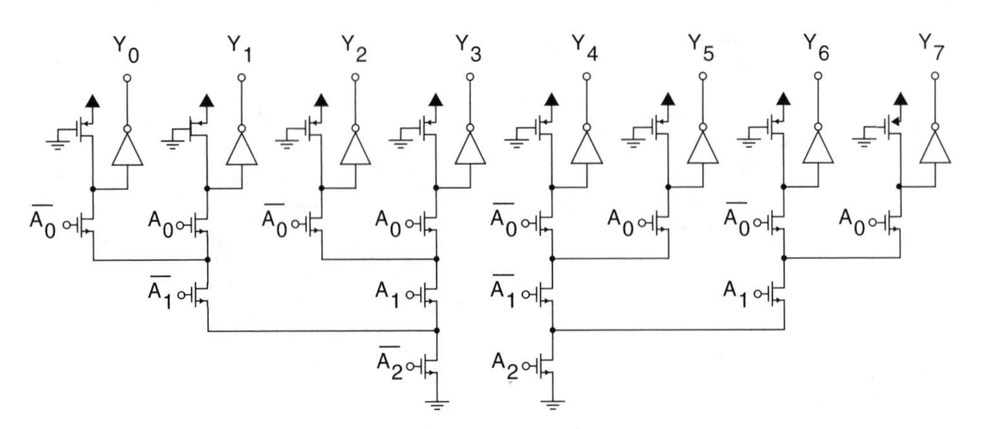

Figure 10.6 A 3-to 1-of-8 decoder based on a tree structure.

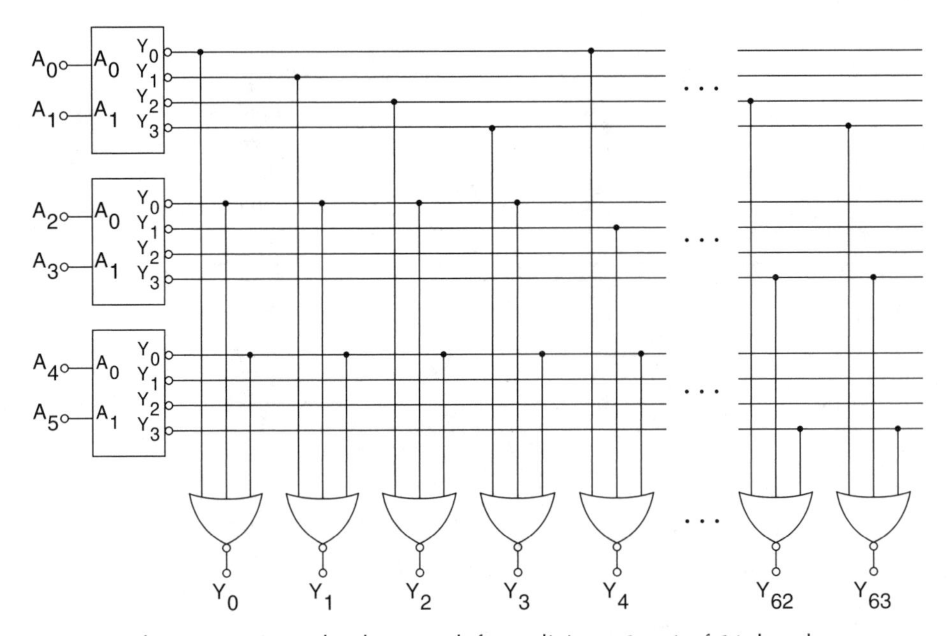

Figure 10.7 A two-level approach for realizing a 6-to 1-of-64 decoder.

When realizing decoders having more than eight outputs, a two-level approach is usually taken. For example, a *6-to 1-of-64 output decoder* can be realized using three *2-to 1-of-4 inverting decoders* followed by 64 3-input *nor* gates as shown in Fig. 10.7. The first level of inverting decoders might be realized using the circuit shown in Fig. 10.8. This circuit has only one output *low* at a given time. Since each of the first-level decoders has only one output low, there

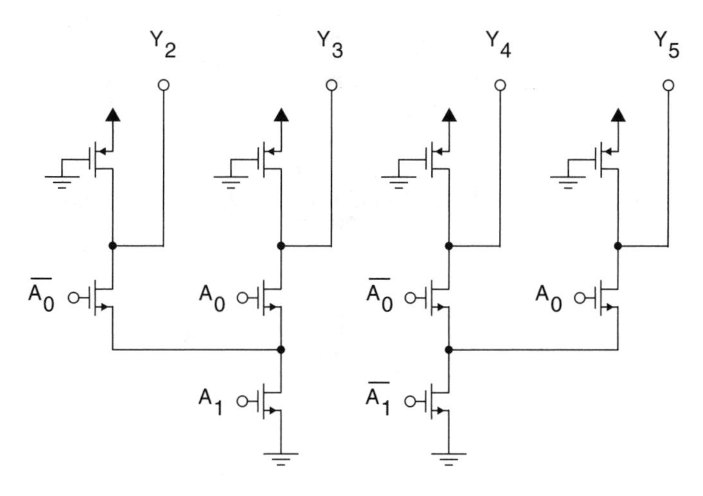

Figure 10.8 A four-output inverting decoder.

will be only a single nor gate that has all inputs *low* and a *high* output. As for the previous circuits, a modern realization would probably use *dynamic loads*. Also, when using a two-level approach, latches might be added to the outputs of the first-level decoders in order to *pipeline* the circuits when *latent time delay* is not a problem and large throughput is desired.

10.2 Barrel Shifters

A *barrel shifter* is a circuit that will shift the contents of a bus a specified number of positions left or right as specified by a control word. This is an important function in computers and many signal-processing ICs. Normally, when shifting to the right, the positions vacated will be filled with values from the left, or if no values are available, then filled with zeros; alternatively, when no values are available, the vacated positions may be filled with the value of the *most significant digit* (i.e., the most left digit). As a third alternative, some shifters may actually rotate the contents of a bus filling the least significant bits (LSBs) with the previous contents of the most significant bits (MSBs) for a *shift left* and vice versa for a *shift right*. The actual functionality is application dependent.

A barrel shifter essentially consists of a number of multiplexors, one for each bit of the bus. A barrel shifter also normally contains a decoder as well that is used to supply the necessary control bits to the multiplexors. For example, assuming a 32-bit bus, a barrel shifter might contain the equivalent of 32 32-input multiplexors and the decoder. For barrel shifters with only a few bits, the required functionality can be realized using a single level of logic as shown in Fig. 10.9. This architecture shifts right either zero, one, two, or three bits, assuming A_0 is the most significant input bit and B_0 is the most significant output bit. Also, the least significant bits are rotated into the vacated most significant positions. For example, if $S_{h1} = 1$ (and

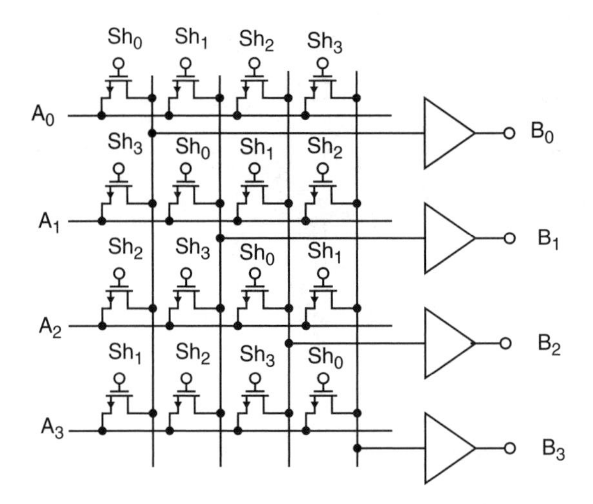

Figure 10.9 A 4-bit shifter with a programmable shift of 0 to 3 bits. The least significant bits are rotated into vacated most-significant positions.

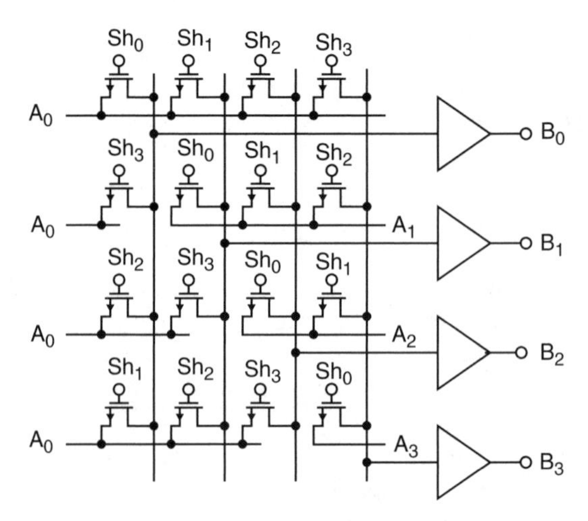

Figure 10.10 A 4-bit shifter with a programmable shift of 0 to 3 bits. The vacated most-significant positions are filled with the most-significant bit of the input.

$S_{hi} = 0$, $i = 2, 3, 4$), then a shift right one position takes place. For this example, $B_3 = A_2$, $B_2 = A_1$, $B_1 = A_0$, and A_3 is rotated into B_0. In a slightly modified architecture, the vacated bits are filled with the most significant bit of the input as shown in Fig. 10.10 (Rabaey, 1996). In this example, assuming $S_{h1} = 1$, then as before, $B_3 = A_2$, $B_2 = A_1$, and $B_1 = A_0$, but now B_0 also equals A_0.

In the 4-bit examples just presented, each output bus is the equivalent of a 4-bit *nor* gate. If this approach was used for a 32-bit barrel shifter, the junction capacitance on each line would be excessively large. For larger barrel shifters, a multistage approach is necessary, similar to what has been previously described for decoders. For example, assume a bus has 32 bits. Then a five-stage approach might be adopted. The first stage would shift the input either left by one bit, right by one bit, or leave it unshifted. The second stage would shift its input either left by two bits, right by two bits, or leave it unshifted. The third stage would possibly shift left or right by four bits, the fourth stage would possibly shift left or right by eight bits, and the fifth stage would possibly shift left or right by 16 bits. The combination of all five stages would allow a shift left or right by any integer between 0 and 31. The details of an architecture that realizes this algorithm are described in Rabaey (1996).

10.3 Counters

Counters are basic building blocks that are often used in logic design, particularly for generating timing signals. There are many ways of designing counters (Taub and Schilling, 1977). Perhaps one of the most common is the *ripple counter* shown in Fig. 10.11 realized using JK flip-flops. Assuming the flip-flops change on the negative edge, then once every negative edge the output of the first flip-flop will change its state. Thus, its output will be a square wave at one-half the frequency of the input clock. Similarly, the output frequency of the second flip-flop will be one-fourth the input frequency, the third flip-flop will toggle at one-eighth the input frequency, and the last flip-flop will toggle at one-sixteenth the input frequency. *Counters that produce an output signal that is an integer submultiple of an input or clock signal are often called dividers.*

The maximum input frequency of this counter is constrained only by how fast the first flip-flop can toggle. This makes the ripple counter one of the fastest available, particularly if a *biphase* flip-flop is used for the first stage. However, the time from when the first flip-flop changes to when the last flip-flop has settled can be quite large. This is because the changes ripple through from flip-flop to flip-flop. This makes the counter a poor choice for use in *synchronous* circuits. However, for cases in which the *ripple delay* is not a problem, this counter can be an excellent choice. An example of this is its use as a frequency divider in *phase-locked loops* used in very high-speed satellite communication systems.

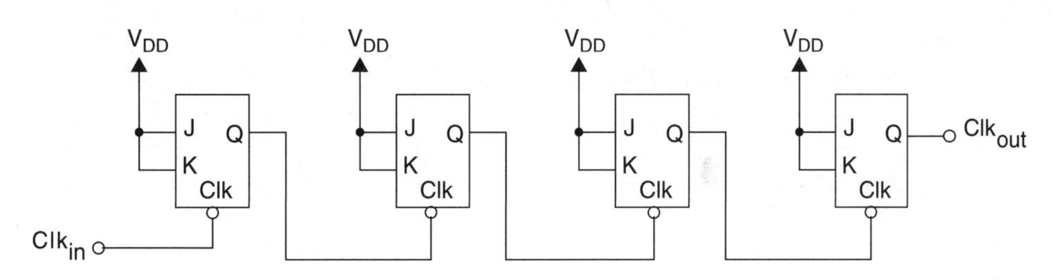

Figure 10.11 A ripple counter.

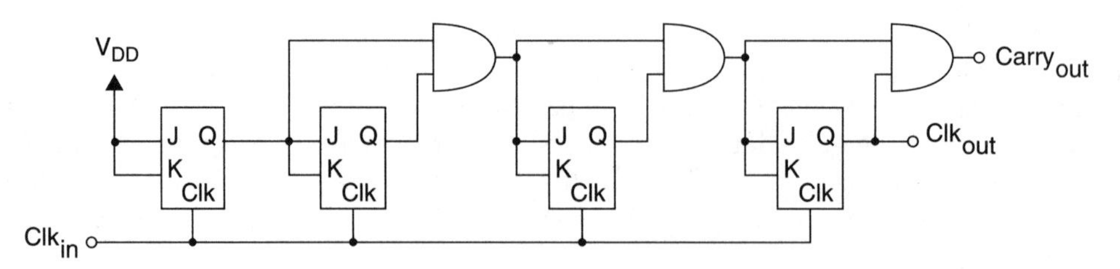

Figure 10.12 A serial-carry synchronous counter.

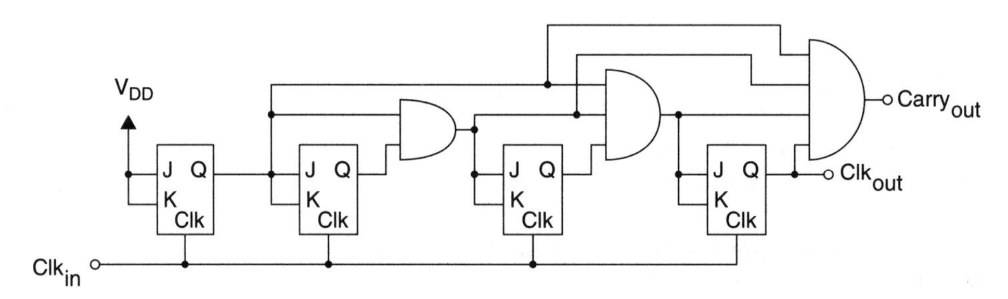

Figure 10.13 A carry-lookahead counter.

An example of a *synchronous counter* is shown in Fig. 10.12. This counter is synchronous since a single clock input is used for all of the flip-flops. The *and* gates are used as a carry-in to each flip-flop. If all of the preceding flip-flops are in the "1" state, then the output of the *and* gate will be one and the next flip-flop will toggle on the next activating clock edge.

This counter is called a *serial-carry counter.* Its major speed limitation is the time it takes the signal to ripple through the *and* gates when all *and* gates are changing.

A third type of counter is shown in Fig. 10.13. This counter is often called a parallel or *carry-lookahead counter.* In this synchronous counter, a separate *and* gate is used for generating every carry-in. Thus, there is only one gate delay from the time the clock changes to when all carry-ins are stable, although the delay can be moderately large for the *and* gates having a large fan-in.

All of the preceding counters were examples of four-bit binary counters, that is, counters with states that change in a *binarily* increasing or decreasing fashion. There are many other possible sequences for counters. An example of a popular nonbinary counter is a *Gray-code counter.* This counter has only one output change state every active clock edge. An example design of a Gray-code counter was presented previously in Example 7.4 of Chapter 7. Another often encountered alternative is to cascade two counters in a *ripple-carry* fashion. For example, the first counter might be a *three-state counter,* whereas the second counter could perhaps be a *seven-state counter* with the overall division ratio being the product of the two, namely 21, for our example.

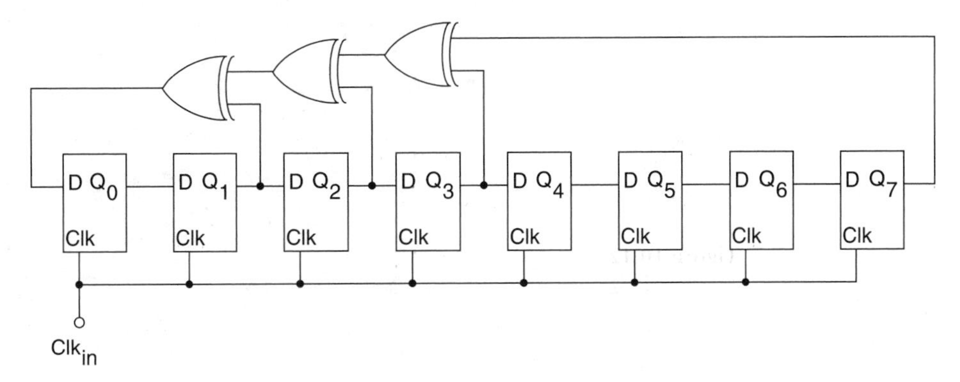

Figure 10.14 An eight-bit maximal-length shift register.

Another example of a commonly used counter is a *maximum-length sequence generator.* This type of counter is also called a *pseudo-random noise generator.* For an n-bit shift register, it cycles through $2^n - 1$ different states without repeating a single state twice. Furthermore, the value of the output in a particular state is very random and has very little correlation with the output of any other state in time. This type of counter is used extensively in *on-chip* testing as we shall see in Chapter 13.

It is realized using a delay line. Depending on the length of the delay line, certain flip-flop outputs are fed through a multi-input *exclusive-or* gate whose output is used as the input of the delay line. For example, an eight-bit *maximal-length* shift register is shown in Fig. 10.14 where a four-input *exclusive-or* circuit is actually implemented as three two-input *exclusive-or* circuits.[1] This circuit will cycle through every possible state, except the "0" state where all flip-flops have zero outputs. If it ever gets into that state, possibly at power-up, then it will never leave it. This situation is called *lockout.* It can be prevented by adding an asynchronous set to one of the flip-flops and setting it at power-up time. Alternatively, the state of all "0"s can be detected, and if it occurs, additional logic can be used to override the output of the *exclusive-or* gate and to input a "1" to the delay line. A third alternative is to realize the counter using *exclusive-nor* gates rather than *exclusive-or* gates. In this case, the *lockout state* is all "1"s, which can be prevented by adding an *asynchronous reset* to one or more flipflops.

There are many possibilities in choosing which flip-flop outputs to use as inputs to the *exclusive-or* (or *exclusive-nor*) gates, when constructing maximal-length shift registers. Some possibilities for shift-register lengths containing 4 to 15 flip-flops are shown in Table 10.1.

[1]This might be an application in which the multiple-output Domino CMOS *exclusive-or* gate presented in the last chapter would be considered.

Table 10.1 Feedback Connections That Result in Maximal-Length Pseudo-Random Shift Registers (the number of flip-flops, n, equals N + 1)

N	*Ex-or* gate inputs (first FF has index "0")
4	$Q_2 \oplus Q_3$
5	$Q_2 \oplus Q_4$
6	$Q_4 \oplus Q_5$
7	$Q_1 \oplus Q_2 \oplus Q_3 \oplus Q_7$
8	$Q_4 \oplus Q_8$
9	$Q_6 \oplus Q_9$
10	$Q_8 \oplus Q_{10}$
11	$Q_2 \oplus Q_3$
12	$Q_1 \oplus Q_9 \oplus Q_{10} \oplus Q_{11}$
13	$Q_0 \oplus Q_{10} \oplus Q_{11} \oplus Q_{13}$
14	$Q_1 \oplus Q_{11} \oplus Q_{12} \oplus Q_{13}$
15	$Q_{13} \oplus Q_{14}$

10.4 Digital Adders

Digital *adders* are one of the most important logic subcircuits and therefore have been discussed many times previously in this book. In this chapter, much of the material discussed previously in different locations is brought together and expanded on. In addition, some techniques used for realizing *fast* multibit adders are introduced (Hwang, 1979; Waser and Flynn, 1982; Ercegovac and Lang, 1985).

Single-Bit Adders

A single-bit *adder* has three inputs, A, B, and *carry-in*, C_{in}, and based on these inputs generates the two outputs, *sum*, S, and *carry-out*, C_{out}, according to the logic equations

$$S = A \oplus B \oplus C_{in} \tag{10.1}$$

for the *sum-generate function*, and

$$C_{out} = C_{in}(A + B) + AB \tag{10.2}$$

for the *carry-generate function*. An example of a *full adder* that realizes this function is repeated from Fig. 4.32 and is shown in Fig. 10.15 with the simplification shown in Fig. 4.35. This circuit is a popular choice for a full adder used in logic-cell libraries.

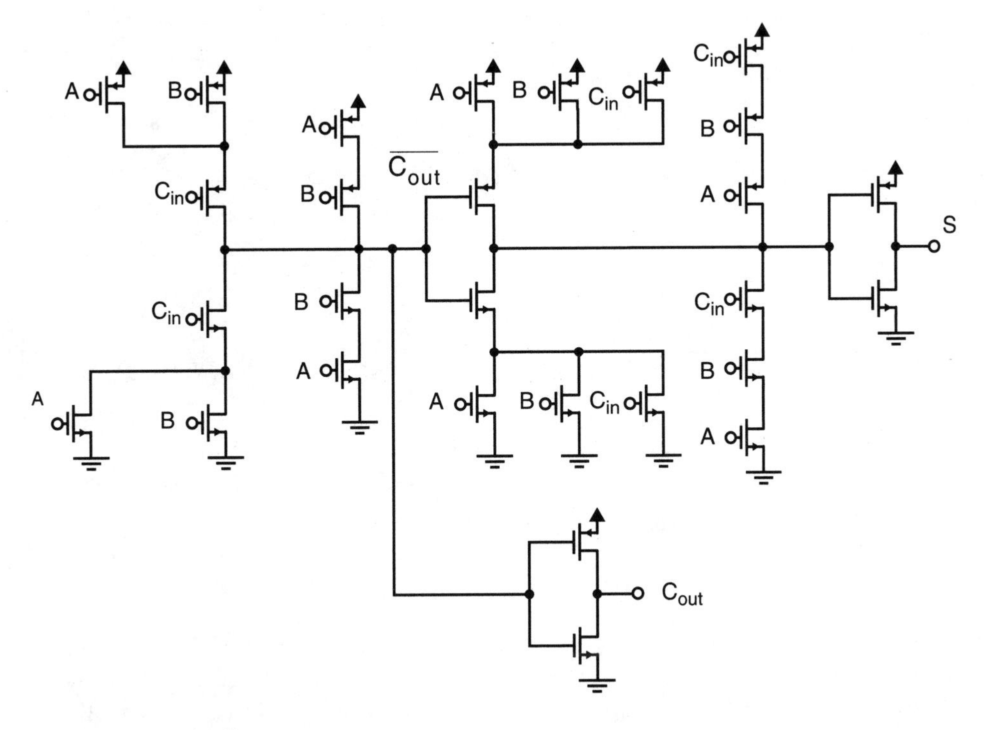

Figure 10.15 A CMOS realization of the full-adder function.

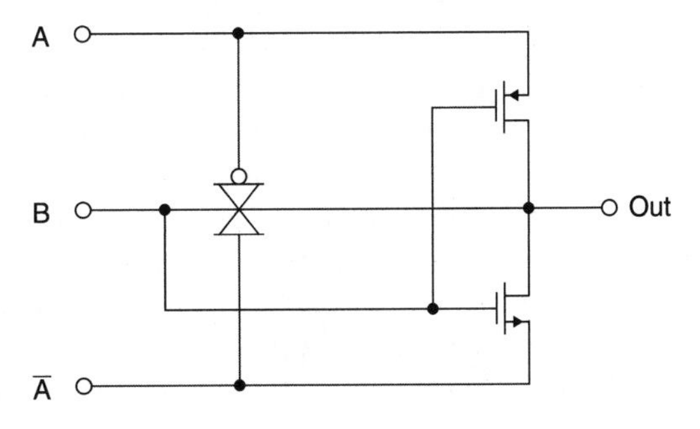

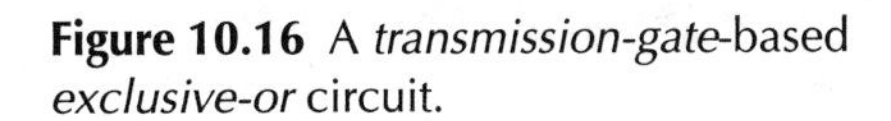
Figure 10.16 A *transmission-gate*-based *exclusive-or* circuit.

Another popular approach is based on using the *exclusive-or* circuit introduced in Problem 5.13 of Chapter 5 and shown in Fig. 10.16. In this *exclusive-or* circuit, when A is *low*, the inverter is disabled and the output is equal to B, which is transmitted through the transmission gate. When A is high, the transmission gate is disabled and the inverter is powered-up. Since its input is B, the output will be $\overline{B}$. Thus, the function realized is Out $= \overline{A}B + A\overline{B}$ as desired. This *exclusive-or* circuit, although simple, is not buffered between the input and output, and

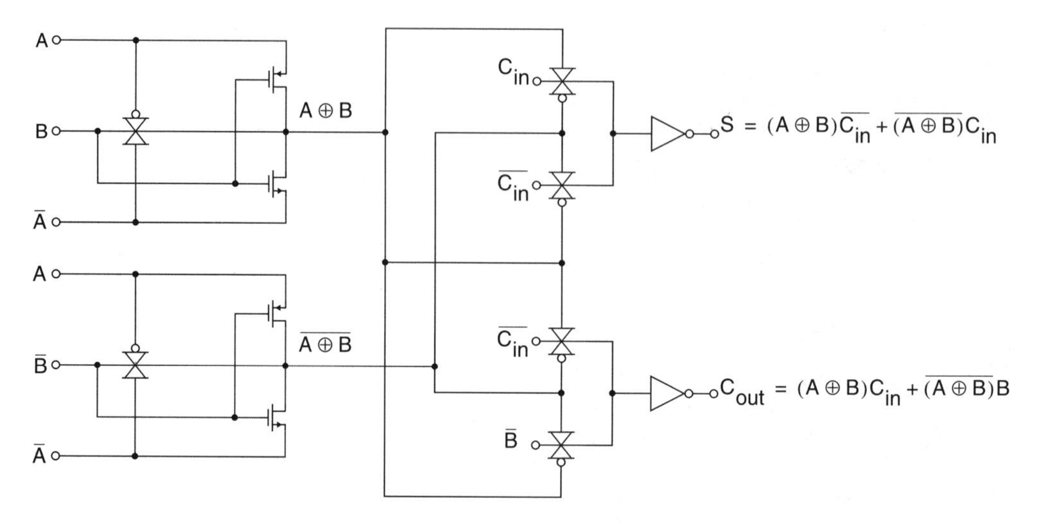

Figure 10.17 A *transmission-gate*-based full adder.

therefore it is not possible to cascade a large number of them. Two of these circuits and two two-input multiplexors with some inverters can be used to realize a full adder based on the realization that $S = A \oplus B \oplus C_{in}$ can be realized using $A \oplus B$ to control a two-input multiplexor having $\overline{C_{in}}$ and C_{in} as the inputs. Furthermore, $A \oplus B$ and $\overline{A \oplus B}$ can be used to generate C_{out} by using a two-input multiplexor having B and C_{in} as the inputs. That is, when $A \oplus B$ is true, then $C_{out} = C_{in}$. When $\overline{A \oplus B}$ is true (i.e., A and B are either both "0" or both "1"), then $C_{out} = B$. Based on these observations, the transmission-gate-based full adder shown in Fig. 10.17 results. Not shown in Fig. 10.17 are the two inverters required to realize $\bar{B}$ and $\overline{C_{in}}$.

A third alternative for a single-bit full adder, based on differential circuit design, was described in Section 9.5 of Chapter 9 and is shown again in Fig. 10.18. It has been modified to have no d.c. power dissipation. This particular implementation, and variations of it, where, for example, the cross-coupled loads are replaced by dynamic Domino-CMOS loads, are the author's preferred realizations.

In addition to the sum-generate and carry-generate functions, there are two other single-bit functions that are often used when realizing multibit adders. These functions are the *generate function*, G, and the *propagate function*, P. The generate function is a "1" if it is known that the carry-output of a single-bit adder is a "1" independent of the carry-in values. This function is given by

$$G = A \cdot B \tag{10.3}$$

The propagate function will be a "1" when the carry-out value is simply the carry-in value irrespective of the values for A or B. Its function is given by

$$P = A + B \tag{10.4}$$

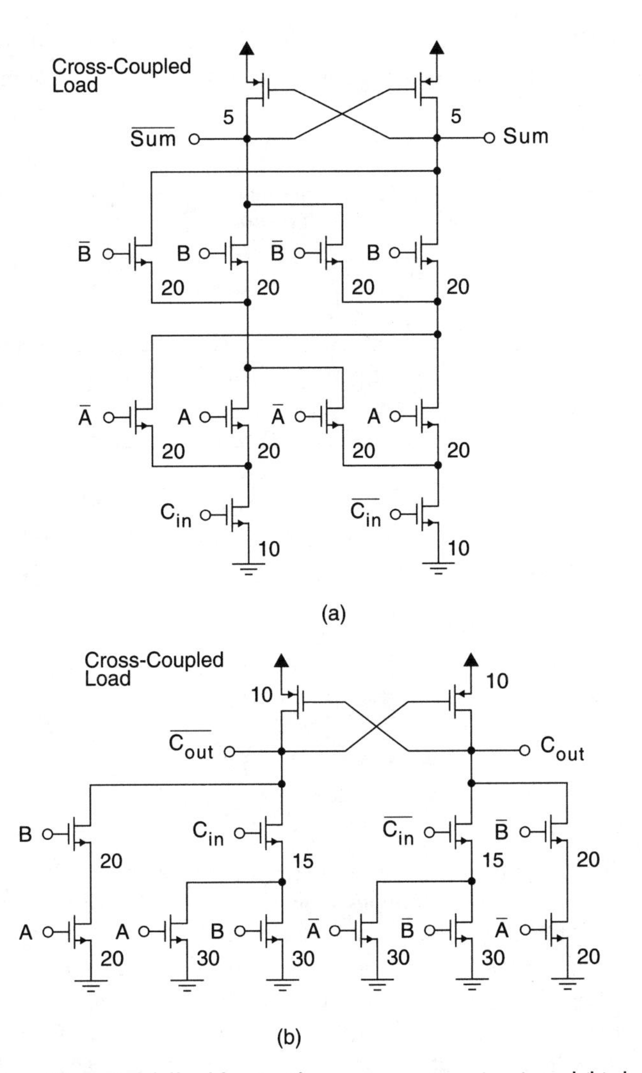

Figure 10.18 A fully differential full adder. (a) The sum-generate circuit and (b) the carry-generate circuit.

An alternative function that can be used in most realizations based on carry-generate and propagate circuits is

$$P = A \oplus B \tag{10.5}$$

This function is not a proper propagate function for when A and B are both "1"s, but in this case, we know the generate function will be a "1", and, in most multibit realizations, this guarantees the correct result, as we shall see. Also, in many realizations, using (10.6) for the propagate function allows the same circuit to be used for generating the sum-out using the additional function

$$S = P \oplus C_{in} \tag{10.6}$$

As a final note, when multibit adders are described, all of the variables will be given subscripts. For example, C_{in} for the ith bit will be labeled C_{in_1}.

Ripple-Carry Adders

The simplest way of realizing a multibit adder is to simply cascade a number of single-bit adders where the carry-out of the ith bit, C_{out_i}, is connected to the carry-in of the next bit, C_{in_i+1}, as we have seen many times previously. This architecture is a very reasonable choice when only a few bits are to be added, particularly when fully differential circuitry is used that is optimized for the carry-propagate function.

Even when fully differential circuits are not used, a *ripple-carry adder* is a reasonable choice if one makes use of the fact that complementing all inputs and outputs of a single-bit adder leaves the functions unchanged. For example, consider the function for the complement of the carry-out.

$$\begin{aligned} \overline{C_{out}} &= \overline{(AB) + C_{in}(A+B)} \\ &= [\overline{AB} \cdot \overline{C_{in}(A+B)}] \\ &= (\bar{A} + \bar{B}) \cdot (\overline{C_{in}} + \bar{A}\bar{B}) \\ &= \bar{A}\bar{B} + \overline{C_{in}}(\bar{A} + \bar{B}) \end{aligned} \tag{10.7}$$

What this means is that the complement of the output can be obtained by using the same circuitry, but simply complementing all inputs. The same is true for realizing the complement of the sum output. Now, if in single-ended realizations of full adders all inputs are complemented, and the inverters at the outputs are eliminated, then the actual noncomplemented outputs are obtained with less delay. This technique can be used to realize fast ripple-carry adders. Assume the symbols used for a regular full adder, a full adder with all inputs and outputs complemented, a full adder with its output inverters eliminated, and a full adder with inputs complemented but outputs uncomplemented (i.e., the output inverters eliminated) are as shown in Fig. 10.19, respectively. Note that the circuits for Fig. 10.19a and b are identical as are the circuits for Fig. 10.19c and d. Further, assume the complemented inputs of all signals are available. This might be the case, for example, if they come from registers having differential outputs. A fast four-bit ripple-carry adder can now be realized as shown in Fig. 10.20, where each block is simply an ordinary full adder with its output inverters eliminated.

In the next few sections, some alternative techniques to ripple-carry adders will be discussed.

Carry-Save Adders

This technique is important when a number of additions are necessary. An example of this might be in a digital-signal processing application or an array multiplier (as described later).

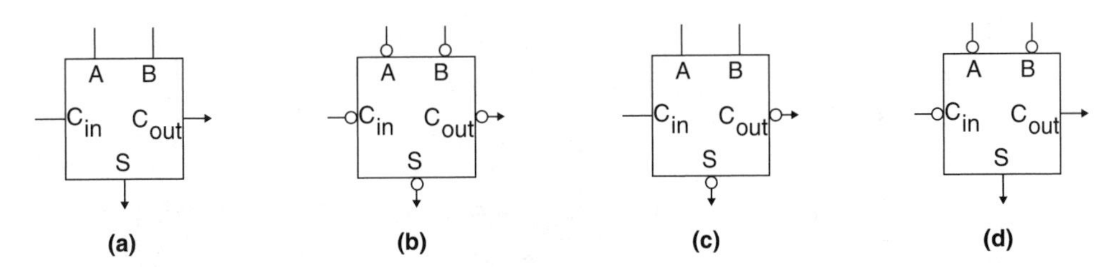

Figure 10.19 The symbols for (a) a full adder, (b) a full adder in which all inputs and outputs have been complemented, (c) a full adder in which the inverters at the output have been eliminated, and (d) a full adder in which all inputs have been complemented and the inverters at the outputs have been eliminated.

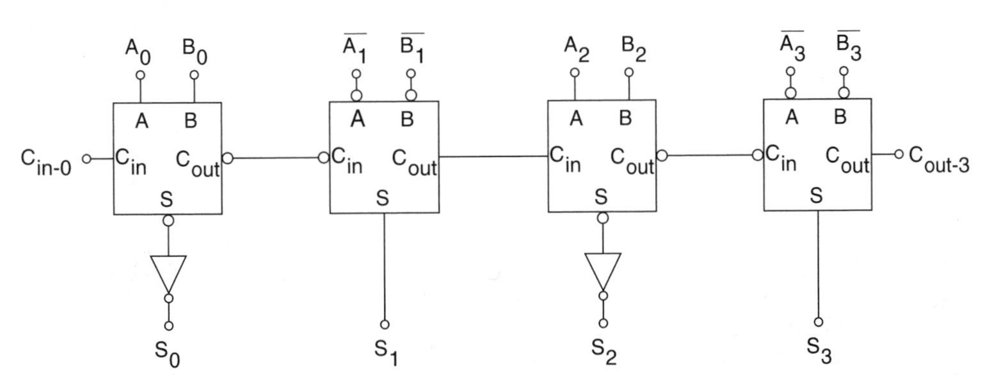

Figure 10.20 A fast four-bit ripple-carry adder based on using complemented inputs.

The basic idea is to save the carry propagates until the very last addition, where a very fast adder is used that is optimized for fast carry-propagate speed. At all levels before the last, the carry-outputs are not rippled through the adders. Rather, they are shifted to a more significant position and added to the carry-ins of the next adder. To be specific, consider the carry-out of the ith bit of an adder. This will be sent to the i+1th carry-in of the next adder. This is shown in Fig. 10.21 for adding four five-bit binary numbers, A, B, C, and D. Notice that the top-level carry-outs are shifted right before being added into the next stage down. The same is true for the carry-outs at the second level down. The critical path for this architecture would be through the two left-most full adders and then the carry propagation through the bottom row of full adders. The delay through this path would normally be minimized through the use of techniques such as carry-propagate adders or carry-select adders to be described shortly.

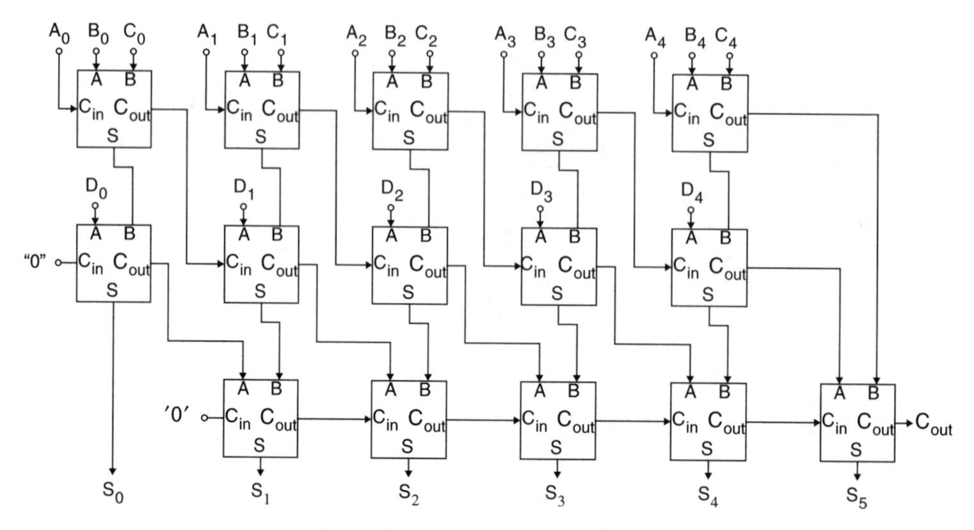

Figure 10.21 A carry-save adder.

CARRY-LOOKAHEAD, CARRY-PROPAGATE, AND CARRY-SELECT ADDERS

There have been many different architectures proposed during the last five or six decades for realizing fast multibit adders. All of these techniques are based on being able to calculate the carry propagation much faster without having to wait for it to ripple through all of the full adders. Many of the algorithms, such as the *carry-lookahead* technique, are based on calculating the carry signal for a number of bits at the same time with circuits that have fewer gates in the signal path. Unfortunately, this often involves using gates that have a large numbers of inputs. The fact that multi-input gates can be very slow has often been largely ignored in arguments for one technique versus another. Another consideration that has seldom been taken into account when evaluating the merits of one technique versus another is the loading on gates that are part of the critical path. The answer to many of these questions is beyond the scope of this text; this section is rather intended as an introduction to some of the concepts involved in realizing fast multibit adders, with additional details given for the *carry-select adder*, normally a good choice.

A carry-lookahead adder is based on calculating the carry-out for a block of bits in parallel to, and separately from, calculating the sum outputs of the block. Using (10.3) and (10.4), at the ith bit, the carry-out is given by

$$C_{out\_i} = G_i + P_iG_{i-1} + P_iP_{i-1}G_{i-2} + \ldots + P_i \ldots P_1G_0 \tag{10.8}$$

Based on this formula, the carry-out of a four-stage adder can be found using the formula

$$C_{out\_3} = G_3 + P_3G_2 + P_3P_2G_1 + P_3P_2P_1G_0 + P_3P_2P_1P_0C_{in\_0} \tag{10.9}$$

Besides the circuits required to find the carry generates and carry propagates for each bit, four *and* gates, one of them being a five-input *and* gate, and a four-input *or* gate are required. Therefore, carry-lookahead realizations require substantial additional circuitry and have gates with very large fan-ins in the *critical path.* Because of these large fan-ins, it is the author's opinion that the critical path for a four-bit block would be substantially slower than the critical path based on a ripple-carry block realized using the differential circuits of Fig. 10.24. For this reason, it is recommended that carry-lookahead techniques not be used in an IC[2] and they will not be discussed further. (They are covered in some detail in Weste and Eshraghian, 1993.)

Most of the other techniques for realizing fast adders are based on generating generate and propagate signals for a block of bits at a time. When this is done for a single-bit full adder, the required equations are as given by (10.3) and (10.4). When calculating the signals for a larger group of bits at a single time, the functionality required is as follows: if the carry-out of the block is guaranteed to be a "1" irrespective of the carry-input to the block, then the block-generate signal, G_{blk}, is a "1"; otherwise, it is a zero; if a block carry-in equal to a "1" would cause the block carry-out to be "1", then the block propagate signal, P_{blk}, is a "1", otherwise it is a "0". The block carry-out can be calculated using the formula

$$C_{out} = C_{in} \cdot P_{blk} + G_{blk} \tag{10.10}$$

The signals G_{blk} and P_{blk} in a *carry-propagate adder* would be found at the same time as the C_{out} of previous blocks is being found. Once the C_{out} of the previous block is available, which is the C_{in} to the current block, then the carry-out of the current block can be found in only one additional gate delay. This is the basic idea behind a carry-propagate adder. There have been many different realizations proposed based on these ideas. A number of popular approaches based on using transmission gates are described in Weste and Eshraghian (1993). Rather than try and describe all of the different approaches that have been proposed, we will describe a single approach based on the use of fully differential circuits to realize a particular version of a carry-propagate adder that is often called a carry-select adder.

A carry-select adder is based on the following realizations: a block carry-generate signal can be found by finding the block carry-output with a "0" carry-in to the block; in addition, the block-propagate signal can be found by finding the block carry-out assuming a "1" carry-in to the block. For example, assuming four-bit blocks, then two four-bit ripple-carry adders can be used, the first having a "0" block carry-in, and the second having a "1" clock carry-in. The block carry-out of the first four-bit adder is the block-generate signal and the block carry-out of the second four-bit adder is the block-propagate signal. Once these are found, the block carry-out can be found using (10.10). In addition, the sum outputs of the block can be found by selecting the outputs of the first or second four-bit adder using a multiplexor controlled by the block carry-in. This is shown symbolically for a four-bit block in Fig. 10.22. Differential realizations

[2]This opinion is controversial and not necessarily widely accepted.

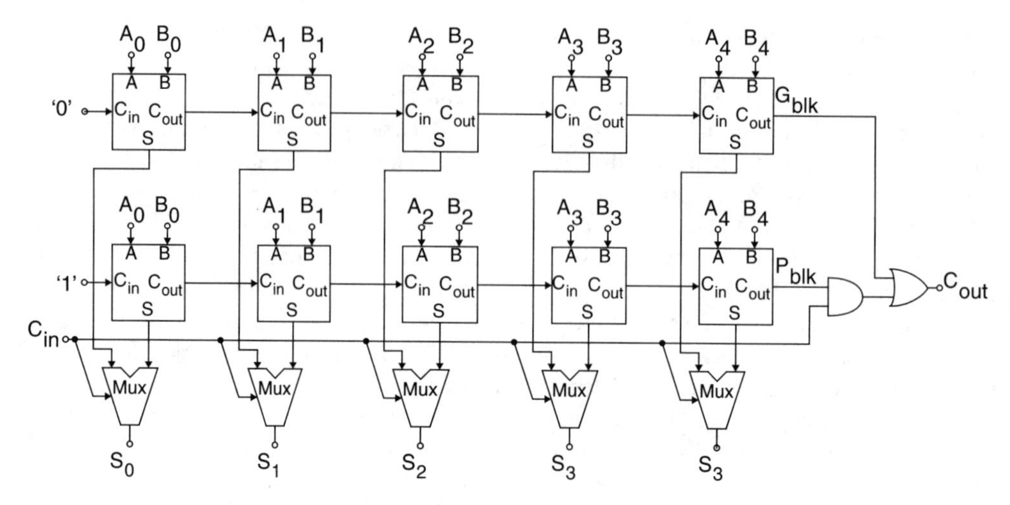

Figure 10.22 A four-bit block used in a carry-select adder.

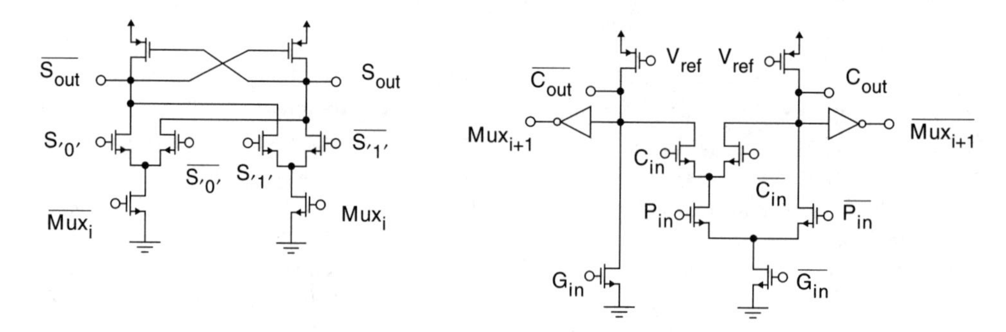

Figure 10.23 Differential realizations of (a) the multiplexor and (b) the logic required for generating C_{out} in Fig. 10.22.

of the multiplexor and the logic required to generate the carry-out of the block are shown in Fig. 10.23. A number of implementation details should be considered:

1. The loading of the carry-in to each block must be minimized. Since four multiplexors are controlled by this signal, inverter buffers are included for this noncritical path. These inverters might be taken fairly small except for the ones driving the multiplexors of the last block. The generation of the carry-out of each block is a minimally loaded critical path and therefore this output is not buffered.
2. The time delay from the block carry-in to the block carry-out is critical. Therefore, the gates in this path might be realized using pseudo-NMOS loads as shown in Fig. 10.23b. Also, the carry-in signal is connected to transistors close to the output. Finally, the sizes of this gate might be taken comparatively larger than that of other gates.

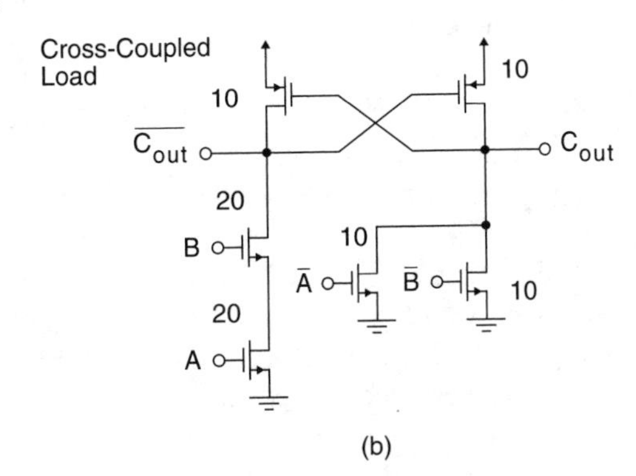

Figure 10.24 A fully differential full adder in which the carry-in is known to be a "0". (a) The sum-generate circuit and (b) the carry-generate circuit.

3. The generation of the carry-out of the first block is also critical. Therefore, a separate additional set of four carry-generate circuits for this block might be used that are not loaded down by sum-generate blocks. These gates might be realized using pseudo-NMOS loads. This would mean two sets of carry-generate circuits for the first block, but this is not a large penalty to pay for the speed increase. Also, the first block does not require two sets of sum-generate circuits or multiplexors, so the total area required for the first block is still less than for the other blocks.
4. In large adders, speed and some area can be saved by taking the blocks larger by one more bit per block, after the first two blocks, to account for the delay of the gates that produce the block carry-outs (Weste and Eshraghian, 1993; Rabaey, 1996). For example, the block sizes might be taken as 4, 4, 5, 6, 7, 6 for a 32-bit adder. For smaller adders, it is not certain this justifies the lack of regularity. For adders larger than 32 bits, it almost certainly should be done.
5. The full-adder blocks that have "0" or "1" inputs can be simplified by removing transistors that are known to be *off* and replacing transistors known to be *on* by simple connections. For example, the sum-generate and carry-generate blocks with a carry-in equal to "0" are shown in Fig. 10.24a and b, respectively. The circuits for the carry-in being a "1"

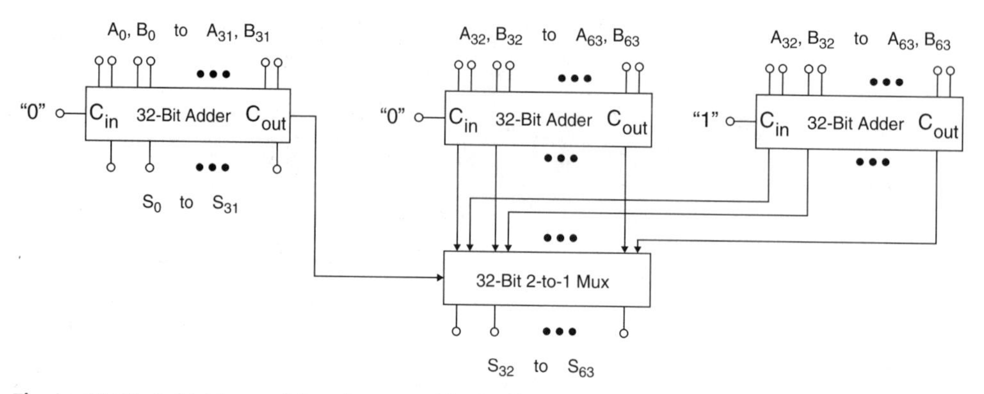

Figure 10.25 A 64-bit conditional-carry adder based on using three 32-bit adders and a 32-bit wide multiplexor.

are similar (see Problem 10.7). Note how the simplifications have also allowed some of the transistor widths to be taken smaller.

One other technique for realizing fast multibit adders will be introduced: that of *conditional-carry adders*. The idea is rather simple and is an extension of the architecture shown in Fig. 10.22. It will be described for a 64-bit adder example. Assume three 32-bit fast adders have been designed using the carry-select techniques just described. One of these 32-bit adders will be used to calculate the least significant bits; the other two 32-bit adders and a 32-bit wide 2-to-1 multiplexor will be used to calculate the most significant bits. One of these latter adders will have its carry-in taken as a "0" and the other will have its carry-in taken as a "1". The outputs of one of these latter adders will be selected by the multiplexor under the control of the carry-out of the 32-bit adder used to calculate the least significant bits. This is shown in simplified form in Fig. 10.25. This 64-bit adder is only slightly slower than a single 32-bit adder by the delay of the 32-bit wide 2-to-1 multiplexor. The trade-off involved is that the area is increased by about 65% over the area by just extending a 32-bit conditional-carry adder to the 64-bit level. In the author's opinion, for most applications, the increase in speed justifies the increase in area.

Digital Subtractors

When implementing digital arithmetic that allows for negative numbers, the most popular strategy by far is to use *2s-complement* arithmetic (Roth, 1985). In this methodology, positive numbers have an additional most-significant bit added that is always a "0". Thus, a 32-bit number can realize positive numbers with the largest magnitude being given by

$$A_{max} = 2^{31} - 1 \tag{10.11}$$

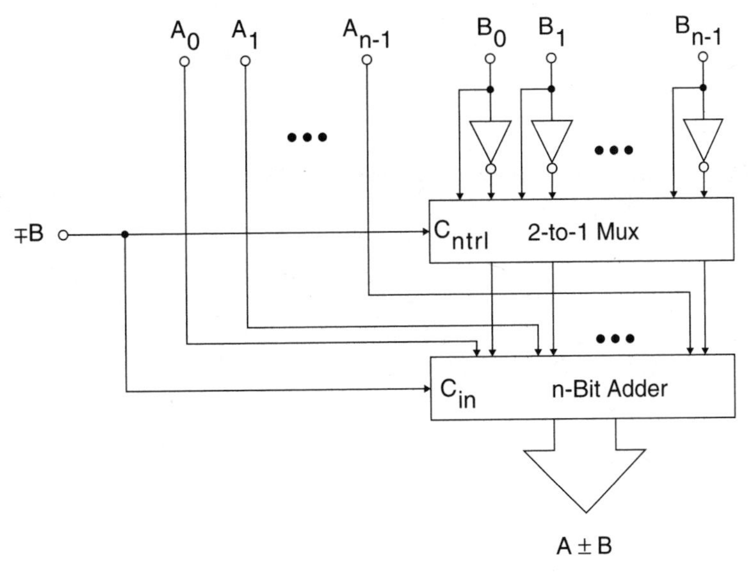

Figure 10.26 Realizing the subtractor/adder using 2s-complement arithmetic.

A negative number is realized by taking a positive number having the same values for all bits, complementing the bits, and then adding "1". Thus,

$$-A = \bar{A} + 1 \tag{10.12}$$

The major advantage of this methodology is that the procedure for adding a positive number to a negative number is the same as when adding two positive numbers.[3] In addition, when implementing a subtractor in a 2s-complement system, based on (10.12) we see it is simply necessary to invert all inputs and set the carry-in of the least significant bit to a "1" to realize the subtraction operation. This is illustrated in Fig. 10.26.

10.5 Digital Multipliers

A digital multiplier (Hwang, 1979; Ma and Taylor, 1980; Waser and Flynn, 1982; Ercegovac and Lang, 1985) is one of the more complicated arithmetic functions introduced in this text. Normally, a multiplication operation of positive numbers is broken down into a number of

[3]Assuming an n+1-bit adder is used to add two n-bit numbers.

conditional additions of a multiplicand based on values of a multiplier. The procedure is started by considering the least significant bit of the multiplier. If this is a "1", then the multiplicand is sent to the next level, otherwise all zeros are input. At the second level, the second least significant bit of the multiplier controls the operation. If it is a "1", then the multiplicand, after shifting it left by one bit, is added to the partial product produced by the first level. The shift left reflects the fact that the second least significant bit of the multiplier is weighted with twice the value as the least significant bit. At each further level, the multiplicand, shifted left by one additional bit, is conditionally added to the partial product produced by the previous level depending on the corresponding bit of the multiplier. This procedure works irrespective of whether the multiplicand is negative assuming *2s-complement arithmetic* is used, assuming that at each level an additional bit is added to account for sign extension. However, without modification, it cannot handle negative multipliers. The procedure in which the multiplier is constrained to be positive is best illustrated through the use of an example.

Example 10.1

For each bit of the multiplier, assumed to be equal to the binary equivalent of 7, give the partial product when multiplied by a multiplicand equal to –11 assuming five bits are used for both the multiplier and the multiplicand.

Solution: The binary value for 11, represented using five bits, is given by 01011. When this is converted to –11, using (10.12), we have

$$\begin{aligned} -11 &= \overline{01011} + 1 \\ &= 10100 + 1 \\ &= 10101 \end{aligned} \tag{10.13}$$

This multiplicand will be conditionally added at each stage depending on the corresponding bit of the multiplier M = 00111. Starting from the least significant bit, and using sign extension at each level, we have.

$$P_0 = 10101 \times 1 = 110101 = -11 \tag{10.14}$$

Notice how six bits have been used to represent the result due to the need for sign extension so overflow does not occur. At the next level, the multiplicand is shifted left and we have

$$\begin{aligned} P_1 &= P_0 + 101010 \times 1 \\ &= 110101 + 101010 \times 1 \\ &= 1011111 \\ &= -33 \end{aligned} \tag{10.15}$$

as expected since $-33 = -11 + 2 \times (-11)$. Again, the result has been sign extended so that now seven bits are used for the partial product. Continuing,

$$\begin{aligned} P_2 &= P_1 + 1010100 \times 1 \\ &= 10110011 \\ &= -77 \end{aligned} \tag{10.16}$$

The partial product is now represented using eight bits. Next,

$$\begin{aligned} P_3 &= P_2 + 10101000 \times 0 \\ &= 110110011 \\ &= -77 \end{aligned} \tag{10.17}$$

using nine bits. This completes the multiplication, except for a final sign extension, as it was assumed the multiplier was positive. The final result is therefore

$$P_4 = S = 1110110011 = -77 \tag{10.18}$$

This algorithm can be realized using a parallel architecture as shown in Fig. 10.27. The full adders that have "0" inputs can of course be simplified into what is normally called half adders, that is, adders having only two inputs. Notice also how the carry-outs at each level are added into the next level down; this is the equivalent of the sign extension described in the previous example.

To handle negative multipliers as well as negative multiplicands, the algorithm must be modified, although the modification is not large. The modification is based on the following fact concerning 2s-complement arithmetic (Ercogovac and Lang, 1985): because of the modular nature of the representation, any 2s-complement number is equivalent to

$$x = \sum_{i=0}^{n-1} M_i 2^i - 2^n \cdot \text{sign}(x) \tag{10.19}$$

Using this fact, and the fact that the $\text{sign}(x) = M_{n-1}$, one can express any 2s-complement number as

$$x = \sum_{i=0}^{n-2} M_i 2^i + M_{n-1}(2^{n-1} - 2^n) \tag{10.20}$$

Notice that when M_{n-1} is "0", then (10.20) is simply the binary sum of the first $n-1$ bits properly weighted. When M_{n-1} is "1", then (10.20) is equivalent to (10.19). Continuing, we have from (10.20)

$$x = \sum_{i=0}^{n-2} M_i 2^i - M_{n-1} 2^{n-1} \tag{10.21}$$

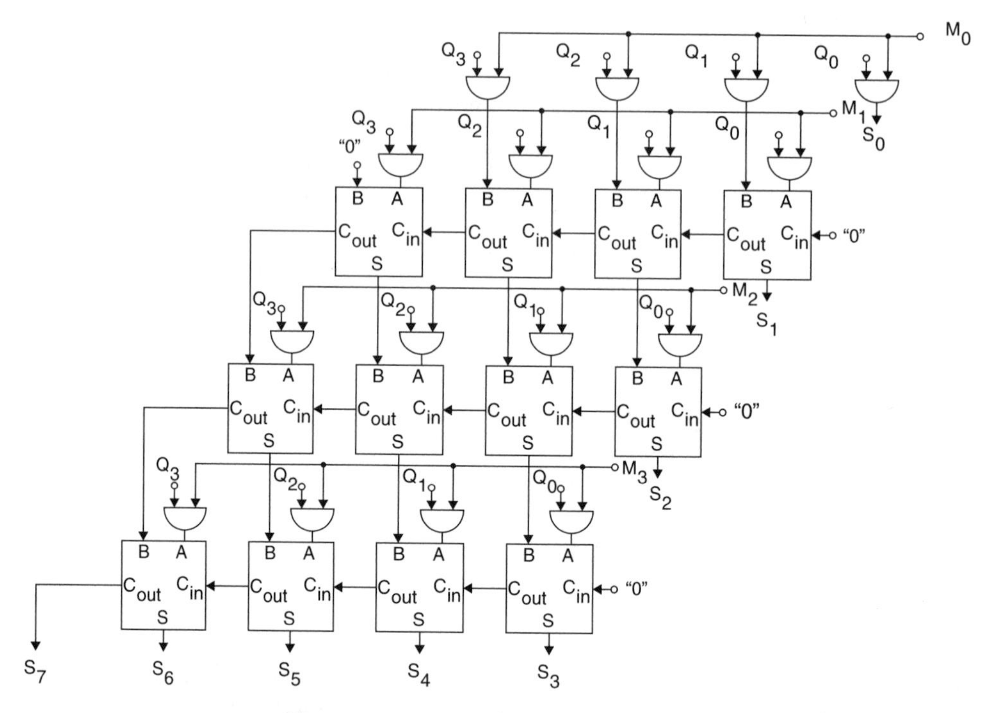

Figure 10.27 A 4×4-bit array multiplier.

Based on this expression, we see both positive and negative multipliers can be handled by proceeding as if the multiplier was positive for the first n–1 bits, but then for the last bit of the multiplier subtracting the multiplicand if it is a "1". Thus, at the last level, the carry-in is set to "1" rather than a "0", and the multiplicand bits are inverted; using fully differential logic, these modifications are minor.

MODIFIED BOOTH MULTIPLIERS

Most modern multipliers are based on an algorithm called the *modified Booth algorithm* (Booth, 1951). This algorithm handles two multiplier bits at each level and also handles negative multipliers in a manner identical to how positive multipliers are handled. It is based originally on the realization that to allow both addition and subtraction operations, then one need only add or subtract the multiplicand at transitions from a "0" to a "1" or a "1" to a "0". For example, consider having a multiplier equal to 63 represented using eight bits. In binary representation, this is equivalent to

$$M = 63 = 00111111 \tag{10.22}$$

Using an algorithm similar to that used in Example 10.1, we see that six additions would be required, one for each bit equal to "1". Now consider expressing 63 as

$$M = 64 - 1 = 01000000 - 00000001 \quad (10.23)$$

Based on this expression, we see the multiplication can be simplified to one addition and one subtraction. Whenever a large number of consecutive "1"s appear in the multiplier, the operation can be simplified into fewer operations using a combination of an addition and a subtraction. Consider now a number of consecutive "0"s followed by a number of consecutive "1"s. For example, consider 56 as a multiplier

$$56 = 00111000 \quad (10.24)$$

Equation (10.24) may be expressed as

$$\begin{aligned} 56 &= 00111000 \\ &= (01000000 - 000010000) \\ &= (64 - 8) \\ &= (2^6 - 2^3) \end{aligned} \quad (10.25)$$

Thus, a multiplication by 56 is achieved by subtracting the multiplicand after shifting it left three positions from the multiplicand shifted left six positions. The *recoding* scheme can be summarized as follows:

1. If the ith bit is a "1" after a string of "0"s, then subtract 2^i where the indexing goes from 0 to $n-1$ for n bits.
2. If the ith bit is a "0" after a string of "1"s, then add 2^i.
3. For $i = 0$, assume the $i-1$ bit was a "0".

This algorithm remains unchanged for handling negative multipliers in a 2s-complement system. The proof of this is left as an exercise for the reader.

In implementations realized using parallel multipliers, this algorithm is modified into what is known as the modified Booth algorithm. In this algorithm, two multiplier bits are processed at each level resulting in only $n/2$ levels. At each level, one of five possible different operations is performed: nothing, add or subtract the multiplicand, or shift the multiplicand left and then add or subtract it. To see how this operation works, consider the following example in which the multiplier bits $i-1$ to $i+1$ are given as $M_{i+1}M_iM_{i-1} = 101$. This would correspond to adding the multiplicand, after shifting it left i bits, because of the 1-to-0 transition at the ith bit and then subtracting the multiplicand after shifting it left $i+1$ bits because of the 0-to-1 transition at the $i+1$th bit. These two operations can be combined into a single operation by simply subtracting the multiplicand after shifting it left i bits. In a similar manner, the operations for other bit combinations are found and summarized in Table 10.2. The operation performed at each level may be summarized as $Z_{i/2} = -2M_{i+1} + M_i + M_{i-1}$ for $i = 0, 2, \ldots, n-2$, assuming n is even and M_{-1} is assumed to be a "0".

Table 10.2 The Recording Operations of the Modified Booth Algorithm

Bit				
M_{i+1}	M_i	M_{i-1}	Operation	
0	0	0	No operation (no transition)	0
0	0	1	Add multiplicand (a 1-to-0 transition at bit i: Q)	+Q
0	1	0	Add the multiplicand (a 0-to-1 transition at bit i and a 1-to-0 transition at bit i+1: 2Q – Q)	+Q
0	1	1	Add twice the multiplicand (a 1-to-0 transition at bit i+1: 2Q)	+2Q
1	0	0	Subtract twice the multiplicand (a 0-to-1 transition at bit i+1: –2Q)	–2Q
1	0	1	Subtract the multiplicand (a 1-to-0 transition at bit i and a 0-to-1 transition at bit i+1: –2Q + Q)	–Q
1	1	0	Subtract the multiplicand (a 1-to-0 transition at bit i: –Q)	–Q
1	1	1	No operation (no transition)	0

Table 10.3 The Recoded Signals Corresponding to the Modified Booth Algorithm

Signal	Logic function	Description
$u_{i/2}$	$\overline{M_i \oplus M_{i-1}}$	A "1" means shift left
$v_{i/2}$	$\overline{M_{i-1}}$	A "1" means subtract
$w_{i/2}$	$M_{i+1} M_i M_{i-1} + \overline{M}_{i+1} \overline{M}_i \overline{M}_{i-1}$	A "1" means no operation

In most implementations, the multiplier bits are recoded into three signals that are distributed through the array based on Table 10.2. These signals are summarized in Table 10.3.

Example 10.2

Design a fully differential logic gate that produces $w_{i/2}$ and $\overline{w_{i/2}}$ assuming fully differential multiplier bits as inputs.

Solution: A possible solution that realizes this function is shown in Fig. 10.28. Note the inclusion of output buffers as the recoded signals need to be supplied to a large number of gates.

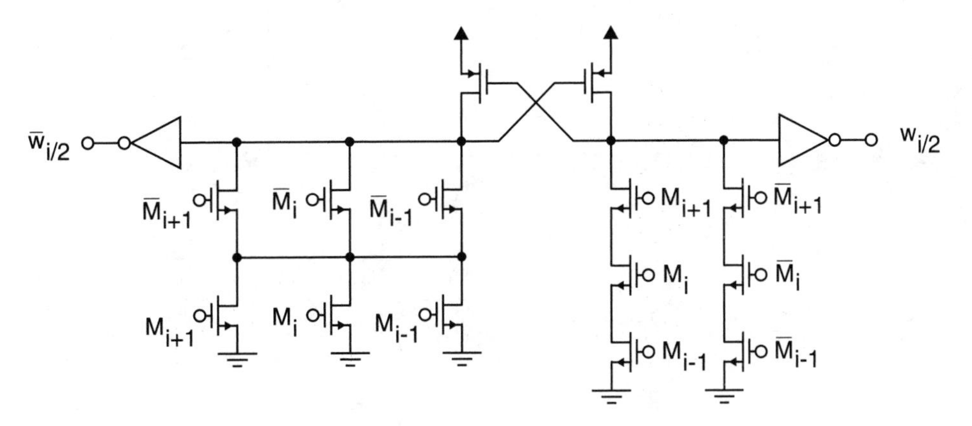

Figure 10.28 A possible fully differential implementation of the recoding circuit that controls the *no-op* function.

When realizing a modified Booth algorithm multiplier, logic circuits need to be added in front of the full adders at each level. These circuits will select the multiplicand or the multiplicand shifted left by one bit depending on the value of $u_{i/2}$, *complement* it if a subtraction is occurring dependent of the value of $v_{i/2}$, and force a "0" to be input when $w_{i/2}$ is a "1" irrespective of the values for the other two *recoded* signals. In addition, when the multiplicand is being subtracted, then a "1" is sent into the carry-in of the least significant bit. This operation will often be realized by an additional half adder at the next level down.

Example 10.3

Design a fully differential logic gate that given multiplicand bits Q_j and Q_{j-1}, and recoded signals $u_{i/2}$, $v_{i/2}$, and $w_{i/2}$, produces an output that can be used for the A_j input of the jth full adder at the i/2 level. Assume the complements of all inputs are available.

Solution: A possible solution that realizes this function is shown Fig. 10.29. Notice that when $w_{i/2}$ is a "1", A_j is forced to a zero irrespective of the other inputs.

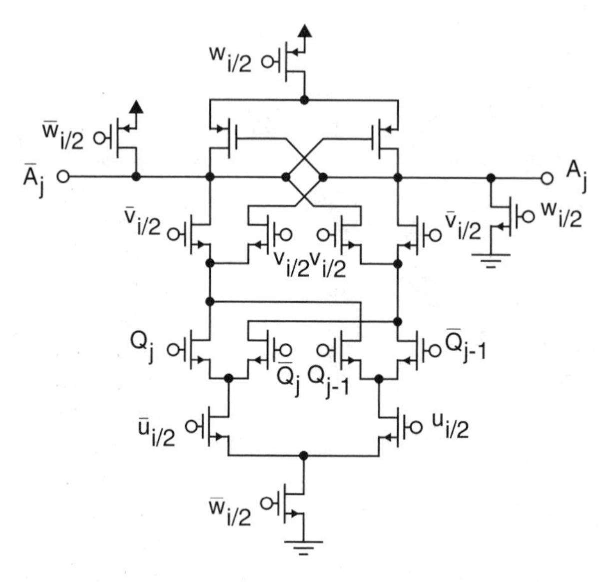

Figure 10.29 A logic gate that can be placed in front of full adders when implementing a modified Booth algorithm multiplier.

When implementing the modified Booth algorithm, the sign bit at each level usually needs to be extended by two bits (Ercegovac and Lang, 1985). The details of this are beyond the scope of the present text. Another modification that is usually used in modern multipliers is *carry-save addition* with a simple but effective modification. In traditional carry-save adders, the carry-outs of each full adder are shifted left and added at the next level down. It is possible to modify this algorithm so the carry-outs are shifted left by two bits and then added two levels down from the level at which they are produced (Iwamura et al., 1984; Oowaki et al., 1987). Similarly, partial-product outputs skip a level and are added in two levels down the array. This algorithm results in dividing the partial-product levels into even and odd levels. The propagation through the odd levels takes place at the same time as, but independent of, the propagation through the even levels. The final partial product of the odd levels is then added to the final partial product of the even levels at the end of the array using an adder designed to have a very fast carry-propagate time using techniques described in the last section. This results in cutting the delay for the signals propagated vertically through the array by almost one-half for large multipliers. In addition, this speed-up is attained with the multiplier array retaining a relatively regular structure. Again, the details of this architecture are beyond the scope of this text and readers are referred to Oowaki et al. (1987) for more details.

10.6 Programmable Logic Arrays

Any acyclic logic function can be realized by two levels of logic gates, if both the inputs and their complements are available. For example, if Karnaugh map reduction is used, and the "1"s are grouped, then the minimized functions consist of *and* gates followed by *or* gates. These can be realized by two levels of *nand* gates. Alternatively, if the "0"s are grouped, then the functions can be realized by a level of *or* gates feeding into *and* gates or equivalently two levels of *nor* gates. Most other logic-function minimization techniques are such that the resulting circuits are to be realized by two levels of *nand* or *nor* gates, at the user's discretion.

A *programmable logic array (PLA)* is a very efficient realization of a number of two-level logic functions. Inside ICs, the *nor-nor* implementation is normally preferred as it results in a fewer number of series transistors in the drive networks of gates and therefore faster operation.

PLAs are very popular in modern digital IC design for a number of reasons. Once the size of the PLA has been determined, then all the layers except one can be laid out irrespective of the logic functions being implemented. This makes it possible to completely automate the generation of layouts for a PLA. In a typical program used to generate PLAs, the user might implement the truth tables required for the logic functions. These would be minimized by the program, which would then also produce and optimize the layout. PLAs are often used in decoders, controllers for microprocessors, and state machines. By using PLAs, not only is a large part of the design automated, but the chances of errors are also reduced. Thus, PLAs are used extensively in modern digital IC design.

In CMOS, there are two common alternatives used for realizing PLAs: pseudo-NMOS approaches and dynamic approaches. The pseudo-NMOS approach is often used for small PLAs because it is compact and simple. Unfortunately, it has d.c. power dissipation. For larger PLAs, it would be too power hungry and slow, hence dynamic circuits are used. In dynamic PLAs, there are again two common approaches: in one approach, both logic levels are evaluated during the same phase; in the second approach, the levels are evaluated on different phases.

Pseudo-NMOS PLAs

A pseudo-NMOS PLA can be used to realize two levels of *nor* gate functions in a compact manner. Shown in Fig. 10.30 is the type of logic function that can be realized by pseudo-NMOS PLAs. As was mentioned, normally the inputs and their complements are available. The complements may be generated using inverters, as is shown. Often they will be available for free if the inputs come from differential circuits such as latches. The outputs of the first level of *nor* gates represent the maxterms of a general logic function. The number of gates at the first level is the same as the maximum possible number of maxterms that can be used. The actual connections of Fig. 10.30 were arbitrarily chosen for illustrative reasons only. In some PLAs, additional inverters at the outputs might be added so that the complement of every logic function is also available.

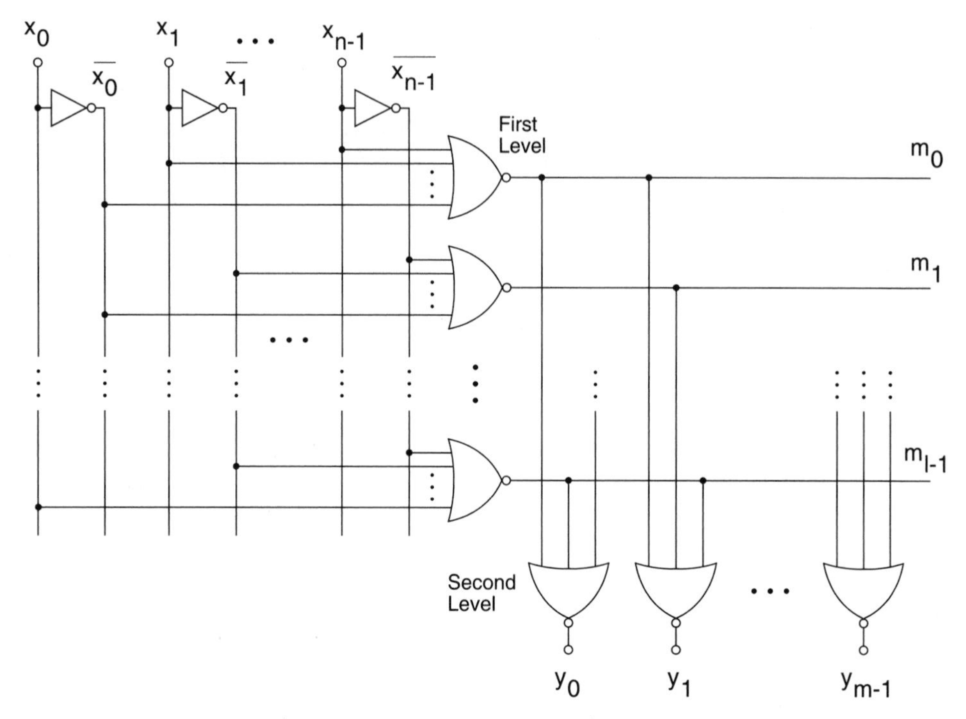

Figure 10.30 A *nor-nor* PLA architecture.

A pseudo-NMOS realization of the logic circuit of Fig. 10.30 is shown in Fig. 10.31. Note that the PLA is laid out in two grids; the *or*-plane and the *and*-plane. The outputs of the first plane run horizontally, whereas the inputs of the first plane run vertically. The opposite is true of the second plane. Distributed through both planes are transistors, but, as shown, all of the transistor gates are connected to the input lines. These connections may or may not be made, depending on the function implemented. These connections are determined by the layout of the *active region* mask. If a connection is not desired, then an active region will *not* be placed in the corresponding gate area. Rather, there will be thick field oxide there. This corresponds to having a transistor that is always turned *off*, or, more exactly, no transistor at all. Alternatively, there will be an active region wherever a transistor connection is desired. Finally, notice that pseudo-NMOS loads are used for the outputs of each plane. The voltages V_{ref} might be generated by a circuit similar to that shown in Fig. 9.3 of Chapter 9. The PLA of Fig. 10.31 can be laid out very compactly, similar to how the schematic has been drawn.

The major limitation on the PLA of Fig. 10.31, besides its d.c. power dissipation, is that the speed significantly degrades for large PLAs. This is primarily due to slow "0" to "1" transitions, which are limited by the *low* currents of the p-channel pseudo-NMOS loads. One possible solution that has been proposed is to use current-mode-type gates in which the voltage swings are limited similar to what was described in Section 9.1.

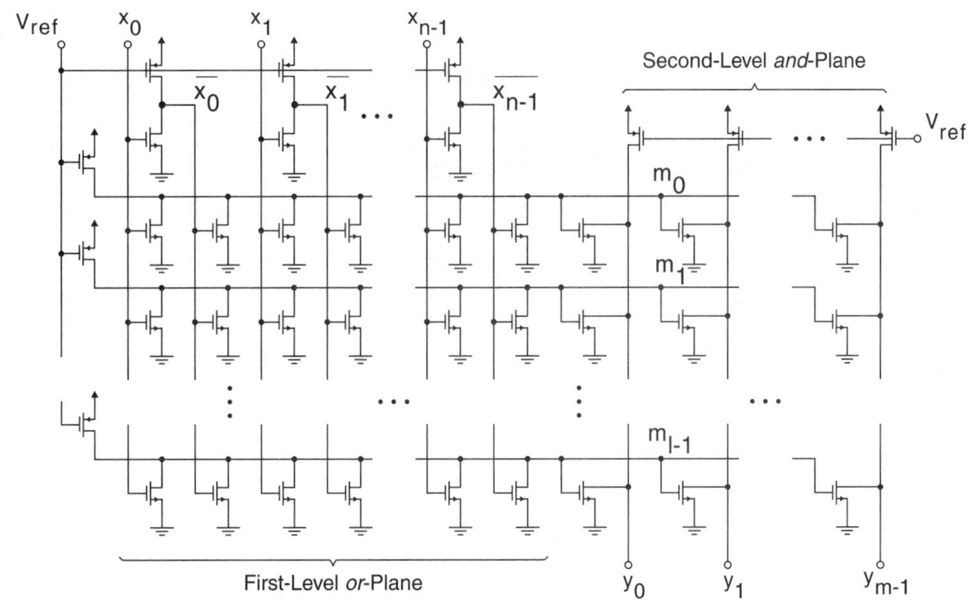

Figure 10.31 A pseudo-NMOS *or-and* or equivalently *nor-nor* PLA.

In this approach, logic gates are used that are similar to bipolar I^2L gates. These are realized by simply adding some *diode-connected* transistors to the output lines of each PLA. In addition, an output buffer would often be used in order to regenerate full-voltage output swings and to drive any large output loads. A simplified schematic showing this approach is given in Fig. 10.32. Also shown in Fig. 10.32 are reasonable device W/L ratios and a circuit for generating V_{ref}. This approach can give roughly a factor of two increase in speed for small-to-moderate-sized PLAs.

For large PLAs, it is necessary to use dynamic techniques with precharging. Not only does this result in faster speeds, but the power dissipation is minimized. As mentioned previously, there are two popular approaches for realizing dynamic CMOS PLAs: self-timed PLAs in which both planes are evaluated at the same time and two-phase PLAs in which the second plane is being evaluated while the first plane is being precharged, and vice versa.

Self-Timed Dynamic PLAs

The realization of dynamic PLAs is complicated by the fact that an odd number of inverters or a latch must normally separate dynamic-precharge gates in order to prevent partial discharge of the latter gates, similar to what was described in Section 9.2 for Domino gates. One possible means to get around this problem with PLAs is to delay the evaluation of the second plane using self-timed clocks until it has been guaranteed that the first plane has settled. This technique is also useful in dynamic decoders.

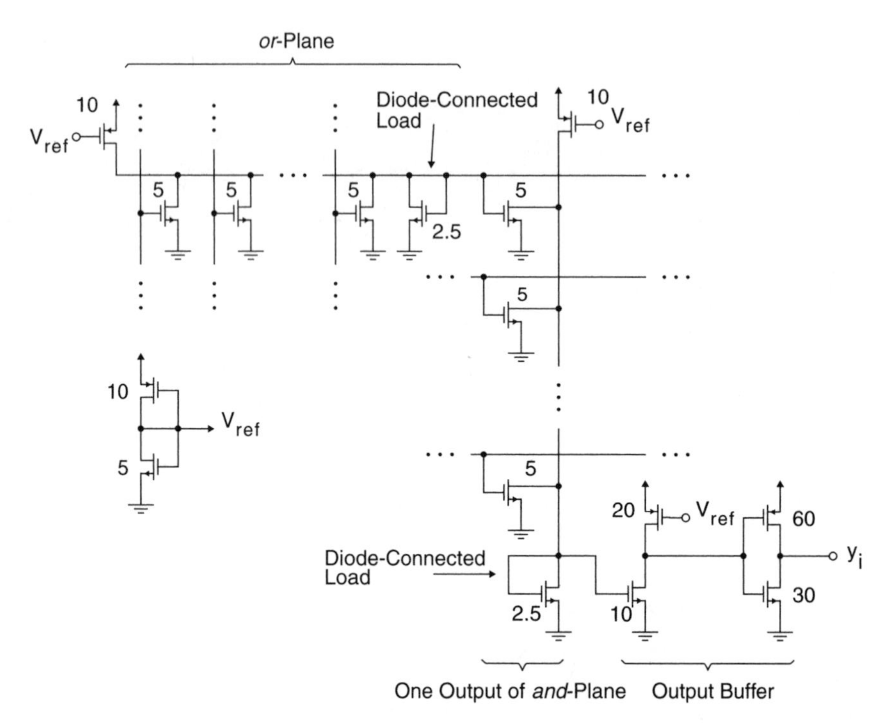

Figure 10.32 A current-mode PLA with reduced voltage swings.

An example of a self-timed dynamic PLA is taken from Nakayama et al. (1989) and is shown in Fig. 10.33. In this diagram, the circuits for generating the complementary inputs are not shown. Between the first *or*-plane and the second *and*-plane is a dynamic buffer. The operation of the self-timed PLA is as follows: during ϕ_1, the outputs of both planes are precharged *high* by p-channel transistors clocked by $\overline{\phi_1}$. Also, at this time, the inputs of both planes are *forced low* to guarantee that all n-channel transistors in the arrays are *off*. The inputs to the first plane are forced low by the *nor* gates that are clocked by ϕ_1. When "ϕ_1" = "1", their outputs are all zero. The inputs to the second plane are forced low by the dynamic buffers that separate the two planes.

The dynamic buffers have two stages; the first stage is precharged high during ϕ_1, whereas the second stage is precharged low: when ϕ_1 goes low and the evaluate phase begins, the nodes of the dynamic buffer do not change immediately. The output of the first stage of the buffer stays high at least until Eval goes high. The signal Eval is produced by the self-timing circuitry. It will not go high until the outputs of the first plane $m_0,...,m_{n-1}$ are stable at their final states. When Eval finally goes high, the output of the first stage of the ith buffer will go low if m_i is high. Otherwise, it will stay high. If the output of the stage goes low, then the output of the buffer, m_i', which was low during the precharge stage, will go high. Otherwise, it will stay low. At the end of $\overline{\phi_1}$, the outputs of the second plane are sensed by dynamic sense amps. Examples of these will be described in the next chapter on memories.

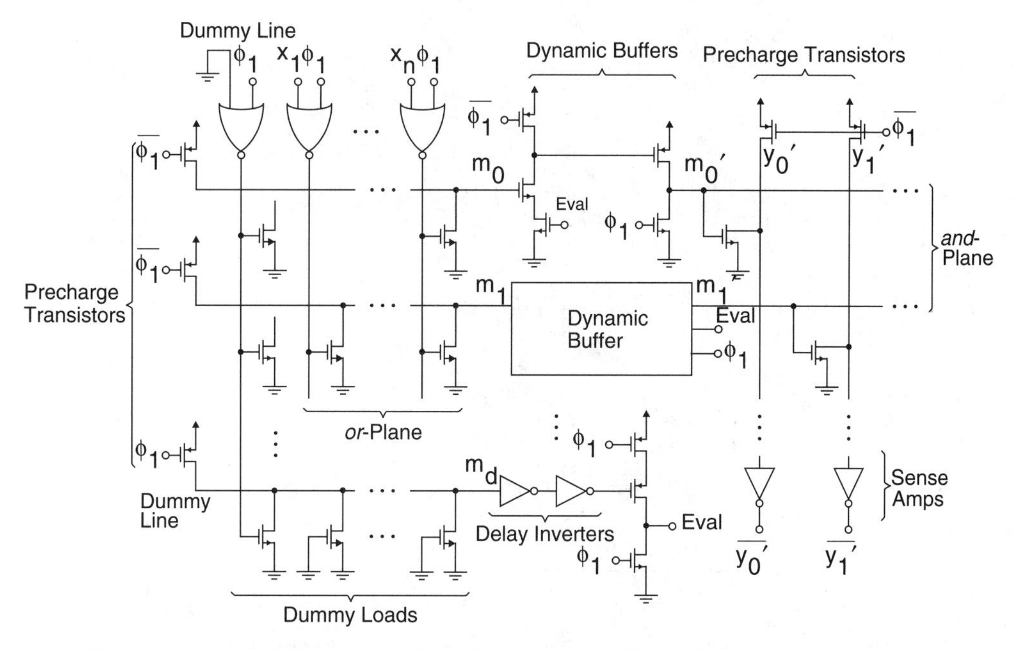

Figure 10.33 A dynamic PLA with precharging and self-timed second-plane evaluate.

The self-generated Eval signal is low during ϕ_1, the precharge phase. During $\overline{\phi_1}$, the evaluate phase, it goes high only after all the outputs of the first plane are stable, as was just mentioned. Until Eval goes high, it is guaranteed that the inputs to the second plane, from the dynamic buffers, are low, thereby guaranteeing that none of the outputs in the second plane will begin to move before the inputs to the buffers are stable. This guarantee is due to the fact that the delay from when ϕ_1 goes low to when Eval goes high is larger than or equal to the worst-case delay through the first plane of the PLA. This is true because the delay from when ϕ_1 goes low to when Eval goes high is generated by a signal path matched to the slowest possible delay path through the first plane of the PLA. The slowest possible delay path through the PLA would be a signal path in which every transistor was present. Thus, the circuit to generate Eval consists of a *dummy line*, matched to the signal path through the PLA, with a number of dummy transistors connected to it equal to the largest number of transistors possible in an actual signal path. In addition, some extra delay is added to the Eval generation circuitry by adding a couple of extra inverters. Finally, it is guaranteed that Eval is low during ϕ_1 by adding the Domino inverter after the delay inverters, as is shown in Fig. 10.33.

The preceding example exemplifies modern self-timed dynamic design techniques and merits careful study.

Two-Phase PLA

An alternative approach for realizing dynamic CMOS PLAs is to *pipeline* the signal paths by placing a latch between the first and second planes. The first plane will now be evaluated on the opposite phase as compared to when the second plane is evaluated.

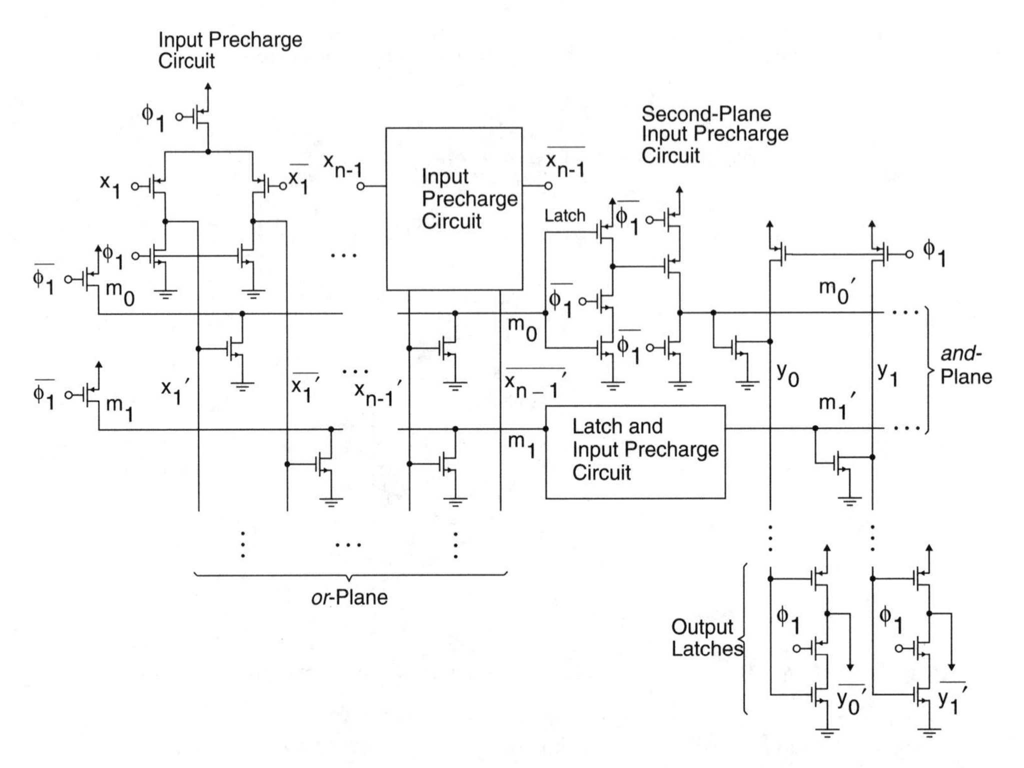

Figure 10.34 A two-phase PLA using NORA logic.

An example of such a PLA that is based on using NORA logic is shown in Fig. 10.34. In this PLA, the first *or*-plane is precharged high during ϕ_1. At this time, the input precharge circuit of the first plane guarantees that all first plane inputs are low. The second plane is precharged high when ϕ_1 = "0" and evaluated when ϕ_1 = "1".

The two planes are separated by true single-phase clock latches and second-plane input precharge circuits. The latches sample the outputs of the first plane during the time ϕ_1 = "0", which is the first plane's evaluate phase. The second-plane input precharge circuits isolate the outputs of the latches from the second plane, at this time, by precharging all second-plane inputs low. At the same time, the second-plane lines (i.e., y_i) are precharged high.

During the second plane's evaluate phase, when ϕ_1 = "1", the latches store the outputs of the first plane and the second-plane input precharge circuits are activated. Every latch that has a "0" stored in it will cause the corresponding second-plane input line to go high. Otherwise, it will be left low. While the second-plane lines are being evaluated, the PLA output latches are in sample mode. They will store the PLA outputs when the second-plane is being precharged, that is, during the time ϕ_1 = "0".

Very often, input latches before the PLA might be included. Indeed the first-stage input precharge circuits are easily modified to make them into combination latch precharge stages. Sometimes a large CMOS inverter might also follow the output latches for buffering.

The dynamic circuits that have just been described for realizing CMOS PLAs are only a few examples out of many possibilities. They are felt to be reasonable choices for PLAs, particularly the latter example. They are also felt to be good examples of nontrivial CMOS dynamic logic design.

10.7 Bibliography

O. Bedrij, "Carry-Select Adder," *IRE Transactions on Electronic Computers*, ED-11, 340–346, 1962.

A.D. Booth, "A Signed Binary Multiplication Technique," *Quarterly Journal of Mechanical and Applied Mathematics,* 4, 236–240, 1951.

M. Davio et al., *Digital Systems with Algorithm Implementation*, John Wiley & Sons, 1983.

Y. Dhong and C. Tsang, "High-Speed CMOS POS PLA using Predischarged OR Array and Charge-Sharing AND Array," *IEEE Transactions on Circuits and Systems - II*, 39(8), 557–564, August 1992.

M. Ercegovac and T. Lang, *Digital Systems and Hardware/Firmware Algorithms,* John Wiley & Sons, 1985.

K. Hwang, *Computer Arithmetic, Principles, Architecture, and Design*, John Wiley & Sons, 1979.

J. Iwamura et al., "A CMOS/SOS Multiplier," *IEEE ISSC Digest of Technical Papers,* 92–93, February 1984.

G.K. Ma and F. Taylor, "Multiplier Policies for Digital Signal Processing," *IEEE ASSP Magazine*, 6–20, January 1980.

M. Mano, *Digital Design,* Prentice Hall, 1984.

M. Mano and C. Kime, *Logic and Computer Design Fundamentals*, Prentice Hall, 1997.

Nakayama et al., "A 6.7 MFLOPS Floating-Point Coprocessor with Vector/Matrix Instructions," *IEEE Journal of Solid-State Circuits*, 24(5), 1324–1330, October 1989.

Y. Oowaki et al., "A 7.4ns CMOS 16x16 Multiplier," *IEEE ISSC Digest of Technical Papers,* 52–53, February 1987.

Pelayo, Prieto, Lloris, and Ortega, "CMOS Current-Mode Multilevel *PLA's*," *IEEE Transactions on Circuits and Systems*, 38(4), 434–441, April 1991.

J. Rabaey, *Digital Integrated Circuits,* Prentice Hall, 1996.

C. Roth, *Fundamentals of Logic Design*, West, 1985.

E. Swartzlander, ed., *Computer Arithmetic—Part I and II*, IEEE Computer Society Press, 1990.

H. Taub and D. Schilling, *Digital Integrated Electronics*, McGraw-Hill, 1977.

S. Waser and M. Flynn, *Introduction to Arithmetic for Digital Systems Designers,* Holt, Rinehart, & Winston, 1982.

N. Weste and K. Eshraghian, *Principles of CMOS VLSI Design,* 2nd ed., Addison-Wesley, 1993.

10.8 Problems

For the problems in this chapter, assume the following transistor parameters:

- npn bipolar transistors:

  $\beta = 100$

  $V_A = 80$ V

$\tau_b = 13$ ps
$\tau_s = 4$ ns
$r_b = 330\ \Omega$

- n-channel MOS transistors:
  $\mu_n C_{ox} = 190\ \mu A/V^2$
  $V_{tn} = 0.7$ V
  $\gamma = 0.6\ V^{1/2}$
  $r_{ds}\ (\Omega) = 5000 L\ (\mu m)/I_D\ (mA)$ in active region
  $C_j = 5 \times 10^{-4}\ pF/(\mu m)^2$
  $C_{j\text{-}sw} = 2.0 \times 10^{-4}\ pF/\mu m$
  $C_{ox} = 3.4 \times 10^{-3}\ pF/(\mu m)^2$
  $C_{gs(overlap)} = C_{gd(overlap)} = 2.0 \times 10^{-4}\ pF/\mu m$

- p-channel MOS transistors:
  $\mu_p C_{ox} = 50\ \mu A/V^2$
  $V_{tp} = -0.8$ V
  $\gamma = 0.7\ V^{1/2}$
  $r_{ds}\ (\Omega) = 6000 L\ (\mu m)/I_D\ (mA)$ in active region
  $C_j = 6 \times 10^{-4}\ pF/(\mu m)^2$
  $C_{j\text{-}sw} = 2.5 \times 10^{-4}\ pF/\mu m$
  $C_{ox} = 3.4 \times 10^{-3}\ pF/(\mu m)^2$
  $C_{gs(overlap)} = C_{gd(overlap)} = 2.0 \times 10^{-4}\ pF/\mu m$

10.1 Design a transistor-level realization of a 16-to-1 *multiplexor* using CMOS. A hierarchical solution may be given.

10.2 Give a hierarchical solution for a three-level 8 to 1-of-256 *decoder.*

10.3 Design a *shifter* that can shift a four-bit bus left by 0, 1, 2, or 4 bits. The vacated bits should be filled with zeros.

10.4 Design a divide-by-8 circuit using serial-carry synchronous logic similar to the realization shown in Fig. 10.12. A gate-level realization is adequate. Estimate the maximum period for the clock signal assuming the JK flip-flops require four unit delays and gates require one unit delay.

10.5 Design a frequency divide-by-8 circuit by cascading a divide-by-2 circuit with a divide-by-4 synchronous divider. Estimate the maximum clock frequency in terms of gate delays and compare to the number found in Problem 10.4.

10.6 Design a *maximal-length sequence generator* in which the output has 8 bits, but internally nine flip-flops and only one *ex-or* gate is used. A gate-level solution is adequate. Would this realization be capable of higher speed than that shown in Fig. 10.14 and if so why?

10.7 Give a design of a simplified single-bit *full adder* where it is known a priori that the carry-in is a "1".

10.8 Give a transistor-level realization of the *adder/subtractor* of Fig. 10.26.

10.9 Give a transistor-level realization for the *ripple-carry adder* of Fig. 10.20 using traditional CMOS logic. Give reasonable widths that optimize the critical ripple-propagate path.

10.10 Design a 4-bit *Hex notation* (where the four bits represent 0 to 15 in unsigned binary) to *binary-coded-decimal (BCD)* (where the four bits can only go up to 9) *converter.* The circuit should have a carry-in (C_{in}); when it is a "1", a "1" should be added to the BCD-encoded output. The BCD-encoded output should have a carry-out (C_{out}) output. If the binary equivalent of the output would be greater than 9, then carry-out should be a "1" and the other four output bits should be equal to the binary equivalent minus "1001" (i.e., minus the binary equivalent of 10). A solution that has only logic gates and/or full adders is adequate.

10.11 Two four-bit 2s-complementary numbers are to be multiplied. The multiplicand A is assumed to be signed and is assumed to be equal to "1101" (i.e., –3 in decimal notation), and the multiplier is assumed to be unsigned and equal to "1110" (i.e., +14 in decimal). Show the steps required using binary additions and one-bit sign extension at each level that are required to multiply the numbers.

10.12 Show an array consisting of full adders and two-input *and* gates only that realizes the multiplication of Problem 10.11 and show the logic values at every node for the example given. (*Hint:* there should be three levels in your array with five full adders at each level not counting the required *and* gates.)

10.13 Assume A and B are six-bit signed 2s-complement numbers that are to be multiplied using the modified Booth algorithm. Assume A = –27 and B = –12, and furthermore assume that B is the multiplier. First give the recoded values for B and then give each step of the multiplication.

10.14 Design a serial (in-time) multiplier capable of multiplying two eight-bit numbers. Once a *reset* input goes low, the multiplication should start; once it has finished (seven clock cycles later), a *finish* output should go high. You may assume any necessary preloading of registers has already been done, but state exactly what preloading is necessary. Give a gate-level and/or full-adder-level solution.

10.15 Give the programming required for a PLA so that when it is placed in the feedback loop of a four-bit clocked register, a Gray-code counter is realized. A *high-level* only description is acceptable.

10.16 Give the programming required for a PLA to realize the *Hex-to-BCD converter* described in Problem 10.10.

10.17 Discuss how to estimate the parasitic capacitances of the reduced voltage swing PLA of Fig. 10.32. Given the currents and voltage swings, how can this be used to estimate the worst-case delay through the PLA?

10.18 Design a 16-bit barrel shifter using four levels of multiplexors that can rotate all inputs an arbitrary number of positions left or right.

10.19 Repeat Problem 10.18, but assume the barrel shifter is to be pipelined using true single-phase clock logic.

CHAPTER 11

Integrated Memories

One of the most important building blocks of modern digital systems is the semiconductor memory. A semiconductor memory is used to store digital data. In this way it is somewhat similar to a digital register, but unlike a register, it is normally not possible to access all storage locations in parallel at the same time. Rather, only one or a few memory locations can be addressed during a single access. In a semiconductor memory, in addition to the data lines, there will be additional address inputs, which, after being decoded, will determine which address location or locations are to be accessed. Also, there will be control lines, which determine the function being performed.

Random-access semiconductor memories can have their individual memory locations randomly accessed either for reading or writing. The *random-access memory (RAM)* is contrasted with a *sequential memory* in which the memory locations can be accessed only sequentially (one after the other) in a particular order. For example, a magnetic tape used to store digital data is considered a sequential memory. Sequential memories are beyond the scope of this text.

Whether a storage cell of a random-access memory is being read from or written into is determined by a control line, often designated $\mathsf{R}/\overline{\mathsf{W}}$, which will be a “1” when a memory location is being read. Some semiconductor memories cannot have their contents changed during normal operation. These are called *read-only memories* (*ROMs*). Their contents are determined either at the time of manufacture or during a special write mode.

A number of variations of both RAMs and ROMs will be covered in this chapter. The first memory described in this chapter will be a static RAM. This type of memory utilizes mostly static logic, particularly for the storage locations. Some of the peripheral circuits may use dynamic logic. This is contrasted with dynamic RAMs, in which capacitors are used to store the memory contents. This results in smaller memories for a given number of locations, but, unfortunately, also means the memory contents must be periodically *refreshed* to prevent leakage currents from corrupting their values.

There are a number of different variations of ROMs as well; some have their contents determined at the time of processing by a particular mask and others can actually have their contents electrically programmed, but will still retain their contents in the event of a power loss, and, thus, are *nonvolatile.* These *electrically programmable ROMs (EPROMs)* also have many variations; once programmed some can never have their contents changed and others can have their contents erased by shining ultraviolet light directly on the integrated circuit through a quartz window in the IC package; more modern variations can have their contents electrically erased.

In this chapter, most of the principles of operation of semiconductor memories will be explained when describing the operation of a static RAM. Many of the subcomponents of static RAMs are common to other memories as well. After static RAMs are described in detail, the other memories will be described in somewhat less detail. Most of the memories that will be described are realized using CMOS technology, the most commonly used technology for semiconductor memories.

This chapter will discuss only briefly architectures for modern large IC memories; rather, simplified and small architectures will mostly be described to illustrate the basic principles. To most fully appreciate these principles, a more in-depth understanding of device modelling and *analog* design aspects is necessary. Much of this may be attained by reviewing Chapter 3 before studying this chapter.

11.1 Static Random-Access Memories

As was mentioned, a *static random-access memory (SRAM)* consists of a number of storage locations. At any given time, any one of these storage locations can be written into or read from. Figure 11.1 shows a simplified block diagram of a static RAM that has 4096 storage locations laid out in a 64 by 64 cell array. The size of this rather small memory was chosen only to illustrate the principles; modern memories are often much larger and would typically consist of a large number of arrays. In the example memory, only one of the 4096 storage cells can be *written to* or *read from* at any given time. For this reason, it is called a 4096 × 1-bit memory. Often a number of similar memories will be connected in parallel so that more than one bit can be accessed at any given time. For example, if eight memories similar to that shown in Fig. 11.1 were connected in parallel, eight bits could be read from or written to at any given time. This memory would be called a 4096 × 8-bit memory.

The I/O signals of the memory can be grouped into three categories: *address lines, data lines*, and *control lines*. For our example, there are 12 address lines that will be used to select one of the 4096 memory locations. Six address lines are used to select a particular row and the other six address lines are used to select a particular column. The memory cell at the intersection of the selected row and column is the cell that will be accessed for reading or writing.

The static RAM has two *decoders*, one for selecting a row of the RAM and one for selecting a column. The inputs to the decoders are the buffered address lines and their inverses. Since, for our example, six address lines are used for selecting the row and six are used for selecting the column, each decoder has n inputs and 2^n outputs where $n = 6$. Each decoder has only one output high at any given access time. The decoder outputs used for selecting a row are called *word lines*.

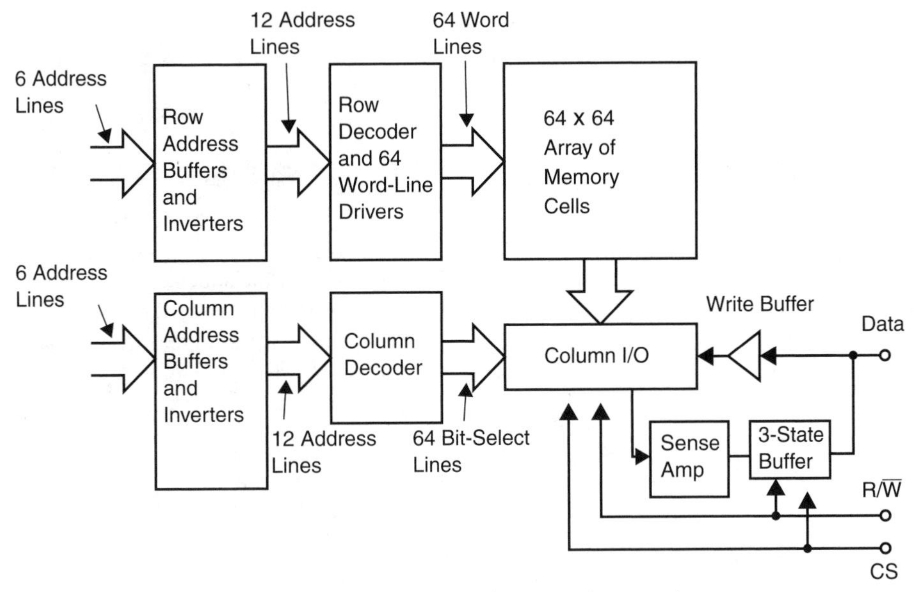

Figure 11.1 A simplified block diagram of a 4096 by 1 RAM.

An expanded but still simplified view of a portion of the array that was shown previously in Fig. 7.11 is repeated in Fig. 11.2. Some of the 64 word lines, W_0, W_1, ..., W_{63}, are shown at the left of the figure. These are outputs of the *row decoder.* When the memory is being accessed, only one of these word lines will be *high* depending on the value of the first six address inputs. The word lines are buffered and then transmitted horizontally through the array. As can be seen from Fig. 11.2, each storage cell in an SRAM array is connected to its differential *bit lines,* which run vertically, by two *access transistors* controlled by the cell's word line. These access transistors operate as *pass transistors.* Since only one word line in an array will be high during access, then one complete row will be selected. Each storage cell in that row will be connected to its particular pair of differential bit lines.

The six address lines going to the *column-address buffers* are responsible for determining which set of bit lines is connected to the *outside world.* Since only one output of the column decoder will be high, then a single pair of differential bit lines will be connected to the differential data lines at the bottom of the array. This connection is through the *column-access transistors,* which function as pass transistors. Thus, the storage cell at the intersection of the high word line and high column line is connected via its cell access transistors and its column access transistors to the array data lines.

What happens next is determined by whether the array has been selected, as controlled by the control line CS, and whether the cell is being read from or written into as determined by $R/\overline{W}$. For example, if a memory is being read, both of the control lines $R/\overline{W}$ and CS (i.e., *chip select*) must be *high.* The signals Data In and $\overline{\text{Data In}}$ in Fig. 11.2 will both be *low.* This allows the state of the storage cell to create a differential voltage across both its bit lines and through the column-access transistors across the data lines at the bottom of the array. The differential

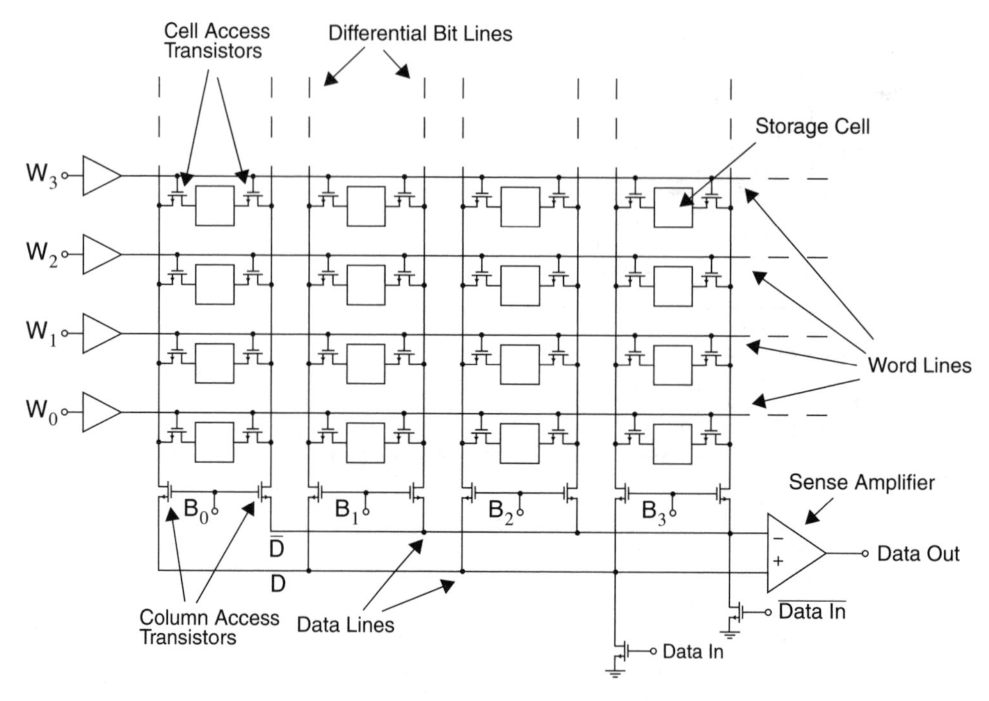

Figure 11.2 A simplified portion of an SRAM array.

voltage across the data lines is sensed by the sense amplifier. The output of the sense amplifier goes to a three-state buffer that will be placed in its low-impedance state and that drives the single Data Out line.

Alternatively, if the memory is being written into, $R/\overline{W}$ will be a "0" and CS will be a "1". This places the three-state buffer into its *high* impedance state so that it does not affect the Data Out line. The state of the Data Out line will be sensed by the write buffer (not shown), which in turn will cause either Data In or $\overline{\text{Data In}}$ to go *high* depending on the value of Data Out. This will also cause one of the data lines at the bottom of the array to go *low*. (They were originally precharged to a high voltage.) This, in turn, causes a large differential voltage across the selected bit lines (which were also originally precharged to *high* voltages), which forces the selected storage cell into the desired state determined by the external value placed on Data Out.

11.2 Static Random-Access Memory Storage Cells

There are many choices possible for the storage cell of a static RAM. In a CMOS circuit in which the static RAM is only a small part of the IC, a typical choice might be similar to that shown previously in Fig. 7.10 and repeated in Fig. 11.3. This storage cell consists of two cross-coupled

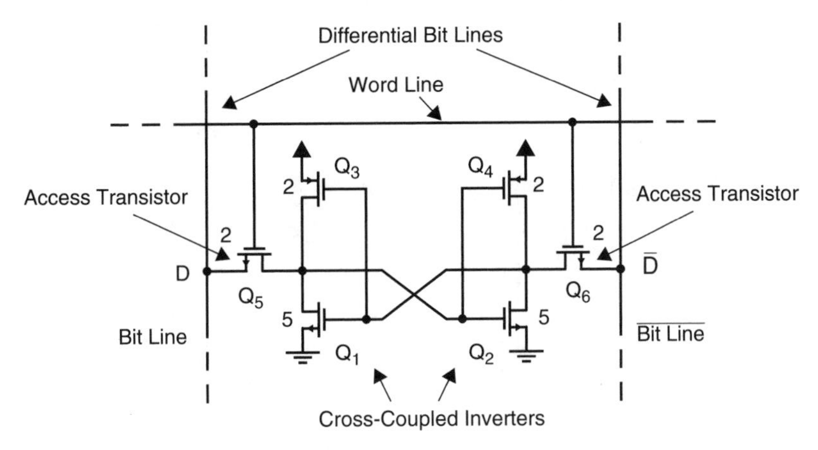

Figure 11.3 A CMOS static RAM storage cell.

CMOS inverters plus a couple of *access* transistors. The access transistors are normally (but not necessarily) n-channel transistors. Also shown in Fig. 11.3 are reasonable choices for the W/Ls of the various transistors. The proper operation of the static RAM cell is dependent on reasonable sizes being chosen for the transistors. This will be discussed more in the next section, although, normally, the relative sizes shown with the access transistors having a W/L around one-third to two-fifths the W/Ls of the n-channel transistors in the cross-coupled inverters is reasonable.

If the word line is at 0 V, both of the access transistors are guaranteed to be *off*. This isolates the memory cell from the bit lines. Similar to the clocked latch, the memory storage cell has two stable states, the output of the left inverter *high* with the right inverter *low*, or vice versa. For consistency, we will arbitrarily assume the output of the left inverter being *high* corresponds to the memory cell storing a "1".

When the memory cell is being written into, one of the bit lines will be precharged to V_{DD}, whereas the other bit line will be held *low* at around 0 V. When the access transistors are then turned *on* by V_{DD} being placed on the word line, the memory cell will then be placed in the same state as the bit lines; when the access transistors are next turned *off*, the cell will remain in this state.

When the memory cell is being read, the first step is to precharge the bit lines *high*. Traditionally, this might be to V_{DD}, although the modern norm is closer to $V_{DD}/2$, for faster operation, as we shall see in the next section.

Next, the appropriate word line will go *high*, which will connect a single memory cell to a set of bit lines. This will cause a low-impedance path to be established between the side of the memory cell that was originally at 0 V and one of the bit lines. This, in turn, causes the bit line's voltage to decrease. The resulting differential bit-line voltage will then be sensed by a sense amplifier once it has reached approximately 50 to 100 mV.

In large static RAMs, the majority of the area is taken up by the memory cells. Thus, great efforts are taken to make its layout as small as possible. The minimum size is severely constrained

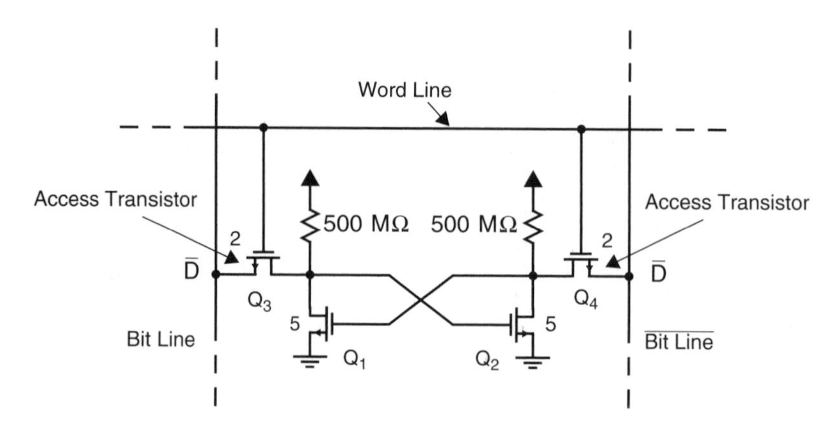

Figure 11.4 A CMOS static RAM storage cell using high-resistivity polysilicon loads.

by the necessity for both n-channel and p-channel transistors in each cell. This in turn requires each memory cell to contain a *well* that must be spaced quite far away from n-channel transistor junctions in the substrate. In modern SRAM cells, this well has been eliminated from the memory cells by using resistive *polysilicon* loads.

A major advance that was developed to reduce the size of static RAM cells was the *double-polysilicon* (or *double-poly*) process. In this process, there are two layers of polysilicon. The top layer of polysilicon can be placed over the bottom layer or over complete transistors without contact being made, unless there are contact holes joining the top layer to the polysilicon or a *diffusion region* below it. In addition, the top layer can be selectively (i.e., only certain areas) doped using ion implantation to make it highly resistive. This allows for the realization of high-value resistors directly above transistors. A static RAM cell based on this double-poly process is shown in Fig. 11.4. The cell still consists of two cross-coupled inverters and two access transistors, but now the loads of the inverters are large-value resistors that are realized by using the second polysilicon layer. Thus, each memory cell consists of only n-channel transistors and polysilicon. Assuming an n-well process is used, which is typically the case, there is no need for any p-channel transistors, or, more importantly, area-consuming wells, anywhere in the memory array. The area required for each memory cell is only that of four n-channel transistors; the resistive loads take up no area as they are realized directly above the n-channel transistors.

Normally, this *four-transistor memory cell* can be realized only in ICs that are processed using a special technology optimized for realizing static RAMs. In other cases, where a static RAM is only a part of a larger IC, as might be the case for a microprocessor or an application-specific IC (ASIC), then the double-poly process is usually not available and the larger *six-transistor cell* of Fig. 11.4 must be used.

The minimum size of the resistive loads of the four-transistor cell of Fig. 11.4 is constrained by the maximum-allowable power dissipation of the memory array.

Example 11.1

Assume the memory array must dissipate less than 50 mW and there are approximately 1 million cells in the array. This means each cell must dissipate less than 50 mW/10^6 = 0.05 μW, which in turn limits the d.c. current of each cell to 0.05 μW/3.3 V = 15 nA, assuming V_{DD} = 3.3 V. At any given time, one of the load resistors will have no current through it, and one will have a current

$$I_{cell} \cong \frac{V_{DD}}{R_L} \tag{11.1}$$

Thus we need

$$\frac{V_{DD}}{R_L} < 1.5\times10^{-8} \tag{11.2}$$

which implies

$$R_L > V_{DD}1.5\times10^{-8} = 500\ \mathrm{M\Omega}) \tag{11.3}$$

This is a minimum size. In modern SRAMs, which can contain tens of millions of memory cells, the polysilicon loads might be as large as a few terraohms (i.e., 10^{12} Ω) resulting in standby currents on the order of a couple of microamperes for the whole IC. One of the consequences of the very large value load resistors is that they have no drive capability. Their only purpose is that when a cell is not being accessed, they can supply more current than the leakage current of an n-channel junction and, thus, preserve the value that was originally written into the storage cell. *The current required to originally write into a cell is supplied only by the peripheral circuits and not by the load resistors.*

To better understand the operations of writing and reading memory storage cells, consider the simplified schematic of a memory array that also shows some peripheral circuits, shown in Fig. 11.5. The schematic shows part of the top and bottom rows and the left two columns of the array. The operation of the memory will be considered for the case first when the memory is being written into and second when it is being read from.

Before writing into the memory, all of the bit lines and the data lines are precharged to V_{DD} by setting $\overline{PR}$ *low* temporarily. This occurs whenever a transition is detected in the address inputs or the CS (chip-select) control line. Next, when writing into the memory, the load transistors at the top of the array are turned *on* by setting $W/\overline{R}$ *low*. The column decoder will cause the data lines that run across the bottom of Fig. 11.5 to be connected to a single pair of bit

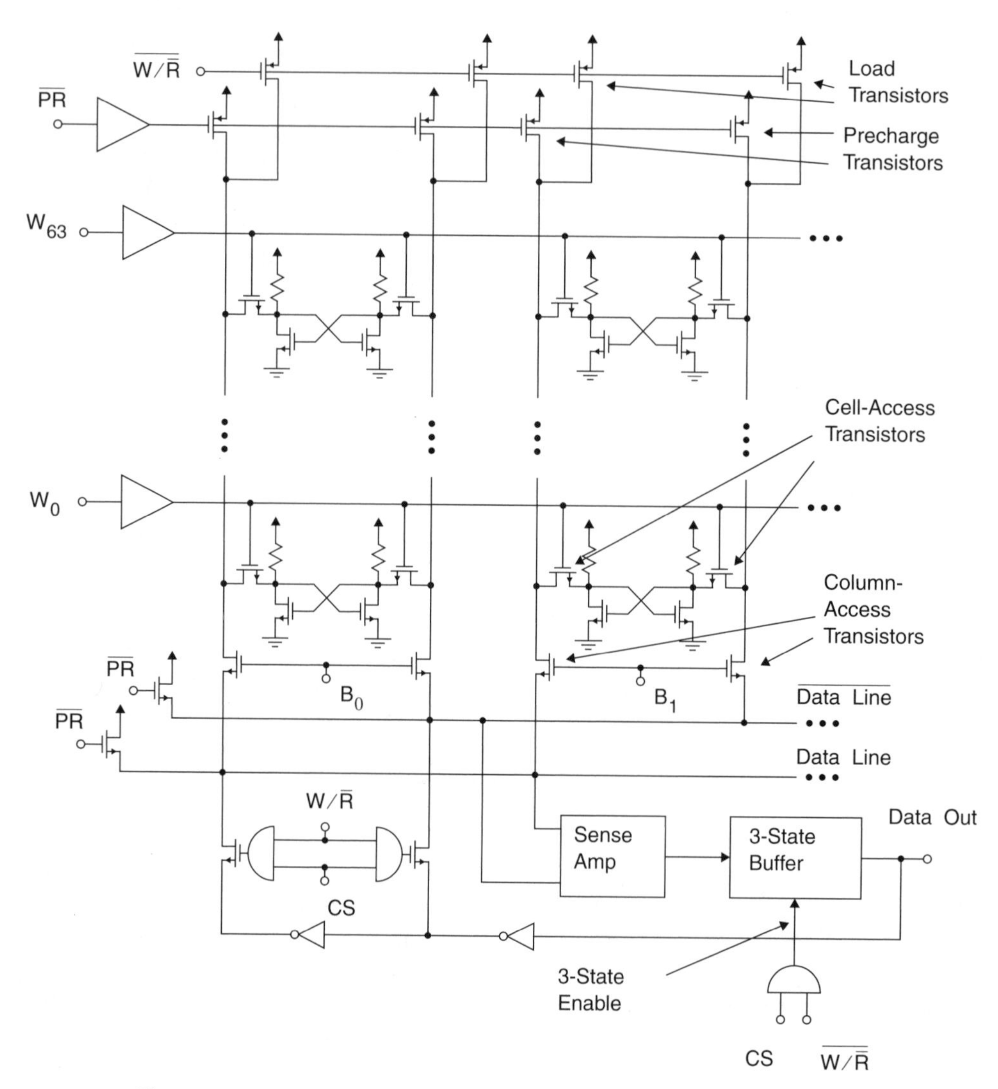

Figure 11.5 An SRAM array with some of the peripheral circuitry shown.

lines that runs vertically through the array. One of the data lines is *pulled low* (remember they were both precharged *high* initially), and this will cause one side of the *selected* bit-line pair to be *pulled low*. The other side will be held *high* by the bit-line load transistor. Thus, a large differential voltage is developed across the selected bit lines. All of the *unselected* bit lines are left at V_{DD} with a very small differential voltage. The row decoder causes a row of memory cells to be connected to the bit lines. The pair of bit lines having a large differential voltage will *write* that voltage into its selected memory cell. The bit lines having small differential voltages will leave their selected memory cells unchanged; *if these bit-line differential voltages are not kept small, the states of the cells connected to them may inadvertently be changed.*

When reading a memory cell, the data lines and all bit lines are again precharged to V_{DD} and initially have very small differential voltages. A row of memory cells is selected by one word line going *high*. This connects the selected row of memory cells to their bit lines, which in turn develops differential voltages across all bit lines. One of these pair of bit lines is connected to the horizontal pair of data lines by the column decoder. This causes a differential voltage to eventually develop across the data lines, which is in turn *sensed* by the sense amplifier.

The p-channel bit-line load transistors, at the top of Fig. 11.5, would always be *on* in a memory that did not have precharging. In a memory with precharging, they are optional but are often included to keep the bit lines precharged if the memory is to be left for long periods of time between accesses. These p-channel loads are often disabled during the read operation.

In the rest of this section, some of the more subtle aspects of reading and writing the memory cells will be considered. These details may be ignored in a first-level course.

The cell- and column-access transistors operate as *transmission gates* and suffer the limitations of n-channel-only transmission gates when transmitting *high* signals. That is, they will be turned *off* unless at least one of the junctions is a transistor threshold voltage, V_{tn}, below the gate voltage. Therefore n-channel-only transmission gates can easily have a 1.5-V drop when transmitting *high* voltages. For example, when writing into a memory cell, the side that is being *driven high* will be charged quickly to $V_{DD} - V_{tn}$ only through the cell-access transistor. When reading a cell's contents, the column-access transistors will not begin to conduct until a bit line has been discharged to $V_{DD} - V_{tn}$, which can represent a significant delay when reading a cell's contents. The voltage drop through the access transistors is the reason why modern memories often precharge buses to only $V_{DD}/2$ rather than to V_{DD} (Nakagome et al., 1991). Also, an alternative that is becoming popular recently is to generate gate *on* voltages greater than V_{DD} for the access transistors using a technique called *charge pumping* (Nakagome et al., 1991; Oto et al., 1983; Witters et al., 1989).

Whenever a cell is connected to a pair of bit lines for reading, its internal voltages may be affected substantially when the word line first goes *high*. This is caused by the large parasitic capacitances of the bit lines, which cannot change voltage immediately. If the initial differential voltage on the bit lines is zero, the bit-line capacitance will cause the internal voltages of the cell to become more equal when the cell is first connected. After some time, the current through the cell to ground will discharge one of the bit lines to establish a large enough differential voltage that can be safely read, but this does take some time. The larger the cell-access transistors are, the larger is the change in internal cell voltages when the cell is first connected to the bit lines. This restricts the widths of the access transistors to less than approximately two-fifths times the width of the n-channel transistors in the cross-coupled inverters of the cell. Otherwise, the equalizing of the internal cell voltages during the read operation can significantly increase the time required to establish a differential bit-line voltage large enough to be correctly determined by the sense amplifier.

It is important to guarantee that before accessing a cell for reading, the differential bit-line voltage is zero; otherwise a large differential bit-line voltage can actually inadvertently change the contents of a cell during a read operation (or even during a write operation for the columns that are not being accessed). For this reason, all bit-line voltages need to be equalized after every write operation; this normally happens at the beginning of every access irrespective of the type (i.e., read or write) as has been previously mentioned. Therefore, when reading or writing a memory cell, the first thing that usually happens, irrespective of whether a memory is being written into or being read from, is all the bit lines are precharged. For the memory shown, the bit lines are precharged to V_{DD}. More sophisticated memories will have the bit lines precharged to approximately $V_{DD}/2$ in order to obtain greater speed, as was also mentioned; for moderate speed memories, V_{DD} precharge is adequate and results in simpler circuitry.

To appreciate the details of the voltage changes of the SRAM cell when it is being written into, consider the simplified circuit of Fig. 11.6a. Initially, it is assumed that the memory cell is in the "0" state. That internal node V_a is at 0 V and V_b is at V_{DD}. It is also assumed that both bit

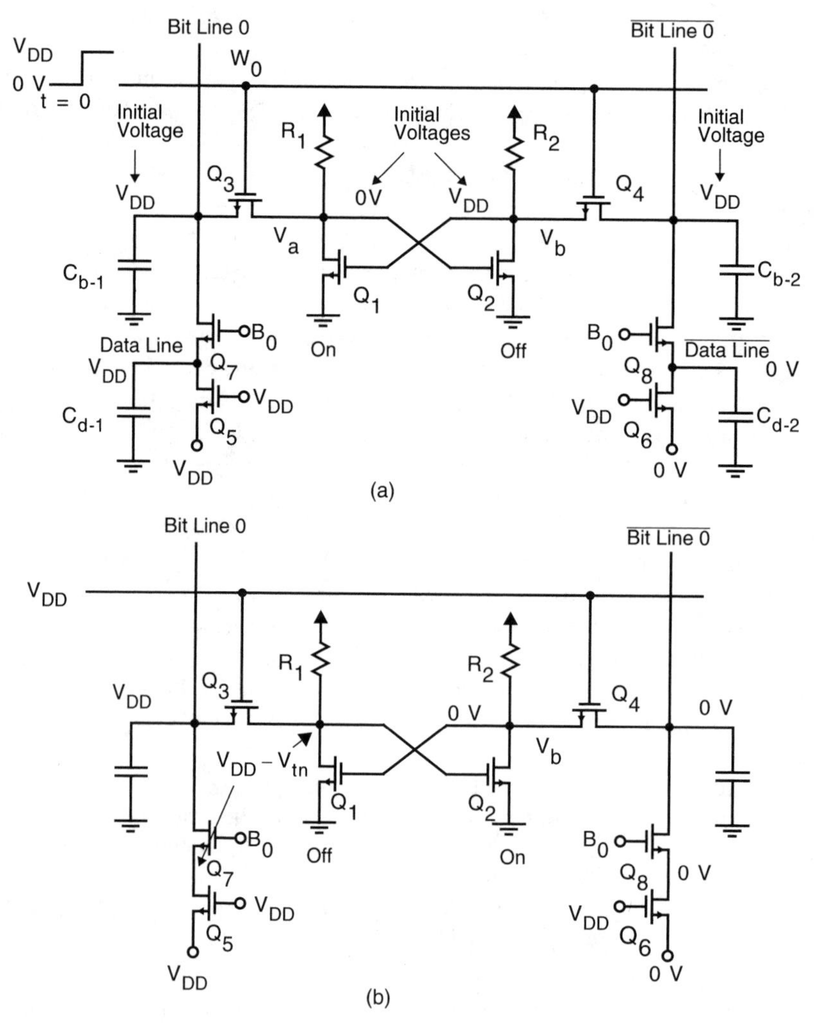

Figure 11.6 The bit-line and memory cell voltages (a) during the write operation just before the word line goes *high* and (b) after the memory cell has changed state to a "1".

lines were initially precharged to V_{DD} and then $\overline{\text{Bit Line 0}}$ was discharged to 0 V before the memory cell is connected to the bit lines.

When the word line goes *high*, both of the access transistors Q_3 and Q_4 will turn *on*. Initially, V_a will not change in voltage very much because Q_1 is much wider than Q_3 and initially V_{G-1} is at V_{DD}. However, very quickly V_b (i.e., V_{G-1}) is discharged from V_{DD} to 0 V through the series connection of Q_4, Q_8, and Q_6. This causes Q_1 to turn *off*. V_a is now charged up to $V_{DD} - V_{tn\text{-}3}$ by Q_3 acting as a transmission gate. Depending on the body-effect parameter, this voltage will be approximately 1.5 V below V_{DD}. Thus, the high voltage stored in a memory cell would normally be closer to $V_{DD}/2$ than V_{DD}. The voltages of the memory cell after changing state to a "1" are shown in Fig. 11.6b.

When reading a cell, the operation is similar with a few differences. First, the bit lines and data lines are precharged to V_{DD}. Next, the appropriate column-access signal and word line will go *high*. This will connect one pair of bit lines to the data line and one memory cell to the bit line. The resulting voltages at this time are shown in Fig. 11.7. It was assumed that the memory cell was initially in a "1" state. A few details should be noted. First, initially the data lines and bit lines do not move due to their large parasitic capacitances. Next, notice that V_b, which initially was 0 V, is pulled up to around $V_{DD}/4$ or so. This voltage is dependent on the ratio of the memory cell-access transistors to the n-channel cross-coupled transistors in the memory cell. The larger the access transistors, the more V_b will be pulled to a higher voltage. This minimizes the difference between V_a and V_b, which makes reading the memory cell slower. This is the major reason for keeping the access transistors only between one-third and two-fifths the size of the cross-coupled transistors; by keeping the cross-coupled transistor widths larger than the access-transistor

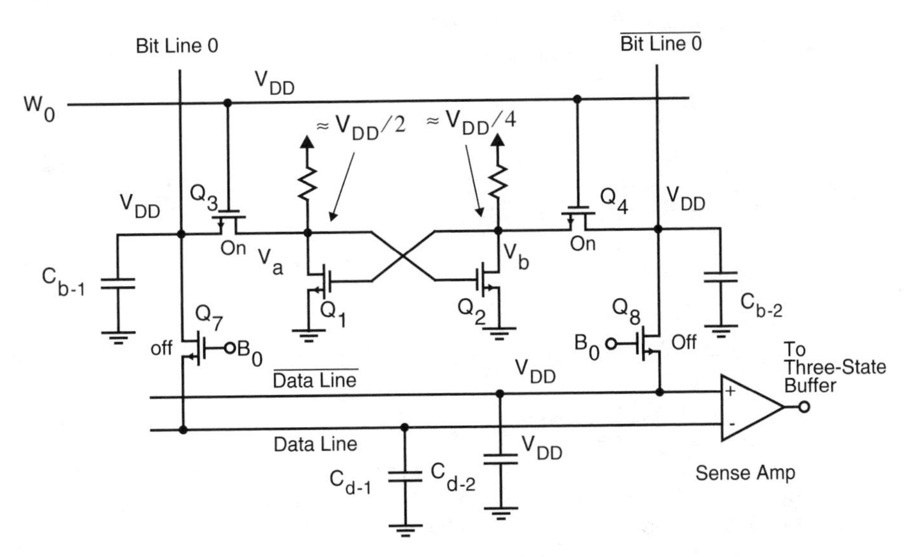

Figure 11.7 The memory cell voltages just after the word line goes *high* during a memory read operation.

widths, it is guaranteed that during a memory read, the low voltage of the memory cell does not become large enough to appreciably turn *on* the cross-coupled transistor that was originally *off*.

For the approximate voltages shown, Q_2 will still be conducting much more than Q_1 just after the access transistors turn *on*. This will discharge $C_{b\text{-}2}$ and, therefore, $\overline{\text{Bit Line 0}}$ more quickly than Bit Line 0. Once $C_{b\text{-}2}$ is discharged to $V_{DD} - V_{tn}$, then (and only then) will Q_8 begin to conduct as well. At this time, $C_{d\text{-}2}$ will start to be discharged, which will cause the voltage on $\overline{\text{Data Line}}$ to decrease. After a differential voltage of around 0.1 V is developed across data lines, then the sense amplifier can be used to detect the value previously stored in the memory cell.

It should be emphasized that before $\overline{\text{Data Line}}$ begins to have its voltage change, $\overline{\text{Bit Line 0}}$ must first decrease from V_{DD} to $V_{DD} - V_{tn}$, as was just mentioned. Taking the body effect into account, then V_{tn} might be on the order of almost 1.5 V (for V_{nn} = 3 V). The time required for this change is significant. Thus, the reading operation of a memory cell can be significantly increased in speed if the bit lines are precharged only to around $V_{DD}/2$ during the read operation. In this case, the differential voltage across the data lines will start to develop immediately after the cell-access transistors are turned *on*. This significantly decreases the time it takes from when the word line goes *high* to when the sense amplifier can be safely latched.

Example 11.2

Using SPICE, simulate the write and read operation of the memory cell of Fig. 11.3 (connected as in Fig. 11.6). Assume the p-channel load transistors (shown in Fig. 11.5) are always *on*. Let the transistors have the following widths (with reference to Fig. 11.6):

Transistors	Widths (μm)
Q_1, Q_2	3
Q_3, Q_4	1
Q_5, Q_6	10
Q_7, Q_8	6

All the lengths should be taken as 0.6 μm.

Solution: Figure 11.8a shows simulated voltages of the internal cell node, bit line, and data line on the side of the cell that was originally sitting at 3 V. For the first 10 ns, the bit lines and data lines are precharged (for 3 ns) and then a “0” is written into this side of the cell. For the next 10 ns, the bit lines and data lines are again pre-

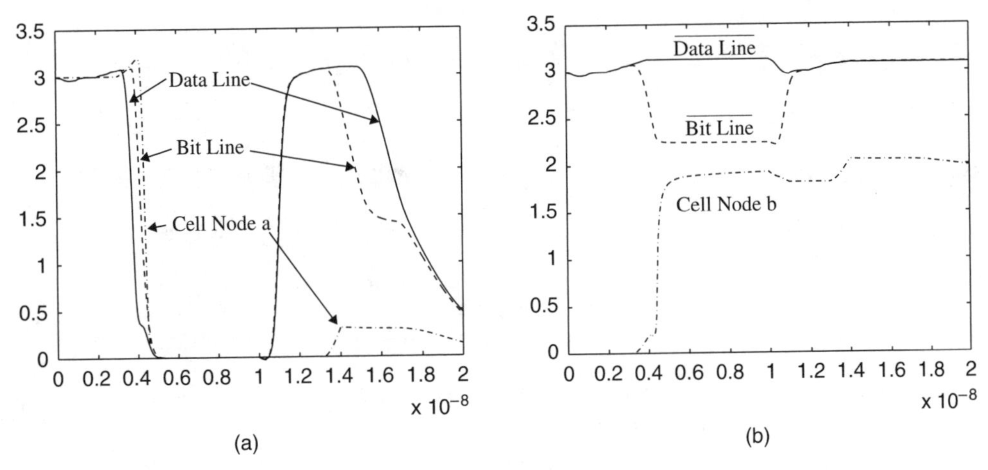

Figure 11.8 Simulated waveforms of writing into and reading from a memory cell. The write operation occurs during the first 10 ns; the read operation occurs during the second 10 ns.

charged (for 3 ns) and then the cell is *read.* Figure 11.8b shows the simulated node voltages of the corresponding nodes on the side of the cell that was initially at a "0".

A few details are worth emphasizing. With respect to Fig. 11.8, note that the write operation is quite fast once precharge has taken place. Second, note that the high voltage written into node b of the cell is less than 2 V (in Fig. 11.8b). Also notice that during the read operation, there is a significant delay from when the bit line starts to discharge to when the data line starts to discharge (in Fig. 11.8a) due to the time it takes before the column-access transistors turn *on.* Finally, note that during the read operation the low voltage of the memory cell (node a in Fig. 11.8a) does not stay at 0 V but is *pulled high* to just less than 0.5 V when the cell-access transistors are turned *on.*

Sense Amplifier

In most large memories, the parasitic capacitances of the bit lines and the data lines are quite large. These parasitics are primarily due to the junction capacitances of the large number of transistors connected to the buses. They are also due to the interconnect capacitances of the buses themselves, which can often be quite long (and therefore have large parasitic capacitances). These parasitics cause the bus voltages to change very slowly, particularly when a memory cell is being read. To shorten the time for reading a memory cell, almost every modern semiconductor RAM uses bus equalization and sense amplifiers. In this approach, the differential bus voltage is first preequalized to be zero. This is normally done when the memory is not being accessed or at the beginning of an access, the same as bus precharging.

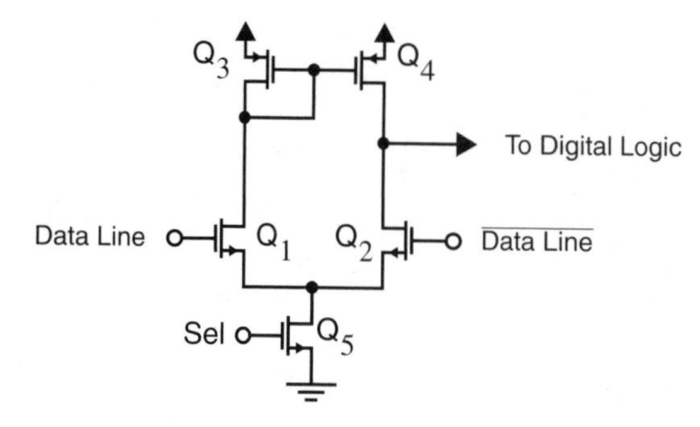

Figure 11.9 A simple sense amplifier.

Next, when the memory cell is being read, a differential bus voltage will slowly develop. Because of the bus capacitances, it can take quite a while before this voltage becomes large enough to drive a logic gate. However, a sense amplifier, which can accurately detect small differential voltages, can be used to detect and amplify bus voltages as small as 50 to 100 mV, after only short delay times. The sense amplifiers are then used to drive the logic gates.

There are many possibilities for sense amplifiers. Most recent designs are based on a differential-input gain stage, similar to what is often used as the first stage of an operational amplifier. For example, a very simple sense amplifier is shown in Fig. 11.9 (Yu et al., 1981). When the sense amplifier is not being used, Sel will be *low* and Q_5 will be *off*. During this time, the sense amplifier is deactivated to minimize power dissipation. When the sense amplifier is being used, Sel goes *high* and Q_5 turns *on*. This will bring the drain voltage of Q_5 *low,* which causes both Q_1 and Q_2 to conduct. Depending on the differential data-line voltage, either Q_1 or Q_2 will be conducting more current. The current of Q_1 is *mirrored* by current mirror Q_3, Q_4 so that it can be compared with the current of Q_2. The output of the sense amplifier will be *high* if the current of Q_1 is larger than that of Q_2. Otherwise, it will be *low.*

The speed and gain of a sense amplifier is very dependent on the transistor sizes chosen and on the d.c. common-mode precharge voltages of the differential input signal. Ideally, for a given precharge voltage, the sizes will be chosen so that Q_1, Q_2, Q_3, and Q_4 are all in the *active region*, when the differential input voltage is zero. Although this constraint is not difficult to meet when the bus precharge voltages are around $V_{DD}/2$, it may be difficult to meet for precharge voltages of V_{DD}, for the simple preamplifier shown. Transistor Q_5 would normally be taken wide so that when it is *on*, it is in the *triode region*, unlike what would be the choice if the circuit was used as the first stage of an operational amplifier.

Currently, a popular preamplifier is an improvement over the simple differential gain stage of Fig. 11.9. This sense amplifier is shown in Fig. 11.10 (Kobayashi et al., 1985). This sense amplifier consists of two of the simple preamplifiers of Fig. 11.9 connected *back to back* in parallel. This makes the circuit completely symmetric and gives a differential output that feeds to a second gain stage. In Kobayashi et al. (1985), the second gain stage was a third differential amplifier similar to the circuit of Fig. 11.9.

Another important concept is used in the sense amplifier of Fig. 11.10. This is the addition of *reset switches* that are used to guarantee the sense amplifier is biased in its high-gain region, with a 0 V output voltage, prior to sensing the input voltage. This resetting also eliminates any

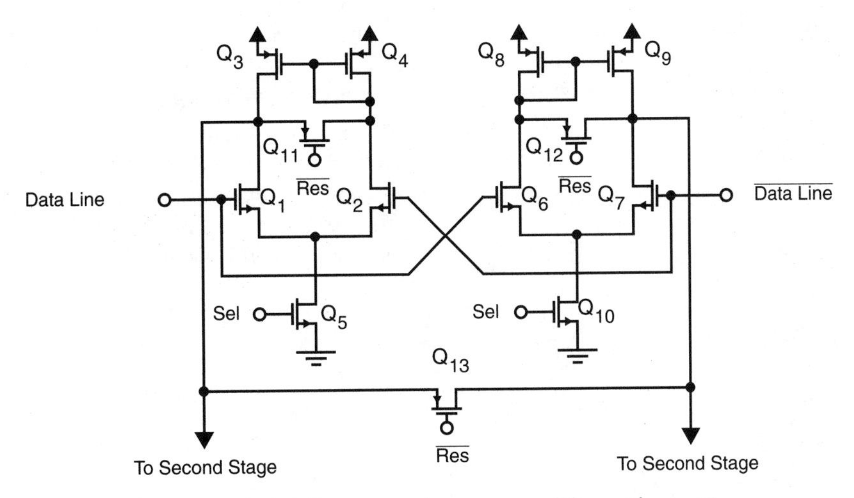

Figure 11.10 A fully symmetric sense amplifier with resetting.

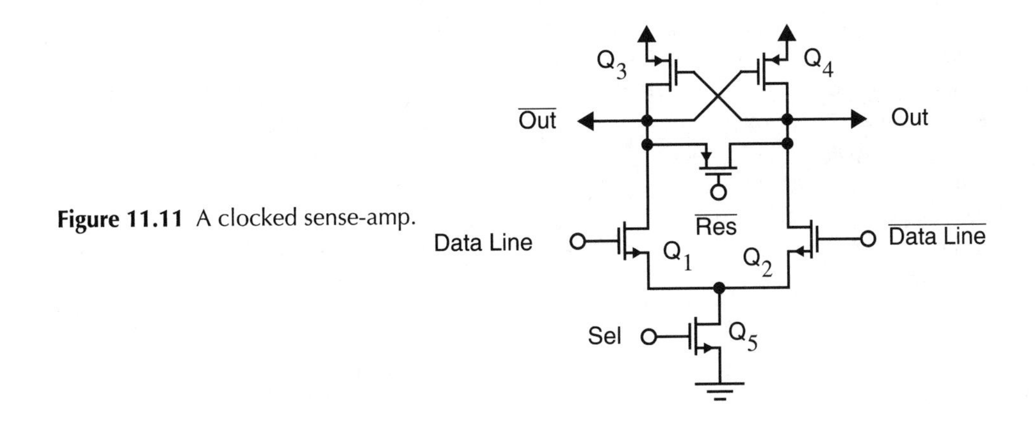

Figure 11.11 A clocked sense-amp.

memory of the previous output value. This concept of *resetting* has been found to greatly increase speed and is widely used at the present time.

The output of a sense amplifier often feeds into a *clocked latch* or an inverter biased using a reset switch into its *high-gain region*. Sometimes, a sense amplifier itself will be a clocked latch, having positive feedback to achieve a larger effective gain. An example of this is shown in the sense amplifier of Fig. 11.11 (Childs and Hirose, 1984). In this sense amplifier, after it is powered up, the differential output is *shorted* by a reset switch for a short while. This reset switch is taken wide enough to guarantee the positive-feedback loop gain through the cross-coupled p-channel load transistors Q_3 and Q_4 is less than unity. Next, when the reset switch is turned *off*, the positive feedback of the cross-coupled load transistors will quickly amplify the originally small differential output voltage into a much larger output voltage.

The previous examples of sense amplifiers are just a few of the variations that have been used for SRAMs. The design of good sense amplifiers is very much an *analog* design issue, which is dependent on an in-depth understanding of transistor modeling, and concepts such as small-signal changes, circuit time constants, and positive versus negative feedback. The importance of these circuits is increasing for applications in addition to memories. In particular, *the use of differential buses being first equalized and then of having their small slowly changing differential voltages quickly sensed by sense amplifiers is becoming prevalent wherever speed is limited by bus signal changes, for example, in high-speed computer chips.*

11.3 Address Buffers and Decoders

Important components of memory circuits are the *address buffers* and *decoders*. The decoder circuits require both the address inputs and their complements. They also represent a significant load on the input address buffers, which are used to obtain the required complements. The large load is due to the typically large number of gates connected to the address-input buffers. An example address-input buffer circuit from Minato et al. (1982) is shown in Fig. 11.12. It is basically a cascade of inverters with the last inverters having n-channel *source-follower pull-ups* as well as p-channel *pull-ups*. The authors state that this gives faster rise times.

The buffered address and its complement are then input to the decoders. Examples of decoders have already been seen in Chapter 10. The decoders used in memories are often straightforward static circuits. They frequently are realized using traditional CMOS logic gates. Because the number of the outputs of the decoders might range from 32 to 512, the decoders usually consist of two or three stages. An example of a two-stage decoder was seen previously (Fig. 10.7 of Chapter 10) and has been repeated here for clarity in Fig. 11.13. This circuit consists of three two-input to 1-of-4 output predecoders. The outputs of these predecoders are the inputs for 64 three-input *nor* gates. The complete circuit realizes a six-input to 1-of-64 output decoder.

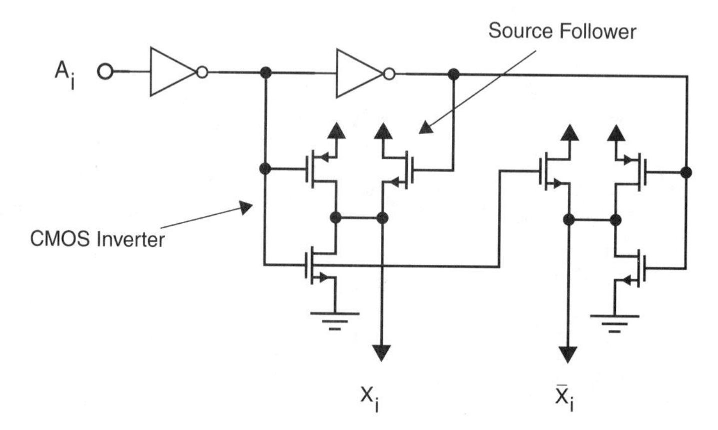

Figure 11.12 An example address-input buffer.

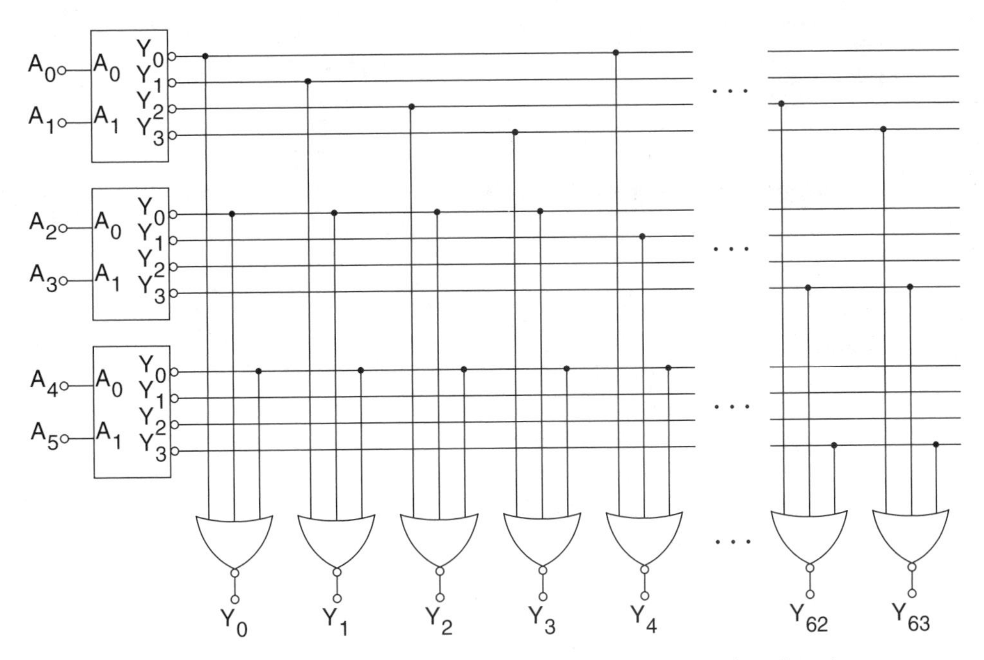

Figure 11.13 A two-level approach for realizing a 6 to 1-of-64 decoder.

More modern memories have started to use improved decoders that take up less space. These often make use of techniques such as *pseudo-NMOS* loads and dynamic logic. An example of a possibility for a decoder that is still static but uses pseudo-NMOS loads to reduce the space and minimize loads is shown in Fig. 11.14. This 8 to 1-of-256 word-line decoder is similar to that of Sasaki (1988). In this decoder, each word line is buffered by an inverter. Eight word-line drivers are grouped into a section. All the word-line buffers of each section are driven by the outputs of a nine-input, eight-output *pseudo-NMOS* logic gate. The inputs to this gate come from three-input *predecoder* gates and from a five-input *nor* gate. There is one five-input *nor* gate per section. Since there are 32 sections, this means there will be 32 five-input *nor* gates. Only one of these five-input *nor* gates will have a high output. All of the others will be *low.* A low input to the section driver guarantees that all of its outputs are *high*, and, therefore, the outputs of its word-line drivers are *low.* It might be noted that in this state, there is no d.c. power dissipated by the section driver. One of the *section-driver gates* will be activated by a high input from the five-output *section-select nor gate.* In addition to the section-select input, each section driver has an additional eight inputs coming from the three-input predecoder *nor* gates. Only one of these inputs will be *high.* This will in turn cause one of the outputs from the section-driver gate to go *low,* which in turn will cause one word-line driver output to go *high.*

It is possible to use pseudo-NMOS logic for the five-input section-select *nor* gates as well as for the three-input predecoder *nor* gates. Possible realizations for these gates are shown in Fig. 11.15a and b, respectively. In these realizations, the section-select gate is actually a six-input gate. The additional input is the block-select ($\mathrm{Blk}_{\mathrm{slct}}$) control signal. When the block is

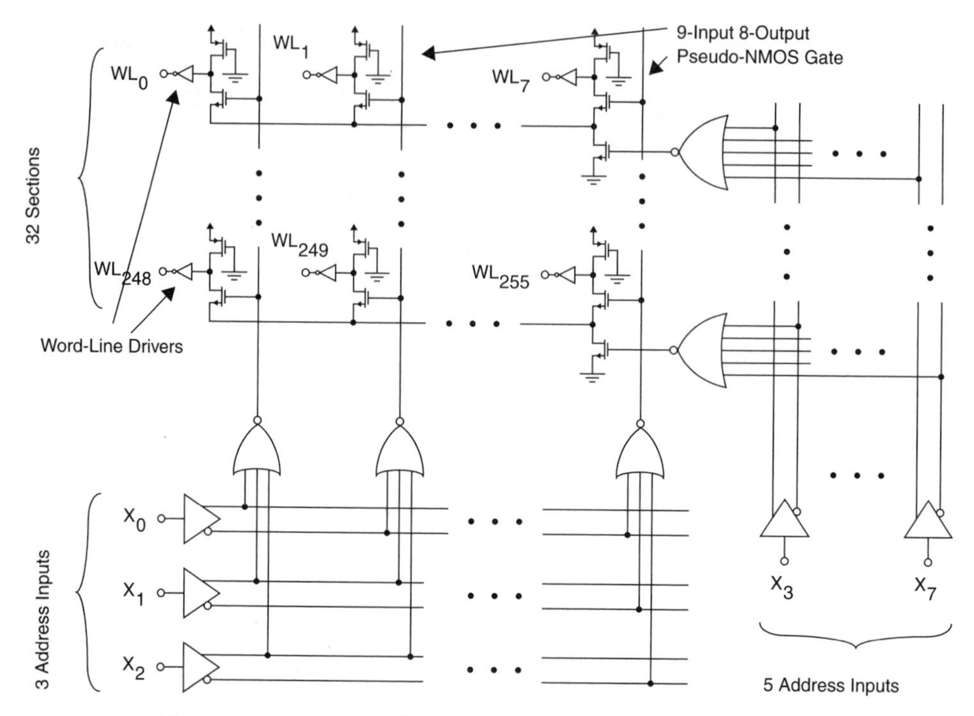

Figure 11.14 An example of an 8 to 1-of-256 word-line decoder.

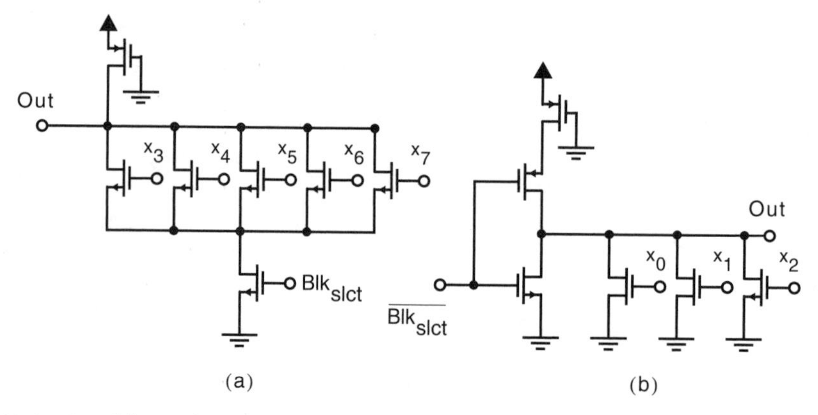

Figure 11.15 Possible implementations of (a) the five-input and (b) the three-input *nor* gates of Fig. 11.14.

not selected, it is guaranteed that all section-select outputs will be *high* so the gates do not dissipate power. Otherwise, the standby power would be too large.

Unfortunately, this causes all of the section-driver gates to be selected. However, this can be overcome by adding an additional $\overline{Blk_{slct}}$ input to the predecoder gates. This guarantees that when the block is not selected, all of the predecoder outputs will be *low,* which guarantees all word-line drivers are *low.* Thus, the proposed decoder has no d.c. power dissipation when in standby mode. There is d.c. power dissipation in the decoders of *selected* blocks, so if low power dissipation is critical, an alternative approach should be taken.

11.4 Dynamic Bus Precharge and Address-Transition-Detect Circuits

Most (but not all) modern *SRAM* circuits rely on *dynamic bus equalization* and possibly precharging of one kind or another. Since most SRAM circuits do not have clock inputs, they must rely on other means to synchronize themselves in order to determine when the dynamic precharging and equalization should occur. Also, many sense amplifiers require timing signals when they are sensing an output. A popular way of generating these timing signals is to use *address-transition-detect (ATD) circuits* (Hardee and Sud, 1981; Childs and Hirose, 1984). Although there are many different ways of realizing this circuit, the circuit shown in Fig. 11.16

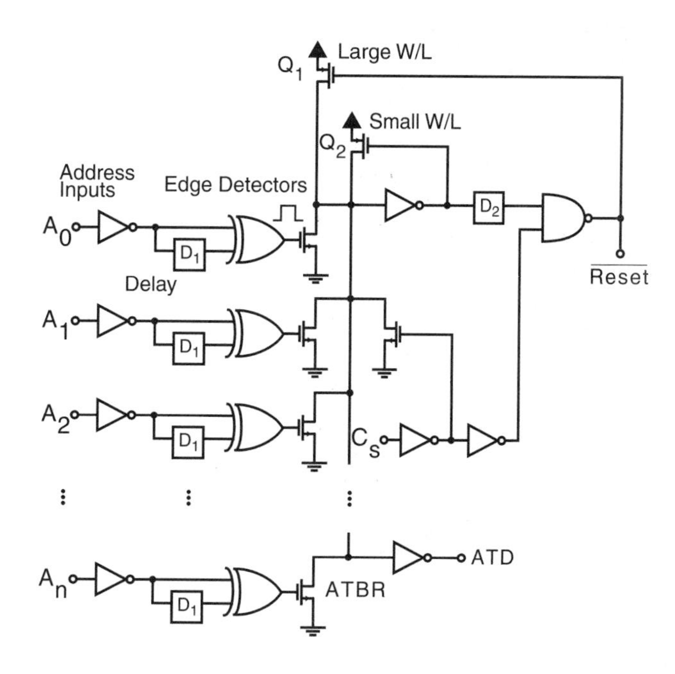

Figure 11.16 A representative address-transition-detect circuit for detecting changes or row address inputs.

is representative. All of the address inputs go to edge detectors. Whenever an address input changes, or the chip-select signal (C_S) goes *high*, the address-transition-detect circuit causes the precharge signal (ATD) to go *high* for a short time, which initiates any internal timing circuits and causes the buses to be precharged. A short time after the ATD has gone *high*, it will automatically go back *low*. A short time after this, new data will be placed on the data lines, assuming the memory is being read.

To see how the ATD circuit works, consider the edge-detector circuitry first. This circuitry is shown symbolically as an *exclusive-or* circuit with one input delayed by time D_1. Normally, both inputs to the *exclusive-or* circuit have the same logic value. However, if an address input changes, then for a short time equal to D_1, the two inputs to the corresponding *exclusive-or* will be different. This causes the *exclusive-or* circuit to output a "1" for duration D_1, whenever the corresponding address input changes. This, in turn, will cause one of the *pull-down* transistors to turn *on,* which causes the previously *high address transition bus* (*ATBR*) to go *low*. This subsequently causes ATD to go *high*. ATD is used to initiate other timing circuits and bus precharges. ATD will stay *high* until the signal passes through the feedback loop composed of an inverter, the delay D_2, and the *nand* gate. Once this happens, Q_1 will turn *on*. Since Q_1 has a large W/L, the bus ATBR will be quickly recharged *high*.

The transistor Q_2 is a moderately high impedance transistor (i.e., small W/L); its function is to keep ATBR *high* when there are no address transitions. When an address transition occurs, Q_2 is easily overcome by the much wider pull-down transistors.

When the memory is not being used, the chip-select signal (CS) is a "0" and the buses are kept precharged, ready for operation. This is guaranteed by having $\overline{CS}$ connected to an ATBR pull-down transistor and also routing $\overline{CS}$ to the *nand* gate, which disables the feedback loop and keeps Q_1 turned *off*.

There are many different possibilities for realizing the edge detectors. Two possibilities from Kobayashi et al. (1985) are shown in Fig. 11.17. Both of these circuits are based on the *exclusive-or*

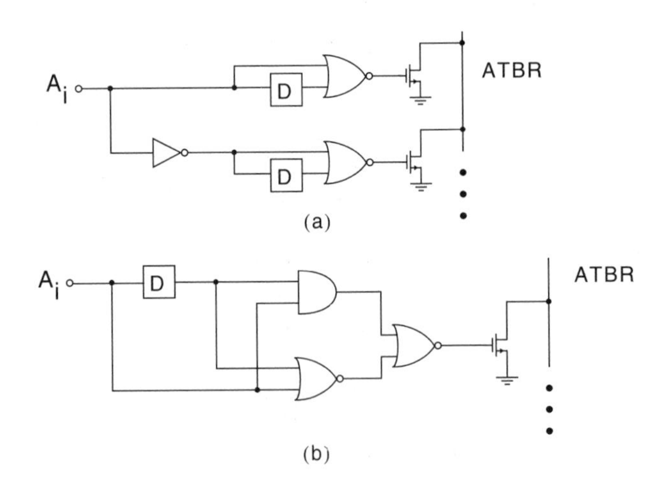

Figure 11.17 Two different possibilities for edge-detection circuits.

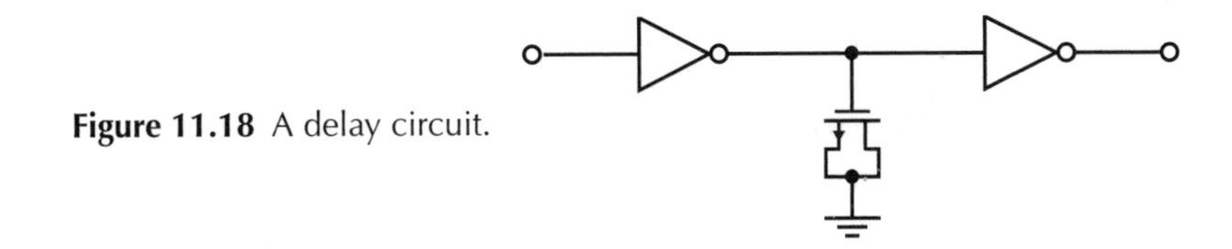

Figure 11.18 A delay circuit.

approach. In Fig.11.17a, the *nor* gates will go *high* at negative-going transitions of its input signal. Since the complement of the address is input to the second *nor* gate, its output will go positive at positive-going transitions of the address input. Thus, one of the *nor* gates will go *high* at any transition and this will cause ATBR to be *pulled low,* thereby activating the address-transition circuitry. The operation of the circuit of Fig. 11.17b is similar. There are other possibilities based upon using transmission-gate realizations of the *exclusive-or* circuit.

The delays required for address-transition-detect circuitry can be realized by a cascade of an even number of inverters. To achieve longer delays, some of the inverters can be loaded by transistors connected as capacitors as is shown in Fig. 11.18.

Often there will be different ATD circuits for the address inputs used to decode the rows as opposed to those used to decode the columns. The row ATD might be used to control the precharge of the bit lines and the data path, whereas the triggering of the column ATD might cause only the data path to be precharged, since, in this case, the values on the bit lines will not be changing. In another variation, a memory may always have its buses in the precharge state unless an address transition occurs. In this case the precharge is deactivated, the decoders (possibly dynamic) are activated, the appropriate word line is pulsed high, and the SRAM contents are sensed and then stored in an output latch (assuming a read operation). The precharge circuitry is then reactivated so the memory is immediately available for the next access (Yamamoto et al., 1985). In addition to this variation, there are many other possibilities; for example, the use of Domino logic for decoders is popular. These other possibilities are beyond the scope of this text.

11.5 Modifications for Large Static Random-Access Memories

At present, SRAM memories have been realized with capacities to store as many as 16 Mbits and more. When realizing memories this large that have to be accessed in very short times, a number of specialized techniques have to be used. Some of these techniques are used to increase the speed. Others are used to minimize power dissipation. Both of these issues are addressed by organizing the memory hierachically into a number of different blocks or arrays. This minimizes the lengths of the word lines and data lines. In addition, at any given time only some of the blocks will be activated, which minimizes power dissipation. For example, in Hirose et al. (1990), the 4-Mbit memory is divided into 32 blocks, with each block having 1024 rows and 128 columns. Also, each block has 16 sense amplifiers, or equivalently one sense amplifier for every eight columns, which reduces the data-line length and therefore parasitic

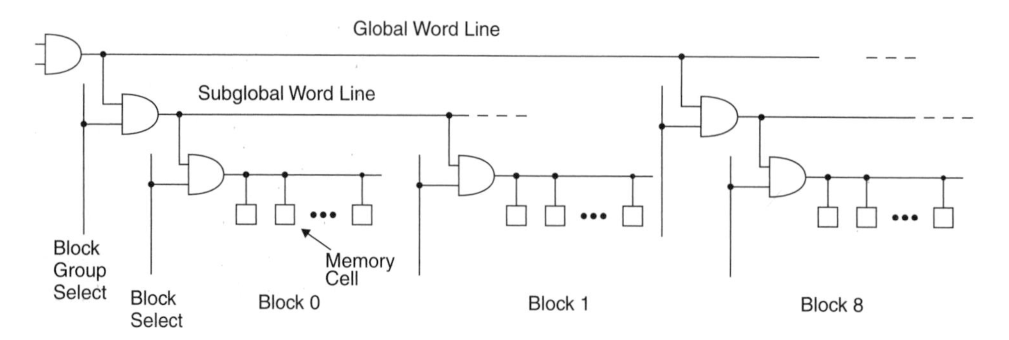

Figure 11.19 The hierarchical word decoding of Hirose et al. (1990).

capacitance in the signal path between the memory cells and the sense amplifiers. Each array or block in this architecture realizes separately a $32K \times 4$ memory.[1] Ten address inputs are used to choose the row and five address inputs are used to choose which block is selected. This latter selection is done using two levels as shown in Fig. 11.19. Only one block group will be selected and the corresponding subglobal word line will go *high*. This allows a single word line of one block in the selected group to go *high*. Each block group consists of eight blocks. Therefore, two address inputs select a group, and three address inputs select a block from within the group. By using a two-level block decoding scheme, the parasitic capacitances of the various lines are minimized. Three address inputs determine which column will be connected to a particular sense amplifier, and two inputs will determine which of four data buses will be connected to the external pins. Four sense amplifiers from each block are connected to external data pins in parallel.

In another example (Goto et al, 1992), the memory is organized into 64 blocks with each block being 1K by 256. In Seno et al. (1993), the memory looks externally like a $4M \times 4$ memory. Internally, the memory is subdivided into four quadrants. Each quadrant will be activated during every memory access to supply one of the four external data bits. However, inside each quadrant there are 32 blocks with each block having 1024 rows and 128 columns. Obviously, many different hierarchical architectures are possible, but the principles involved for each architecture are similar.

11.6 Dynamic Random-Access Memories

When realizing very large random-access memories, it is critical to minimize the area of the memory cells. For large memories, this is possible by using a capacitor to store the logic values. These memory cells are much smaller than SRAM memory cells, but require peripheral circuits that are considerably more complex; however, in a memory IC, since the storage cells

[1]This memory can actually be configured using a control pin so that each array is a $128K \times 1$ memory.

make up the great majority of the IC, the total number of memory cells in a dynamic RAM is considerably greater for given a similar silicon area. These memories, called dynamic memories, trade off the additional complexity and slower speed for a greater number of memory locations. For general-purpose memory inside most computers, this has proven to be a reasonable choice; at present, the great majority of computers use dynamic RAMs for general-purpose storage. This has resulted in dynamic RAMs being the most often produced IC. Due to its large volume, it has historically been the driving force behind the development of new processing technologies. At present, 1-Gbit dynamic RAMs are just being reported and 256-Mbit dynamic RAMs are in production. Ten years previous to this time, having over 1000 million transistors in a single IC was almost inconceivable.

The cost of building a competitive production line capable of producing state-of-the-art dynamic RAMs is large, approximately $3 billion, and increasing. There are few U.S. companies left willing or able to invest this much capital into a market place in which profits might not be returned for 5 years and profit margins are cyclic. Largely because of the high cost and pressure for short-term profits from large stock investors, such as pension funds and mutual funds, most U.S. companies have stopped producing dynamic RAMs, with a few notable exceptions such as IBM. Irrespective, the total volume of production of dynamic RAMs is so great that this remains an area of great importance.

Dynamic Memory Cells

The memory cell used in a *dynamic RAM (DRAM)* is shown in Fig. 11.20. It consists of a single capacitor, C, and an access transistor, Q_1. The capacitor is the storage element. In a modern 256-Mbit dynamic RAM based on a 0.25-μm technology, it would usually be 20 to 40 fF. To realize a capacitor of this size in as small an area as possible (i.e., around 0.7 μm^2 for a 0.25-μm process), the capacitor is usually a three-dimensional structure and realized in either a trench (Taguchi et al., 1991) or a stacked structure overtop the memory cell (Mori et al., 1991; Oowaki et al., 1991). In the latter case, a plate of the capacitor actually acts as an electrical shield to protect the bit line from capacitively coupled noise from adjacent bit lines. A 256-Mb dynamic RAM

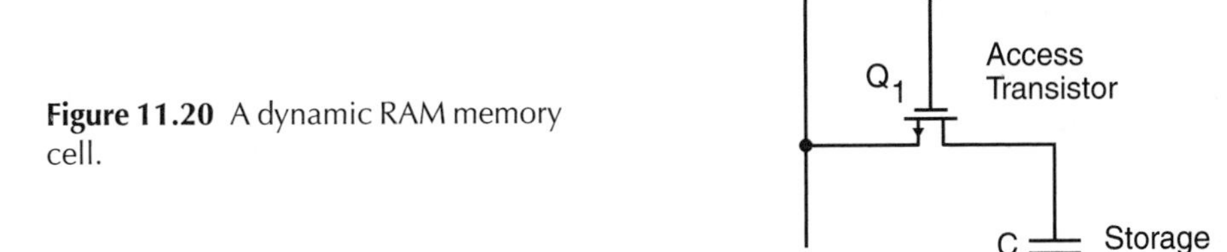

Figure 11.20 A dynamic RAM memory cell.

is described in Sugibayashi et al. (1993), and 1-Gb dynamic RAMs are described in Sakashita et al. (1996) and Yoo et al. (1996).

When a "0" is stored in a DRAM cell, the storage capacitor will be discharged to around 0 V by the bit line being held at a low voltage just before the access transistor is turned *off* when the word line goes *low*. When a "1" is to be stored, the bit line is held at a high voltage of perhaps 2.2 or 3.3 V just before the access transistor is turned *off*. This will leave the storage capacitor charged to a high voltage around V_{DD}.

Reading the contents of a dynamic RAM cell is a more complicated and error-prone procedure than reading the contents of an SRAM cell. The major reason for this is that a typical storage capacitor may be only one-tenth the size of the parasitic capacitance of the bit line. When a DRAM cell is being read, the bit line is first precharged to a high voltage, perhaps 3.3 V. Next, the DRAM cell is connected to the bit line. If the cell had contained a "1", the bit-line voltage will not change when the cell is connected to the bit line. However, if the cell had contained a "0" and was originally discharged, when it is connected to the bit line, the bit line will be discharged to a slightly smaller voltage. The difference between these cases must be detected. This difference is usually small, possibly as small as 0.2 V. This small change is difficult to detect since there is not a set of differential bit lines as was the case for SRAMs. A third reason for the difficulty of determining the state of dynamic RAM cells is this voltage change is of the same order as noise being coupled through the substrate.

To differentiate between the cases of storing a "1" and storing a "0", it is necessary to generate a comparison voltage, which is then used as a second input to a differential sense amplifier. This comparison voltage is usually generated by adding some extra *dummy cells* that are always discharged to 0 V. During read operation, a dummy cell will be connected to two bit lines matched to, and in close proximity to, the actual bit line being read. The voltage change on these reference bit lines will be one-half the change on the bit line being read assuming a "0" had been stored. Thus, a comparison voltage approximately half way between the possible final voltages of the bit line being read is produced.

Another problem with DRAMs is that every time a DRAM cell is read, it will be charged to a voltage very close to the precharge voltage of the bit lines, perhaps 3 V or so. Thus, the read out is *destructive*; every time a DRAM cell is read, it will be charged *high* very close to the "1" state irrespective of its original contents. The correct value must be restored. Therefore, after each read operation, the contents of the memory cell must be rewritten by either keeping the bit-line voltage *high* if a "1" had been stored or by *pulling* the bit line *low* to around 0 V if a "0" had been previously stored.

A third problem with DRAMs is that an isolated capacitor can store charge for only a finite time. Eventually the leakage current of the junctions of the access transistor will corrupt the stored value. For example, if n-channel access transistors are used, which predominantly is the case, then a stored "1" would eventually be corrupted to a "0" if it were left long enough. Thus, periodically, every cell of a DRAM must be read and refreshed. Normally, no cell may be left without being refreshed for longer than 4 ms, or so.

A simple solution to the above problems is to use many dynamic sense amplifiers throughout the memory, perhaps one for each bit line (or two, or four, bit lines). These sense amplifiers will be described next.

DYNAMIC RANDOM-ACCESS MEMORY SENSE AMPLIFIERS

An example of a DRAM dynamic sense amplifier is shown in Fig. 11.21. During normal operation, the sense amplifier outputs, D and $\overline{D}$, are precharged to *high* and equal voltages, in this case V_{DD}, by Q_1 and Q_2. Also, the reference capacitor, C_{dummy}, is discharged to 0 V. Next, also before sensing occurs, the storage capacitor, C_S, is connected to its bit line. This bit line has parasitic capacitance $C_{p\text{-}1}$, which is approximately equal to 12 C_S for this example. At the same time, the reference capacitor, C_{dummy}, is connected to a *pair* of matched bit lines, which together have a parasitic capacitance $C_{p\text{-}2}$ approximately equal to 24 C_S. This will create a differential voltage difference between the single bit line on the left and the pair of bit lines on the right.

In the next phase, the dynamic p-channel load transistors of the sense amplifier, Q_1 and Q_2, are turned *off*. The sense amplifier is then connected to the single bit line on the left and the two bit lines on the right. This will charge the sense amplifiers outputs to the initially small difference between the single bit line on the left and the two bit lines on the right. During the actual sensing, current is supplied to the cross-coupled transistors, Q_3 and Q_4, by turning Q_5 *on*. Because Q_3 and Q_4 are connected in a positive-feedback arrangement, this will amplify the initially small differential voltage to a much larger voltage. Indeed, if the device sizes are chosen carefully, and a controlled *turn-on rate* is used for Q_5, one output of the sense amplifier will be left at a high voltage near V_{DD} and the other side will be discharged almost to ground. In some sense amplifiers, some small pseudo-NMOS loads might be included in parallel with Q_1 and Q_2 to guarantee that after sensing, the differential output voltage is as large as possible.

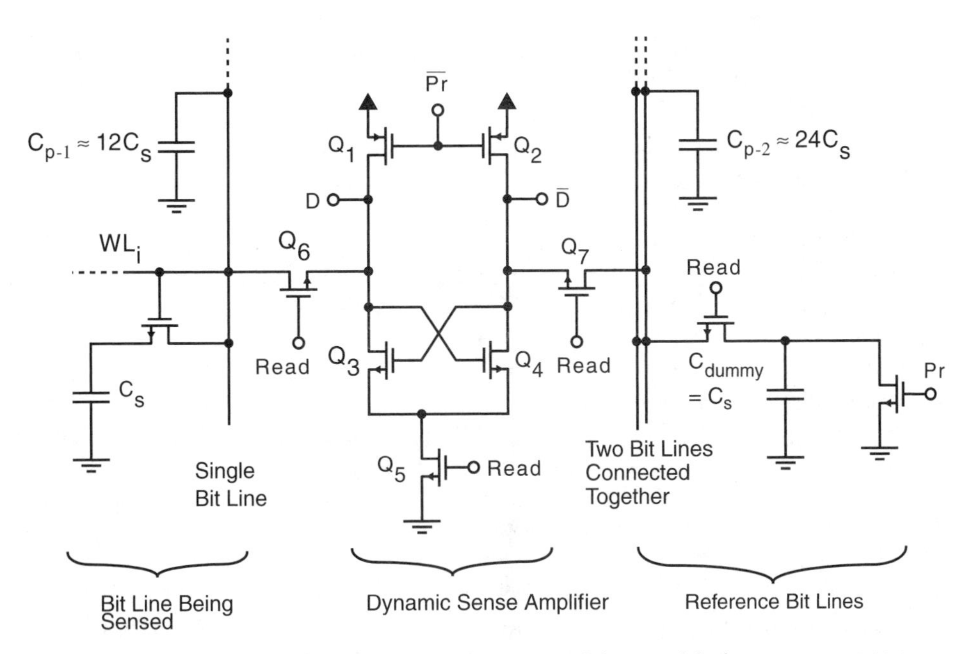

Figure 11.21 A dynamic cross-coupled sense amplifier suitable for use in a DRAM.

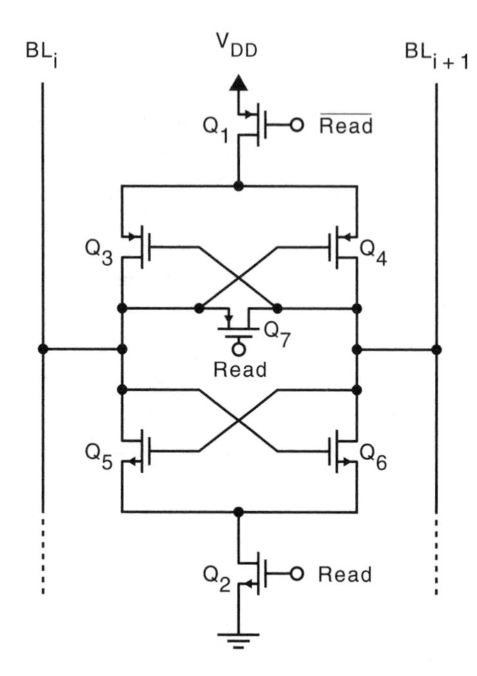

Figure 11.22 An alternative DRAM sense amplifier.

Irrespective, the positive feedback of the sense amplifier regenerates a large enough voltage so that the voltage level of the storage capacitor, C_S, is restored to either 0 V if a "0" had been previously stored or $V_{DD} - V_{tn}$, if a "1" had been previously stored,[2] which automatically refreshes the sense amplifier.

Even if a DRAM is not being read, its storage cells must be periodically refreshed by reading all cells every so often or their contents will be lost. However, this takes the memory out of usage for only a small percentage of the time. To see this, consider the following example. Normally, there will be a sense amplifier for every four bit lines, which might contain a total of 256 memory cells, for example. Every refresh phase, one of these 256 memory locations will be sensed and automatically refreshed. If one assumes the memory cycle time is 20 ns and that each 1 ms all 256 memory cells must be refreshed, then the total amount of time used for refreshing is $256 \times 2 \times 10^{-8} = 5.12\ \mu s$. Thus, the memory cannot be used for regular operation $(25.6/1000\ \mu s) \times 100\% = 0.25\%$ of the time while it is being refreshed. With properly designed memory controllers, this causes only a negligible degradation of system throughput.

Most of the original DRAMs had sense amplifiers similar to that just described. However, more modern DRAMs might have a sense amplifier similar to that shown in Fig. 11.22 (Lu and Chao, 1984). In this scheme, the sense amp consists of two cross-coupled n-channel transistors

[2]This assumes a voltage drop equal to V_{tn} across the access transistors. In modern DRAMs the high voltage level of the word lines may be actually greater than V_{DD} eliminating this voltage drop. This high word-line voltage is obtained using charge-pump circuits to be described later.

and two cross-coupled p-channel transistors. Normally, both bit lines will be precharged to $V_{DD}/2$. Also, a dummy capacitor will be precharged to $V_{DD}/2$. During sensing, the storage capacitor of the memory cell being read will be connected to one bit line and the dummy cell will be connected to the other bit line, which is used as a reference. Next, the sense amplifier is turned *on* by turning Q_1 and Q_2 *on*. This takes the initially small differential voltage and amplifies it to a full differential voltage of $\pm V_{DD}$ refreshing the memory cell at the same time.

The precharging of the bit lines to $V_{DD}/2$ can be achieved by precharging one bit line to V_{DD}, discharging the other bit line to ground, and then connecting them together. This is the reason for Q_7. Connecting the two bit lines together before sensing the memory cell's value also eliminates any memory from previous operations, a critical requirement. Similarly, the precharging of the dummy capacitor to $V_{DD}/2$ can be achieved by using two dummy capacitors, one precharged to V_{DD} and the other discharged to ground, and then connecting them together. This will *equalize* their voltages to a voltage very close to $V_{DD}/2$. Only one of these dummy capacitors is then connected to the appropriate bit line during the actual sensing operation. The appropriate bit line is one that does not have the storage cell being read connected to it.

Level-Boosted Word Lines

Also, most DRAM memories using the scheme just described also use a word-line voltage that has a high voltage that is larger than V_{DD}. This gives a large effective gate voltage for the access transistor (i.e., Q_1 in Fig. 11.20) when a dynamic RAM memory cell is being accessed. The *voltage-level boost* is achieved using a *charge-pump driver.* A simplified schematic of a voltage-level-boost word-line driver is shown in Fig. 11.23 (Fuji et al., 1989). The charge pump that is used to boost the voltage level is realized by transistors Q_1, Q_2, Q_3, and Q_4. Transistor Q_2 is used to realize a capacitor. It normally will have a large W and L. When P_r is *high,* both Q_1 and Q_4 are *on* and Q_3 is *off.* This precharges capacitor Q_2 close to V_{DD}.[3] Next, when P_r goes *low,* Q_1 and Q_4 turn *off* and Q_3 turns *on.* This connects the bottom plate of Q_2 to V_{DD}. Assuming capacitor Q_2 is large enough so that all of its charge cannot be immediately lost, the voltage change in the bottom plate of capacitor Q_2 from ground to V_{DD} will cause the top plate voltage, V_{BOOST}, to be boosted to a voltage above V_{DD}. This voltage is then used as a boosted power-supply voltage for a level translator (transistors Q_5 to Q_{10}) and the word-line driver (transistors Q_{11} and Q_{12}). With the proper dimensioning of sizes, this circuit can easily supply a word-line voltage that is 1.5 to 2 V above V_{DD}. *It is predicted that using charge pumps to generate on-chip voltages outside of the range of* V_{DD} *will become more common as 3 V and smaller power-supply voltage becomes prevalent.*

This concludes the description of some of the special circuits commonly used in dynamic RAMs. In the next section, some of the special circuits used in read-only memories (ROMs) will be described.

---

[3]Actually to $V_{DD} - V_{tn}$ for the realization shown, although modified realizations can boost the voltage to V_{DD}.

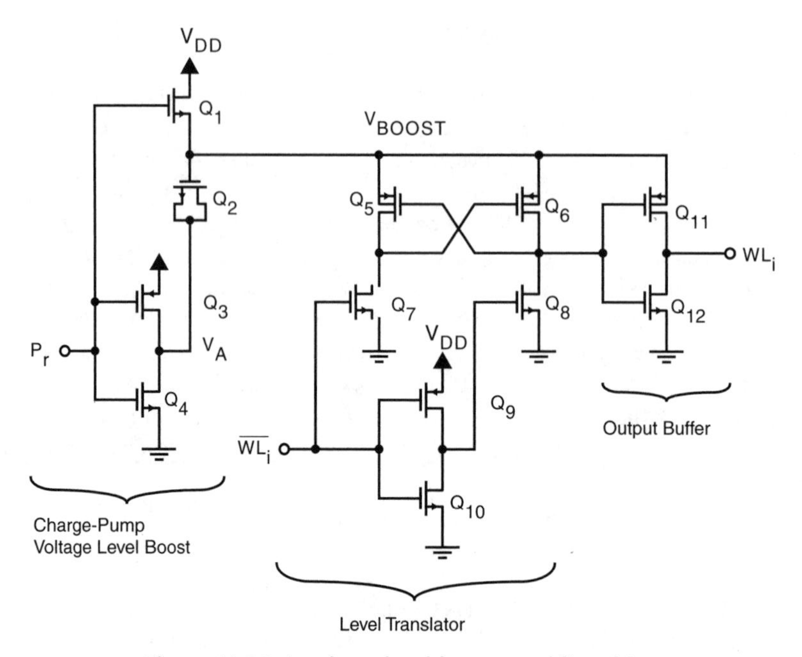

Figure 11.23 A voltage-level-boost word-line driver.

11.7 Read-Only Memories

Many times, the desired contents of a memory are known at design time; in this case it is often desired that the contents be programmed at design time so that the memory does not have to be *down-loaded* during actual usage. Also, it is often desired that the memory be *nonvolatile*, that is, its contents be preserved, even when the power supplies are turned off. In these cases, as long as the contents of a memory never need to be changed, one would normally use a *read-only memory* (*ROM*). Not only are these memories nonvolatile, they also are denser for a given technology, and are potentially faster and consume less power. A ROM might be used to store the μ-code that programs the μ-controller of a microprocessor IC. Another example might be a *ROM lookup table*. ROMs can also be used to implement *acyclic logic*. In this application, they are useful when almost all *min-terms* are present.

There are many types of ROMs that can be used for nonvolatile storage of data. The first type that will be described is called a *mask-programmed ROM*. This type of *ROM* has its contents permanently programmed at the time of processing. The actual contents are determined by the layout of a mask. For example, the mask that defines the active regions can be used to either create or not create a transistor at every storage location. The former case might correspond to a "0", whereas the latter might correspond to a "1". This type of ROM is often used when a ROM is included as a part of a larger IC. It might also be used for ROMs that are to be produced in very high volume.

There are also many types of ROMs whose contents can be programmed by the user after all the processing has been completed. One example of this is called a *fusible-link ROM* or a *writable ROM (WROM)*. This type of ROM is usually realized using a bipolar process. There will be a bipolar transistor at every memory location having its base connected to a word line and its emitter possibly connected to a bit line by a fuse. If a “0” is desired at a particular memory location, the fuse is *blown*. Conversely, a *nonblown* fuse corresponds to a “1”. This type of ROM will not be described in any greater detail as it is waning in popularity.

A more popular type of *user-programmable ROM* is based on an MOS technology, where a transistor is included at every storage location that has an adjustable threshold voltage that is electrically programmable. When a “1” is desired, the transistor threshold voltage will be programmed so the transistor never turns *on* when the appropriate word line goes *high*. The contents of ROMs of this sort can often be erased. For example, ultraviolet light might be used to erase the cells, in which case the ROMs are called *EPROMs* for *electrically programmable read-only memories.*[4] If the contents can be electrically erased, then the memories might be called *EEPROMs* or E^2PROMs for *electrically erasable programmable read-only memories.*[5] This latter type is gaining in popularity. These types of programmable ROMs will be described after first describing mask-programmable ROMs.

MASK-PROGRAMMABLE READ-ONLY MEMORIES

A mask-programmable ROM is similar to a RAM except for some important differences. The circuitry for writing into a memory cell is not included and the contents of each memory cell is permanently fixed by the layout of a particular mask. For example, a “0” could be programmed into a specific memory location by including a transistor at that memory location, whereas a “1” would be programmed by not having a transistor. This could be determined by the mask that defines the active regions. By having an active region at a particular storage location, that is a region that does not have thick *field oxide*, and by having a polysilicon word line cross that region, a transistor will be formed. Conversely, if there is thick field oxide at a given storage location, a transistor will not be formed.

A simplified architecture for a mask-programmable ROM array is shown in Fig. 11.24, where the location of a transistor in the memory array corresponds to a “0”. During operation, all bit lines are precharged to a high voltage. Also, the single-output data line is precharged *high*. Next, a single word line will be *pulsed high* depending on some address inputs. Also, a single bit line will be connected to the output data line depending on the remaining address inputs. If a transistor had been included at the juncture of the chosen word line and the chosen bit line, the output data line will be *pulled low*, otherwise it will be left *high*. The value of the data line will be compared to a *reference voltage*, approximately half way between the two possible final values for the data line, by a sense amplifier, which determines the memory output. This architecture should be compared to the architecture given previously in Fig. 11.15 for an SRAM.

---

[4]Sometimes the acronym EPROM is used to designate *erasable programmable read-only memory*. There is not a consensus on the exact meaning of the acronym EPROM.

[5]Digital IC designers, more than perhaps any other professionals (except perhaps medical doctors), seem to like to use acronyms. This often makes it difficult for others to understand their vocabulary.

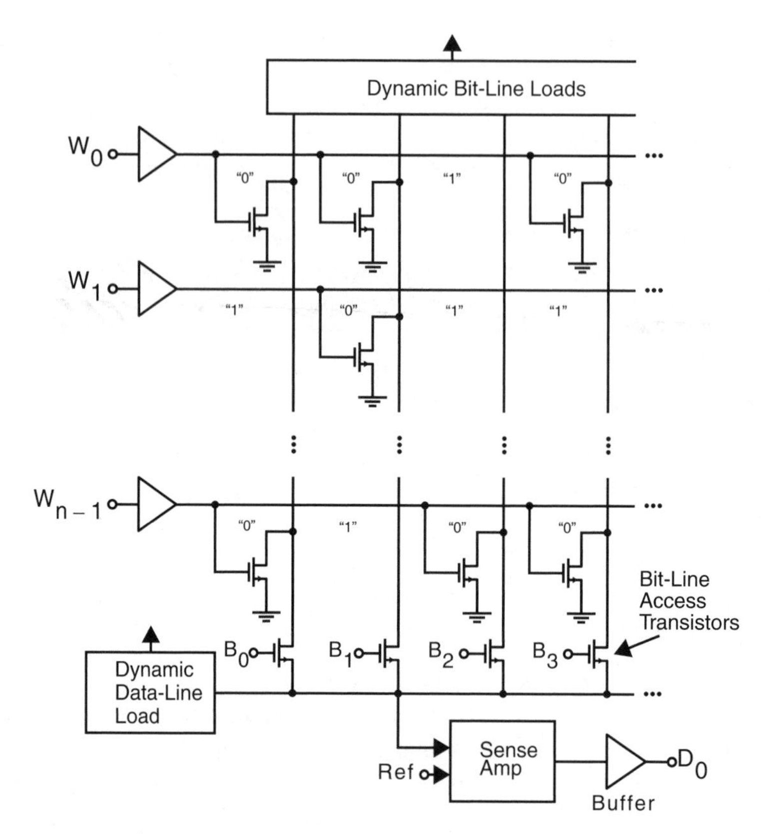

Figure 11.24 A ROM with some peripheral circuitry.

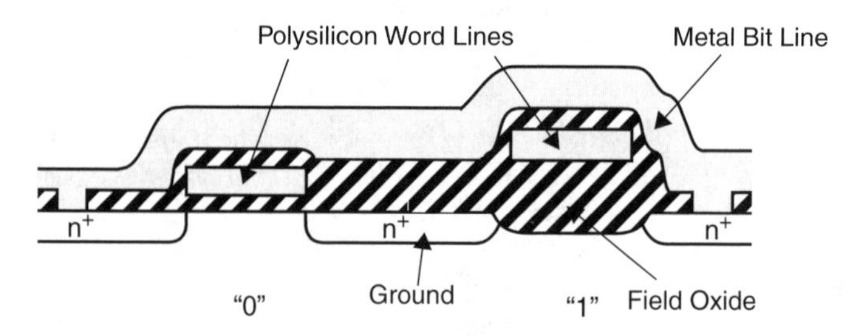

Figure 11.25 A cross section of ROM memory cells taken along the bit-line axis.

A cross-sectional view of two adjacent memory locations is shown in Fig. 11.25. This cross section is taken along the bit-line axis, which corresponds to the vertical axis in Fig. 11.24. The memory cell on the left stores a "0", since a transistor is realized, whereas the memory cell on the right stores a "1", since there is field oxide under the polysilicon word line and the transistor will never turn *on*.

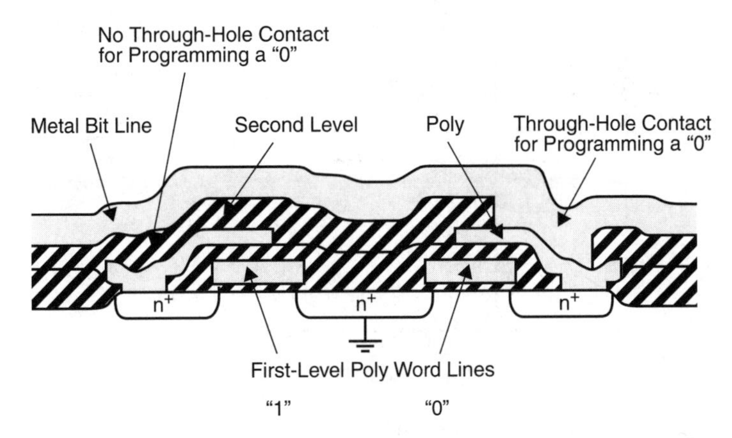

Figure 11.26 An alternative method for realizing ROMs based on programming the contact mask.

As an aside, it should be noted that *a polysilicon word line should always be strapped by a parallel metal line, placed above it, perhaps in second-level metal, to decrease the resistance of the word line.* The metal line runs in parallel directly above the polysilicon line and the two are connected together by metal vias. Usually, there will be one metal via for each memory cell or couple of memory cells. If the *metal strapping line* is not included, it would take an excessively long time for the word-line signal to propagate through a memory array due to the high resistance of a distributed RC transmission line realized by the polysilicon only.

Another possibility for realizing mask-programmed ROMs is to have transistors everywhere, but to have a contact from the transistor drain to the bit line only if a "0" is desired. A cross section of two memory cells based on this scheme is shown in Fig. 11.26. The ROM is based on a double-poly process. The second level of poly is used to realize buried contacts to the transistor drains. The metal bit line (Al) runs above the second level of poly. If a "0" is desired, a contact between the second level of poly and the Al bit line is formed. Otherwise they are isolated by deposited SiO_2. Because second-level poly is used for buried contacts to the transistor drains, the actual programming contacts can be made partially over the transistor gates. This makes the cells only three-fourths the size of what they would be without buried contacts (Masuoka et al., 1984). Furthermore, because the programming mask is used near the end of the processing steps, most of the processing can be done independently of the contents of the ROM, the time required from the programming step to when the IC is finished is minimized resulting in short *turnaround* times. Most of the processing can be finished before programming; once the contents of the memory have been decided on, and the contact mask has been produced, only a few processing steps need be completed before the IC processing is complete.

Sensing Read-Only Memory Cells

To preserve space, most ROMs have only a single bit line per memory cell. When this is the case, it is necessary to also generate a reference voltage so that a differential sense amplifier has a reference input. This reference input is ideally half way between the values generated by

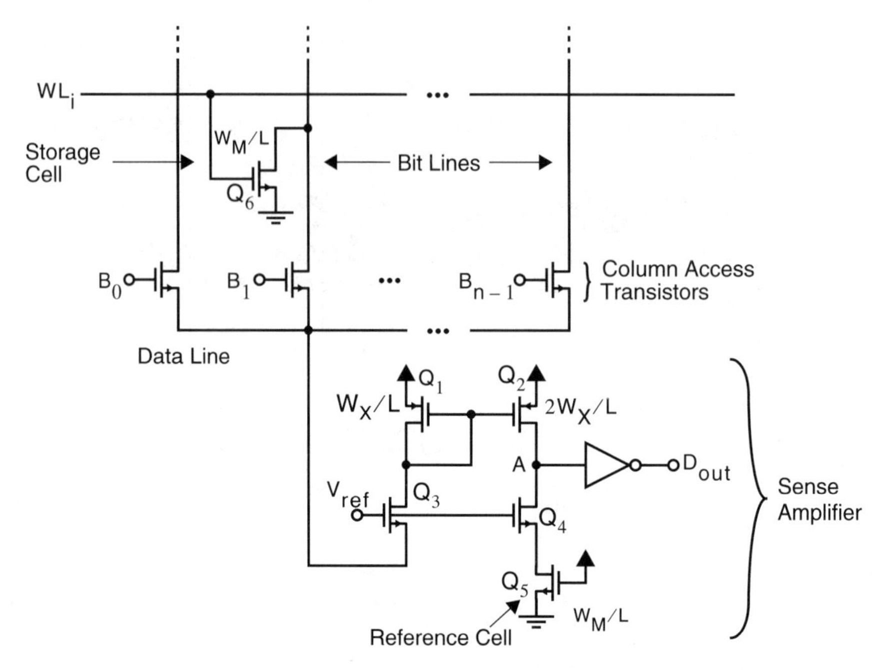

Figure 11.27 A simple sense amplifier for a ROM Memory Array.

actual memory cells for the cases of a "1" and a "0" being stored. There are many different approaches possible for generating the reference input to the sense amplifier. A moderately simple possibility is shown in Fig. 11.27. In this case, the data line is fed into common-gate transistor Q_3. The purpose of this transistor is to limit the voltage changes of the data line. Next, the current from the accessed bit line, which flows through Q_3, is amplified by a factor of two by p-channel current mirror Q_1 and Q_2. This current is then compared to the current generated by reference-cell transistor Q_5. If the contents of the memory location being read is "0", then a conducting transistor will be present that is connected between the accessed bit line and ground, for example, Q_6, if WL_i is *high* and B_1 is *high*. Once this current is amplified by two by current mirror Q_1, Q_2, then it will be twice the current of reference transistor Q_5 (assuming both transistors are in saturation). This will cause node A to go to a high voltage and D_{out} to go *low*. Alternatively, if there were no conducting transistor connected to the accessed bit line, then Q_3 and also current mirror Q_1, Q_2 would not be conducting. In this case, reference transistor Q_5 will *pull* A *low* and D_{out} will be *high*.

For higher speed operation, albeit with higher power dissipation as well, it is possible to add bias currents to common gate transistors Q_3 and Q_4. The bias current added to Q_4 should be twice the size of the current added to Q_3. This will further limit the voltage swings of both the data line and the accessed bit line.

A slightly better sensing scheme can be realized by adding a complete *dummy bit line* as well as a reference cell. This will match the parasitic capacitance, as well. For an example, the reader is referred to Nakayama et al. (1991). Indeed it is even possible to include two dummy

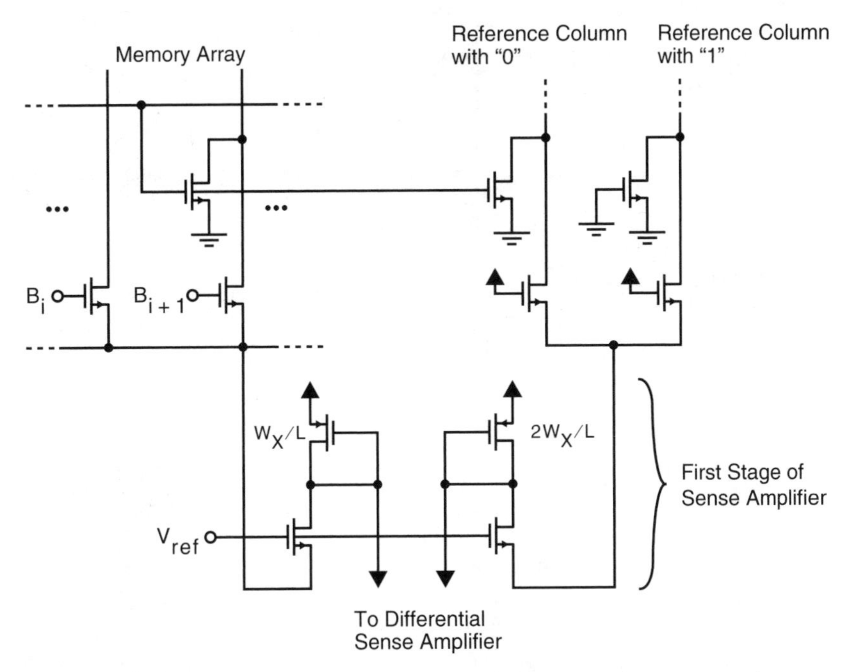

Figure 11.28 A ROM sense amplifier using two dummy columns as a reference.

bit lines as is shown in Fig. 11.28. One dummy column is programmed with "0"s and the other dummy column is programmed with "1"s. The current from both these columns is added, goes through a common gate transistor, and feeds into a diode-connected p-channel load. Also the current from the selected bit line of the memory array goes through a common-gate transistor and a diode-connected p-channel load, but with only half the width as the load for the dummy columns. This guarantees that the output voltage of the load for the dummy columns will be in between the two possible output voltages of the load used for the selected bit line. Also, all parasitic capacitances are matched so any dynamic effects will also be matched.

A third alternative is shown in Fig. 11.29. This architecture is based on having two differential bit lines for every column. A "1" is programmed into a memory cell by having the drain of the transistor in the memory cell connected to the left bit line, whereas a "0" is programmed by having the drain connected to the right bit line. Thus, both bit lines would normally be precharged *high* to approximately $V_{DD}/2$, and during operation one of the bit lines will be *pulled low.* The resulting differential voltage is then sensed by the first stage of the sense amplifier. The voltage swings on the data lines and bit lines are minimized by the common-gate connected input transistors of the first stage of the sense amplifier as was done in previous ROM architectures. The main reason for this differential bit-line architecture being fast is that during a read operation, the stored signal can be safely determined after only a short time when a small differential voltage has developed across the pair of bit lines, as opposed to waiting a much longer time for a large bit-line voltage to develop, which is required for safe sensing in

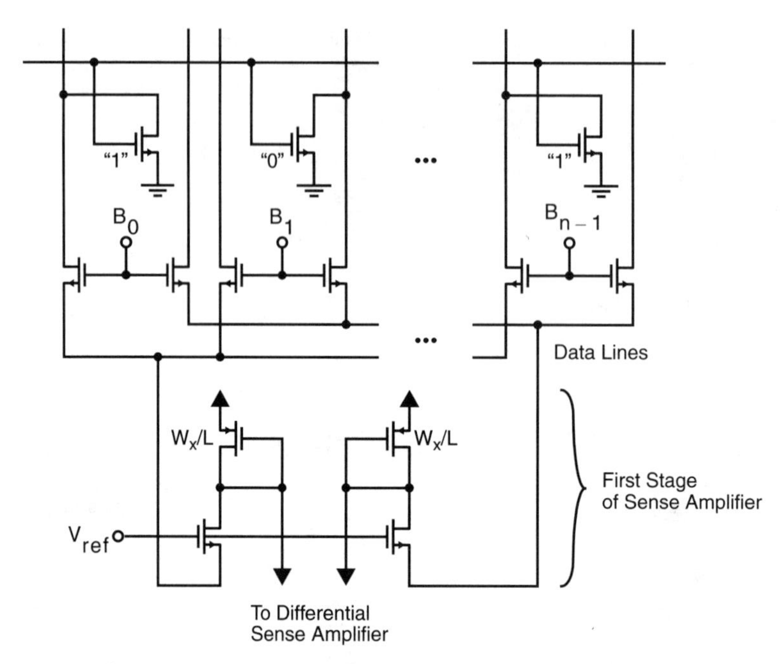

Figure 11.29 A ROM with a differential bit-line architecture.

nondifferential designs. Although the memory array requires more area, it still is much denser than a static RAM array given similar technologies. The additional area required is felt to be a penalty that is well justified in many high-speed applications.

Programmable Read-Only Memories

In the last section, ROMs were described that were programmed during IC processing by the geometry of photolithographic masks. For many applications, this procedure has many drawbacks. First, there is a large *turnaround time* required from submitting the desired contents to the time of receiving the memory. This time includes time for producing the mask and time for processing the ICs, as well as scheduling and delivery times. Since the first two steps are often done by different companies, the wait can be large. A second problem is that there are large costs involved each time the contents of the ROM are changed. This makes mask-programmable ROMs suitable only for high-volume applications or applications where the contents are known with a high level of confidence well ahead of time. For other applications, it is desirable that the user can program the contents of the ROM well after all processing has been completed. This type of ROM is called a *programmable ROM* or *PROM*. There are many different techniques for realizing a PROM, but all involve some penalty such as lower speed or larger area in return for the greater flexibility.

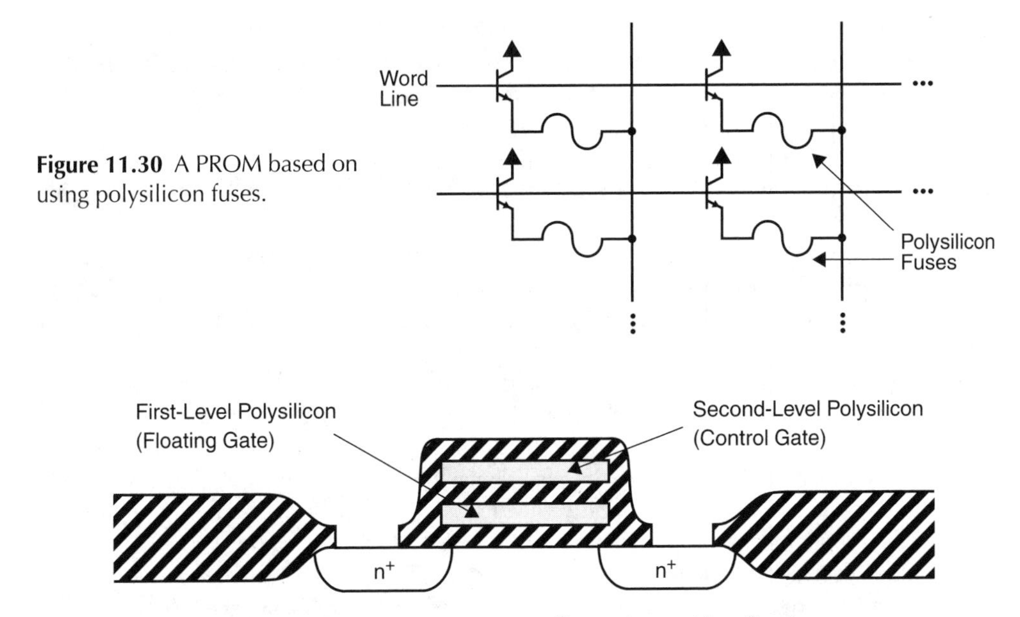

Figure 11.30 A PROM based on using polysilicon fuses.

Figure 11.31 An EPROM memory-cell transistor with a floating gate.

Most PROMs are realized in ICs that are completely dedicated to realizing the memory, although this is slowly changing; more and more often PROMs are being realized on an IC that contains many other functions such as a digital signal processor or perhaps a microprocessor.

There are many different ways a ROM can be designed so that it is user programmable. Some of the earlier PROMs were based on the use of *fuses*. These were typically realized using bipolar technology and polysilicon fuses (Wallace, 1980). More recently, there have also been CMOS versions (Metzger, 1983). A simplified version of part of the memory array of a bipolar fusible PROM is shown in Fig. 11.30. For this technology, intact fuses are used to represent "1"s, whereas *blown* fuses represent "0"s. The bit lines will have loads connected to ground. When a word line is *pulsed high*, the connected bit lines will change to a high voltage, whereas the unconnected bit lines, due to *blown* fuses, will remain at low voltages. By decoding the value of the desired bit line, the memory is read.

As time goes on, fusible PROMs are becoming less popular and a different type of PROM, namely *electrically programmable PROMs* or *EPROMs* are becoming more popular. This type of PROM is programmed by changing the effective threshold voltage of an MOS transistor. The most common way of realizing an MOS transistor, with an electrically programmable threshold voltage, is by realizing a *floating-gate* transistor (Frohman-Bentchkowsky, 1971; Salsbury et al., 1977). The basic structure is shown in Fig. 11.31. It is realized using a double-poly process. The first-level poly is used to create a gate that is electrically isolated, that is, it is left unconnected and floating. On top of this floating gate is placed a control gate that is electrically coupled to the word line. The floating gate is used for long-term storage of charge. If it is charged heavily negative with electrons, then it will make the threshold voltage of the transistor very positive so that no channel forms, even when the control gate to source voltage is equal to V_{DD}. If the floating-gate charge is removed, then the effective transistor threshold voltage, V_{tn}, will be small enough so that a channel will be present when the control-gate-to-source voltage is V_{DD}. The symbol for a floating-gate transistor is shown in Fig. 11.32.

The memory array of an EPROM based on floating-gate technology is very similar to a normal *ROM* memory array, and is shown in Fig. 11.33. All of the memory-cell transistors that have floating gates that store negative charge will have threshold voltages larger than V_{DD} and will never turn *on* when the appropriate word lines go *high*. Those without negative charge will turn on and hence will pull their bit lines *low* when their word lines are *driven high*. Thus, the negatively charged cells correspond to "1"s, whereas the uncharged or erased cells correspond to "0"s.

There are many ways for charging and discharging the floating gate (Muller et al., 1977). Originally, the most popular way of charging the gate with electrons was by injecting *hot electrons* from the drain end of the channel to the gate. This is achieved by putting the control voltage of the gate to a quite high voltage and placing the drain at a moderately high voltage. The high voltage of the control gate is capacitively coupled to the floating gate, placing it also at a high voltage, although not as high a voltage as the control gate. This causes a channel to form. The high drain voltage causes current to flow along the channel. The high electric fields cause electrons to be generated in the channel, near the drain, with energies in excess of the Si/SiO^2 energy barrier (3.2 eV). These are injected into the floating gate. After some time, enough electrons will be injected into the floating gate to seriously reduce the current flow and give the transistor an effective threshold voltage greater than V_{DD}, thus programming the cell to a "1". Originally, a 3.5-μm process would require V_{GS} = 25 V, V_{DS} = 16 V, and about 50 ms of programming time (Muller et al., 1977). During programming the maximum current might be 2 mA. A more modern 0.6-μm CMOS process would require only 10 μs to program the cell with V_{GS} = 12 V and V_{DS} = 6 V (Nakayama et al., 1991). Most EPROMs that use hot-electron injection to program the floating gate need *off-chip* voltages supplied during program. These

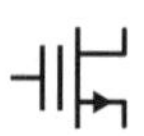

Figure 11.32 The symbol for the floating-gate transistor.

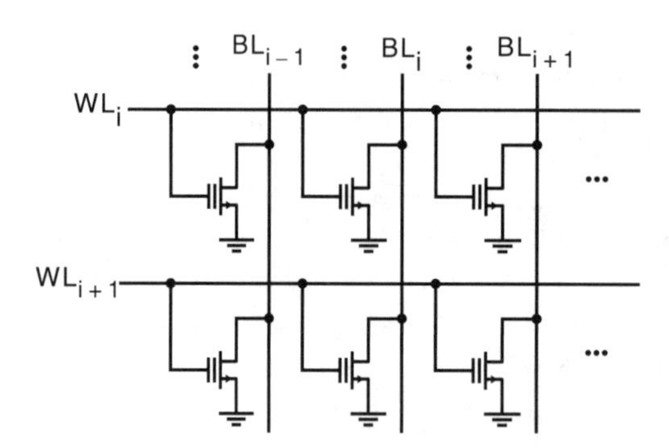

Figure 11.33 Part of a memory array of an EPROM.

large voltages are not easily generated *on-chip* because of the large drain current required during programming. Thus, most EPROMs are programmed by plugging them into dedicated programmers that have the requisite, nonstandard, power-supply voltages.

An alternative method for introducing negative charge into the floating gate is based on *tunneling* by *Fowler–Nordheim field emission* (Lenzlinger and Snow, 1969). This tunneling to the floating gate can occur without drain current if the oxide thickness between the floating gate and the drain is very small. Thus, EPROM cells that made use of Fowler–Nordheim tunneling originally looked as is shown in Fig. 11.34. The floating gate was extended over the n^+ diffusion at the drain end. Also, in the region of the extension, the oxide separating the floating gate from the drain is very thin, on the order of 0.01 μm or less. For programming, the control gate is taken to a high voltage of around 10 to 12 V for a modern process. Both the drain and the source are left at 0 V. This causes an electric field of around 10^7 V/m to exist across the very thin oxide. This field is large enough for the electrons to tunnel to the floating gate, charging it negatively, which in turn causes the threshold voltage, V_{tn}, to become greater than V_{DD}.

One of the major advantages of Fowler–Nordheim injection over hot-electron injection is that large drain currents are not required. This makes it much easier to generate the high voltage required for programming using an on-chip charge-pump circuit. This is usually necessary if a memory is to be programmed without physically removing it from its circuit board, as most systems do not have the large voltages needed for programming available.

In some of the more modern 0.2 μm and smaller technologies, the normal gate oxide thickness is approaching dimensions as small as 0.01 μm, and it is no longer necessary to have special thin oxide regions to use Fowler–Nordheim injection for programming. For the above reasons, Fowler–Nordheim injection is gradually replacing hot-electron injection as the desired method for programming EPROM cells. Fowler–Nordheim injection does require good control of processing in order to be reliable.

Once programmed, a floating gate of an EPROM cell transistor will store its charge for decades until it is erased. There are a variety of different methods for erasing EPROM cells. Originally, the most popular method was to use ultraviolet light. UV EPROMs were packaged in ceramic packages that had quartz windows overtop the ICs. To erase them, high-intensity ultraviolet light was shone through the window onto the chip. This excited the electrons enough to escape from the floating gates, causing the threshold voltages to become more negative, which in turn erased all cells to the "0" state. UV EPROMs have two serious drawbacks; they

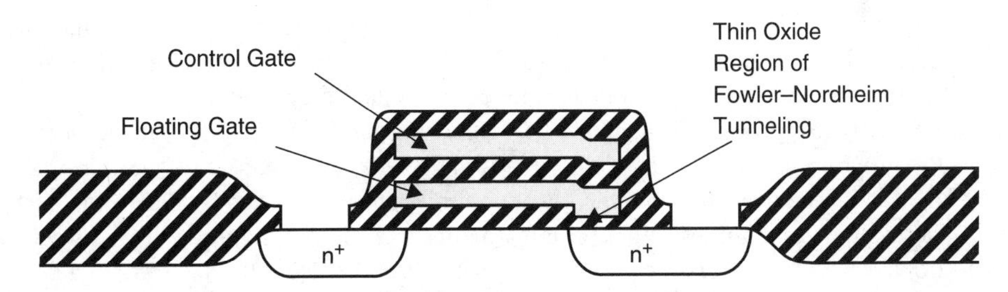

Figure 11.34 An EPROM cell that uses Fowler–Nordheim tunneling for charge transfer to floating gate.

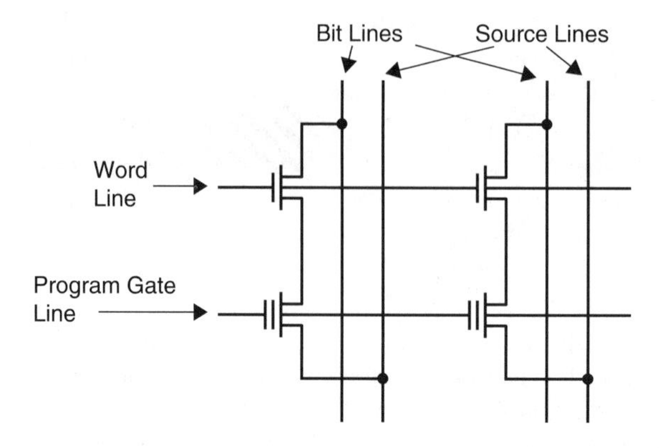

Figure 11.35 Two adjacent memory cells of a common EEPROM memory array.

are expensive due to the special ceramic packaging and they have to be removed from their circuit boards to be erased.

More recently, UV EPROMs are being replaced by electrically erasable EPROMs. These are EPROMs that can be erased electrically without removing them from the actual circuits in which they are normally used. There are two popular types of electrically erasable EPROMs, EEPROMs and *flash EPROMs*. EEPROMs can have individual bytes erased, whereas for flash EPROMs, the smallest unit that can be erased is usually a complete memory array or sector. In return for this lack of flexibility, flash EPROMs have simpler memory cells and therefore can realize larger memories.

Both types of memories normally rely on Fowler–Nordheim tunneling for erasure. When a cell is to be erased, the voltage of the control gate will be placed at a negative voltage with respect to the drain voltage. The source voltage may be the same as the drain voltage, at some interim voltage between the gate and the drain voltage, or it may be left floating depending on the particular memory technology. Alternatively, the drain voltage might be left floating and the gate-to-source voltage made negative. In all cases, the floating gate voltage is coupled to a negative voltage and the electrons tunnel to the drain (or possibly source) causing the threshold voltage to become more negative, which again erases the EPROM cell to the "0" state.

Most EEPROMs, where individual bytes can be erased, are based on a two-transistor memory cell as shown in Fig. 11.35 (Johnson et al., 1980). Each memory cell consists of a floating-gate transistor in series with a normal n-channel enhancement transistor. During a normal read operation, the program gate line will be held at a high voltage (around V_{DD}) and the individual memory cells will be selected using the word line. Also, during programming and erasing of the cells, individual cells are selected using the word line and the additional select transistors. When a cell is being programmed, both the source line and the bit line will be *held low*, whereas the program gate line will be pulsed to a high voltage (perhaps 12 V or more). When erasing a cell, both the word line and the bit line will be pulsed to a high voltage (i.e., 12 V or more), whereas the source lines might be left floating and the program gate line is *held low*. The additional select transistors are needed for each cell to deselect all transistors in a column that are not to be erased.

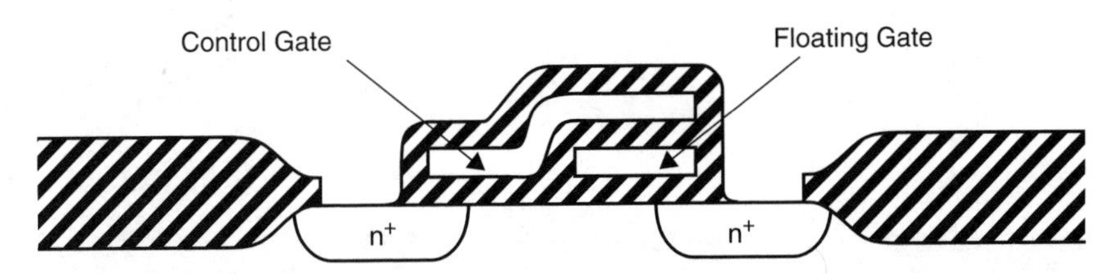

Figure 11.36 A Flash EEPROM cell with an overlapped control gate.

A more recent development has been a flash EPROM (Pavan et al., 1997), which has single transistor floating-gate cells, often similar to UV EPROMs, but which can also be electrically erased. However, because select transistors are not present for flash EPROMs, it is necessary to erase a minimum of all the transistors in a column at any one time. Normally, all transistors of a sector, where a sector might be a 16K memory array in a multiarray memory, will be erased at any given time.

When electrically erasing the cells of a flash EPROM, it is necessary to guarantee that the floating-gate transistors never be erased long enough to result in threshold voltages less than 0 V and therefore become depletion transistors. If this accidentally happens, the memory cells will pull their appropriate bit lines *low* even when they are not selected and their control gates are left *low*. One way of guaranteeing that the transistor threshold voltages never become negative during erasure is by checking transistor threshold voltages after every erase pulse (Nakayama et al., 1991). Once the threshold voltages have become more negative than a predetermined value, erasure is stopped. Alternatively, a modified memory cell as shown in Fig. 11.36 can be used (Samachisa et al., 1987). In this memory cell, the control gate overlaps the floating gate to effectively form a series enhancement transistor. When the floating gate is erased, it does not matter if positive charge is left on the floating gate giving negative threshold voltage in that region of the transistor. The threshold voltage under the control gate region of the memory cell is still guaranteed to be positive. Thus, nonselected memory cells will not be conducting during a read operation. The 20% increase in cell size seems a small penalty to pay for the greatly simplified erase procedure. Another penalty incurred is a greater word-line capacitance. The effect of this penalty was not mentioned in (Samachisa et al., 1987), but it is somewhat alleviated by approximately 50% greater cell current during the read operation as compared to the cell of Fig. 11.34. Overall, the author's opinion is that the cell of Fig. 11.36 is a reasonable choice that should be seriously considered.

NAND ARCHITECTURES

A recent development in memory architecture has been the *NAND* architecture for flash EPROMs (Masuoka et al., 1987). This architecture can also be used for EEPROMs. In an EPROM based on the NAND architecture, a number (usually 8 or 16) of floating-gate transistors are connected in series between a bit line and ground, and separated from both by standard MOS access transistors, as shown in Fig. 11.37. The major and significant reason for considering this architecture to realize a flash EPROM is that the layout area is very small. The reason

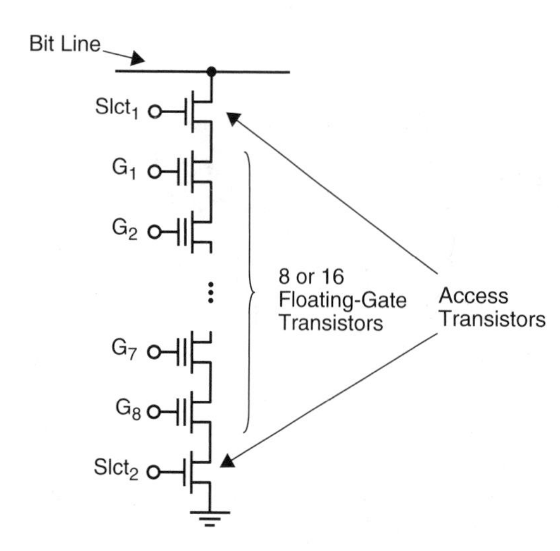

Figure 11.37 The NAND architecture used for some flash EPROMs.

for this is that no contacts are required in the transistor string, which allows the gate stripes to be placed very close to each other. It is claimed that this results in over a 40% savings in area.

Transistors that store "0"s are programmed to have negative threshold voltages, whereas transistors that store "1"s are programmed to have positive threshold voltages. At any given time, only one transistor in the string would be accessed. When a particular storage cell is to be read, the bit line is first precharged *high* (this corresponds to a "1"), and then the gates of the access transistors and all storage transistors that are not being read are set *high* to V_{DD}. The gate of the transistor that is being accessed is kept at 0 V. If this transistor has a negative threshold voltage (corresponding to a "0"), it will be conductive and the bit line will be discharged to ground through the transistor string, albeit somewhat slowly because of the large number of series transistors. If the transistor being accessed was storing a "1", then its threshold voltage will be positive, it will not be conductive with 0 V on its gate, and the bit line will not be discharged.

Fowler–Nordheim tunneling is used for both erasing and programming. During erasing, every node except the gates are connected to a voltage much larger than V_{DD}. The gates are kept at 0 V. This causes all transistors with floating gates to become depletion devices with negative threshold voltages (again, corresponding to "0"s). When a cell is to be programmed to a "1", then the access transistor going to ground is kept *off* (i.e., $Slct_2$ = "0"), the string is connected to the bit line (which is kept at 0 V), and the gate of the transistor being programmed is connected to a voltage much larger than V_{DD}. The gates of the transistors not being programmed to "1"s will have their gates connected to a voltage only moderately higher than V_{DD}. This changes the threshold voltage of the accessed transistor to be greater than 0. At present, the popularity of NAND-architecture EPROMs is increasing quickly.

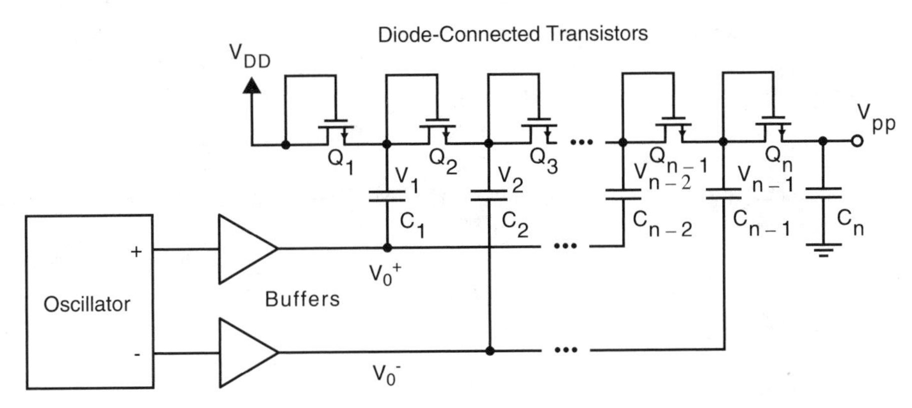

Figure 11.38 A charge pump suitable for generating a large on-chip voltage.

Charge Pumps

Many of the modern EEPROM and flash EPROM circuits generate the higher than normal voltages required using on-chip charge-pump circuits. This is possible only when the current load on the output voltages of the charge pumps are very low. This is normally the case when Fowler–Nordheim tunneling is used for both programming and erasure. A typical charge-pump circuit, suitable for generating a large positive voltage, is shown in Fig. 11.38 (Oto et al., 1983).

The oscillator produces two complementary square-wave outputs, which are then buffered in order to drive the large load of the charge-pump capacitors. To understand how the charge pump works, consider the steady-state case with only a capacitive load on V_{pp}, so that there is no d.c. current through the charge pump. When V_0^+ goes to 0 V, V_1 is clamped to $V_{DD} - V_{tn}$ by diode-connected transistor Q_1. This will charge C_1 to $V_{DD} - V_{tn}$. When V_0^+ next goes positive, Q_1 turns *off* and no charge can escape through it. This *pumps* V_1 up to $V_{DD} + V_{C\text{-}1}$ or to $2V_{DD} - V_{tn}$. At the same time $\overline{V_0}$ will be at 0 V and V_2 will be charged through diode connected Q_2 to $V_1 - V_{tn}$ or to $2V_{DD} - 2V_{tn}$. Similarly, as one continues down the chain, the most positive voltage across capacitor i is given by $i \times (V_{DD} - V_{tn})$. Thus, the ideal output voltage, in the absence of a d.c. load is given by $n \times (V_{DD} - V_{tn})$. If there is any d.c. load current at all, the output voltage will be much less. Also, the transistor body effect causes the transistor threshold voltages, V_{tn}, to become quite large for transistors near the output of the charge pump. This also limits the output voltage. A third limitation is due to leakage currents of the junctions, which can become appreciable at higher temperatures due to the large reverse-bias voltage across the junctions. Irrespective of these limitations, a charge pump consisting of 10 diode-connected transistors can easily generate a charge-pumped output of 20 V when the d.c. load current is minimized. This technology normally requires thicker field oxides than normally used for a given minimum channel length in order to prevent undesired substrate inversion.

The actual output voltage of the charge pump is very difficult to determine accurately, so normally on-chip charge pumps have some voltage regulation circuitry included. For example, Oto et al. (1983) compared the output voltage, after scaling it with a voltage divider, to a desired reference voltage. The difference was then fed back to control the frequency of the oscillator. If the difference was negative, the oscillating frequency was increased, if it was positive, the frequency was decreased.

In this section on programmable ROMs, only the currently most popular methods for programming the cells have been covered. There are many other mechanisms, such as triple-poly transistors, or metal nitride oxide semiconductor (MNMOS) transistors that will not be covered as they are currently less popular. Also, whenever there are high voltages on an IC, special care must be taken in designing logic and devices connected to the high voltages if reliable functionality without breakdown is desired. The reader is referred to the *IEEE Journal of Solid-State Circuits* and the *IEEE Journal of Electronic Devices* for details concerning these and other issues.

11.8 Bibliography

L. Childs and R. Hirose, "An 18ns 4K×4 CMOS SRAM," *IEEE Journal of Solid-State Circuits*, 19(5), 545–551, October 1984.

D. Frohman-Bentchkowsky, "A Fully-Decoded 2048-Bit Electrically Programmable MOS ROM," *Proceedings of IEEE International Solid-State Circuits Conference,* 80–81, February 1971.

S. Fuji et al., "A 45-ns 16-Mbit DRAM with Triple-Well Structure," *Proceedings of IEEE International Solid-State Circuits Conference,* 248–249, February 1989.

H. Goto et al., "A 3.3V 12-ns 16-Mb CMOS SRAM," *IEEE Journal of Solid-State Circuits*, 27(11), 1490–1496, November 1992.

K. Hardee and R. Sud, "A Fault-Tolerant 30ns/375mW 16k×1 NMOS Static RAM," *IEEE Journal of Solid-State Circuits*, 16(5), 435–443, October 1981.

T. Hirose et al., "A 20-ns 4-Mb CMOS SRAM with Hierarchical Word Decoding Architecture," *IEEE Journal of Solid-State Circuits*, 25(5), 1068–1074, October 1990.

W. Johnson et al., "A 16Kb Electrically-Erasable Nonvolatile Memory," *Proceedings of IEEE International Solid-State Circuits Conference,* 152–153, February 1980.

Y. Kobayashi et al., "10μW Standby Power 256K CMOS SRAM," *IEEE Journal of Solid-State Circuits*, 20(5), 935–940, October 1985.

M. Lenzlinger and E.H. Snow, "Fowler–Nordheim Tunnelling into Thermally-Grown SiO_2," *Journal of Applied Physics,* 40, 278–283, January 1969.

N. Lu and H. Chao, "Half-V_{DD} Bit-Line Sensing Scheme in CMOS DRAMS," *IEEE Journal of Solid-State Circuits*, 9(4), 451–454, August 1984.

F. Masuoka et al., "An 80ns 1Mbit MASK ROM with a New Memory Cell," *IEEE Journal of Solid-State Circuits*, SC-19(5), 651–657, October 1984.

F. Masuoka et al., "New Ultra-High Density EPROM and Flash EEPROM Cell with NAND Structure Cell," *IEDM Technical Digest,* 552–555, 1987.

L. Metzger, "A 16K CMOS PROM with Polysilicon Fusible Links," *IEEE Journal of Solid-State Circuits*, SC-18(5), 562–565, October 1983.

Minato et al., "A Hi-CMOSII 8K×8K Bit Static RAM," *IEEE Journal of Solid-State Circuits*, 17(5), 793–798, October 1982.

R. Muller et al., "An 81920Bit Electrically-Alterable ROM Employing a One-Transistor Cell with Floating Gate," *IEEE Journal of Solid-State Circuits*, SC-12(5), 507–514, October 1977.

S. Mori et al., "A 45-ns 64 Mb DRAM with a Merged Match-Line Test Architecture," *IEEE Journal of Solid-State Circuits*, 26(11), 1486–1492, November 1991.

Y. Nakagome et al., "An Experimental 1.5V 64-Mb DRAM," *IEEE Journal of Solid-State Circuits*, 26(4), 465–472, April 1991.

T. Nakayama et al., "A 60-nsec 16-Mb Flash EEPROM with Program and Erase Sequence Controller," *IEEE Journal of Solid-State Circuits*, 26(11), 1600–1605, November 1991.

Y. Oowaki et al., "A 33-ns 64-Mb DRAM," *IEEE Journal of Solid-State Circuits*, 26(11), 1498–1507, November 1991.

D. Oto et al., "High-Voltage Regulation and Process Considerations for High-Density 5V Only $E^2PROM's$," *IEEE Journal of Solid-State Circuits*, 18(5), 532–538, October 1983.

P. Pavan et al., "Flash Memory Cells—An Overview," *Proceedings of the IEEE,* 85(8), 1248–1271, August 1997.

B. Rossler and R.G. Muller, "Erasable and Electrically-Programmable Read-Only Memory Using the n-Channel SIMOS One-Transistor Cell," *Siemans Forsch. U. Entwickl-Ber,* 4, 345–351, November 1975.

N. Sakashita et al., "A 1.6GB/s Data-Rate 1-Gb Synchronous DRAM with Hierarchical Square-Shaped Memory Block and Distributed-Bank Architecture," *IEEE Journal of Solid-State Circuits*, 31(11), 1645–1655, November 1996.

K. Sasaki et al., "A 15ns 1-Mbit CMOS SRAM," *IEEE Journal of Solid-State Circuits*, 23(5), 1067–1072, October 1988.

Salsbury et al., "High-Performance MOS EPROM's Using a Stacked Gate Cell," *Proceedings of IEEE International Solid-State Circuits Conference,* 186–187, February 1977.

G. Samachisa et al., "A 128K Flash EEPROM Using Double-Polysilicon Technology," *IEEE Journal of Solid-State Circuits*, 22(5), 676–683, October 1987.

K. Seno et al., "A 9-ns 16-Mb CMOS SRAM with Offset-Compensated Current Sense Amplifier," *IEEE Journal of Solid-State Circuits*, 28(11), 1119–1124, November 1993.

T. Sugibayashi et al., "A 30-ns 256-Mb DRAM with a Multidivided Array Structure," *IEEE Journal of Solid-State Circuits*, 28(11), 1092–1098, November 1993.

M. Taguchi et al., "A 40-ns 64-Mb DRAM with 64-b Parallel Data Bus Architecture," *IEEE Journal of Solid-State Circuits*, 26(11), 1493–1497, November 1991.

R.K. Wallace, "A 35 nsec 16K PROM," *Proceedings of IEEE International Solid-State Circuits Conference,* 148–149, February 1980.

J. S. Witters et al., "Analysis and Modelling of On-Chip High-Voltage Generator Circuits for Use in EEPROM Circuits, *IEEE Journal of Solid-State Circuits*, 24, 1372–1380, October 1989.

Yamamoto et al. "A 256K CMOS SRAM with Variable-Impedance Data-Lines," *IEEE Journal of Solid-State Circuits*, 20(5), 924–928, October 1985.

J. Yoo et al., "A 32-Bank 1 Gb Self-Strobing Synchronous DRAM with 1 GByte/s Bandwidth," *IEEE Journal of Solid-State Circuits*, 31(11), 1635–1644, November 1996.

K. Yu et al., "HMOS-CMOS—A Low-Power High-Performance Technology," *IEEE Journal of Solid-State Circuits*, 16(5), 454–459, October 1981.

11.9 Problems

For the problems in this chapter, assume the following transistor parameters:

- npn bipolar transistors:

  $\beta = 100$

$V_A = 80$ V
$\tau_b = 13$ ps
$\tau_s = 4$ ns
$r_b = 330\ \Omega$

- n-channel MOS transistors:
  $\mu_n C_{ox} = 190\ \mu A/V^2$
  $V_{tn} = 0.7$ V
  $\gamma = 0.6\ V^{1/2}$
  $r_{ds}\ (\Omega) = 5000L\ (\mu m)/I_D\ (mA)$ in active region
  $C_j = 5 \times 10^{-4}\ pF/(\mu m)^2$
  $C_{j\text{-}sw} = 2.0 \times 10^{-4}\ pF/\mu m$
  $C_{ox} = 3.4 \times 10^{-3}\ pF/(\mu m)^2$
  $C_{gs(overlap)} = C_{gd(overlap)} = 2.0 \times 10^{-4}\ pF/\mu m$

- p-channel MOS transistors:
  $\mu_p C_{ox} = 50\ \mu A/V^2$
  $V_{tp} = -0.8$ V
  $\gamma = 0.7\ V^{1/2}$
  $r_{ds}\ (\Omega) = 6000L\ (\mu m)/I_D\ (mA)$ in active region
  $C_j = 6 \times 10^{-4}\ pF/(\mu m)^2$
  $C_{j\text{-}sw} = 2.5 \times 10^{-4}\ pF/\mu m$
  $C_{ox} = 3.4 \times 10^{-3}\ pF/(\mu m)^2$
  $C_{gs(overlap)} = C_{gd(overlap)} = 2.0 \times 10^{-4}\ pF/\mu m$

11.1 Assume an SRAM IC has $2^{24} \cong 16.8$ million cells, that $V_{DD} = 3.3$ V, and the total power dissipation must be less than 100 mW. What is the maximum size of the resistive loads of the memory cells?

11.2 Assuming the minimum channel lengths of transistors is 0.5 μm and a very simple set of design rules similar to those described in Section 2.3 of Chapter 2, estimate the layout of a memory cell similar to that shown in Fig. 11.3. Assume a layout in which the well regions are shared by two adjacent rows (as are the junctions of the access transistors that are connected to the bit lines).

11.3 Repeat Problem 11.2 but for the memory cell shown in Fig. 11.3. Assume the resistor loads are realized on top of the transistors and do not take up any additional area.

11.4 There are 64 static RAM memory cells connected to a single word line. Each memory cell has two access transistors, with W/L = 2 μm/0.5 μm, whose gates are connected to the word line. Estimate the parasitic capacitance of the word line, assuming each metal line is 1500 μm long, 1.2 μm wide, and 0.5 μm thick. The SiO_2 under each word line is 0.5 μm.

11.5 Estimate the rise time of a word-line driver, which is an inverter with a p-channel having W/L = 40 μm/0.5 μm, and an n-channel transistor having W/L = 2 0 μm/0.5 μm. Assume the word line has the characteristics described in Problem 11.4. You may ignore junction capacitance and assume the input to the buffer is a perfect negative-going step.

11.6 Repeat Example 11.2, but assume all transistor lengths are 0.5 μm and the following widths:

Transistors	Widths (μm)
Q_1, Q_2	2
Q_3, Q_4	1
Q_5, Q_6	5
Q_7, Q_8	4

11.7 Using SPICE simulate the sense amplifier shown in Fig. 11.11. Assume during the first period, the input is +0.05 V and during the second period, the input is –0.05V. The load capacitance should be 0.4 pF and the transistor lengths should be 0.5 μm. The following transistor lengths should be used:

Transistors	Widths (μm)
Q_1, Q_2	4
Q_3, Q_4	4
Q_5	8

11.8 Design a 9 to 1-of-512 word-line decoder using an architecture similar to that shown in Fig. 11.14. Give a hierarchical solution. Device sizes are not required.

11.9 Give a transistor-level realization of an edge-detection circuit. Try to optimize area and speed.

11.10 Give a transistor-level realization of the hierarchical word decoding shown in Fig. 11.19. To increase efficiency and speed, different types of gates may be substituted as long as the same basic function is realized.

11.11 Assume the access transistors of a dynamic RAM cell are $1\ \mu m \times 0.5\ \mu m$ and the storage capacitance is 35 fF. Further assume the bit-line capacitance due to the interconnect adds 50% additional capacitance to the bit-line capacitance due to the access transistor junctions. Estimate how many storage cells can be connected to one bit line without the parasitic bit-line capacitance becoming greater than 12 times the storage capacitance.

CHAPTER 12

GaAs Digital Circuits

12.1 Introduction

So far, all of the digital integrated circuits considered have been realized using silicon, which has a valence of four. There are other semiconductor materials that can also be used to realize digital integrated circuits. Many of these alternative semiconductors are based on crystals composed of two different atoms, one having a valence of three and one having a valence of five. Examples of these III–V compound semiconductor materials include *gallium arsenide* (GaAs), *indium phosphide* (InP), *aluminum arsenide*, (AlAs), and *indium arsenide* (InAs), as well as others. Of these alternatives, semiconductors based on GaAs have become the most important commercially, at present.

The primary use of GaAs integrated circuits is for very high-speed applications; historically, commercially available GaAs circuits have operated at about twice the maximum speed of commercially available Si circuits. They are also somewhat more resistant to radiation than silicon-based ICs.

There are a number of reasons why GaAs-based integrated circuits have been faster than Si-based ICs. The velocity of electrons in n-doped regions of GaAs is greater than that of electrons in n regions of silicon, especially at lower electric fields. At a somewhat typical electric field of 1 V/μm, electrons in GaAs have slightly less than double the velocity of electrons in Si. Plots of carrier velocities versus electric fields taken from Sze (1981) are shown in Fig. 12.1. At higher fields, the velocity advantage of electrons in GaAs over electrons in Si is not as large because velocity saturation starts to occur in GaAs for fields larger than 0.3 V/μm. However, this fact is partially offset by another effect. Most often, large electric fields occur across short transistor channels often having lengths of 0.5 μm or less. For these short channel lengths, an effect called *ballistic transport* or *transient overshoot* is responsible for electrons achieving higher peak velocities very close to where the electrons are injected. It is responsible for significantly larger velocities for electrons in the channels of transistors having short channel lengths; it becomes significant particularly at channel lengths of 0.1 μm or smaller.

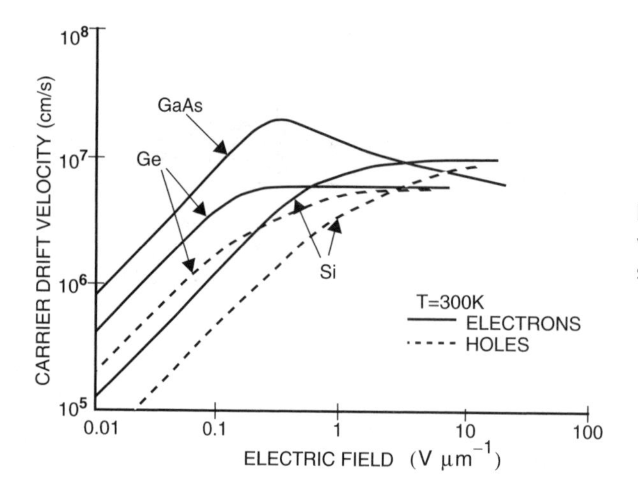

Figure 12.1 The measured drift velocity of carriers in some semiconductors from Sze (1981).

It should be noted that holes in p-doped GaAs have substantially lower velocities than holes in p-type Si. Thus, p-type devices are almost never used in GaAs integrated circuits.

Another important advantage of GaAs ICs over Si ICs is that at room temperatures, a GaAs substrate is quite high impedance, as high as 10^6 to 10^8 Ω-cm in GaAs as compared to 10^3 to 10^5 Ω-cm in Si. This helps minimize parasitic capacitances, particularly at the junctions of transistors and second for interconnect lines, as long as they are not too close to other lines. This is particularly significant as dimensions shrink in modern technologies. Together, these two advantages have resulted in the fastest GaAs digital ICs being about twice as fast as the highest-speed Si digital ICs. This two-to-one advantage has held fairly constant over the past 10 years, and is predicted to continue in the near future.

This is not to say that GaAs technology does not have some disadvantages, the major one being cost. The cost per wafer is quite high since the volume of production of GaAs ICs is much less than that of silicon ICs, and is predicted to remain so in the future. This is exacerbated by the GaAs wafers being smaller than Si wafers, perhaps 4–6 in. in diameter compared to 8 in. for silicon. Also, the yield for GaAs ICs is lower than for silicon ICs. Other disadvantages of GaAs technology include larger thermal resistivity (about 3 to 1 over Si) and sometimes harmful substrate effects, as will be described later.

Altogether, these disadvantages has relegated GaAs technology to the most critical highest speed and usually low-volume applications. However, in these most critical applications, GaAs technology is becoming somewhat entrenched and is expected to be an important technology in the near future.

12.2 GaAs Processing and Components

The most common semiconductor device found in GaAs technology is the n-channel MESFET. This acronym stands for metal-semiconductor field-effect transistor. The device is quite similar to an n-channel JFET (Long and Butner, 1990), except a Schottky diode is used to isolate the gate

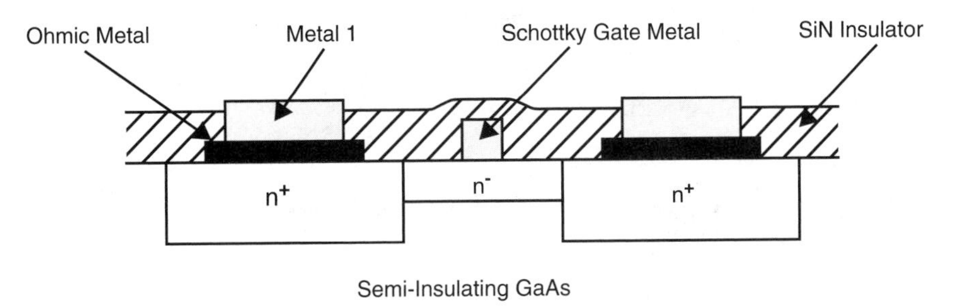

Figure 12.2 A typical cross section of an ion-implanted GaAs MESFET.

from the channel rather than a pn junction. To date, it has not been possible to realize reliable transistors similar to MOSFETs with gates insulated from the channel by high impedance materials such as SiO_2 or SiN. This is due to the large number of surface states in GaAs and to the poor quality of deposited SiO_2 (as opposed to thermally grown SiO_2 in silicon).

A cross section of a typical GaAs MESFET is shown in Fig. 12.2. The MESFET is a non-self-aligned transistor realized using ion implantation (Rode et al., 1982) to realize depletion MESFETs. In this rather mature process, the n^- channel is implanted first, followed by the n^+ source and drain regions. On top of the source and drain regions, an ohmic metal, such as a *gold–germanium–nickel* alloy (Au/Ge/Ni), is applied by thermal evaporation through contact openings in a *silicon nitride* (SiN) insulator to form connections to the transistor junctions. The metal used to form the gate makes a Schottky barrier with GaAs. An example alloy might be *titanium–platinum–gold* (Ti/Pt/Au). Often the Schottky metal might be used for the first level of interconnect, commonly called Metal 1. This layer is often deposited directly on top of the ohmic metal used to contact the junctions in order to lower the resistivity of the junction contacts. To ensure that reliable Schottky junctions form under the base metal, without any shorts forming between the base metal and either of the junctions, it is necessary that the base metal never overlap the n^+ regions. Since this requirement is achieved without the use of self-aligned techniques, there might be up to 2 μm of n^- channel separating the Schottky gate region from the n^+ source and drain regions in a conservative 1985 process. This large spacing increases the series source and drain resistances besides making them somewhat unpredictable and very sensitive to mask alignment. More modern self-aligned processes have helped to minimize this variability (Rochi, 1990). In one of these processes, the gate might be deposited first and used as a mask for the n^+ source and drain regions (Rochi, 1990).

Most currently available processes can realize both depletion and enhancement transistors. A depletion transistor might have a threshold voltage between –1.5 and –0.5 V. An enhancement transistor might have a threshold voltage between 0.0 and 0.2 V. To reliably realize enhancement transistors, good control over the threshold-voltage reproducibility is essential.

There are many alternative advanced processes for realizing higher speed transistors than MESFETs. In one variation, sometimes called *high electron mobility transistors (HEMTs)* (Rochi, 1990; Mimura et al., 1980), separate layers are deposited, using *molecular-beam epitaxy (MBE)*, for producing carriers in the channel and through which the carriers flow. The layer for producing the carriers might be an n^- aluminum–gallium–arsenide layer, whereas

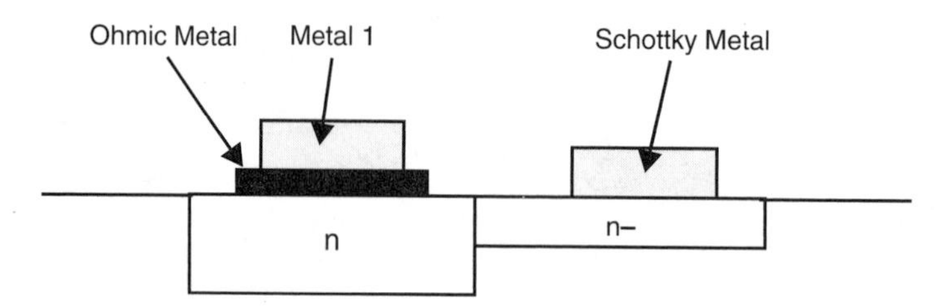

Figure 12.3 A possible realization of a Schottky diode.

the conduction layer might be an undoped GaAs layer. Since conduction occurs in an undoped layer, very few collisions occur between carriers and therefore higher electron mobilities exist. This effect is particularly pronounced at lower temperatures. The higher mobilities result in larger unity gain frequencies and a more reliable realization of enhancement transistors. Irrespective of the differences, the transistors are still modeled similar to JFETs.

Besides transistors, modern GaAs processes can realize a variety of other devices; these might include Schottky diodes, high-quality thin-film resistors, capacitors, and even on-chip inductors.

A cross section of a Schottky diode is shown in Fig. 12.3. These diodes can be turned on and off very quickly since there is no minority charge storage when they are conducting. Thus, a *bridge* of diodes is often used to realize high-speed switches. The diodes are also used extensively for level shifts. When used in this application, it is important to consider the voltage drop of the series impedance due to the n^{-} channel as well as the drop of the diode itself.

Most GaAs processes have the capability of realizing thin-film resistors that are often used for loads in low-noise amplifiers. A typical material that is often used to realize thin-film resistors is nickel chromium (NiCr). This material has very good temperature stability and excellent matching between adjacent resistors. A typical resistivity might be 50 to 100 $\Omega/\Box$. Other possible materials that might be used to realize thin-film resistors include cermet (CrSiO) having resistivities of 50 to 500 $\Omega/\Box$ and tantalum nitride (TaN) having a resistivity of around 100 $\Omega/\Box$.

There is also a variety of ways of realizing capacitors on a GaAs IC. The most common way is two layers of metal separated by a dielectric such as silicon nitride (SiN), silicon dioxide (SiO_2), or a polyimide. Sometimes a sandwich of two materials, such as SiN and SiO_2, might be used. These capacitors are often called *metal–insulator–metal (MIM)* capacitors. Typical thicknesses for the dielectric in 1985 might be 0.1 to 0.2 μm. This is considerably thicker than typical capacitor dielectrics found in a modern silicon process and results in smaller capacitances per unit area. For example, a 0.1-μm dielectric layer of SiO_2 would result in 0.34 fF/μm^2. A 10 pF capacitor would require an area of 3.0×10^4 μm^2, which is quite large. Besides MIM capacitors, *interdigitated* metal fingers and reverse-biased diodes have also been used to realize capacitors.

Spiral on-chip inductors are often used in high-speed analog circuits operating at 1 GHz and up. A 10-nH inductor operating at 1 GHz might have a Q-factor of 3–5, whereas a 3-nH inductor operating at 2 GHz might have a Q-factor of 10–15. The major limitation on the Q-factor is

Table 12.1 Representative Parameters for a 0.5-μm Ga_As Process

Parameter	Typical size
Transconductance (per μm of transistor width for $V_{GS} = 0$ V)	0.15 mS
Gate-source capacitance (per μm of transistor width for $V_{GS} = 0$ V)	1.6 fF
Gate-drain capacitance (per μm of transistor width for $V_{DS} = 2$ V)	0.16 fF
Pinch-off voltage	–1.2 V
Diffusion substrate capacitance (per μm of transistor width)	0.2 fF
Wiring capacitance of first metal on oxide (per μm of transistor width)	0.1 fF
Wiring capacitance of air-bridge interconnect (per μm of transistor width)	0.05 fF
Drain-source impedance (per μm of transistor width for $V_{GS} = 0$ V and $V_{DS} = 2$ V)	75 kΩ
NiCr sheet resistance	50 Ω/□
MIM capacitance	0.3 fF/μm²
Transistor unity current gain frequency	15 GHz

due to the resistance of the conductive layer. Self-resonance is usually not a problem in the 1 GHz to 10 GHz frequency range due to the high-impedance substrate.

Another component, unique to GaAs, that is found in some processes, is *air-bridge* interconnect. This interconnect is basically metal suspended in the air on metal pillars that might be separated by 20 μm or so. Since most of the interconnect is suspended in the air, it has very little parasitic capacitance, perhaps one-half the capacitance of the first-level metal sitting directly on a dielectric; for this reason, *air-bridge interconnect* is usually the principal interconnect used in forming inductors, when it is available.

Table 12.1 lists a number of representative parameters for a typical 0.5 μm gate-length process capable of realizing MESFETs with unity-gain frequencies (f_ts) of 15 GHz. A state-of-the-art process might be capable of realizing FETs with a 0.1 μm gate length and f_ts of 60–80 GHz and up.

12.3 MESFET Modeling

Most MESFET models are based on JFET models with modifications to account for second-order differences between silicon JFET's and GaAs MESFETs. Common large-signal equations used for the JFET in SPICE are

$$I_D = \beta\frac{W}{L}[2(V_{GS}-V_t)V_{DS}-V_{DS}^2](1+\lambda V_{DS}) \quad \text{for } V_{DS} < V_{GS}-V_t \tag{12.1}$$

which is used in the triode region, and

$$I_D = \beta\frac{W}{L}(V_{GS}-V_t)^2(1+\lambda V_{DS}) \quad \text{for } V_{DS} > V_{GS}-V_t \tag{12.2}$$

which is used in the active region.

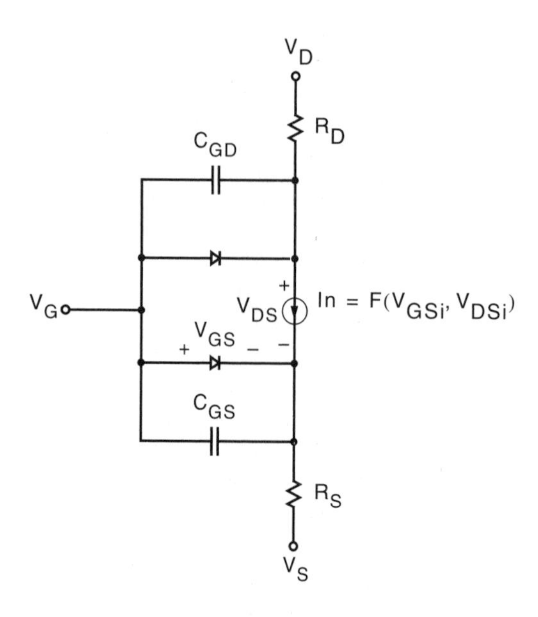

Figure 12.4 A commonly used large-signal model for GaAs MESFETs.

Both of these equations apply only when the MESFET is conducting, that is for $V_{GS} > V_t$. These equations are similar to those used for modeling silicon MOSFETs, but there is an important difference for MESFETs; if the gate-source junction or gate-drain junction becomes forward biased by more than 0.7 V or so, gate conduction occurs. This is modeled by adding *Schottky diodes* to the equivalent circuit as is shown in Fig. 12.4 (Long and Butner, 1990). Normally, these will be reverse biased under normal operation, but will conduct for large gate voltages.

The diode model is the same as that used for silicon JFETs; but one important difference is that a transit time of zero is used for the diodes. This is because in GaAs MESFETs, the diodes are Schottky diodes, which do not have appreciable *minority-carrier charge storage* when they conduct. The only charge storage is due to the diode depletion capacitance, which makes the diodes much faster to turn *off*. For this reason, some GaAs logic gates are often designed so that the gates conduct for "1" inputs without serious deleterious effects occurring.

Also note that in the model of Fig. 12.4 resistances, R_S and R_D, are included in series with the junctions. Since the MESFETs are lateral devices, these resistances can be large and have substantial effects, particularly in reducing the effective transconductance of the transistors.

Another important difference between silicon JFETs and GaAs MESFETs is that MESFETs typically operate partially in the *velocity saturation region*, particularly for large drain-source voltages. This has two major effects. First, it can cause the I_D versus V_{DS} curve to flatten for drain-source voltages less than $V_{GS} - V_t$. This effect is sometimes called the *early saturation effect* and can be taken advantage of in some applications. Second, the velocity saturation can change the relationship between $V_{GS} - V_t$ and I_D, in the active region, from a squared relationship to one that is somewhere between linear and squared. These effects are often modelled empirically. For example, in one commonly used modification, called the *Curtice model* (Curtice, 1980), I_D is given by (12.2) multiplied by a hyperbolic tangent. In this model, we have

$$I_D = \beta(V_{GS}-V_t)^2(1+\lambda V_{DS})\tanh(\alpha V_{DS}) \tag{12.3}$$

which is used for all regions where V_{DG} is greater than $-V_t$. The parameter α is empirically derived by curve fitting and might have a value on the order of 2 to 5 V^{-1}. There are many other models that have been proposed to more accurately model the effects of velocity saturation (Long and Butner, 1990, p. 16).

Another important difference between the JFET model of Fig. 12.4 and a GaAs MESFET is that the gate capacitance goes down substantially when the transistor turns *off* and the channel disappears. This is not modeled by the diode models used in Fig. 12.4, where a substantial depletion capacitance remains even after the channel is gone. Also, the division of the gate capacitance between C_{GD} and C_{GS} is quite complicated and poorly modeled by a typical JFET model. Despite all these difficulties, the JFET-based model, with an appropriate choice of parameters, is adequate for the transient simulation of GaAs digital circuits realized with MESFETs with an accuracy of 10 to 20%.

12.4 MESFET Second-Order Effects

There are many second-order effects, some of which are not easily modeled, that are responsible for MESFETs exhibiting differences other than those predicted by simple JFET-based models. Most of these second-order effects can be attributed to either the high-impedance substrate, which is only semi-insulating, or to traps in the substrate.

Back-Gating

One second-order effect that has received wide attention in the technical literature is called *back-gating* (Kocot and Stolfe, 1982). This effect occurs when an adjacent transistor has junctions at much lower voltages than the transistor being considered. Because the substrate is only semi-insulating, this causes the MESFET threshold voltage to become more negative, which in turn causes a decrease in current of the transistor under consideration. This phenomenon is very similar to the body effect in MOSFETs. Back-gating is very process dependent and has been greatly minimized in recent years. In addition, there are many isolation techniques that virtually eliminate it, such as using p^+ isolation regions or reverse-biased Schottky diodes as shields. Even when these are not used, back-gating can be eliminated by using larger spacings around transistors that have junctions with very negative bias currents. There is no back-gating effect when adjacent transistors are biased positively with respect to the transistor under consideration.

MESFET Hysteresis

Another second-order nonideality of MESFETs exhibits itself as a hysteresis in the I–V curves of a MESFET, which might be obtained using a curve tracer. This effect is often called *drain-lag*

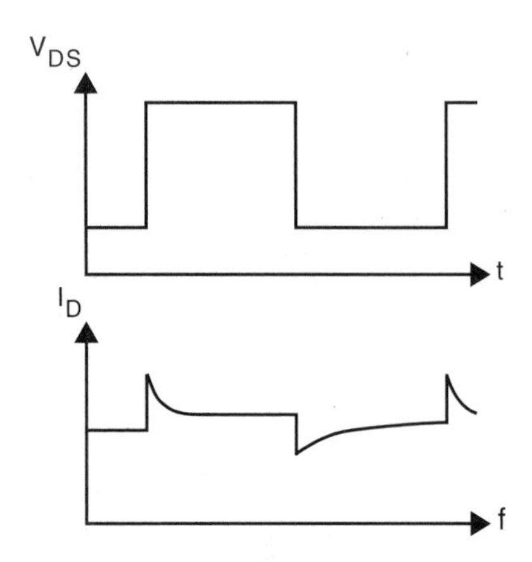

Figure 12.5 The effect of changing V_{DS} on I_D showing the hysteresis of the GaAs MESFET.

(White and Thurston, 1976). One way this effect is noticed occurs when the drain-source bias voltage of a MESFET is suddenly increased by a substantial amount (for example a volt or more). Initially, the current will noticeably increase due to the widening of the channel caused by capacitive coupling between the drain and the substrate under the channel. Next, over a time-frame of a microsecond to a millisecond, electrons injected into the substrate are trapped by deep-level traps that provide additional negative charge, which in turn causes the channel to narrow again. The opposite occurs when the drain-source voltage is suddenly decreased, but the time constant is longer here as it takes longer for traps to release than to capture electrons. Figure 12.5 shows some typical waveforms for I_D as V_{DS} changes with time.

Another way this effect manifests itself is by causing the output impedance of the MESFET (i.e., r_{DS} when the MESFET is in the active region) to be frequency dependent. This effect causes the output impedance to decrease substantially at higher frequencies, perhaps by as much as a factor of three. The effect begins to occur somewhere between 100 Hz and 1 MHz, typically.

This transistor hysteresis is one of the major limitations on the applicability of GaAs MESFETs for wideband analog circuits, but is not too detrimental to most digital circuits.

There are other second-order effects that can be important in certain applications, but these normally are not too detrimental to logic circuits. These additional effects include frequency-dependent transconductance and subthreshold drain-current conduction, even when the gate-source voltage V_{GS}, is less than the threshold voltage. These effects will not be covered here and are described in Long and Butner (1990) for the interested reader.

12.5 Logic Design with MESFETs

There are many different logic families that can be implemented using MESFET transistors. Which family is chosen usually involves a trade-off between speed, power dissipation, noise immunity, and area. Other important considerations include level-shift circuitry that might be required for d.c. compatibility between input and output voltage levels and the need for multiple power-supply voltages. Some of the more popular logic families will be described in this section.

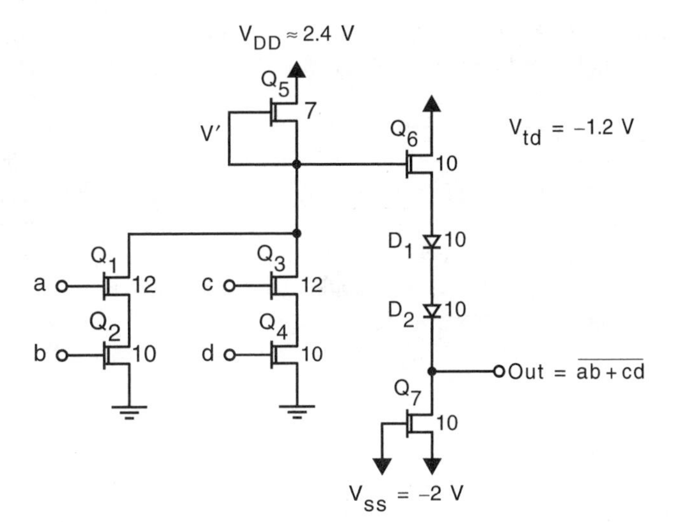

Figure 12.6 A typical buffered-FET logic gate.

BUFFERED-FET LOGIC

One of the first popular GaAs logic families is often called *buffered-FET logic (BFL)*. A typical BFL logic gate is shown in Fig. 12.6. In this logic family, only *depletion* transistors and Schottky diodes are used. The depletion transistors might have threshold voltages of −1.2 V for a modern process. For higher speed logic, more negative threshold voltages might be used, perhaps −1.5 to −2 V; for smaller power dissipation, threshold voltages of −1 to −0.5 V might be used; the modern trend is toward the latter.

The basic logic circuit looks very similar to NMOS logic gates described in Chapter 4. There will be a network of driver transistors having a logic gate input connected to each transistor gate. In addition, there will be a load transistor connected as a current-source load, that is, with its gate connected to its source. The source of the load transistor is connected to the second-stage level-shift network. If the *driver network* is high impedance, the output of the first stage, i.e., V′ in Fig. 12.6, will be at a high voltage. Normally, this will be equal to V_{DD}, which might be around 2.4 V when transistors with −1.2-V threshold voltages are used. If the drive transistor network is low impedance, then the output voltage of the first stage will be low, perhaps around 0.4 V. For the circuit of Fig. 12.6, this will be the case if inputs a and b are both high and/or if inputs c and d are both high.

It is usually not possible to have more than two transistors connected in series, such as Q_1 and Q_2, when realizing *and* functions, although it is common to see up to five transistors connected in parallel when realizing *or* functions.

Because the transistors in the driver network are all depletion transistors, the typical input voltages might be −1.2 V or less for a "0" input and +0.6 V for a "1" input. To realize output voltages with these values, it is necessary to level shift the output of the first stage to more negative voltages. This is accomplished using the second stage composed of source-follower Q_6, two level-shifting diodes, D_1 and D_2, and current-source-connected Q_7. Also, to obtain output

"0" voltages less than 0 V, a second negative power supply, V_{SS}, is required. Its value might be around –2 V for depletion transistors having $V_{td} = -1.2$ V.

When both Q_6 and Q_7 of the buffer stage are in the active region, they will both have 0-V gate-source voltages as they are both the same widths and have the same currents. In addition, there is no appreciable body effect in GaAs to cause shifts in the threshold voltages, as in MOS circuits. Thus, all of the level shift is achieved in the voltage drop across D_1 and D_2. This might be 0.8 V each, if the finite series resistance of the diode is taken into account.

For a BFL gate to work correctly, the widths of the drive transistors need to be wider than that of the load transistor. Ideally, the widths will be chosen so that when V' is changing from a "0" to a "1", the current available for charging parasitic capacitances will be the same as the current available for discharging them during "1"-to-"0" transitions. During the "0"-to-"1" transition, the available current is from that of the load transistor Q_5, as the driver network is high impedance. During the "1"-to-"0" transition, the available current is the difference between the current from the driver network and the load transistor. For the magnitude of the charging current to equal the magnitude of the discharging current, the driver network must sink approximately twice that of the load-transistor current during "1"-to-"0" transitions. This will approximately be the case for the transistor widths shown in Fig. 12.6. Note that during "1"-to-"0" transitions, the driver transistors that are *on* will have a gate-source voltage greater than 0 V, i.e., around 0.6 V, so they need not be taken twice as wide as the load transistor. Note also that when driver transistors are taken in series, the top transistor will sometimes be taken wider than the bottom transistor, as is the case for Q_1 as compared to Q_2. Otherwise, the logic gate would have a different threshold voltage when Q_1 is changing with Q_2 *on*, as compared to when Q_2 is changing with Q_1 *on*. This would result from Q_2 acting as a source-degeneration resistor for the former case. This in turn would cause Q_1 to have a smaller effective transconductance, which in turn would increase the gate threshold voltage. By taking Q_1 slightly wider, this effect is approximately cancelled. For the latter case of Q_1 *on* and Q_2 changing, the drain-source impedance of Q_1 is in series with the larger drain-source impedance of Q_5 and does not have an appreciable effect.

The bias current for the level-shift network can be chosen dependent on the fan-out requirements. A typical value for general purpose gates might be equal to two times the current of the load transistor when the output is low. The preferable choice is a trade-off between speed and power dissipation. For larger fan-outs, the buffer stage can be biased at much higher currents at the expense of excess power dissipation.

The number of diodes used in the level-shift network is dependent on the threshold voltage of the transistors. For $-2\text{ V} < V_t < -1.5\text{ V}$, three level-shift diodes might be used. For $-1.5\text{ V} < V_t < -1.0\text{ V}$, two diodes might be used, whereas for a threshold voltage of around –0.5 V, one level-shift diode would normally be used.

BFL logic is one of the faster GaAs logic families, but it is also one that is most wasteful of power. Making the rough approximations that a $V_{DD} = 1.2\,V_{SS}$, and that the level-shift network is biased at 1.5 times the current of the load transistor, then the average power dissipation of a BFL gate is approximately given by

$$3.5 I_L V_{DD} = P_D \tag{12.4}$$

where I_L is the load-transistor current when the output is low. For a 0.5-μm technology and a 5-μm wide load transistor, I_L might be around 0.7 mA. Using $V_{DD} = 2.4V$ gives an average power dissipation of 6 mW per gate. This would limit a chip to around 800 gates if a maximum total IC power dissipation of 5 W is assumed. *Thus, although BFL is fast, it is limited to medium-scale integration levels.*

12.6 Capacitively Enhanced Logic

To reduce the power dissipation of BFL logic, a GaAs logic family called *capacitively enhanced logic (CEL)* was developed (Livingstone and Mellor, 1980). This logic family is very similar to BFL except the buffer-level-shift stage is replaced by a capacitively enhanced low-power level-shift network. A typical gate is shown in Fig. 12.7.

The level-shift network is simply a couple of forward-biased diodes that are biased at low currents by current-source-connected Q_6, which is quite narrow. Since the current through the level-shift network is very small, no source-follower buffer is required, as is the case for BFL. However, the load transistor Q_5 is normally taken slightly wider since it must supply the d.c. current of the level-shift network. Thus, in Fig. 12.7, Q_5 is taken 9 μm wide as opposed to having a 7 μm width in the BFL gate of Fig. 12.6.

Since there is very little d.c. current in the level-shift network, it would normally be high impedance and have little drive capability during transients. This is alleviated by including a capacitor in parallel with the level-shift network, as is shown in Fig. 12.7. During transients, the current for driving the succeeding gates comes primarily from the parallel capacitor. Thus, the logic is called capacitively enhanced.

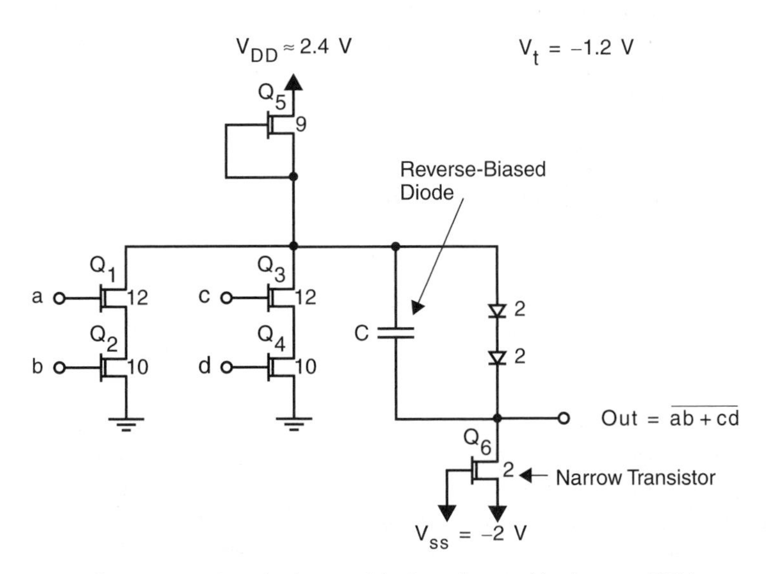

Figure 12.7 A typical capacitively-enhanced logic gate (CEL).

The average power dissipation of a CEL gate is approximately given by

$$P_D = 1.1 I_L V_{DD} \tag{12.5}$$

or less where I_L would be the current of a load transistor of an approximately equivalent BFL logic gate. This makes the typical power dissipation about one-third that of an equivalent BFL gate. For fan-outs of two or less, the speed is comparable to that of a BFL gate. Thus, a common design methodology is to normally use CEL logic gates for most gates in an IC, but to use BFL logic gates for places where a large fan-out must be driven. Since the logic levels of the two families are compatible, the two types of gates can be freely mixed inside an IC to obtain an optimum trade-off between power dissipation and large drive capability. There is a slight area penalty to pay for CEL gates because the capacitor takes a large area, but overall this methodology is usually preferable to using only the logic families.

Example 12.1

Assuming a 5-μm MESFET has $I_D = 0.7$ mA when $V_{GS} = 0$ V and $V_t = -1.2$ V, estimate the average power dissipation of the CEL gate shown in Fig. 12.7. Compare this to the average power dissipation of the BFL gate of Fig. 12.6.

Solution: Using

$$I_D = \beta \frac{W}{L}(V_{GS} - V_t)^2 \tag{12.6}$$

gives

$$\frac{\beta}{L} = \frac{I_D}{W(V_{GS} - V_t)^2} = 97\ \text{Am}^{-1}\text{V}^{-2} \tag{12.7}$$

For the circuit of Fig. 12.7, the current in the level shift network is determined by Q_6. The current through Q_6 is given by

$$I_{D\text{-}6} = 2 \times 10^{-6} 97(-1.2)^2 = 0.28\ \text{mA} \tag{12.8}$$

Thus, the level-shift network dissipates $(V_{DD} - V_{SS})$ 0.28 mA or 4.4 V × 0.28 mA = 1.23 mW. The driver network has an average power dissipation given by $(I_{D\text{-}5} - I_{D\text{-}6})V_{DD}/2$ since it dissipates power half of the time. Using (12.6), we have $I_{D\text{-}5}$ = 1.26 mA when it is in the active region (i.e., the output of the gate is low). Thus, the driver network dissipates on the average 1.2 mW and the total power dissipation of the CEL gate is 2.4 mW.

Proceeding in a similar manner for the BFL gate, we have the power dissipation of the buffer stage given by $(V_{DD}-V_{SS})\, I_{D\text{-}7}$ where $I_{D\text{-}7} = 1.4$ mA using (12.6). Thus, the level-shift network dissipates 6.2 mW. The driver network dissipates $I_{D\text{-}5}(V_{DD}/2)$ on the average or 0.98 mA × 2.4 V/2 = 1.2 mW to give a total power dissipation of 7.3 mW, which is approximately three times the dissipation of a CEL gate.

Despite the lower power dissipation of CEL gates, the average dissipation of around 2 mW per gate still limits an IC to around 1500 gates or so, which is not enough for very large-scale integrated (VLSI) circuitry. The next GaAs logic family to be discussed, direct-coupled logic or DCL, is somewhat slower but has much less power dissipation and therefore allows the realization of VLSI circuits in GaAs.

Direct-Coupled Logic

Direct-coupled logic (*DCL*) uses GaAs enhancement transistors as driver transistors. This allows the voltage levels corresponding to both the "0" and "1" logic levels to be greater than 0 V. This, in turn, allows the level-shift stage to be eliminated. A typical DCL gate is shown in Fig. 12.8. It is very similar to a depletion-load NMOS gate discussed earlier in Chapter 4.

The basic logic gate is composed of two types of transistors, *enhancement* transistors and *depletion* transistors. The threshold voltage of the depletion transistor might be around –0.6 V typically. The depletion transistors are used exclusively for current-source-connected loads (where the transistor gate is connected to the source).

The driver transistors are enhancement transistors. They might have nominal threshold voltages between 0 and 0.3 V, with 0.2 V being a typical choice. To realize DCL gates with a reasonable yield, highly predictable processing is required, where tight uniformity on the threshold voltage is maintained across the wafer with very small standard deviations, perhaps

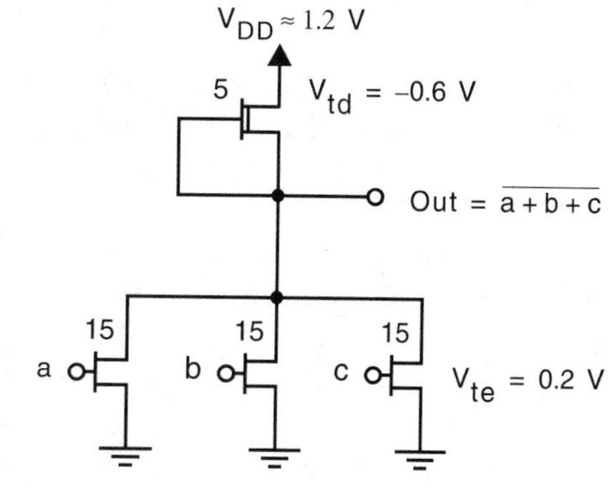

Figure 12.8 A typical direct-coupled logic gate.

of 30 mV or smaller. Also, it is important to minimize the series source resistance of the enhancement transistors in order to maximize the gain at the gate-threshold voltage. For these reasons, a self-aligned technology is often required for realizing DCL gates.

DCL logic gates are *ratioed gates*, where the driver transistors must be wide enough to sink at least twice the load current, when the output voltage is around the gate-threshold voltage, during "1"-to-"0" transitions. A ratio of 3 to 1 for the width of the enhancement transistors, as compared to the depletion load transistors, is considered to be around the minimum necessary.

During the operation of a DCL gate, when all the inputs are low, the output will be high, usually high enough in voltage to forward bias the gates of the transistors of succeeding DCL gates. Thus, a typical "1" output voltage of a DCL gate might be 0.7 to 0.8 V when MESFET transistors are used. When HEMT transistors are used, the typical output "1" voltage might be 0.8 to 0.9 V because of the larger gate-to-source voltages typical of HEMT transistors with conducting gates. This gives slightly better noise margins when HEMT devices are used.

When any input is high, the output voltage is low. The output voltage for a "0" output is dependent on the width of the driver devices; when wider drivers are used, the output low voltage will be less. For the 3:1 ratio of driver-to-load widths shown in Fig. 12.8, a typical output low voltage might be as large as 0.2 to 0.3 V. This implies that for typical input voltages corresponding to "0" inputs, the driver transistors of succeeding gates may not be completely turned off; this is particularly true because of subthreshold conduction that occurs for V_{GS} only slightly less than V_{te}. Normally, for a fan-in of three or less, this conduction has no ill effects; however, for large fan-in, when there is a large number of parallel enhancement driver transistors, the subthreshold currents of the input transistors may be large enough to keep the output low even when all inputs are "0". This limits the maximum fan-in of DCL gates, particularly at low temperatures. It is possible to have larger fan-ins if wider driver transistors are used, but this increases the capacitive load on preceding gates. In most applications an engineering compromise is necessitated that is application and process dependent.

It is not possible to reliably realize DCL gates with two series transistors, or equivalently, a single double-gate transistor, as is possible for BFL and CEL gates. Otherwise, the noise margins would be compromised too much. Thus, sometimes there will be more gate delays for functions realized using DCL gates than for functions realized using BFL or CEL gates because of the requirement for small fan-ins. This is one of the limitations of DCL logic that must be accepted.

Another major limitation of DCL gates is that they have very little drive capability, particularly for "0"-to-"1" transitions similar to NMOS gates, but also for "1"-to-"0" transitions because of the limited effective gate-source voltage of the enhancement driver transistors. There are many variations of DCL logic where *push–pull* buffers are added when large loads need to be driven. An example of this, called superbuffer FET Logic (SBFL) was presented in Tanaka et al. (1985).

A typical example of a DCL inverter, with an output buffer added, is shown in Fig. 12.9. The buffer consists of Q_3 and Q_4. During output "0"-to-"1" transitions, Q_3 is turned off and Q_4 operates as a source follower. Since it is also moderately wide, it can source larger currents during the transition than a simple depletion load. During output "1"-to-"0" transitions, Q_3, acting as a common-source amplifier, can sink a reasonably large current since it is also fairly wide. Also, at this time, V' will be low and Q_4 is turned off, so all of the drain current of Q_3 is

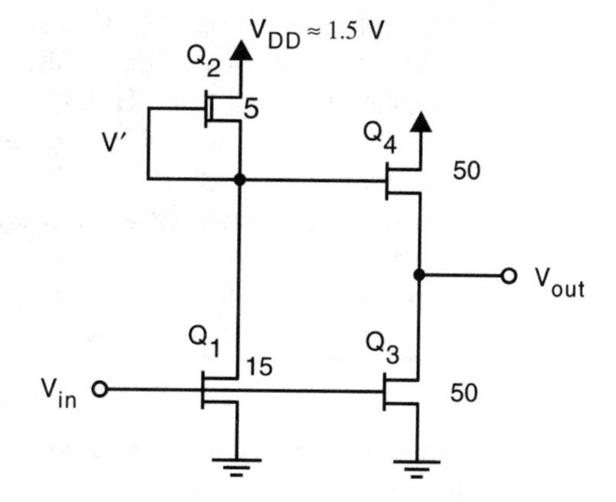

Figure 12.9 A DCL inverter with a push–pull output buffer added for driving large capacitive loads.

used to discharge output load capacitances. This should be contrasted with a normal DCL gate where part of the enhancement driver transistor's drain current is used to sink the depletion load current and is therefore unavailable for discharging the output load capacitance.

When push–pull buffers are used, it is necessary to increase the power-supply voltage somewhat to account for the voltage drop of source-follower Q_4 when it is conducting. If this larger power-supply voltage is also used for all ordinary DCL gates that do not have large loads, and therefore do not require output buffers, the overall power dissipation is increased. Alternatively, one can use two power supplies, perhaps 1 to 1.2 V for ordinary DCL gates, and 1.5 to 1.7 V for DCL gates having push–pull buffers. Unfortunately, this would often increase the system cost and complexity substantially; which choice is preferable is application dependent and an engineering compromise is again necessary.

Because DCL logic has less drive-current capability, it is normally much slower than either BFL logic or CEL logic, perhaps only one-third to one-half the speed when transistors having similar lengths are used; however, its power is substantially less.

Example 12.2

Consider the gate of Fig. 12.9. If, $\beta/L = 97 \text{ Am}^{-1} \text{ V}^{-2}$ and $V_{td} = -0.6$ V as in Example 12.1, then the depletion load, when it is in the active region, will have a current given by

$$I_L = W\frac{\beta}{L}(-V_{td})^2 = 0.17 \text{ mA} \tag{12.9}$$

The average power dissipation of the gate is then given by

$$P_D = \frac{I_L V_{DD}}{2} = 0.13 \text{ mW} \tag{12.10}$$

assuming the gate's output will be low half the time and high half the time. Thus, the average power dissipation is one-eighteenth that of a CEL gate and one-fifty-sixth that of a BFL gate. This allows GaAs to be used for VLSI circuits having more than 100,000 gates in a single IC. *Because of the low power-supply voltages, the small parasitic capacitances of GaAs, and the high speed, DCL GaAs gates presently have the best power-delay product of any logic family in any technology.* This should prove to be a major driving force in increasing the popularity of GaAs for digital ICs.

However, in addition to the lower speed of DCL gates compared to other GaAs logic families, the DCL family suffers from other major problems; it has low noise margins and is very prone to errors. When a DCL gate is turned off, the load current is diverted from the driver to the transistor gate input of a succeeding logic gate. This means the ground path changes, and with the high speeds involved, there is considerable ground bounce due to both resistive and inductive effects. If this is considered along with the small noise margins, it is seen that DCL is very prone to noise errors. A single error in a digital IC at any time makes the chip practically unusable. For many years, experts in GaAs digital IC design felt that DCL would never be used in *real* ICs intended to be used in actual products. This has proven not to be the case, largely due to an elegant solution to the problem that was developed by Vitesse Semiconductor.[1] The solution is an excellent example of a very simple solution to a difficult problem, which after the fact is obvious, but typically is found only by the most imaginative of designers. Vitesse Semiconductor added an extra layer of metal that completely covered the entire IC and was only used as a ground plane. This lowered the ground impedances enough so that errors due to ground bounce were virtually eliminated. The additional cost incurred was the extra noncritical ground layer and an additional mask for the contacts required. It also eliminated leakage currents due to photoconductive effects that could otherwise occur in GaAs ICs. With this modification, GaAs digital ICs are a very reasonable alternative for high-speed VLSI circuits.

There is one other GaAs logic family that is very popular for digital ICs that must run at the very fastest possible speeds, that of fully differential source-coupled logic, which will be described in the next section.

Source-Coupled Logic

GaAs *source-coupled logic (SCL)* is the GaAs version of bipolar fully differential current-mode logic described in Chapter 8. An example of a GaAs carry-generate circuit that could be used as part of a full adder is shown in Fig. 12.10. The similarity between this gate and the bipolar realization of the same function that was shown previously, i.e., Fig. 8.12 in Chapter 8,

[1]The individual who originally proposed this solution is unknown by the author.

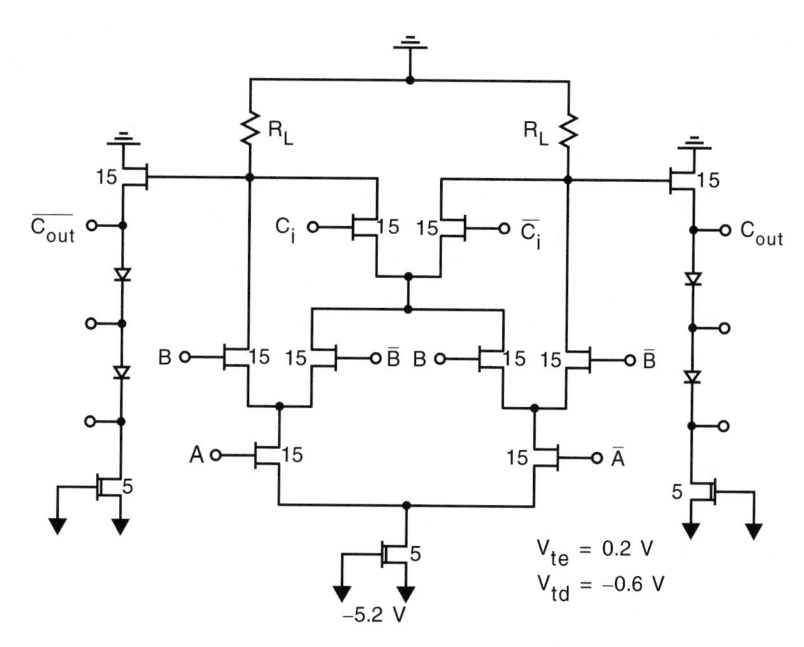

Figure 12.10 A source-coupled logic realization of a carry-generate circuit.

is obvious.[2] Indeed, even the voltage levels and voltage swings are similar. The values for the resistive loads are chosen so that the differential voltage swing is 0.8 to 1.0 V, typically. This is process dependent and, in some realizations, the simple current-source-connected depletion bias transistors might be replaced by regulated current sources. The design of these is beyond the scope of this book.

SCL logic is growing in popularity for highest speed applications for a number of reasons. First, it is very fast. In some applications, it is the fastest GaAs logic family to date. The reasons for this are the same as the reasons for the high speed of current-mode logic. First, the voltage swings can be chosen to be small, as low as 0.4 to 0.5 V. This minimizes the time to charge or discharge parasitic capacitances. Second, for many arithmetic functions, such as full adders and parity circuits, it is often possible to eliminate inverters that would be required in single-ended logic designs. An example of this is seen in Fig. 12.10 where the delay from the carry-in, C_i, to the carry-out, C_{out}, is almost the same as the delay of a single buffered inverter. This results in very fast ripple-carry adders due to the minimizing of the logic delays. A similar example is shown in Fig. 12.11, where a multiplexer circuit is shown. In this circuit, when S = "1", the output X will be equal to the input A, otherwise if S = "0", $X = \bar{B}$. The delay from the signal inputs, A or B, is approximately the same as that of a simple buffered inverter having a small voltage swing; furthermore, this remains the case even if one desires to have the inputs inverted.

[2]Part of the reason for the similarity is that the author designed both gates.

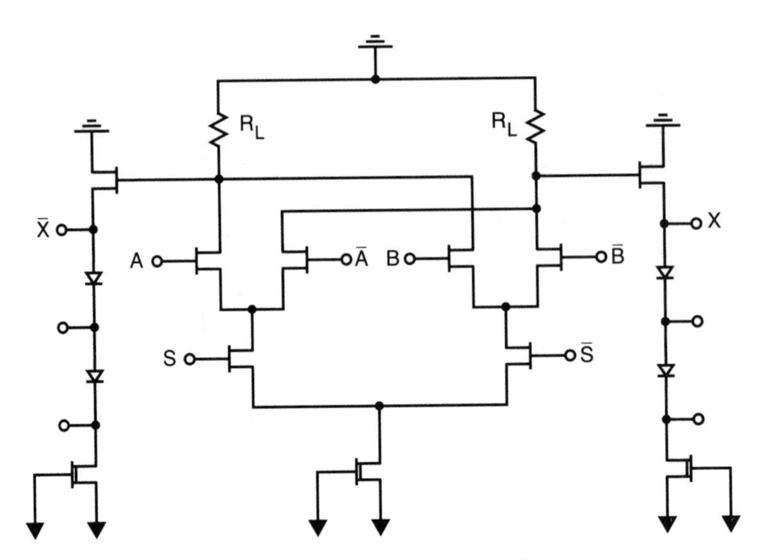

Figure 12.11 A source-coupled logic multiplexor gate.

Another reason for the popularity of the SCL family is that it is relatively immune to noise errors. Again, the reasons for this are the same as the reasons for the noise immunity of silicon bipolar current-mode logic. First, because all the signals are differential, noise inputs are often common-mode signals that have minimal effects on the differential voltages. Second, as long as both load resistors are connected together before they are connected to ground, as is shown in Figs. 12.10 and 12.11, then there are minimal changes in the currents through ground paths or the power supplies when the output values of the gates change. For very high-speed logic, this feature is critical to good noise immunity.

The SCL family does have some serious limitations. First, since all signal interconnects are differential, each signal path requires two wires and thus the routing area is increased. This is not usually a problem for short interconnects between adjacent gates, but can cause a significant area penalty for long buses composed of many signals. Second, the number of gate inputs is limited, typically to about three or so. To get around this limitation, additional gate delays and power dissipation often result. Finally, level-shift buffer networks are required to shift from logic at one level to a lower level. This results in additional power dissipation, although it does help to drive larger fan-outs. Despite these limitations, the popularity of SCL circuit realizations is predicted to increase for the highest speed applications primarily due to its high speed and noise immunity.

Dynamic GaAs Logic

Recently, a number of different approaches for realizing GaAs dynamic logic gates have been presented. For example, GaAs *capacitively coupled Domino logic (CCDL)* was described in Hoe and Salama (1991), with a typical gate shown in Fig. 12.12. Also, GaAs *two-phase dynamic FET logic (TDFL)* was presented in Nary and Long (1992) with a typical gate shown in Fig. 12.13.

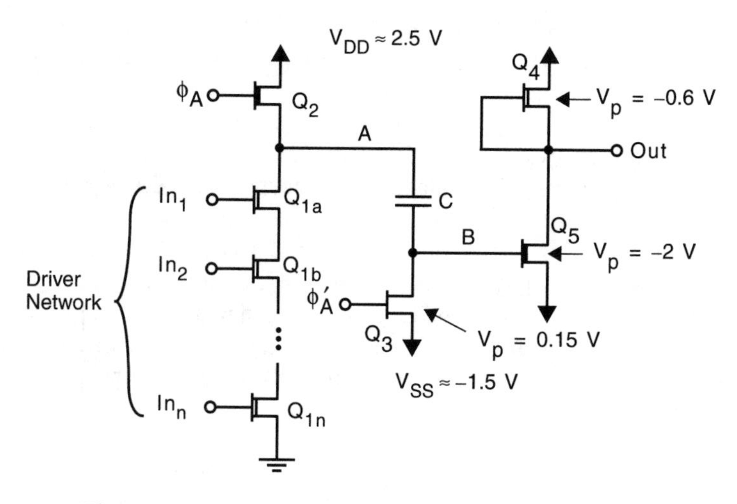

Figure 12.12 A capacitively coupled Domino logic gate.

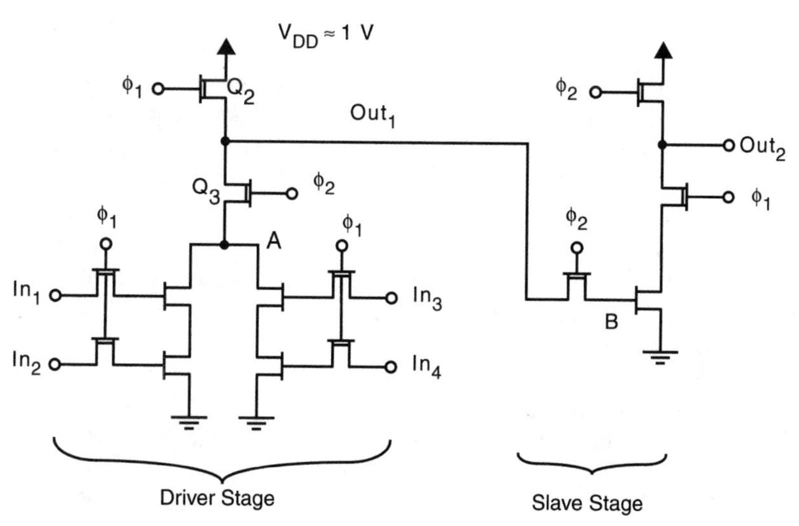

Figure 12.13 A GaAs two-phase dynamic FET logic gate.

The approach of Fig. 12.12, CCDL, is somewhat similar to Domino-CMOS logic that was presented in Chapter 9. In CCDL, during precharge phase ϕ_A, node A is precharged to V_{DD} (approximately 2.5 V) and node B is discharged to V_{SS} (approximately −1.5 V). Note that ϕ_A' is simply a level-shifted version of ϕ_A. Thus, during the precharge time, the coupling capacitor C is precharged to $V_{DD} - V_{SS}$, or about 4 V. Next, during $\bar{\phi}_A$, precharge transistors Q_2 and Q_3 are turned off. At this time, node A will discharge to ground if the driver network is low impedance, otherwise it will be left at a high voltage, although not necessarily V_{DD}, due to charge sharing in the driver network. If node A is discharged to ground, node B will also be pulled low enough to cause the inverter output to go high. Otherwise, it will be left low, the same value it has during the precharge phase, ϕ_A.

The driver stage of CCDL gates does not dissipate d.c. power, however, the *inverter* does dissipate d.c. power, around two times or more the amount of a typical DCL gate. This is primarily due to the dual power-supply voltages used. However, the driver network can include many transistors in series, so the gates can realize more complicated logic functions than DCL gates. A serious limitation is that three different types of transistors are required: an *enhancement* transistor and *depletion* transistors having two different threshold voltages. In addition, two power supplies and clock drivers (and level shifters) are required. Also, a fairly large *level-shifting* capacitor is needed. Finally, the logic has a similar limitation as Domino-CMOS logic in that inverting functions are difficult to realize unless differential logic is used. Because of these limitations, it is expected that CCDL logic will be limited to a few very specific applications, although the principles developed may be used in other future dynamic logic families.

In the alternative TDFL approach, described in Nary and Long (1992), a clocking scheme very similar to one used originally for two-phase PMOS logic gates (Penney and Lau, 1972) is proposed. In this approach, each successive stage is clocked using complementary waveforms. In addition, each stage has dynamic storage. Together, these characteristics make the logic suitable for pipelined circuits. In addition, the logic does not dissipate d.c. power, so it is suitable for high-speed very low-power circuits. Indeed, it has been reported (Penney and Lau, 1972) that TDFL gates would typically have 10 to 15 times less power dissipation than CMOS or BiCMOS logic gates, making this possibly the lowest power logic family to date.

To understand the operation of TDFL logic, consider the representative gate shown in Fig. 12.13. Two stages are shown with *complementary* clock waveforms being used for each stage. The first stage includes a *driver network* with four inputs; the second stage is simply a *clocked inverter.* Consider the first stage when ϕ_1 is high and ϕ_2 is low. At this time, the output of the first stage, Out_1, is *precharged high.* Also, the input signals are coupled through *pass transistors* to the gates of the enhancement transistors in the driver network. Also, since ϕ_2 is low and Q_3 is turned off, the driver network is isolated from the output of the first stage. Next, ϕ_1 goes low and ϕ_2 goes high. This stores the input signals on the gates of the enhancement transistors of the driver network and isolates the first stage from preceding stages, which are now free to change. Also, node A is connected to the output of the first stage, Out_1, which is also connected to node B, the input of the succeeding stage. If the driver network of the first stage is low impedance, then the output of the first stage will be pulled low, otherwise it will be left high assuming the capacitance at its output node, Out_1, is *large enough.*

TDFL has the advantage that only a single power supply, perhaps 1.0 V, is required, if *off-chip* clock drivers are used. Otherwise a second *on-chip* power supply is needed for the clock drivers as the voltage levels of the clock signals might be between 0 and –1.5 V for depletion transistors having nominal threshold voltages of –1.1 V. Also, the clock waveforms should be nonoverlapping or at least cross at a voltage of –0.7 V or lower. In addition, clock signals should be distributed from the end of the pipeline to the beginning, similar to what is preferable in all pipelined circuits.

Perhaps the major limitation of TDFL logic is the need for the capacitance at the output of each stage to be much larger than the capacitance at adjacent nodes (nodes A and B in

Fig. 12.13). This would not normally be the case unless there is a very large *fan-out*. If the output capacitance is not significantly larger, then the output can be pulled *low* when it is connected to adjacent nodes, if they had been previously discharged, even though the driver network is high impedance during the evaluation phase. This would seriously degrade NM_H when the output should be left at a "1". To prevent this, it is usually necessary to add additional capacitance to the output nodes of the gate, which unfortunately slows the gates and does increase the dynamic power dissipation somewhat. Despite this limitation, *TDFL is expected to be used in applications where extremely low-power dissipation and high speed is required.*

12.7 GaAs Logic Family Comparison

The advantages and disadvantages of the various GaAs logic families are summarized in Table 12.2. Based on this table and on the previous description of the various logic families, a number of guidelines as to which family should be used can be made.

First, GaAs logic should be used only in applications in which silicon technology is either too slow or an excellent speed–power product is necessary. When GaAs logic is used, it is expected that for most applications DCL is the preferable choice, primarily due to its low power dissipation. It can therefore be used in applications in which a large number of gates are required on a single IC. When the absolutely fastest speed is required, then SCL logic should normally be chosen, although a combination of CEL and BFL gates might be a consideration. Where extremely low-power dissipation and moderately high speed (compared to silicon but not as compared to the other GaAs logic families) is necessary, then TDFL or DCL should be chosen.

12.8 Heterojunction Bipolar Technology

One of the most exciting new technologies that uses III–V semiconductors and related crystals is the realization of *heterojunction bipolar transistors*. In this technology, bipolar transistors,

Table 12.2 A comparison of Different GaAs Logic Families

Logic family	BFL (Fig. 12.6)	CEL (Fig. 12.7)	DCL (Fig. 12.8)	SCL (Fig. 12.10)	TDFL (Fig. 12.13)
Speed	Very good	Good	Poor	Excellent	Very poor
Power dissipation	Very poor	Poor	Good	Poor	Excellent
Speed–power product	Poor	Good	Excellent	Moderate	Excellent
Number of power supplies	2	2	1	1	1+
Noise margins	Good	Good	Poor	Adequate	Good
Fan-in	Good	Good	Poor	Very poor	Good
Fan-out	Good	Moderate	Very poor	Moderate	Very poor

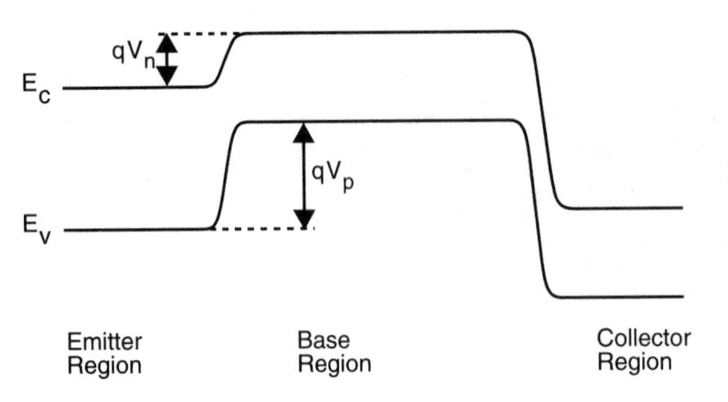

Figure 12.14 An energy-band diagram of a heterojunction bipolar transistor.

very similar to silicon bipolar transistors, are realized, but different semiconductor materials are used for the emitter and base regions. The semiconductor material used for the emitter region has a wider bandgap potential than that of the base region. This gives the process designer an additional degree of freedom that is not possible with silicon technology. With silicon technology, it is necessary to dope the base region with a considerably lower concentration than is used for the emitter region in order to achieve a reasonable transistor current gain or β. In *heterojunction bipolar technology* (*HBT*), this is no longer necessary due to the difference in the bandgap energies of the base and emitter regions. It is now possible to dope the base region equal to or greater than the emitter without the transistor current gain suffering excessively; this results in a number of improved transistor characteristics. To see why this is so, consider a heterojunction bipolar transistor realized using a p^+ GaAs base, and an emitter region formed of a $Ga_{0.7}Al_{0.3}As$ compound (commonly referred to as *algas* or *AlGaAs*). The bandgap potential of the GaAs base is 1.42 eV, whereas that of the AlGaAs emitter is about 1.8 eV. An energy-band diagram corresponding to these materials is shown in Fig. 12.14. It can be seen that the junction potential opposing the diffusion of electrons across the base-emitter junction, qV_n, is considerably less than the potential opposing the diffusion of holes, qV_p. This remains true even for a heavily doped base and a lightly doped emitter region. This difference in energies results in large βs even when the base is heavily doped.

The heavily doped base region results in a number of desirable characteristics. Most of these are due to the fact that the heavily doped base results in a small intrinsic base resistance, r_b. This in turn results in a higher f_{max}, the unity-gain frequency for power gain. Furthermore, HBT transistors have large unity-current-gain frequencies due to the high mobility of III–V compounds. Together, these result in very fast transistors, as fast as 100 GHz f_t at present. In addition, the low emitter doping results in low base-emitter depletion capacitances. Finally, the small r_b also minimizes one of the major sources of noise in bipolar circuits.

GaAs HBT digital circuits exhibit many of the same desirable features as silicon-based bipolar circuits when compared to FET digital circuits. These include a better ability to drive large capacitive loads due to the larger transconductance and the possibility of designing logic circuits having small voltage swings due to large transconductances and good V_{BE} matching between transistors.

A typical cross section of an AlGaAs HBT transistor is shown in Fig. 12.15 (Asbeck et al., 1984). Alternative processing techniques might result in very different cross sections. The

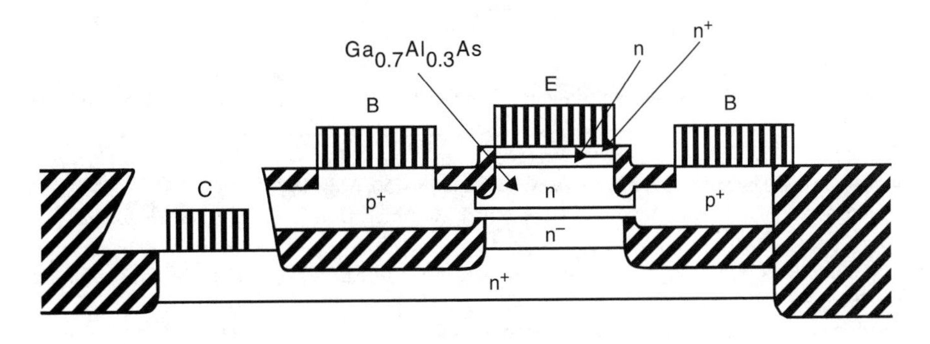

Figure 12.15 A self-aligned HBT transistor cross section.

details of the different possible processes and the trade-offs between them are beyond the scope of this text. For an example of a fairly conservative AlGaAs process the interested reader is referred to TRW (1991), as an example. In this process, a transistor realized as four parallel transistors, each having an emitter area of 2×10 μm, and biased at a total current of 8 mA, which is 50% of the maximum bias current, had an f_t of 19.4 GHz and an intrinsic base resistance of only 2.6 Ω. More modern, smaller, dimensions would have resulted in considerably higher frequencies.

One of the nice features of HBT technology is that the transistors do not seem to exhibit many of the defect-related limitations of MESFET transistors such as hysteresis, back-gating, and frequency-dependent transconductance and output impedance. The transistors also have a very large output impedance and intrinsic gain. Indeed the transistors described in Asbeck et al. (1984) have an early voltage of approximately 700 V and an intrinsic gain of 18,500! This is substantially larger than that reported for silicon bipolar transistors. The major limitations of GaAs HBT transistors, as compared to silicon bipolar transistors, are a large base-emitter voltage, on the order of 1.2 V, and increased self-heating effects due to smaller thermal conductivity of GaAs as compared to silicon.

The recommended design of HBT logic gates is identical to the design of silicon bipolar current-mode logic gates that were described in detail in Chapter 8, except a couple of additional constraints should be considered. The most important difference is that the base-emitter junction voltage, under forward bias conditions, might be around 1.2 V for an AlGaAs process. This increases the power dissipation and limits the number of levels of the current-mode logic circuits, which in turn limits the fan-in to three or four. A more recent HBT technology based on *indium phosphide* materials (i.e., an n InP substrate and collector, a p^+ InGaAs base, and an n InP emitter) (Schur, 1989) results in smaller base-emitter voltages, perhaps 0.9 V, but also decreased output impedances, which fortunately does not have much effect on digital applications. It is predicted that this technology will be extremely important for future, ultra-high-speed digital circuits. At present, this technology is just reaching the maturity level where circuits with around 1000 transistors can be realized. Also, at present, InP-based circuits hold the speed record for the fastest digital gates ever realized.

12.9 Bibliography

P. Asbeck et al., "GaAs/GaAlAs Heterojunction Bipolar Transistors with Buried Oxygen-Implanted Isolation Layers," *IEEE Electronic Device Letters,* EDL-5, 310–312, August 1984.

W.R. Curtice, "A MESFET Model for Use in the Design of GaAs Integrated Circuits," *IEEE Transactions on Microwave Theory and Technology,* MTT-28, 448–456, May 1980.

D. Hoe and A. Salama, "Dynamic GaAs Capacitively-Coupled Domino Logic," *IEEE Journal of Solid-State Circuits*, 26(6), June 1991.

C. Kocot and C.A. Stolfe, "Backgating in GaAs MESFETs," *IEEE Transactions on Electronic Devices,* ED-29, 1059–1064, July 1982.

A Livingstone and P. Mellor, "Capacitive Coupling of GaAs Depletion-Mode FETS," *Proceedings of the Institute of Electrical Engineers,* 127, Pt. 1, 297–300, October 1980.

S. Long and S. Butner, *Gallium-Arsenide Digital Integrated Circuit Design,* McGraw-Hill, 1990.

T. Mimura et al., "High-Electron-Mobility Transistor Logic," *Japanese Journal of Applied Physics,* 19, L225–L227, 1980.

M.R. Namordi and W.A. White, "A Low-Power Static GaAs MESFET Logic Gate," *Proceedings of GaAs IC Symposium,* 21, 1982.

K. Nary and S. Long, "GaAs Two-Phase Dynamic FET Logic: A Low-Power Logic Family for VLSI," *IEEE Journal of Solid-State Circuits*, 27(11), October 1992.

W. Penney and L. Lau, *MOS Integrated Circuits,* Van Nostrand Reinhold, 1972.

M. Rochi, *High-Speed Digital IC Technologies,* Artech House, 1990.

A. Rode et al, "A High-Yield GaAs MSI Digital IC Process," *Proceedings of the IEEE International. Electron-Modelling Conf. (IEDM)*, 162, 1982.

A.S. Sedra and K.C. Smith, *Microelectronic Circuits*, 3rd ed., Oxford University Press, 1991.

M. Schur, "Submicron GaAs, AlGaAs/GaAs and AlGaAs/InGaAs Transistors," in *Submicron Integrated Circuits,* R.K. Watts, ed., Wiley-Interscience, 1989.

S.M. Sze, *Physics of Semiconductor Devices,* 2nd ed., John Wiley & Sons, 1981.

K. Tanaka et al., "Super-Buffer FET Logic (SBFL) for GaAs Gate Arrays," *Proceedings of the 1985 Custom Integrated Circuits Conference,* 425–428, May 1985.

M.H. White and M.O. Thurston, "Characterization of Microwave Transistors," *Solid-State Electronics,* 13, 523–542, 1976.

N. Yokoyama et al., *ISSCC Digest of Technical Papers,* 218–219, February 1981.

12.10 Problems

For the problems in this chapter, assume the following transistor parameters:

- $V_{td} = -1.2$ V
  $V_{te} = 0.1$ V
  $\beta/L = 100$ A m^{-1} V^{-2}
  $L_{min} = 0.5$ μm
  $V_D = 0.9$ V
  $\lambda = 0.1$ V^{-1}

C_{GS} = 1.6 fF per micrometer width

C_{SB} = C_{DB} = 0.2 fF per micrometer width

12.1 For V_{DD} = 2 V, V_{SS} = –1.5 V, and V_{td} = –0.8 V, and the following transistor widths, estimate the power dissipation of the gate in Fig. 12.6. Assume defaults for all other parameters.

Transistors	Widths (μm)
Q_1, Q_3	6
Q_2, Q_4	5
Q_5	3.5
Q_6, Q_7	5

12.2 Prove equation (12.3), which gives the power dissipation of the gate in Fig. 12.6.

12.3 For the parameters of Problem 12.1, and assuming inputs b = "1", c = "0", d = "1", estimate node voltages for the cases a = "0" and a = "1" for the gate shown in Fig. 12.6.

12.4 For V_{DD} = 2 V, V_{SS} = –1.5 V, and V_{td} = –0.8 V, and the following transistor widths, estimate the power dissipation of the gate in Fig. 12.7. Assume defaults for all other parameters and that the widths of the diodes is 2 μm.

Transistors	Widths (μm)
Q_1, Q_3	6
Q_2, Q_4	5
Q_5	5
Q_6	2

12.5 Prove equation (12.4), which gives the power dissipation of the gate in Fig. 12.7.

12.6 For the parameters of Problem 12.4, and assuming inputs b = "1", c = "0", d = "1", estimate node voltages for the cases a = "0" and a = "1" for the gate shown in Fig. 12.7.

12.7 For V_{DD} = 1.5 V, V_{td} = –0.7 V, and V_{te} = 0.1 V, and assuming all *drive transistors* are 6 μm wide and the *load transistor* is 2 μm wide, estimate the power-dissipation of the gate in Fig. 12.8. Assume defaults for all other parameters.

12.8 For the gate of Fig. 12.8 and assuming the gate is driving a similar gate, estimate the node voltages for the cases of the output high and the output low.

12.9 For the gate of Fig. 12.10, assuming $R_L = 3\ k\Omega$ and the voltage drop across the diodes is 1 V, find node voltages for the cases A = "1", B = "0", C_i = "1", and for the same inputs but with C_i = "0". Give your assumptions for the input voltage levels.

12.10 What is the power dissipation of the gate of Fig. 12.10?

12.11 Design a GaAs *biphase* D flip-flop using BFL logic. Give reasonable device widths where the maximum width is constrained to be less than 20 μm.

12.12 Design a static RAM memory cell using DCL logic.

12.13 Design a full adder using BFL and/or CEL logic. Give reasonable transistor sizes. Discuss the justifications for your design.

12.14 Design a full adder using DCL logic. Give reasonable device sizes.

12.15 Compare the design obtained in Problem 12.13 to that obtained in Problem 12.14 in terms of number of transistors (and transistor sizes), power, and speed.

12.16 A divide-by-two circuit is required in a phase-locked loop to control the carrier frequency of a millimeter-wave communication system. Speed is of the most importance. Give a GaAs design.

12.17 A serial-to-parallel interface circuit is required to help test some high-speed logic circuits. Assume the input is to be sampled at the negative edge of a clock signal, and that every eight clock signals eight successive serial bits are to be stored in eight latches. Give a hierarchical design assuming DCL.

CHAPTER 13

Digital System Testing

This chapter deals with how to design digital integrated circuits so they can be tested. As digital integrated circuits continually get larger, it is essential that IC designers seriously consider how an IC is to be tested, both during the prototype design and debugging phase, and during production. These considerations must start from the time of the first conception of the overall system design. Designing an IC so that it is testable is a nontrivial task, but it is one that the designer must succeed at or all other efforts are for naught. It is this author's opinion that almost every digital IC should include circuitry whose sole function is to help with the on-chip testing. Readers should not assume that because this chapter is last in the book the importance of testing is minimal; rather, *for digital ICs intended for realizing real systems, testing may be one of the most important considerations.*

A reason why modern ICs, and even more so complete systems, are difficult and sometimes impossible to test is twofold: first, it is difficult to access internal nodes during testing; second, as circuits become larger and more complicated, the number of test vectors required to exercise all aspects of a digital circuit with a high probability of not missing any faults is growing exponentially. This results in large testing times, incomplete testing, and ultimately loss of profit. The only solution is the typical engineering solution to every problem resulting from complexity: *"divide and conquer."*

Through the years, and especially in the past few years, a number of techniques have been developed that make the problem if not easy, at least tractable. Designers who are either unaware or ignore these techniques place themselves in peril.

When applying the cardinal law of designing for testability, *"divide and conquer,"* the system is separated into smaller functional blocks, test vectors are applied to the inputs of these blocks, and the outputs are compared to the expected outputs of a functionally correct block.

The expected outputs are often obtained from computer simulations. This process is repeated until there is a high probability that the circuits have no faults. A 100% confidence level is almost never attained; the closer one comes to this ultimate goal, the more successful the design is.

A number of issues must be considered when making a design testable:

1. How should an IC be subdivided so that each subcircuit is individually testable? If an IC is subdivided into only a few subcircuits, it is easier to get the test vectors to their inputs, but a greater number of test vectors is required to properly exercise each subcircuit.
2. How can test vectors be delivered to the inputs of subcircuits and what test vectors should be used? If these test vectors must be shifted in from *off chip*, through a serial shift register, the time required for inputting them becomes prohibitive; for this reason, much research into generating test vectors *on chip* for each subcircuit has taken place.
3. How can the results from the testing of the subcircuits be compared to known correct values? Do they need to be shifted serially off chip, which is very time consuming, or can they be compared to known correct outputs locally at the outputs of each subcircuit? If so, how should the comparison be done without having to store excessive numbers of known correct outputs?
4. How can the test process be controlled? Normally, this would be done by automatic test equipment (ATE) in partially relying on the results of computer simulations obtained during the design phase. Often, an on-chip controller might be used. Hopefully, the amount of software that must be developed for each chip is kept to a minimum. This is an area in which some recent developments of an IEEE standard approach, designated *ANSI/IEEE Standard 1149.1* (IEEE, 1990)—*the Standard Test Access Port and Boundary-Scan Architecture*, is rapidly gaining in popularity and is having a major impact on the how testing is done.

In this chapter, some of the more traditional approaches for *built-in-self-test (BIST)* will be described first followed by a brief description of *boundary-scan* techniques. Testing is an area of extensive on-going research and development and without writing a complete book dedicated to just this subject, the most that can be accomplished in a single chapter is the presentation of an introduction and overview.

13.1 Conservative Design Principles

One of the most important principles to follow when designing for testability is to use conservative design principles; a number of principles are recommended:

1. Static logic circuits should be used where the clock can be stopped without internal logic values being lost, at least for the majority of the circuitry in an IC. When dynamic circuits are used, then *bleeder* transistors, that is small *pseudo-NMOS* loads, should be added to allow for stopping the clocks. When purely dynamic circuitry is used, then extensive simulations should be done at design time, and extensive considerations of

potential subtle error sources, such as charge sharing, clock feedthrough, and leakage currents, should be expended.

2. A single global clock signal should be used for the complete IC. It should be distributed in a direction opposite to that for the propagation of signals. Together these principles help minimize potential problems with race conditions. If it is necessary to generate clock signals that are different from the global clock signal, then this should be done using a single flip-flop, and consideration should be given to the edge on which the new clock signal is effective.
3. Master–slave D flip-flops should be used. These are less sensitive to input-signal glitches than SR, JK, or toggle (T) flip-flops. They also do not require as small rise and fall times for clock waveforms compared to edge-sensitive flip-flops. Also, it is easier to modify D flip-flops to include circuitry for testing than it is for other types.
4. A liberal number of small test pads (perhaps 20 μm by 20 μm) should be distributed throughout the IC. Much consideration should be given to which nodes have these test pads connected to them. They are intended primarily for use during the debugging phase, particularly in the event of some unforeseen error or subtle timing problem. They are not very useful for production testing. After the initial design and debugging, they can be removed and the layout can be automatically compacted using CAD tools; however, this should not be done until a large degree of confidence in the design has been gained.
5. Careful consideration should be given to the layout of the power-supply interconnects and even more so to the ground-line returns. *Glitches* on the ground lines can cause subtle errors that are difficult to detect and may cause errors only at temperature and/or process extreme *corners*. Excessive currents through narrow interconnect lines can cause metal migration that results in poor long-term reliability. This can be particularly troublesome around I/O buffers, which should have separate power-supply and ground interconnect and, ideally, I/O pads, from the rest of the chip. This helps prevent corrupting the power-supply and ground voltages everywhere. In addition, a number of I/O pads should be used for the power-supply, and particularly for ground, interconnects.
6. Substrate connections and well ties should be used copiously throughout an IC. I/O buffers should be surrounded by guard rings to help prevent latch-up. In addition, on-chip bypass capacitors with low Q-factors (i.e., finite series resistances) should be placed throughout the chip. *The low Q-factors are necessary to prevent resonances with the inductance of bonding wires.*

The use of conservative design principles can make debugging ICs considerably easier. When realizing state-of-the-art ICs, it may be necessary to use more advanced, and therefore riskier, techniques for critical circuits in the IC; however, it is still generally possible to use conservative techniques for the majority of logic in an IC.

13.2 Scan-Design Techniques

Perhaps the major hurdle to exhaustive testing of sequential circuits is the huge number of test vectors required for every possible state of an IC to be exercised. Normally, the number is so large for typical ICs that a testing procedure that checks all states cannot even be contemplated due to the time required for the testing. For this reason one should include built-in testing with some type of *scan-design* circuitry (Eichler, 1974; Funatsu et al., 1975; Williams and Parker, 1983; Eichelberger and Williams, 1978; McCluskey, 1984; Bennetts, 1984) that allows smaller blocks inside an IC to be individually tested. This circuitry is used to divide the IC into either registers or strictly *acyclic* logic (sometimes called *combinatorial* logic) having no memory. These two different types of logic circuitry can be tested individually; furthermore, the number of test vectors required is greatly reduced.

The testing is accomplished by operating the circuits in two different modes: regular operation, sometimes called *data mode*, or *test mode*. In *test mode*, all the internal registers of an IC are configured as shift registers. These can be then tested by shifting appropriate codes through them. While this is being done, they are also used as an input path through which test vectors are shifted into the internal sections of an IC. After test vectors are shifted in, then the outputs of the acyclic logic portions of the IC are captured by latching them into the registers. These logic outputs are then shifted serially out of the IC and compared to known good vectors.

By using the on-chip registers configured as shift registers to input test vectors to the on-chip acyclic logic, the number of additional pins required for testing is minimized. The major limitation of this technique is the large time required to shift in a single set of test vectors and the time to shift out the outputs of the acyclic logic circuitry.

Level-Sensitive Scan Design

A possible system diagram of a sequential logic circuit that uses scan-design techniques is shown in Fig. 13.1 where the two different latches used in the reconfigurable D flip-flops are shown separately (Williams and Parker, 1983). This particular implementation was invented by researchers from IBM in the early 1970s (Eichler, 1974; Eichelberger and Williams, 1978) and has been used extensively by IBM ever since then, particularly for gate arrays. It was called *level-sensitive scan design (LSSD)*.

In LSSD, the master latches have two clocks; the slave latches are operated using a third clock. A possible implementation is shown at the logic level in Fig. 13.2. This is basically a two-input D flip-flop. If nonoverlapping clocks Clk_1 and Clk_3 are used, then D_1 is the input. Alternatively, Clk_1 can be held at a "0", while nonoverlapping clocks Clk_2 and Clk_3 are toggled, then D_2 is the input. In an alternative realization, a single clock might be used for the master latch with a multiplexor added to select between two possible inputs, D_1 and D_2. In this latter configuration, it is necessary that the multiplexor switch states only when the master clock is low, assuming cross-coupled *nor* gates are used to realize the latches.

The procedure for testing using LSSD registers proceeds as follows with respect to Fig. 13.1.

1. First, the shift registers are tested by using the flip-flops in shift-register mode. This is achieved by keeping Clk_1 at a "0" and applying two *nonoverlapping* clock signals to clock inputs Clk_2 and Clk_3. This serially shifts in the input signals that are sequentially input to the signal Scan-In. Some example input sequences that might be used include all "0"s, all "1"s, and/or a sequence of "1"s and "0"s, such as "10110010110010 . . .," or

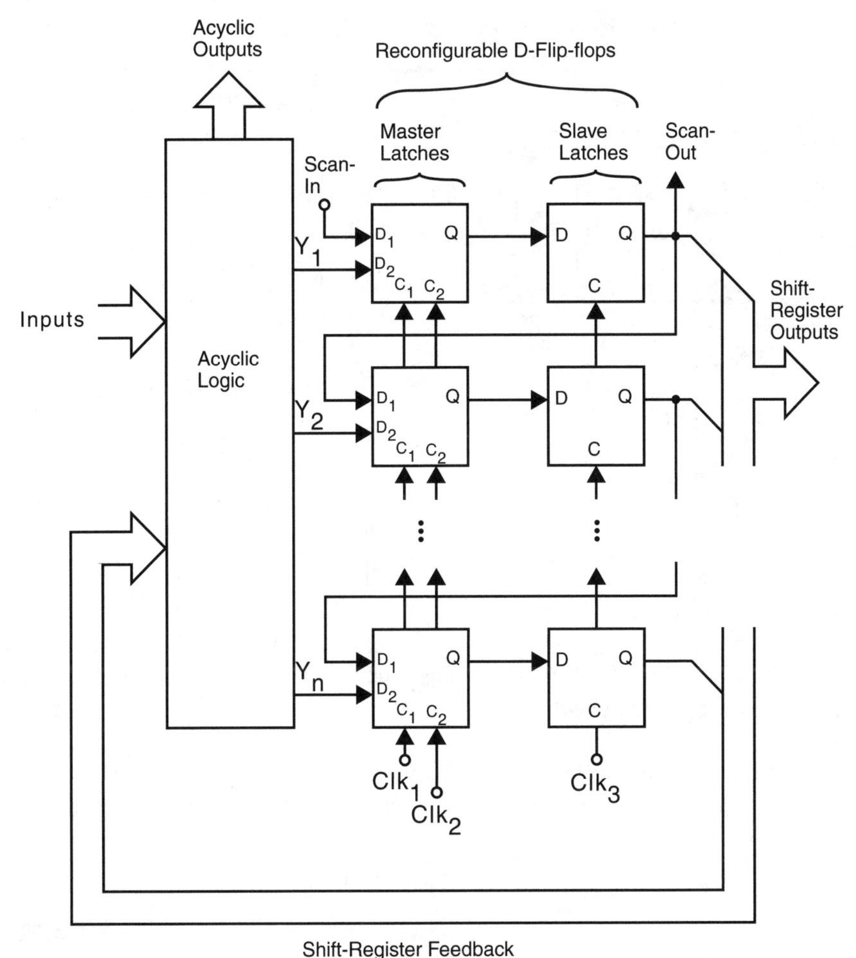

Figure 13.1 A system diagram showing the use of level-sensitive scan-design techniques (Bennetts, 1984).

perhaps a nearly random but known sequence. After the flip-flops have been found functional, it is time to test the *acyclic logic*.

2. The first test vector is shifted in serially, from Scan-In, again using nonoverlapping clocks Clk_2 and Clk_3, while Clk_1 is held low. This step may have (more likely, would have) occurred at the same time the shift registers were tested. The first test vector for the *acyclic logic* is shifted in until the required bits are stored in the *slave latches,* which are connected to the inputs of the acyclic logic.
3. The outputs of the *acyclic logic* are strobed into the master latches by holding clocks Clk_2 and Clk_3 low, and strobing Clk_1 high and then back low again.

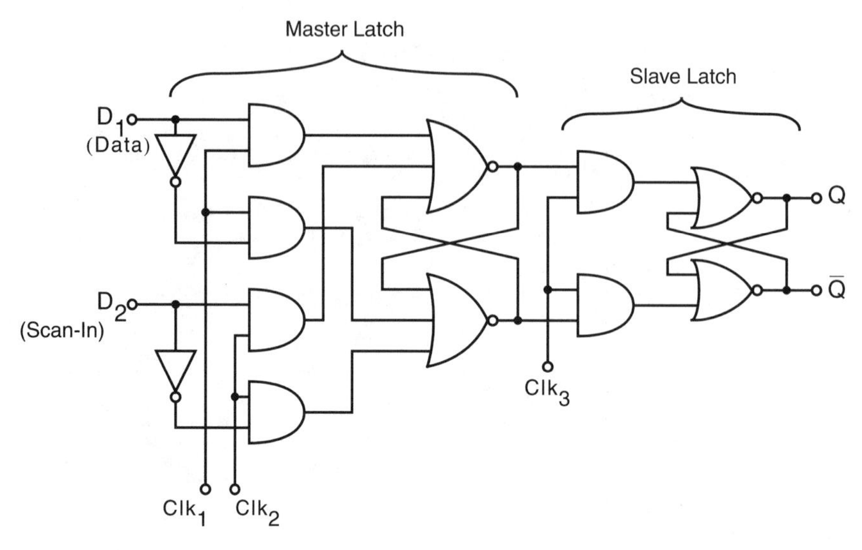

Figure 13.2 A possible realization of an LSSD reconfigurable D flip-flop.

4. The test results from the *master latches* are serially shifted off chip through Scan-Out using nonoverlapping clocks Clk_2 and Clk_3, while Clk_1 is held low. They are compared off-chip to the expected results.

This completes the testing for the first acyclic test vector. This process is repeated numerous times until the acyclic logic has been adequately exercised. Obviously, this can take some time.

LEVEL-SENSITIVE SCAN DESIGN IMPLEMENTATION ISSUES

There are a number of important issues that must be dealt with for the successful implementation of LSSD techniques.

Level-Sensitive Design. First, the state of the logic system must be a function of input logic levels only, independent of gate delays, rise and fall times, and skews between signals. This will be the case for the system diagram of Fig. 13.1, if all clocks are nonoverlapping and have frequencies low enough so that all circuits have settled in one period.

Glitch-Free Clock-Gating. It is important that when clocks are stopped, shorter pulses are never encountered. A possible *gating* circuit that guarantees this is shown in Fig. 13.3. During normal operation, $\overline{\text{Disable}}$ will be a "0". This effectively disconnects *nor* gate 2 from the circuit and *nor* gate 1 acts as an inverter. When the clock is to be turned off, $\overline{\text{Disable}}$ goes to a "1". If this occurs when the signal Clk is a "0", the output of *nor* gate 2, which was a "0", will immediately go to a "1" which will keep Clk low from then on. However, if $\overline{\text{Disable}}$ goes to a "0" when Clk is a "1", the output of *nor* gate 2, which is a "0", will not go to a "1" until Clk goes to a "0". This effectively disables the clock gating until the end of a period of the Clk signal.

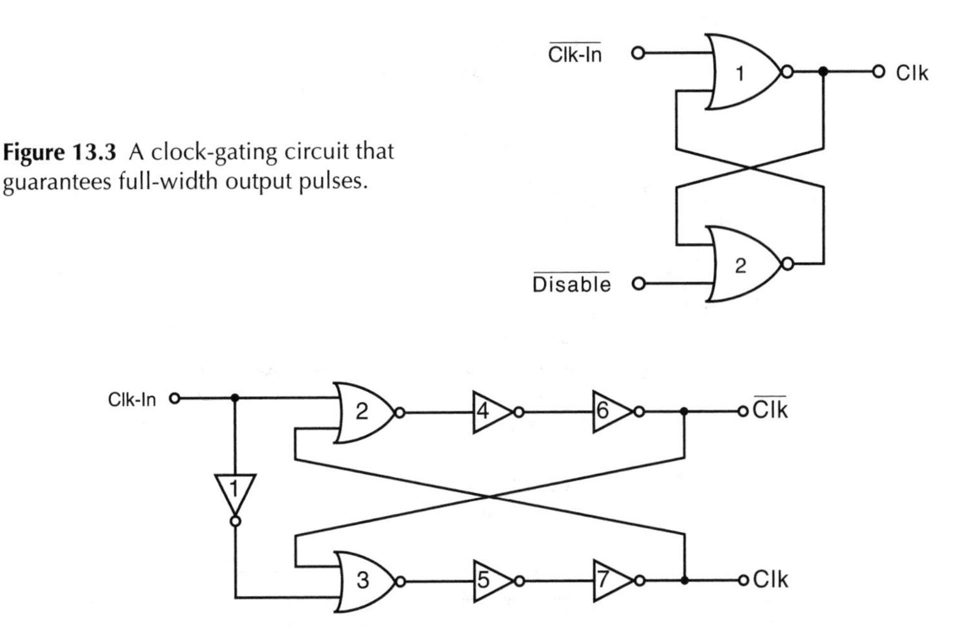

Figure 13.3 A clock-gating circuit that guarantees full-width output pulses.

Figure 13.4 A nonoverlapping clock generation circuit.

Nonoverlapping Clocks. A closely related issue is that all clocks should be *nonoverlapping*. A typical circuit that is often used to generate two nonoverlapping clock waveforms was described in Chapter 7, and is shown again in Fig. 13.4. This circuit guarantees that when Clk-In changes value, first the clock output that was high comes low, and only then will the other output go high. The nonoverlap time is equal to the delay of a *nor* gate added to the delay of two cascaded inverters. Also, if the threshold voltage of the first inverter is lower than that of the *nor* gates, the circuit will have hysteresis and will generate clean output waveforms even when the input waveform has slow rise and fall times. The inverters at the outputs of the *nor* gates also play the role of output buffers; the widths of the transistors in them should be increased the closer they are to the outputs.

A possible circuit that incorporates both the glitch-free clock gating and the nonoverlapping clocks is shown in Fig. 13.5. In this circuit, if Slct is a "1", Clk_1 is enabled, whereas if Slct is a "0", Clk_2 is enabled. When Clk_1 is enabled, cross-coupled *nor* gates 1 and 5 are used to generate the nonoverlapping clocks Clk_1 and Clk_3. When Slct is low, cross-coupled *nor* gates 3 and 5 are used to generate nonoverlapping clocks Clk_2 and Clk_3. Although the circuitry looks large, most of the area is taken by the inverters used to buffer the outputs of the *nor* gates, which are required to drive the long clock lines. Also, the circuitry need only be included once per IC. Furthermore, very efficient CMOS implementations are possible with few p-channel transistors required when cross-coupled differential latches are used.

Clock-Waveform Distribution. Another related issue is the distribution of the clock waveforms, as was mentioned in the previous section. Notice that in the system of Fig. 13.1, the

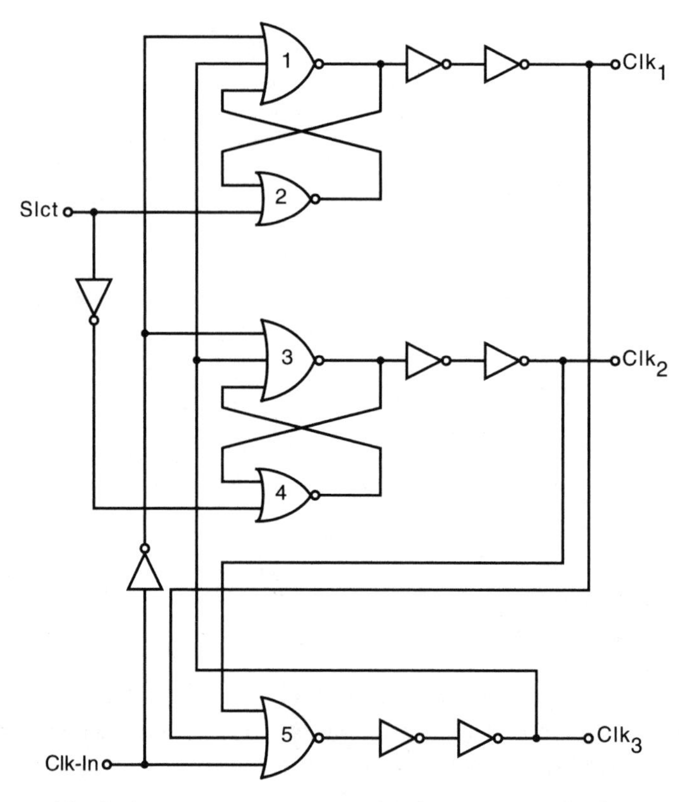

Figure 13.5 A possible clock-generation circuit suitable for generating the three clocks required in LSSD circuits.

clock waveforms are routed from the Scan-Out flip-flop to the Scan-In flip-flop. That is the clock waveforms are routed in the direction opposite to the direction of signal propagation. This makes the circuitry less sensitive to clock delays and skew due to unequal loading as it is distributed along the shift register from output to input. This routing is different than that shown in Fig. 12 of Williams and Parker (1983).

Example 13.1

Design an efficient CMOS realization of the LSSD clock-generation circuit of Fig. 13.5.

Solution: A possible implementation is shown is Fig. 13.6. Note that only a single p-channel transistor is required for each gate output despite the fact that the circuit is fully static, although it is a ratioed-logic circuit. Further note that the disable circuitry

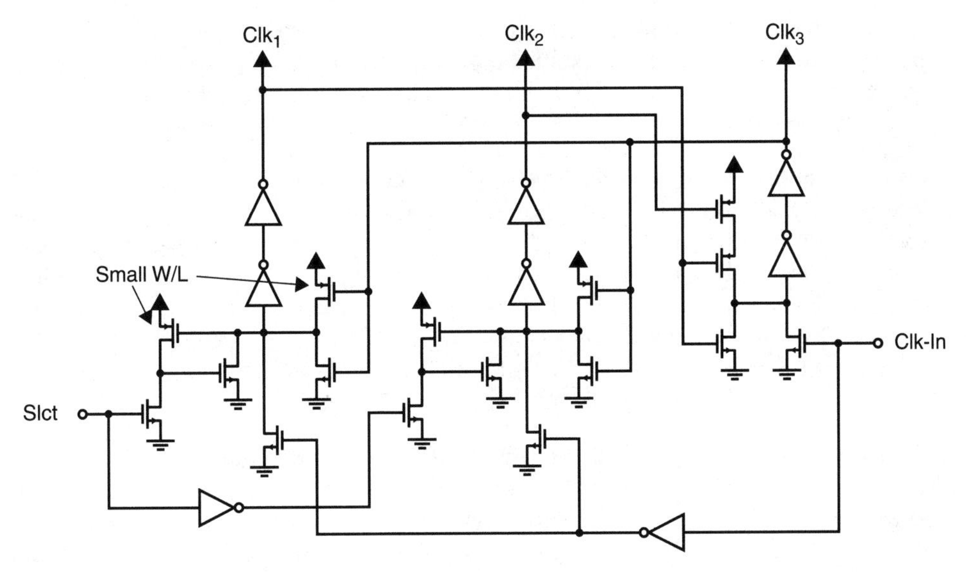

Figure 13.6 An efficient CMOS realization of the LSSD clock-generation circuit of Fig. 13.5.

requires only an extra three transistors per output, excluding the inverter connected to the Slct signal. This circuit is intended for use in LSSD testing, but it also illustrates many useful principles for realizing clock-generation circuits for other testing techniques to be described later.

Testing Time. Perhaps the major shortcoming of LSSD design techniques is the large number of clock cycles required to shift in each test vector and shift out the results of a single test vector. Despite this fact, it has been used extensively by companies such as IBM for medium to small complexity ICs. For larger ICs, it cannot be used without modification due to the shifting time, however scan-path techniques are now used, along with other techniques, in almost every IC that has been designed with built-in self-test.

Related Test-Design Techniques

Scan/Set Testing. There are a number of test-design techniques that have been proposed that are similar to or use techniques just described for scan design. For example, Funatsu et al. (1975) describe a technique called *scan-path* design that was developed at the same time or slightly before LSSD techniques, but uses a different, slightly less robust, clocking scheme. Another related technique has been denoted the *Scan/Set* technique (Stewart, 1977). In this technique, the shift-register flip-flops used for shifting in test vectors or shifting out test-result vectors are separate from the shift registers actually used during regular system operation. In addition to the shift registers, multiplexors are required that select system inputs from either the regular inputs or from the additional test-shift registers. An advantage of *Scan/Set* techniques is that output vectors can be stored and shifted out during regular system operation. A disadvantage is that often additional circuitry is required as compared to LSSD techniques if good fault coverage is desired.

Random Access Test. Another related technique, called *random-access scan* (Ando, 1980), allows each flip-flop to be individually addressed, both for loading in test vectors and reading test results, in a manner similar to that used for accessing the memory locations of a static RAM, as described in Chapter 11. This technique has a larger overhead than LSSD, which is required for the decoders and additional address lines going to every flip-flop. It does have a major advantage that it is much more useful during the debugging phase of IC design. Also, it considerably cuts down on the time required for shifting in test vectors and shifting out test results when only a small part of an IC is being tested.

Random-Access LSSD. In yet another related design-for-test technique, a number of separate LSSD paths are included in a single IC. These can be individually accessed using a decoder. This greatly shortens the time for serially shifting in test vectors and shifting out the test results when only a part of the IC need be tested. In some implementations, the input shift registers can be loaded or read in parallel at the expense of additional pins needed for testing. *Most modern large ICs, that use LSSD techniques would use some form of this technique to reduce the shifting-time overhead during debugging.*

13.3 Localized Test-Vector Generation and Test-Output Compression Techniques

A number of techniques have been developed that are designed to reduce the time required for shifting in test vectors and shifting out test results without greatly increasing the overhead of additional circuitry and/or interconnect. Most of these techniques are based on generating test vectors locally. In addition, the test results for a large number of clock cycles are often compressed into a single *signature* vector that can then be serially shifted out in much less time. When both on-chip test-vector generation and output-vector compression are included, the IC is often said to include *built-in self-test (BIST)*.

The precursor of most of these techniques was the *signature-analysis* technique, originally developed by Hewlett Packard, for board testing (Nadig, 1977). In this technique, the contents of a register are modified by a large number of sequential samples of a circuit output node (or nodes). Ideally, the resulting signature stored in the register will be different for correct and faulty circuits. Before describing the details of the test-vector generation techniques and the related signature-analysis techniques, it is first necessary to discuss the basis for most of the currently popular techniques, namely maximal-length linear-feedback shift registers.

Maximal-Length Linear-Feedback Shift Registers

A *maximal-length linear-feedback shift register (LFSR)* is a logic circuit that generates *pseudo-random* digital words. These shift registers are often denoted maximal-length LFSRs or some-

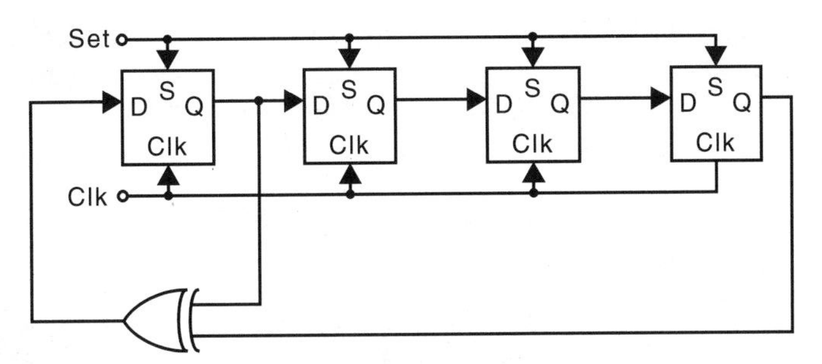

Figure 13.7 A four-stage maximal-length LFSR suitable for generating pseudo-random digital words.

times more simply just LSFRs. An example of a four-bit LFSR is shown in Fig. 13.7. The input to an LFSR is the output of a multi-input *exclusive-or* circuit, which is often, but not necessarily, implemented by a cascade of two-input *exclusive-or* circuits. The inputs to the *exclusive-or* circuits are selected outputs of the D flip-flops in the shift register. If appropriate shift-register outputs are first *exclusive-ored* and then fed back, and if the shift register starts in a state that is not all "0"s, then it can be shown (Bardell et al, 1987) that the sequence generated has a number of desirable properties. First, if the shift register has n stages, the sequence will go through $2^n–1$ distinct states. For a four-bit shift register, this would imply 15 states; for a 16-bit shift register, this would imply 65,535 states. Thus, the states are said to be of *maximal length.* Furthermore, the output of any particular flip-flop is a *pseudo-random sequence.* It will have $2^{n–1}$ "0"s and $(2^{n–1}$-1) "1"s randomly distributed. The required inputs to the *exclusive-or* gates are found by applying *finite-field theory* to the *residue classes modulo g*, where *g* is a primitive polynomial (Wang, 1989). The coefficients of *g* are represented by the feedback connections from the shift-register outputs to the *exclusive-or* gates. Wherever a connection exists, it represents a "1" coefficient, whereas no connection represents a "0". The *exclusive-or* gates represent additions in this particular *finite field.* The states of the shift register represent the coefficients of a second polynomial that keeps changing every clock period. Assuming *g* is *primitive* (see Wang, 1989), it can be shown that polynomials represented by the stored coefficients will rotate through $2^n–1$ distinct states. Luckily, it is not necessary for us to show this in order to design a maximal-length LFSR; indeed, it is not really necessary to know anything about finite-field *theory*; rather, we need only look up a table in which the coefficients of primitive polynomials of a binary number base have been listed (Forney, 1970). However, if it is necessary to accurately predict the probability of detecting faulty circuits using *built-in self-test* techniques that make use of LFSRs, then at least a minimal understanding of finite-field theory is necessary. Furthermore, it is also the basis for many other digital systems such as error-control coding of digital systems (Rao and Fujiwara, 1989; Viterbi and Omura, 1979), spread-spectrum communication systems (Lee and Messerschmitt, 1994), and encryption systems, and therefore is an important field to study for many digital-system designers. Unfortunately,[1] this

[1]This is an area that the author has always found fascinating, but unfortunately has never had time to investigate.

Table 13.1 Suitable Shift-Register Connections for Generating Maximal-Length LFSRs

N	Flip-flop outputs	N	Flip-flop outputs	N	Flip-flop outputs
2	1.2	13	1,3,4,13	24	1,2,7,24
3	1,3	14	1,6,10,14	25	3,25
4	1,4	15	1,15	26	1,2,6,26
5	2,5	16	1,3,12,16	27	1,2,5,27
6	1,6	17	3,17	28	3,28
7	1,7	18	7,18	29	2,29
8	2,3,4,8	19	1,2,5,19	30	1,2,23,30
9	4,9	20	3,20	31	3,31
10	3,10	21	2,21	32	1,2,22,32
11	2,11	22	1,22	33	13,33
12	1,4,6,12	23	5,23	34	1,2,27,34

subject is well outside the scope of this book, and the interested reader is referred to Rao and Fujiwara (1989) and Viterbi and Omura (1979) for further study.

LFSR's are often used to generate pseudo-random test vectors on chip. The outputs of each stage of an LFSR are used as an input to the circuit. The circuit is then clocked for a large number of cycles while the outputs are being monitored. After many clock cycles, if the outputs have all occurred as expected, the circuit is deemed operational; if not, it is deemed faulty.

An alternative configuration for maximal-length shift registers is to replace the *exclusive-or* gates by *exclusive-nor* gates. This results in a maximal-length LFSR as long as the shift register is initialized to any state other than all "1"s. This is easy to guarantee by resetting the flip-flops at system start-up as opposed to setting them for the former case. The complexity is about the same in both cases. Table 13.1 shows suitable feedback connections for realizing maximal-length LFSRs having sizes between 2 and 34 flip-flops (Forney, 1970). The ordering is from the input to the output [as opposed to the ordering in Forney (1970) which is from the output to the input]; also, the index of the first flip-flop is assumed to be "1".

Example 13.2

Design an eight-bit maximal-length LFSR to be used for generating eight-bit test vectors.

Solution: From Table 13.1, we see that the outputs from flip-flops 2, 3, 4, and 8 are to be used as inputs to the *exclusive-or* gates. A possible realization is shown in Fig. 13.8, where *exclusive-nor* gates are used, which means the shift register should be started from the state where all the flip-flop outputs are "0"s.The realization is shown where two input gates have been used. An alternative realization

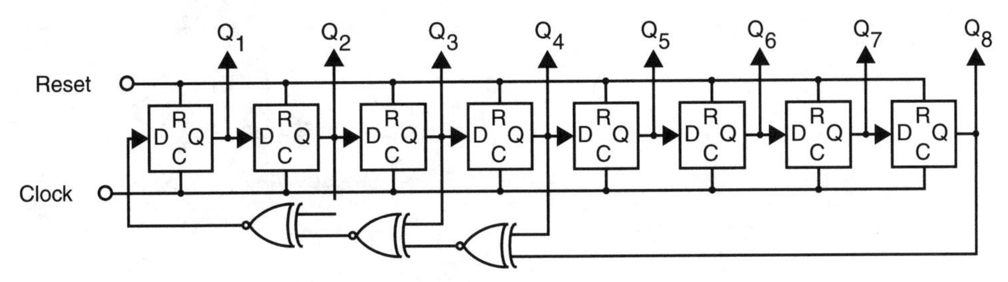

Figure 13.8 An eight-bit maximal-length LFSR based on using *exclusive-nor* gates in the feedback path.

could use a single four-input *exclusive-nor* gate. As an aside, when high-speed operation is desired, one might consider implementing a nine-bit LFSR since this would only require a single two-input *exclusive-nor* gate, which would have much less delay than three two-input gates as shown. In this case, only eight of the flip-flop outputs would be used as the test-vector inputs. Slightly more area would be required for the extra flip-flop, but this would often be a reasonable trade-off given the higher clock rate that would be possible.

Signature Analysis

A related built-in self-test technique is that of *signature analysis*. In its original form, it was used to check the signature of a single node of a circuit board for a large number of clock cycles at a time (Nadig, 1977). The circuit used to do this was a maximal-length LFSR with an additional *exclusive-or* gate added to modify the input to the shift register as is shown in Fig. 13.9, for the four-bit case. An additional *exclusive-or* gate has been included in front of the input to the shift register. One input to this gate is the output from the *exclusive-or* feedback network. The other input is the test input from the external circuit. Every time the shift register is clocked, its input will be dependent not only on the fed back signal but also on the test-input signal. Its final state will therefore be dependent on the sequence of test inputs being sampled every clock period.

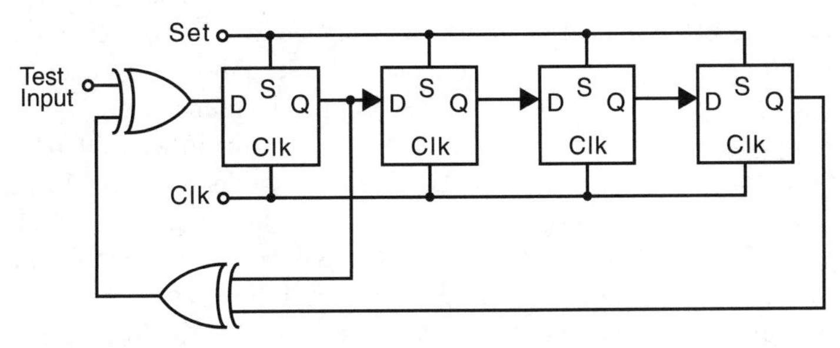

Figure 13.9 A four-stage single-input signature-analysis circuit.

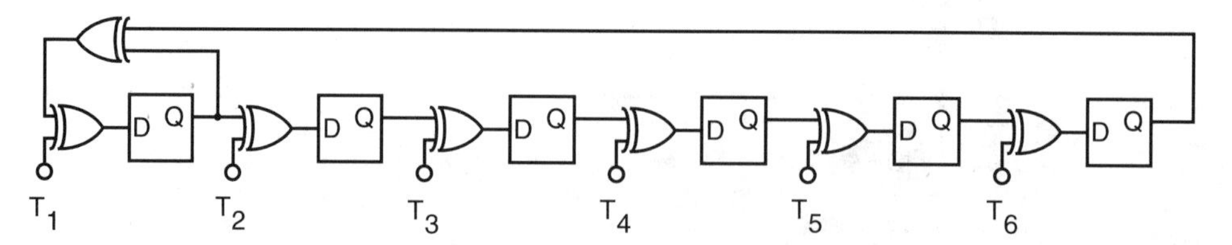

Figure 13.10 A six-stage six-input signature-analysis circuit.

Ideally, if these test inputs are not correct, the LFSR will not end in the same state as it would when testing a functionally correct circuit.

The technique can be extended to produce a signature for N parallel inputs, when using N-bit registers by including *exclusive-or* (or *exclusive-nor*) gates in front of each flip-flop, as is shown in Fig. 13.10 for the case of N equal to six. The necessary set and clock inputs have not been shown for simplicity. Now, the input to each flip-flop is dependent not only on the output of the previous flip-flop, but also on the corresponding bit of the parallel output word being compressed.

It has been shown (Bhavsar and Heckelman, 1981; Carter, 1982) that for large LFSRs that are clocked a large number of times, the probability of a functionally incorrect circuit producing the same signature as a correct circuit is approximately given by $1/(2^N)$ where N is the number of flip-flops.

After the LFSR has been clocked a large number of times, the signature produced must then be compared with the expected result. To do this, the LFSR might be reconfigured into an ordinary shift register whose state is then shifted out serially. Alternatively, an ordinary parallel shift register might be included. After the signature has been produced, it can be latched into the parallel shift register and shifted out serially. The former procedure was proposed in Konemann et al. (1979) where a reconfigurable shift register was described that could either be used to produce random inputs and produce signatures of test outputs, or be used for LSSD testing or during regular system operation. This technique was called built-in logic-block observation and is described in the next section.

Built-in Logic-Block Observation

Built-in logic-block observation (BILBO), combines LSSD test techniques with maximal-length LFSR input-vector generation and multiple-input signature analysis. This method will be described using a four-bit shift register as a simple example, although it should be remembered that during actual use, lengths of at least eight bits or more and preferably 16 bits or more should be used. A four-bit BILBO register is shown in Fig. 13.11. This implementation is not exactly the same as that described in Konemann et al. (1979), but works based on the same principles and is more easily implemented using complex fully-differential logic gates. Depending on the control inputs, C_1, C_2, and C_3, the BILBO register can be configured as a maximal-length LFSR, as a parallel-input signature-analysis register, as a shift register for shifting in test inputs or shifting out compressed test outputs, or as latches to be used during regular system operation of a state machine. It can also be set to the all-"0"state, which is useful for initializing the LFSR

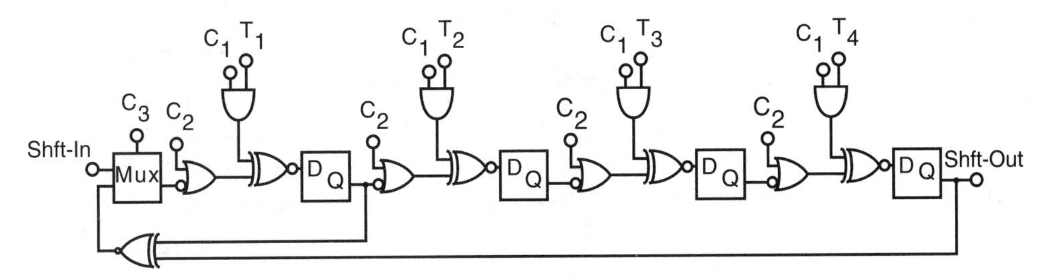

Figure 13.11 A built-in logic-block observation register.

given that *exclusive-nor* gates are used in the feedback path. The various different configurations will be individually described.

During regular system operation, C_1 is a "1" and C_2 is a "1". The value for C_3 is immaterial. By C_2 being a "1", the *or* gate outputs are guaranteed to be all "1"s. This disables the shift register operation. Since C_1 is a "1", the outputs of the *and* gates, which are fed into the *exclusive-nor* gates, will all be equal to their T_i inputs. The other inputs to the *exclusive-nor* gates, from the *or* gates, are all "1"s. Therefore, the outputs of the *exclusive-nor* gates are all equal to the T_i inputs, and the shift register acts as a parallel latch.

If C_1 is a "0", then the outputs of the *and* gates are all guaranteed to be zero. This disables the parallel inputs. If C_2 is also a "1", then the outputs of the *or* gates are all "1"s and the outputs of the *exclusive-nor* gates are therefore all "0"s. This can be used to load "0"s into all the flip-flops, which can be used to prepare the register for operation as an LFSR.

Alternatively, if C_2 is a "0", with C_1 still a "0", then the two inputs to each *exclusive-nor* gates will be a "0" and the inverted output of the previous latch in the shift register. Thus, the outputs of the *exclusive-nor* gates will be equal to the outputs of the previous latches in the shift register. This mode is used both for shift-register mode and for generating *maximal-length* LFSRs used for generating test vectors. The first mode will be the case if C_3 is a "1", in which case the output of the multiplexor is equal to Shft-In; the second mode will be the case if C_3 is a "0", in which case the output of the multiplexor will come from the *exclusive-nor* feedback network.

The final mode is used for *signature-analysis* compression. In this mode, C_1 is a "1", C_2 is a "0", and C_3 is also a "0". The outputs of the ith *exclusive-nor* gate will now be equal to $Q_{i-1} \oplus T_i$. This is slightly different than the signature analysis of Fig. 13.10 in that now *exclusive-nor* gates are used in the feedback, whereas the equivalent of *exclusive-or* gates are used for the parallel inputs, but the difference is immaterial.

The different modes are summarized in Table 13.2. The d values in the table denote do not care situations. Simplified circuits for the different cases are shown in Fig. 13.12a through e.

The use of BILBO registers for BIST is quite straightforward. The BILBO registers are placed between acyclic-logic blocks. A typical configuration is shown in Fig. 13.13. In this case, two acyclic-logic networks are separated by BILBO registers. During regular system operation, the registers are configured in *latch mode* and the whole system operates as a state machine. During *BIST mode*, the acyclic networks are tested in turn. First, acyclic network #1 is tested. The first step is to load "0"s into BILBO register #1 by setting C_1=0 and C_2=0 and clocking the registers once (this could alternatively be accomplished by adding asynchronous

Table 13.2 The Different Modes of a BILBO Register

C_1	C_2	C_3	*and*-Gate output	*or*-Gate output	*exclusive-nor*-Gate output	Mode
0	0	0	0	Q_{i-1}	Q_{i-1}	Shift-register mode
0	0	1	0	$\overline{Q}_{i-1}$	Q_{i-1}	Maximal-length LFSR
0	1	d	1	0	0	Load-"0"s mode
1	0	0	T_i	$\overline{Q}_{i-1}$	$Q_{i-1} \oplus T_i$	Signature-analysis mode
1	0	1	T_i	$\overline{Q}_{i-1}$	$Q_{i-1} \oplus T_i$	Not used
1	1	d	T_i	1	T_i	Parallel-latch mode

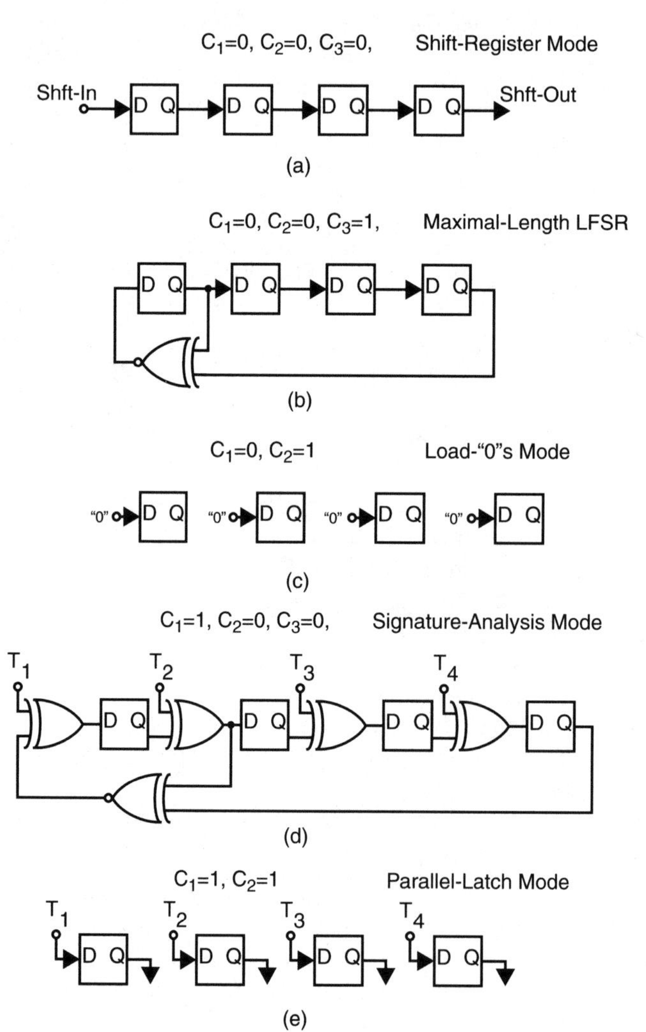

Figure 13.12 The various modes of a BILBO shift register from Table 13.2.

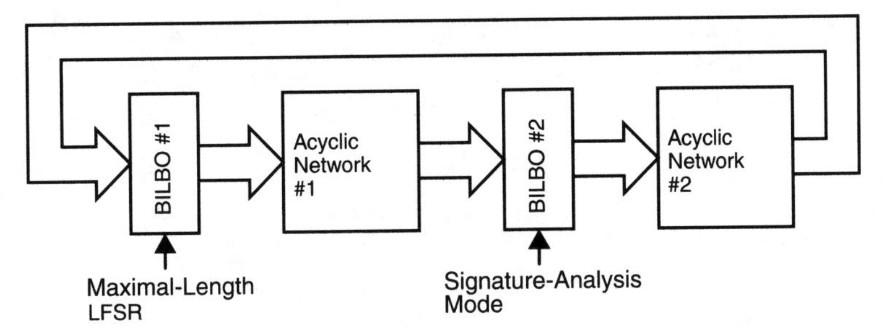

Figure 13.13 A typical configuration when using BILBO for BIST.

resets to the latches). Next, BILBO #1 is configured as a maximal-length LFSR (C_1=0, C_2=0, and C_3=1) and BILBO #2 is configured as a signature-analysis network (C_1=1, C_2=0, and C_3=0). The system is then clocked for a large number of cycles, on the order of a couple of hundred. The next step is to configure BILBO #2 as a shift register (C_1=0, C_2=0, and C_3=0) and the compressed test results are shifted out. The results are then compared with the known correct results.

After testing acyclic network #1, acyclic network #2 is tested. BILBO #2 is first reset and then configured as a maximal-length LFSR. BILBO #1 is now configured in signature-analysis mode, the system is clocked a couple of hundred times, and then the contents of BILBO #1 are shifted out and compared to the known correct results.

In the event that one of the acyclic networks is found to be nonfunctional, the BILBO registers can be configured as ordinary shift-registers and LSSD-like testing can be used to try and debug the circuits. Also, before BIST is begun, the BILBO registers themselves can be partially tested by configuring them as shift registers and shifting through a number of "1"s followed periodically by a number of "0"s.

Example 13.3

Design a CMOS fully differential realization of the logic required in front of a typical flip-flop of a BILBO register (i.e., the *and* gate, the *or* gate, and the *exclusive-nor* gate of Fig. 13.11). Assume differential inputs are available.

Solution: A fully differential realization of the logic is shown in Fig. 13.14. This realization is a single compound gate, which is possible because only noninverting gates feed into the *exclusive-nor* gate of Fig. 13.12. This realization was designed by combining a differential *or* gate with a differential *and* gate with an *exclusive-nor* gate. It is a good example of how complex functions can sometimes be very efficiently realized using differential logic.

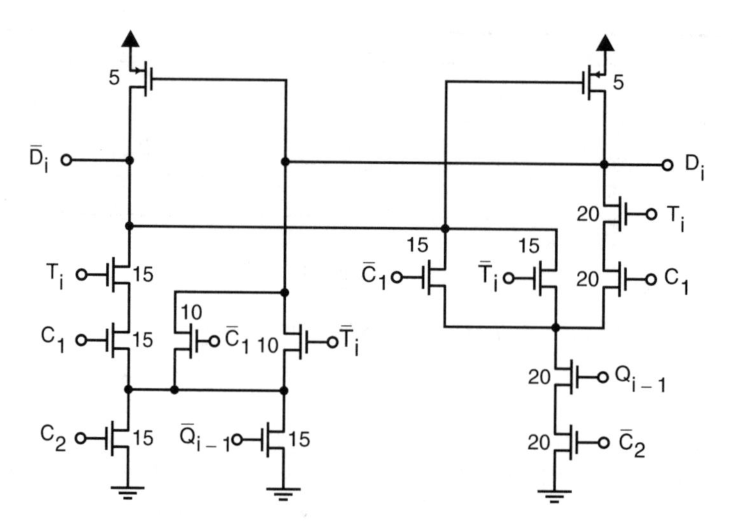

Figure 13.14 A possible realization of the logic used in a BILBO register.

13.4 Boundary-Scan Testing

Another very important recent development in IC testing and board-level testing involves some standardized techniques and architectures for testing digital systems that are commonly called *boundary-scan* testing. This *IEEE/ANSI Standard* 1149.1-1990 (IEEE, 1990) is called the *IEEE Standard Test Access Port and Boundary-Scan Architecture.* It was developed by the *Joint Test-Action Group (JTAG)* between 1985 and 1990 in response to testing problems brought about by the increasing use of surface-mount technology and very large-scale integrated circuits. For this reason, it is also often referred to as the *JTAG 1149.1 Standard.* There is a possibility that this development will eventually revolutionize the testing of digital integrated circuits, and is thus an area that digital IC designers should be familiar with.

Standard 1149.1 describes an architecture for some specialized circuitry that would be included in a digital IC. It greatly enhances the testability of not just the IC, but a complete board as well. The circuitry includes a simple controller, a number of registers, and an instruction set that when included in a digital system allows the system to be broken into separate modules that can be individually tested. Including this test system in an IC requires the use of only four (or optionally five) pins, does not take up a very large area, and greatly enhances the testability of an IC. The additional circuitry is primarily intended to isolate the IC from the rest of the system and to allow control of the inputs and outputs of the system logic (i.e., the logic that is separate from the additional testing logic). It is not intended to replace BIST, rather, it is expected not only to make it easier to isolate the IC for LSSD-like testing, but also to allow automatic test equipment (ATE) to more easily interface with BIST circuitry.

At present, a number of ICs are already on the market that include this standardized testing circuitry; it is possible that in the future a majority of digital ICs will include it. Also, many modern testers being developed have specialized hardware and software specifically included to make use of 1149.1 testing circuitry. For these reasons the importance of digital designers becoming familiar with boundary-scan techniques can not be overemphasized. Unfortunately, due to size constraints, it is not possible to completely describe boundary-scan testing in this section; rather, an overview will be presented. IC designers are strongly encouraged to also consult some of the excellent references that describe boundary-scan testing for a more complete treatment (Maunder and Tullos, 1990; Maunder, 1991; Parker, 1992).

Boundary-Scan Architecture

Figure 13.15, copied from Fig. 1.3 of Parker (1987), shows a simplified diagram of an IC that includes boundary-scan testing. The additional circuitry that must be included to allow for boundary-scan testing includes a *test-access-port (TAP) controller* and a number of registers. The additional registers must include (but are not limited to) an instruction register, a bypass register, and at least one boundary-scan data register. Additional registers can be included optionally. The additional pins are used for control and for serially shifting in and out test data. These four pins are the *test clock* (*TCK*), the *test-mode select input (TMS)*, the *serial test-data input (TDI)*, and the *serial test-data output (TDO)*. Also, a fifth pin can be optionally included to allow for synchronous reset of the test circuitry (*TRST*).

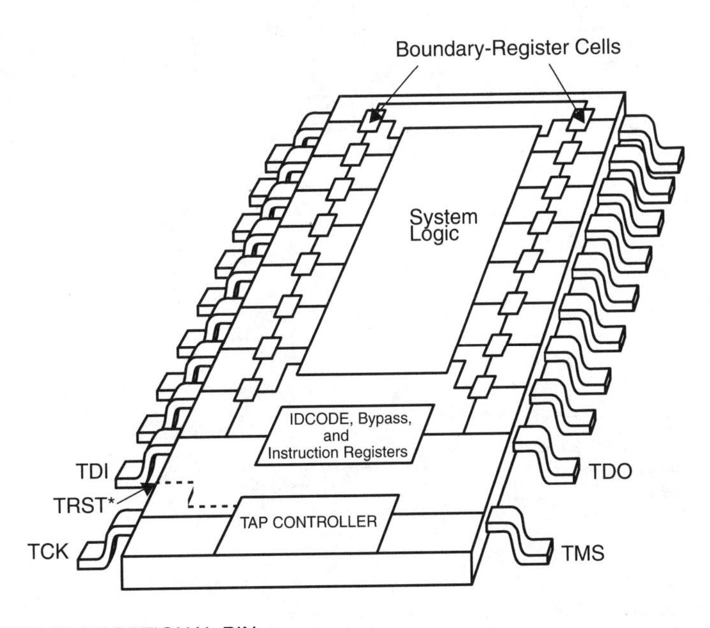

Figure 13.15 A simplified architecture of an 1149.1-compliant integrated circuit. From Parker (1987).

All of the registers can be connected between the serial input and output pins (TDI and TDO). The different testing modes are determined by the logic value on the test-mode select (TMS) pin, the contents of the instruction register, and the state of the TAP controller. During normal system operation, TMS will be held high and all of the boundary registers are transparent. When TMS is set low, testing is initiated.

The controller is a simple 16-state machine that allows instructions to be serially loaded into the instruction register, and also allows testing data to be serially loaded into, or out of, the boundary-scan registers. It also allows data to be shifted into the IC, shifted quickly through the single-cell bypass register, and out of the IC into another IC, thereby allowing a number of ICs to be *daisy-chained* together to allow easy access to an IC further along the chain. Finally, it allows data to be serially shifted into, or out of, optionally included data registers, typically used for testing and/or debugging the IC.

An IC with boundary-scan testing will typically have a boundary-scan register cell placed at every IO pin. Each cell will consist of two or more master–slave D flip-flops, and a number of two-input single-output multiplexors. An example of a boundary-scan cell that might be placed at an input or output pin of an IC is shown in Fig. 13.16.

When it is placed at an IC input pin, the off-chip signals from the input pad would be connected to the node PI. During typical system operation, Mode, a control signal coming from the TAP controller, will be a "0", and the input from off-chip, PI, will be connected directly to the on-chip system logic input, PO, through the multiplexor labeled X_2. The other multiplexor, X_1, is used to either route the off-chip input signals to flip-flop D_1, or to daisy-chain all D_1 flip-flops into a serial shift register for shifting in test vectors or shifting out test results. The latter case is used to determine what the inputs from off chip are. This is useful both for checking the interconnect between chips and for storing the outputs of other chips during testing that do not contain boundary-scan testing circuitry. After these signals are stored in the D_1 flip-flops, by setting the select input, SI (another output signal of the TAP controller, which controls the mul-

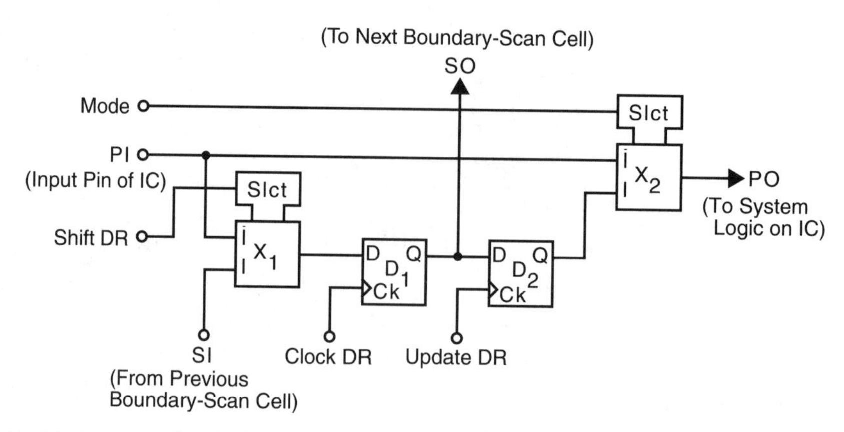

Figure 13.16 An example of a boundary-scan cell that might be placed at an input pin of an IC.

tiplexor X_1) to a "0" and clocking the ClockDR signal, then ShiftDR would be changed to a "1" and the results would be shifted off chip through the now serially connected D_1 flip-flops. The second flip-flop, D_2, is used to store test vectors when the internal logic is being tested. The test vectors would first be serially shifted into the D_1 flip-flops, from test-port input pin TDI, while ClockDR is used with ShiftDR set to a "1". After the shifting operation has finished, the test vectors are latched into the second flip-flop, D_2, by clocking the UpdateDR signal. During test, Mode would be a "1", which disconnects the signal path going from the IC pad to the on-chip system-logic input at PO, and instead connects the output of flip-flop D_2 to the input of the on-chip system logic through multiplexor X_2.

The same boundary-scan cell could be used at an output pin. In this case, an internal system-logic output would be connected to the node PI, and the output pin would be connected to the node PO. Again during regular system operation, Mode would be a "0", which connects PI to PO through multiplexor X_1 with only the additional delay of one multiplexor. When the internal logic of the IC is being tested, the selector of multiplexor X_1 (i.e., the ShiftDR signal) would be a "0". This allows the system-logic output to be stored in flip-flop D_1 on the rising edge of the clock-signal ClockDR. Next, ShiftDR would typically change to a "1", which again connects all the D_1 flip-flops into a serial shift register, which allows the results of the test to be serially shifted off chip through the TDO output pin of the test port. The second flip-flop is included to allow for the control of the values on the output pads, when testing off-chip logic or interconnects, irrespective of the output values of the on-chip logic. During this mode, the desired values would first be shifted into the IC, from the test-port input pin TDI, through the D_1 flip-flops. Once they had been shifted in, they would then be latched into the D_2 flip-flops by clocking the UpdateDR signal. Also, the Mode signal would be set to a "1", which places the value stored in flip-flop D_2 on the output pin, rather than the output of the on-chip system logic. This mode is intended primarily for the testing of interconnect between ICs with boundary scan, but can also be used for supplying test inputs to other ICs that do not have boundary-scan test circuitry added.

There are other possibilities for boundary-scan cells than that shown in Fig. 13.16. For example, every bidirectional pin might have two cells similar to that shown in Fig. 13.16. Other alternatives include very simple cells used for monitoring only without placing a multiplexor between I/O pins and the IC system, and more complicated cells for output pins that can be used to check for shorts at the output pins. The interested reader is referred to Parker (1992) for more details.

The on-chip controller must be able to decode a number of required instructions, and perhaps also decode a number of additional optional instructions. Some of the required instructions include an instruction used to sample or load the boundary-scan cells without affecting the values going to the rest of the system (SAMPLE/PRELOAD), an instruction (BYPASS) that places the bypass register between TDI and TDO, which is intended to allow for fast access to a particular IC, and an instruction used for testing the interconnect between ICs (EXTEST). Some of the optional instructions include an instruction intended to help test the system logic on the IC in an LSSD like manner (INTEST) while the output pin values are deterministically controlled, an instruction intended to allow for the easy testing of on-chip logic using built-in self-test (RUNBIST) with circuit techniques such as BILBO, while the output-pin values are deterministically controlled, instructions used to identify an IC (IDCODE and USERCODE), an instruction used to force all three-state pins to high impedance (HIGHZ), and an instruction that allows for controlling output pin values without incurring large scan times (CLAMP) each time the instruction is asserted, as is the case when EXTEST is asserted. The reader is again referred to Parker (1992) for many of the details concerning these instructions.

Boundary-Scan Testing Methodology

A simple example of a typical test using *boundary scan* might go as follows

1. Initialize the controller and the test-access port.
2. Load the instruction register with SAMPLE/PRELOAD.
3. Shift a test vector into the boundary-scan cells.
4. Load the instruction register with INTEST.
5. Capture the results of the test in boundary-scan cells.
6. Shift the test results out of the boundary-scan register.
7. Repeat steps 2 through 6 as many times as necessary.
8. Shift in a *safe* pattern to be stored in the boundary-cell output flip-flops (i.e., D_2 in Fig. 13.16).
9. Go back to the reset state.

Each of these steps involves going through a number of states of the controller. The progress through these states is determined by the value on test pin TMS each time the test clock, TCK, changes. For example, loading the instruction register involves having TMS low for one period, high for two periods, low for N+1 periods during shifting, where N is the number of cells in the instruction register, high for two periods, low for one period, and finally high for one period, before step 3 is started. The reason for the complicated procedure is primarily due to the decision to keep the number of pins dedicated to testing small. As an aid to determining the required signals, TAG has included a state-transition diagram, shown in Fig. 13.17, as a part of the standard.

The path just specified starts at Test-Logic-Reset, and then in order goes through Run-Test/Idle, Select-DR-Scan, Select-IR-Scan, and Capture-IR, before the instruction word is shifted in. Next, the state Shift-IR is occupied for N clock cycles while shifting in the instruction word, and then goes through the states Exit1-IR, Update-IR, and Select-DR-Scan before starting step 3. A similar procedure might be followed when scanning in a test vector, but would now go through the path with state labels finishing with DR. The actual data register placed between pins TDI and TDO at this time would be determined by the instruction word shifted in previously. Once again the reader is referred to Parker (1992) for additional details about the transition through the various states. Obviously, if boundary-scan testing is to be used effectively, some kind of external tester is needed that can automate the generation of the instruction words, test vectors, complicated signal generation, and verification of the test results, in a moderately high-level manner. Fortunately, this is becoming available with modern IC testers.

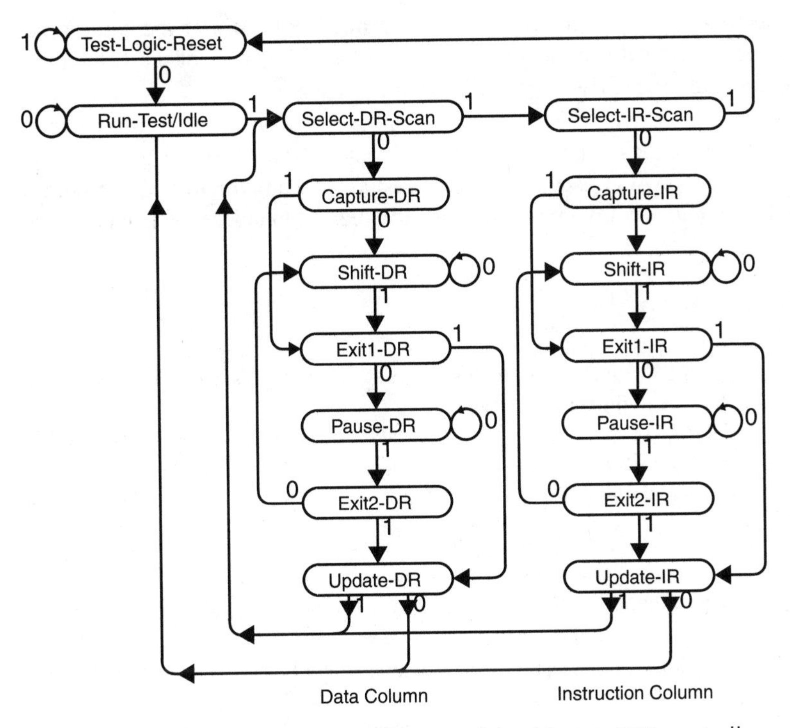

Figure 13.17 The state transition diagram of the 16-state TAP controller.

Another necessary requirement for the widespread use of boundary-scan testing is the easy generation of all the circuit building-blocks that need to be included in an IC, in a process independent manner. This is now possible using a combination of high-level hardware descriptions, such as that based on VHDL and described in Parker (1992), and process-independent layout languages.

As another example of using boundary-scan testing, consider its use in conjunction with optionally included BIST circuitry. A typical procedure might go as follows:

1. Initialize the controller to the TEST-LOGIC-RESET state.
2. Load the instruction register with the SAMPLE/PRELOAD instruction.
3. Shift in a *safe* pattern to be stored in the boundary-scan cells.
4. Shift in the RUNBIST instruction. This puts a BIST register between test pins TDI and TDO, as specified in the RUNBIST instruction.
5. Proceed to the RUN-TEST/IDLE state and remain there while the BIST is being performed.
6. Capture the results (preferably the compressed results) by proceeding through the path of the transition table that includes CAPTURE-DR and shift out the results.
7. Return to the TEST-LOGIC-RESET state.

It is possible to perform this procedure on a number of ICs at the same time, which can greatly decrease the total time required for board testing if most of the ICs on the board have been designed with boundary-scan and BIST included.

The foregoing is only a very brief description of boundary-scan testing techniques, and has been included to give the reader a feeling for the capabilities of these testing techniques, but is far from adequate to allow the user to design in boundary-scan circuitry. To design boundary-scan ICs the reader should consult the references, particularly Maunder (1991) and Parker (1992).

13.5 Bibliography

H. Ando, "Testing VLSI with Random-Access Scan," *Digest of Papers Compcon 80, IEEE Publication 80CH1491-OC,* 50–52, February 1980.

P. Bardell et al., *Built-In Test for VLSI—Pseudorandom Techniques,* John Wiley & Sons, 1987.

R.G. Bennetts, *Design of Testable Logic Circuits,* Addison-Wesley, 1984.

D.K. Bhavsar and R.W. Heckelman, "Self-Testing by Polynomial Division," *Proceedings of the 1981 IEEE Test Conference,* 208–216, 1981.

W.C. Carter, "Signature Testing with Guaranteed Bound for Fault Coverage," *Proceedings of the 1982 IEEE Test Conference,* 75–82, 1982.

E. Eichelberger and T. Williams, "A Logic-Design Structure for LSI Testability," *Journal of Design Automation Fault-Tolerant Comput.,* 2(2), 165–178, May 1978.

E.B. Eichler, "Method of Level-Sensitive Testing a Functional Logic System," U.S. Patent No. 3783254, January 1, 1974.

G.D. Forney, Jr., "Coding and Its Application in Space Communications," *IEEE Spectrum,* 7, 47–58, June 1970.

S. Funatsu et al., "Test-Generation Systems in Japan," *Proceedings of the 12th Design Automation Symposium,* 114–122, June 1975.

IEEE, "IEEE Standard Test Access Port and Boundary-Scan Architecture," *IEEE Standard 1149.1-1990,* IEEE Standards Board, 345 East 47th St., New York, NY 10017, May 1990.

B. Konemann et al. "Built-In Logic-Block Observation Techniques," *Proceedings of the 1979 IEEE Test Conference,* 37–41, 1979.

E. Lee and D. Messerscmitt, *Digital Communication,* 2nd ed., Kluwer, 1994.

C. Maunder, *The Board-Designer's Guide to Testable Logic Circuits,* Addison-Wesley, 1991.

C. Maunder and R. Tullos, The Test-Access Port and Boundary-Scan Architecture, IEEE Computer-Society Press, 1990.

E. McCluskey, "A Survey of Design for Testability Scan Techniques," *VLSI Design,* December 1984.

H. Nadig "Signature Analysis-Concepts, Examples and Guidelines," *Hewlett Packard Journal,* 15–21, May 1977.

K. Parker, *Integrating Design and Test: Using CAE Tools for ATE Programming,* IEEE Computer Society Press, 1987.

K. Parker, The Boundary-Scan Handbook, Kluwer, 1992.

T. Rao and E. Fujiwara, *Error-Control Coding for Computer Systems,* Prentice Hall, 1989.

J. Stewart, "Future Testing of Large LSI Circuits Cards," *Digest Papers 1977 Semiconductor Test Symposium, IEEE Publication 77CH1261-7C,* 6–17, October 1977.

A Viterbi and J. Omura, *Principles of Digital Communication and Coding*, McGraw-Hill, 1979.
N. Wang, *Digital MOS Integrated Circuits,* Appendix, Prentice Hall, 1989.
T. Williams and K. Parker, "Design for Testability—A Survey," *Proceedings of the IEEE*, 71(1), January 1983.

13.6 Problems

13.1 Give a transistor-level realization of the LSSD D flip-flop of Fig. 13.2.

13.2 Give a schematic of a basic storage cell to be used with the LSSD method for testing an IC that is based on CMOS transmission gates.

13.3 Give timing diagrams for all nodes of the circuit of Fig. 13.3 for the cases when (a) Disable goes *low* when Clk is low and (b) Disable goes *low* when Clk is *high*.

13.4 Design a clock-gating circuit similar to Fig. 13.3, but one in which the output is buffered by an inverter and in which all positive pulses are guaranteed to be full length.

13.5 Design a *good* transistor-level realization of the clock-generation circuit of Fig. 13.4.

13.6 Give timing diagrams for all *nor-gate* outputs of the circuit of Fig. 13.5 assuming Slct goes from *high* to *low* when Clk_2 is *high* and then goes back *high* two periods later when Clk_1 is high.

13.7 Give reasonable widths for all transistors (including those in the inverters) of the circuit of Fig. 13.6.

13.8 Give a logic-gate design of the control and decoder circuitry required in order to individually access any one of four LSSD chains.

13.9 Design a 32-bit maximal-length LFSR. Would using a 33-bit maximal-length LFSR and only outputting 32 bits be faster? Why?

13.10 Give a transistor-level design for the first flip-flop of Fig. 13.7 that incorporates the *exclusive-or* gate into the *master* stage of the flip-flop.

13.11 Give a transistor-level design for the first flip-flop and the three *exclusive-nor* gates of Fig. 13.8. Try to optimize speed even at the expense of d.c. power dissipation.

13.12 Design an eight-bit BILBO register to be used in testing. The inputs are Serial-In (S_{in}), Data-In ($D_{in}(0...7)$ which has eight bits), and Control-In ($C_{in}(0...1)$ which has two bits). The outputs are Data-Out ($D_{out}(0...7)$ which has eight bits) and Serial-Out (S_{out}). There is also a clock input. Depending on the two control bits, the register should be configured as shown. For the last case use $D(0) = D_{out}(2) \oplus D_{out}(4) \oplus D_{out}(6) \oplus D_{out}(7)$. A similar polynomial should be used when the register is configured as a signature analysis register. Only a logic-diagram-level solution is required.

C(0)	C(1)	Configuration
0	0	An eight-bit shift register
0	1	All eight bits are set to zero
1	0	An eight-bit signature analysis register
1	1	A pseudo-random noise generator

13.13 Give a reasonable transistor-level design for the boundary-scan cell of Fig. 13.16.

INDEX

D

E

F

G

H

I

J

L

M

N

O

P

V

W